C0-AZS-548

Electron Tomography

Three-Dimensional Imaging with the Transmission Electron Microscope

Electron Tomography

Three-Dimensional Imaging with the Transmission Electron Microscope

Edited by

Joachim Frank

Wadsworth Center for Laboratories and Research
New York State Department of Health
Albany, New York

and

Department of Biomedical Sciences
School of Public Health
State University of New York at Albany
Albany, New York

Plenum Press • New York and London

Library of Congress Cataloging-in-Publication Data

Electron tomography : three-dimensional imaging with the transmission electron microscope / edited by Joachim Frank.
p. cm.
Includes bibliographical references and index.
ISBN 0-306-43995-6
1. Three-dimensional imaging in biology. 2. Transmission electron microscopes. I. Frank, J. (Joachim), 1940- .
QH324.9.T45E43 1992
578'.45--dc20 91-45408
CIP

ISBN 0-306-43995-6

A Division of Plenum Publishing Corporation
233 Spring Street, New York, N.Y. 10013

Printed in the United States of America

Contributors

David A. Agard • Department of Biochemistry and Biophysics and The Howard Hughes Medical Institute, University of California at San Francisco, San Francisco, California 94143-0448

Kaveh Bazargan • Focal Image Limited, London W11 3QR, England

José-María Carazo • Centro Nacional de Biotecnología and Centro de Biología Molecular, Universidad Autonoma, 28049 Madrid, Spain

Hans Chen • Department of Biochemistry and Biophysics and The Howard Hughes Medical Institute, University of California at San Francisco, San Francisco, California 94143-0448

Joachim Frank • Wadsworth Center for Laboratories and Research, New York State Department of Health, Albany, New York 12201-0509; and Department of Biomedical Sciences, School of Public Health, State University of New York at Albany, Albany, New York 12222

Peter W. Hawkes • Laboratoire d'Optique Electronique/CEMES du CNRS, 31055 Toulouse Cedex, France

Zvi Kam • Department of Chemical Immunology, Weizmann Institute of Science, Rehovot 76100, Israel

Michael C. Lawrence • Electron Microscope Unit, University of Cape Town, Rondebosch 7700, South Africa. *Present address:* CSIRO Division of Biomolecular Engineering, Parkville, Victoria 3052, Australia

ArDean Leith • Wadsworth Center for Laboratories and Research, New York State Department of Health, Albany, New York 12201-0509

Pradeep K. Luther • The Blackett Laboratory, Imperial College, London SW7 2BZ, England

Bruce F. McEwen • Wadsworth Center for Laboratories and Research, New York State Department of Health, Albany, New York 12201-0509

Michael Radermacher • Wadsworth Center for Laboratories and Research, New York State Department of Health, Albany, New York 12201-0509

John W. Sedat • Department of Biochemistry and Biophysics and The Howard Hughes Medical Institute, University of California at San Francisco, San Francisco, California 94143-0448

James N. Turner • Wadsworth Center for Laboratories and Research, New York State Department of Health, Albany, New York 12201-0509; and Department of Biomedical Sciences, School of Public Health, State University of New York at Albany, Albany, New York 12222

Ugo Valdrè • Centro di Microscopia Elettronica, Departmento di Fisica, Universita Degli Studi di Bologna, 40126 Bologna, Italy

Terence Wagenknecht • Wadsworth Center for Laboratories and Research, New York State Department of Health, Albany, New York 12201-0509; and Department of Biomedical Sciences, School of Public Health, State University of New York at Albany, Albany, New York 12222

Christopher L. Woodcock • Department of Zoology and Program in Molecular and Cellular Biology, University of Massachusetts, Amherst, Massachusetts 01003

Elmar Zeitler • Fritz-Haber-Institut der Max-Planck-Gesellschaft, W-1000 Berlin 33, Germany

Preface

Some physicists may be drawn to biology because of the challenge that lies in the vast complexity of biological matter; what attracted me initially was the curious paradox that makes electron microscopy of macromolecules possible—*phase contrast*, the contrast that arises not *despite*, but *because* of, the imperfections of the objective lens. It is the capricious nature of such details that carries the promise of future problems finding totally unexpected (and sometimes surprisingly simple) solutions. Once engaged in electron microscopy, as a student I was in awe of the wide range of forms in which living matter is organized, but I was also frustrated by the central limitation of the instrument—that it renders these structures only in the confusing, highly ambiguous form of projections.

Three-dimensional information about an object is usually obtained in a cumbersome way, by a process that does not leave the object intact, namely by cutting and slicing, and by stacking or geometrically relating the resulting images. Consider the origins of anatomy, which set itself the task of making a three-dimensional image of the body with all its organs. It started as a heretical undertaking because it required dissection, intrusion into the body, violating its sanctity which was being upheld by the Roman Church. Because of the need for dissection, the teaching of anatomy in the Middle Ages was a clandestine operation performed by candlelight in a windowless hall, with the corpse lying on a table that was specially designed to hide it rapidly, in case the authorities stormed the premises. Perspective anatomical drawings and three-dimensional models emerged as the result of an intense visual, tactile, and visceral effort on the part of the scholar. Centuries after this type of three-dimensional imaging with the scalpel was begun, computerized axial tomography (CAT) was invented, a miraculous tool to look inside a living body without a single cut.

This book deals with a similar revolution (albeit on a different time scale) in the study of the cell's ultrastructure, brought about by the application of tomographic techniques to electron microscopy. For a long time, structural information about cell components had to be inferred from images of thin sections, the thickness being limited by the path length of 100-kV electrons in biological matter. The limitations of sectioning are well known: it produces distortions and material loss, and additional errors arise in the attempt to stack the section images to form a three-dimensional representation. Organelles of complex shape have proved difficult or impossible to study in this way. The problem is solved by increasing the voltage to the range of 400 to 1000 kV, thereby increasing the penetration thickness, and using a *series of views* rather than a single one to generate a "true" three-dimensional image. Again, an inside look is obtained into the structure, which remains intact during the investigation.

Similar techniques have been developed for macromolecular assemblies that are in a much smaller size range and require no increase in voltage. Thus, electron tomography has filled a large gap: for the first time, all hierarchies of structural organization, ranging from the level of atomic structure (explored by x-ray crystallography) to the architecture of the cell (explored by confocal scanning light microscopy) can now be studied by quantitative three-dimensional imaging techniques that require no symmetry or order. Although this book deals only with the mid-level of structural organization in this vast logarithmic range, the challenges posed by the explosive increase in the amount of data, and the need to make them accessible in some "nested" way are becoming evident. Clearly, the revolution in the biology of the cell will not be complete until a system of data storage, retrieval, and visualization is found that is capable of mapping out the intrinsic complexity of the cell's components—the cell as a walk-in world, one of the momentous challenges of computational biology.

This book emerged as the result of a long and sometimes tedious interaction with the contributors. I was lucky to find authors that were not only experts in their fields but also enthusiastic to cooperate and share my vision. I am very grateful for their patience and endurance. Special thanks go to Michael Radermacher and Bruce McEwen, who discussed with me the concept of the book. I also wish to acknowledge valuable suggestions by Pawel Penczek and Terry Wagenknecht, who helped me read and reconcile the contributions. Finally, I thank Amelia McNamara of Plenum for initiating an endeavor that allowed me to illuminate this stimulating topic from many directions.

Joachim Frank

Contents

PART III: METHODS

PART IV: APPLICATIONS

1

Introduction: Principles of Electron Tomography

Joachim Frank

1. WHAT IS ELECTRON TOMOGRAPHY?

Tomography is a method for reconstructing the interior of an object from its projections. The word *tomography* literally means the visualization of slices, and is applicable, in the strict sense of the word, only in the narrow context of a single-axis tilt geometry: e.g., in medical computerized axial tomography (CAT-scan imaging), the detector-source arrangement is tilted relative to the patient around a single axis. In electron microscopy, where the beam direction is fixed, the specimen holder is tilted around a single axis (Fig. 1). However, the usage of this term has recently become more liberal, encompassing arbitrary geometries. In line with this relaxed convention, we will use the term *electron tomography* for any technique that employs the transmission electron microscope to collect projections of an object and uses these projections to reconstruct the object in its entirety.

Excluded from this definition, more or less for historical reasons, are techniques that make use of inherent order or symmetry properties of the object. These latter, more established techniques are covered by a large body of literature that goes back to the crystallographic methods pioneered at the Medical Research Council (MRC) in Cambridge, England.

Joachim Frank • Wadsworth Center for Laboratories and Research, New York State Department of Health, Albany, New York 12201-0509; and Department of Biomedical Sciences, School of Public Health, State University of New York at Albany, Albany, New York 12222

Electron Tomography: Three-Dimensional Imaging with the Transmission Electron Microscope, edited by Joachim Frank, Plenum Press, New York, 1992.

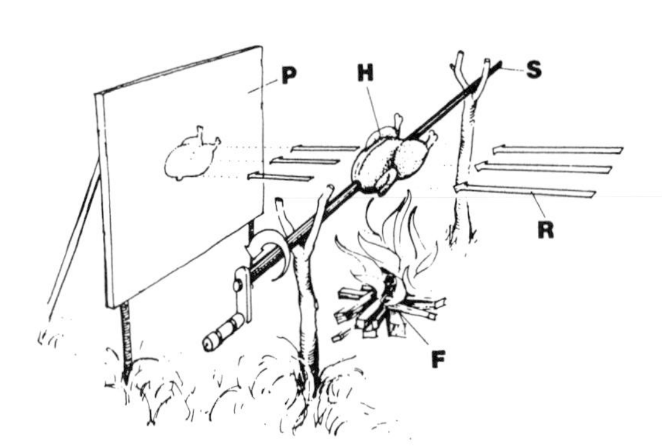

FIGURE 1. Principle of tomographic data collection. The figure was used to illustrate the collection of projections in the electron microscope with single-axis tilting. The grill models the deterioration of the object in the beam. The symbols, originally referring to a German text, have the following meanings: The chicken H on the skewer S is grilled in the fire F. As the chicken is rotated in equal increments $\Delta\theta$, the parallel illumination R produces a series of exposures on the plate P. (Reproduced from Hoppe, 1983, by permission of Verlag Chemie, GmbH, Weinheim, Germany.)

Even more recently, the terms *3D imaging* or *3D electron microscopy* have come into use as general terms to denote the capabilities of the instrument combined with the necessary computational tools to obtain a three-dimensional image of the object's interior. For instance, a new series of Gordon Conferences was started in 1985 under the title "Three-Dimensional Electron Microscopy of Macromolecules," with the intention of providing a forum for scientists approaching the study of biological structure with both crystallographic and noncrystallographic techniques. [The term *3D electron microscopy* may actually sound misleading since it conjectures an instrument with true 3D imaging performance. Such an instrument was actually conceived (Hoppe, 1972; Typke *et al.*, 1976) but never advanced beyond the blueprint stage.]

2. A HISTORICAL PERSPECTIVE

Three-dimensional imaging techniques are now commonplace in many areas of science, and it is difficult to recall that they have emerged only within the past 20 years; before that time, computers were simply too slow to be useful in processing three-dimensional data on a routine basis, although much of the mathematical theory was well developed.

We may consider Plato's simile of the cave as a precursor to the reconstruction problem: here our ignorance of the essence of reality is depicted by the situation of a man in a cave who watches shadows on the walls of his domicile; the shadows are all he sees of the world outside, and, because of the scantness of the information he receives, his comprehension of reality is severely limited. Similarly, a single projection, sometimes actually called a "shadowgraph," of an object is totally insufficient to establish its three-dimensional shape. If we were prevented from changing the angle of view, we would be in a similar situation as the man in the cave, although without the existential ramifications.

The history of tomography (see also the brief account by Herman and Lewitt, 1979) is a history of intellectual challenges in a number of unrelated fields of science. As Elmar Zeitler recounts in Chapter 4, the same mathematical solution to the reconstruction problem that was found by Radon (1917) has had to be rediscovered numerous times. Two Nobel Prizes are directly related to three-dimensional reconstruction: one that was shared by A. Cormack and G. N. Hounsfield in 1979 for the development of computerized axial tomography (see recent recollections by Cormack in OE Reports, 1990) and one, in 1982, to Aaron Klug, in part for his pioneering work in the 3D reconstruction of molecular structures from their electron micrographs.

Klug traces the origins of three-dimensional reconstruction in electron microscopy in his Nobel lecture (Klug, 1983). His laboratory, the Molecular Biology Laboratory of the Medical Research Council, is the appropriate starting point for a brief history of 3D imaging in electron microscopy. The predisposition of this institute for initiating quantitative structure research with the electron microscope is obvious, considering its historic role in the development of protein crystallography under Max Perutz's leadership.

DeRosier and Klug (1968) considered the problem of reconstructing the helical structure of the T4 phage tail from its projection (Fig. 2). To put their contribution into perspective, we must skip ahead and give a basic outline of the principle underlying 3D reconstruction. According to a fundamental mathematical theorem, the measurement of a projection yields a single central plane of the object's three-dimensional Fourier transform. The Fourier transform, an alternative representation of the object, is a breakdown of the object's density distribution into sine waves. The Fourier transform constitutes a complete description of the object in the sense that, if we know the strengths (amplitudes) and phase shifts of all sine waves traveling in all possible directions and having wavelengths down to $d/2$, then the object is completely known to a resolution of d. The projection theorem thus suggests a recipe for reconstructing the object from its projections: by tilting the

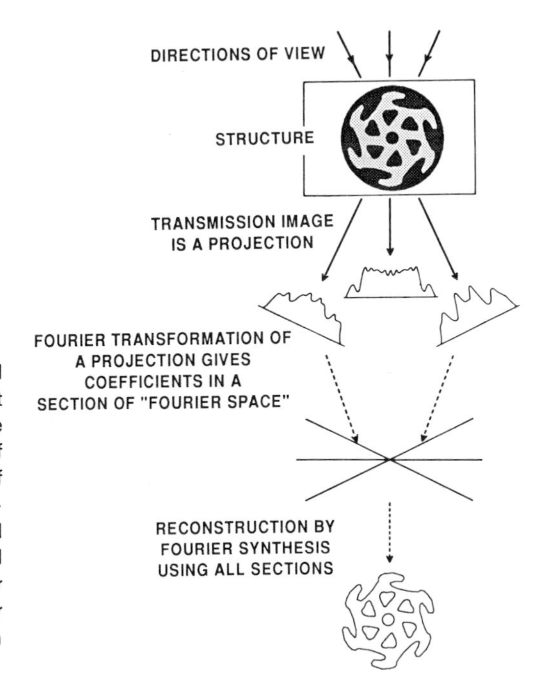

FIGURE 2. Principle of three-dimensional reconstruction: the projections of the object furnish different central sections of the object's Fourier transform. If the number of projections is sufficient (making use of symmetries where possible), then the complete Fourier transform can be regenerated by interpolation, and from this the original object can be retrieved by inverse Fourier transformation. (Reproduced from DeRosier and Klug, 1968, by permission of Macmillan Journals, Ltd.)

object through a range of $\pm 90^\circ$, we effectively sample its Fourier transform on a bundle of planes all intersecting one another on a single axis. It is clear that the angular spacing must be close enough to prevent information loss, in particular, far away from the axis where the planes are maximally spaced and where the information on sine waves with the smallest wavelengths is situated.

The application of this method to electron microscopy poses a problem because the tilt range is normally restricted for several reasons, the most important of which is the need to support the specimen on some type of grid that obstructs the electron path at high angles. Therefore, the angular range in commercial instruments does not exceed $\pm 60^\circ$. Special tilt stages have been designed that push the range to $\pm 85^\circ$ (Chalcroft and Davey, 1984). When the object is contained in a thick plastic section, the increased path length of electrons traversing the section at high angles also becomes a problem. Most recently, a high-voltage electron microscope stage with 360° rotation capability was developed for objects that can be placed into an ultrathin glass capillary (see Chapters 7 and 13). However, apart from these special cases, the experimental restriction to $\pm 60^\circ$ applies, which means that in the general case of an object without symmetry, a significant portion of the Fourier transform simply cannot be measured.

In contrast, when an object does possess symmetries, then the measurement of any projection yields other symmetry-related projections simultaneously. Another way of saying this is that, in this case, only part of the Fourier transform needs to be known for the entire Fourier transform to be generated. Among symmetric objects, those with helical symmetry, such as the T4 phage tail studied by DeRosier and Klug (1968), have a special position in that a single projection may be sufficient to generate the entire Fourier transform.

As early as 1970, Crowther and co-workers at the MRC formulated the approach to be used for reconstructing objects with or without symmetry with great clarity, and they also derived a general formula linking resolution, object size, and number of projections. The first particle with icosahedral symmetry was reconstructed in 1970 (Crowther *et al.*, 1970b). Subsequently, Henderson and Unwin (1975) developed the reconstruction of single-layer, "two-dimensional" crystals in the general crystallographic framework (see Amos *et al.*, 1982).

It is now necessary to illuminate the substantial contributions to the field by another group closely linked to crystallography: the group of Walter Hoppe at the Max-Planck Institute in Munich. Hoppe envisioned the prospect of 3D reconstruction in electron microscopy as early as 1968 (Hoppe *et al.*, 1968); however, he saw the true challenge of electron microscopy in imaging objects not amenable to crystallographic techniques. Consequently, he pursued almost exclusively the development of methods aimed at reconstructing objects lacking symmetry or crystalline order. Progress in this direction was initially slow because many tools of data processing had yet to be developed or adopted from other fields. The reconstruction of the fatty acid synthetase molecule in 1974 (Hoppe *et al.*, 1974) represents a significant achievement which marks the beginning of electron tomography in the proper sense of the term. At that time, essentially all important tools were in place: the use of correlation functions for the alignment of projections, the Smith-Cormack scheme of 3D reconstruction (Cormack, 1964; Smith *et al.*, 1973), and the first sophisticated image-processing software system of modular design dedicated to electron microscopy applications ("EM"; see Hegerl and Altbauer, 1982).

However, work in several other laboratories during that same period pointed to the deleterious effects of radiation damage, which made the quantitative interpretation of images taken with the standard imaging conditions questionable, and cast serious doubts on the significance of 3D information obtained by multiple exposure of the same object. According to Unwin and Henderson (1975), high-resolution information (at least to 7 Å) is preserved when the total dose is kept below 1 e/Å^2. Thus, it became apparent that tomography would produce biologically significant results only under two rather narrowly defined circumstances: (a) when applied to macromolecular structures, only those data collection schemes are acceptable that make use of multiple occurrences of the same molecules, by extracting different projections from different "repeats" of the molecule; (b) when applied to cell components in an entirely different size range where resolution requirements are more modest (50–100 Å), and specialized higher-voltage microscopes must be used for increased penetration, much higher accumulated radiation doses may be acceptable. In fact, these types of objects rarely exist in "copies" with identical structure, thus excluding any approach that uses averaging implicitly or explicitly.

In hindsight, it must be seen as unfortunate that Hoppe's leading laboratory in 3D reconstruction of noncrystalline objects invested its main efforts in an area that does not fall in either category: the 3D reconstruction of single macromolecules (or complex assemblies such as the ribosome) from a tilt series, in the course of which the molecule receives a radiation dose that exceeds the limit found by Unwin and Henderson (1975) by a large factor. (The arguments put forth by Hoppe (1981) attempting to justify 3D electron microscopy of individual macromolecules are not convincing.)

Meanwhile, the general theory of 3D reconstruction was advanced by a number of papers; among these, the works of Bates's group (Lewitt and Bates, 1978a, 1978b; Lewitt *et al.*, 1978), Zwick and Zeitler (1973), Colsher (1976), and Gilbert (1972) should be mentioned for their relevance to our subject matter. Tomography in all areas of science proceeded at such a rapid rate that, in 1975, the Optical Society of America decided to organize a topical meeting on 3D reconstruction in Stanford, California*). This meeting brought together contributors from a wide range of fields, such as geology, radioastronomy, radiology, and electron microscopy. An overview of various implementations and applications was compiled by Herman (1979).

Two hurdles had to be overcome before electron tomography could gain any practical importance as a tool for studying macromolecular or subcellular structure. In the former case, a data collection method had to be found that allowed different projections to be collected from different molecules. Such a method was conceptualized in 1978 (Frank *et al.*, 1978) but not worked out until much later (Radermacher *et al.*, 1987a, b).

In the latter case, the study of subcellular structures in three dimensions,

* *Image Processing for 2D and 3D Reconstruction from Projections: Theory and Practice in Medicine and the Physical Sciences*, OSA Topical Meeting, Stanford University, 4–7 August 1975.

progress hinged upon availability to the biologist of high- or intermediate-voltage electron microscopes equipped with precision tilt stages and supported by sophisticated image-processing resources. Centers with this degree of organization and sophistication did not emerge until the beginning of the 1980s when the National Institute of Health's Biotechnology program started to support three high-voltage microscopes dedicated to the biological sciences*). Thus, for entirely different reasons, the pace of electron tomography development was rather slow in both principal areas of study already outlined, especially considering the state of the art that already existed when Hoppe *et al.*'s fatty-acid synthetase study was published. However, perhaps the most important factor determining the pace of 3D imaging with the electron microscope has been the increase in speed and memory of computers. It must be realized that tomography poses computational problems of such magnitude that, until recently, only groups with access to mainframes were able to make significant progress. Equally important as a time factor was the slow development of flexible image-processing software capable of handling the numerous combinations of operations that are encountered in the analysis of electron microscopic data.

3. THE PRINCIPLE OF 3D RECONSTRUCTION

The principle of 3D reconstruction becomes clear from a formulation of the fundamental relationship between an object and its projections. An understanding of the basic concept of the Fourier transform is needed for this formulation. A brief introduction is provided in the following. For a more detailed introduction, the reader is referred to the specialized literature such as Andrews (1970). A compilation of definitions and formulas for the case of discrete data is provided in the appendix to Frank and Radermacher (1986).

The Fourier transform provides an alternative representation of an object by breaking it down into a series of trigonometric basis functions. For mathematical expediency, complex exponential waves of the form $\exp[2\pi ikr]$ are used instead of the familiar sine and cosine functions. The argument vector is $r=(x, y, z)$, and $R=(X, Y, Z)$ is the so-called spatial frequency, which gives the direction of traveling and the number of full oscillations per unit length. From such waves the object can be built up by linear superposition:

$$o(r)=\sum_i c_i \exp[2\pi iR_i r] \tag{1}$$

with the complex coefficients c_i. The 3D Fourier transform may be visualized as a 3D scheme ("Fourier space") in which the coefficients c_i are arranged, on a regular grid, according to the position of the spatial frequency vector. Each coefficient c_i contains the information on the associated wave's amplitude (or strength),

$$a_i=|c_i|$$

* University of Colorado in Boulder, Colorado; University of Wisconsin in Madison; and New York State Department of Health in Albany, New York.

and phase (or shift with respect to the origin),

$$\varphi_i = \arctan \frac{\mathrm{Im}\{c_i\}}{\mathrm{Re}\{c_i\}}$$

The projection theorem offers a way to sample the Fourier transform of an object by measuring its projections. According to this theorem, the 2D Fourier transform of a projection of the object is identical to a central section of the object's 3D Fourier transform. Thus, by tilting the object into many orientations, one is, in principle, able to measure its entire Fourier transform. Obviously, the projections must be collected with a small angular increment and, ideally, over the full angular range. Then, after the Fourier summation in Eq. (1) is performed, the object can be retrieved.

The angular increment is evidently determined by two parameters (Fig. 3): (i) the mesh size of the Fourier space grid and (ii) the size of the region, in Fourier space, that needs to be filled. These quantities are in turn determined by object diameter and resolution:

1. The mesh size must be smaller than $1/D$, where D is the object diameter.
2. The region in Fourier space for which data must be acquired is a sphere with radius $1/d$, where d is the resolution distance, i.e., the size of the smallest feature to be visualized in the reconstruction.

According to these essentially geometrical requirements, the minimum number of (equispaced) projections works out to be (Crowther *et al.*, 1970a; Bracewell and Riddle, 1967).

$$N = \frac{\pi D}{d}$$

Reconstruction methods may be classified according to the way in which projections are collected or according to the way in which the object is retrieved

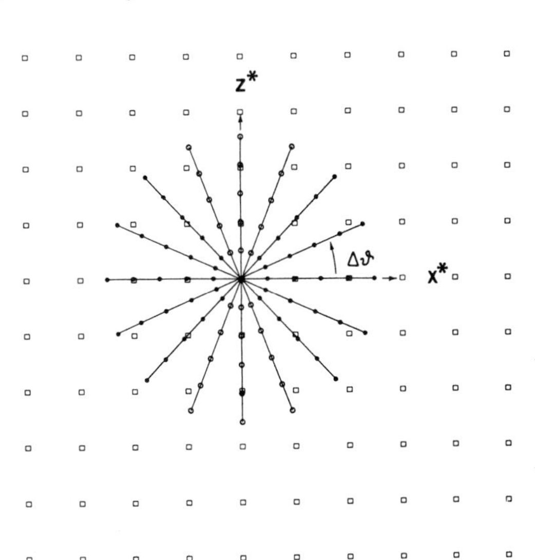

FIGURE 3. Relationship between the polar grid on which Fourier samples are obtained with the Cartesian grid on which data are required for Fourier inversion.

from its measured projections. The former relates to the experiment, while the latter relates to the mathematical and computational aspects of reconstruction as discussed in Chapters 4, 5, and 6. It is the purpose of this introduction to distinguish between the two experimental data collection geometries that have gained practical importance in electron microscopy: single-axis and conical tilting.

Single-axis tilting (Fig. 4*a*) is simply achieved by rotation of a side-entry rod in the electron microscope, whereas conical tilting (Fig. 4*c*) involves either a second tilt capability around an axis perpendicular to the first or a rotation capability in the inclined plane defined by the first tilt. It is easy to see that conical tilting provides a much wider coverage of Fourier space if the maximum tilt angle is the same in both cases (Fig. 4*b*, *d*). However, the price to be paid for this information

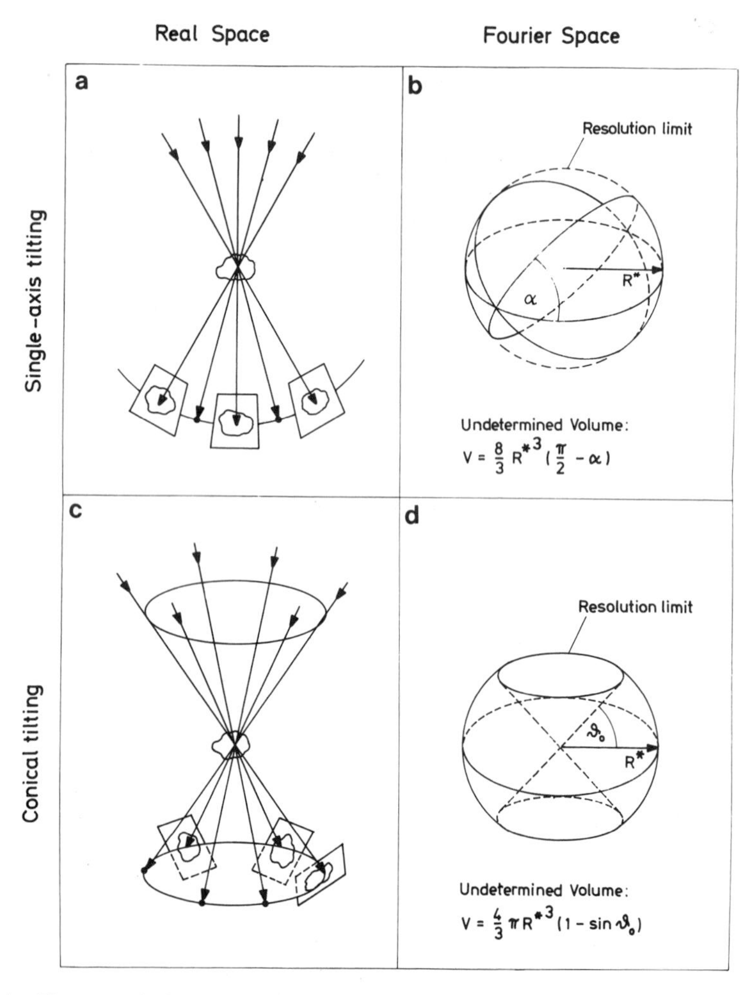

FIGURE 4. The two principal data collection geometries and the resulting coverage of 3D Fourier space: (a) single-axis tilting; (b) Fourier space sampled with single-axis tilting: the missing region has the shape of a wedge; (c) conical tilting; (d) Fourier space sampled with conical tilting: the missing region has the shape of a cone. (Reproduced from Radermacher, 1980.)

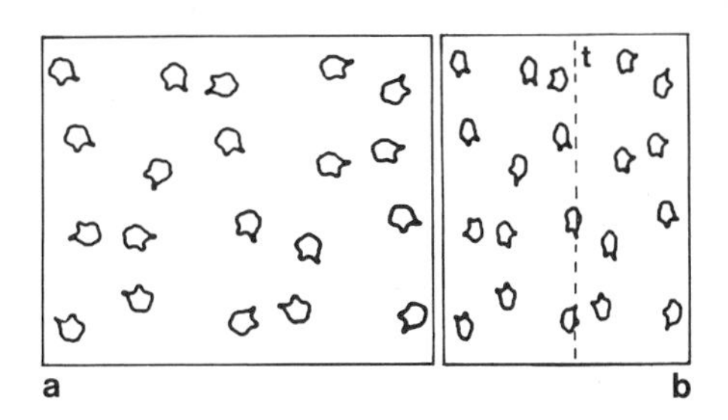

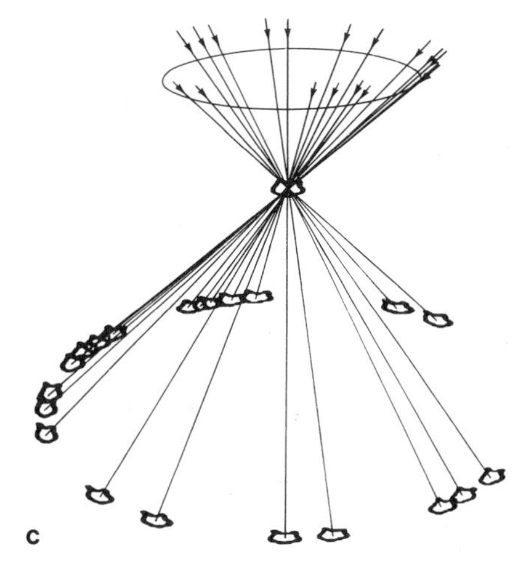

FIGURE 5. Random-conical data collection using a single exposure. The molecule is assumed to settle on the grid in a defined orientation: (a) specimen grid without tilt; (b) specimen grid tilted by 50° (direction of tilt axis indicated). In the experiment, an image of the tilted specimen (b) is recorded first and, subsequently, an image of the untilted specimen (a). This latter image is needed to determine the azimuthal angles in the equivalent conical geometry (c). (From Radermacher *et al.*, 1987b, reproduced by permission of Blackwell Scientific Publ. Ltd.)

gain is a more than twofold increase in total dose (Radermacher and Hoppe, 1980; Frank and Radermacher, 1986). Because of this disadvantage, and the more elaborate experimental procedure, conical data collection has not been used in experimental protocols according to which a single biological structure is multiply exposed.

Conical data collection, however, has found widespread application in reconstructions of a single macromolecule from its "repeats" which are uniquely oriented with respect to the support grid. It is easy to see that such a structure, due to its remaining rotational freedom, presents a conical series of views with random azimuths when its support grid is tilted by a fixed angle (Fig. 5; Frank *et al.*, 1978; Radermacher *et al.*, 1987b). Since this "random-conical" reconstruction, being

TABLE 1. Experimental Data Collection Geometries and Their Areas of Application

Geometry	Single axis	Random conical
Type of structure	Cell component	Macromolecular assembly
Preparation	Plastic embedding	Negative staining or ice embedding
Example	Kinetochore, Golgi apparatus	Ribosome
Reference	McEwen & Frank, 1990	Radermacher *et al.*, 1987a
Number of views	50–100	100–500 or more
Total dose	1000 e/Å	<10 e/Å
Practical resol. achievable	50 Å	20 Å

based on an averaging approach, is restricted to structures that occur in identical form, it cannot be applied to a large range of subcellular structures whose precise shapes vary even though the different variants of a given structure follow a common architectonic principle.

The different experimental situations and parameters of application are summarized in Table 1. (For an overview, see Frank, 1989.)

4. HOW THIS BOOK IS ORGANIZED

The book is organized into four parts. The first relates to the physics of imaging in the electron microscope. The second is an exposition of the theory pertinent to an understanding of three-dimensional reconstruction in electron microscopy. The third part covers the current working methods. The last part represents an attempt to give an up-to-data account of biological results achieved with the technique.

In the beginning chapter on image formation, Peter Hawkes explores the conditions that must be satisfied if the observed projections are to be regarded as simple integrals of an object function. For it is of course always possible to apply the reconstruction procedure "blindly" to a set of experimental images and obtain an "object function." Whether this reconstructed function has any meaning, or even a tractable relationship to the physical object, is investigated by Hawkes, which sets the stage for the rest of the book.

Pradeep Luther covers much of the experimental data presently known about material loss and section shrinkage in the electron beam, as well as methods to measure these radiation-damage effects. Knowledge of the behavior of the specimen is of crucial importance in planning an experiment that requires multiple exposure of the same specimen.

In the third chapter, which opens the part of the book devoted to the mathematics of reconstruction, Elmar Zeitler gives an authoritative account of a whole class of approaches to reconstruction by the use of orthogonal function series, which are quite elegant and sophisticated. Even though these techniques have gained little practical use in the past, they elucidate the fundamental relationships between the three-dimensional object and its projections. Furthermore, the recent successful work by Vogel and Provencher (1988) demonstrates that seemingly intractable problems, such as the assignment of angles to randomly spaced projections, may become surprisingly simple if formulated in the appropriate function space.

Weighted back-projection methods have a special position in the practical implementation of 3D reconstruction, mainly because of their mathematical tractability and computational speed. It is therefore appropriate that a chapter be devoted to an expose of the mathematical foundation of these methods. Michael Radermacher summarizes the rationales and important formulas of weighted back-projection methods for regular and general geometries. This treatment effectively covers the major portion of all electron tomography being done today: on the one hand, the reconstruction based on single-axis tilt data with equal angular increments, which underlies most of the results reviewed by Chris Woodcock and Bruce McEwen, and, on the other hand, the so-called random-conical reconstruction technique, which has dominated the reconstruction of single, asymmetric macromolecular assemblies, covered by Terence Wagenknecht.

Because of its practical importance, the random-conical reconstruction technique in principle deserves a separate chapter in a book of the present scope. However, this subject has been treated in depth in a recent article by Michael Radermacher (1988), to which the reader is referred instead.

One of the important recent developments in electron tomography is in the field of restoration. Due to the absence of information in a sizable angular range, the issue of restoration is of eminent interest and of crucial importance in ensuring a practical role for tomography in the structural biology of the future. Next to the issues addressed by Hawkes and Luther, the problems associated with missing data and the various ways of approaching these problems are the features that distinguish electron tomography of biological structures from tomography in other fields of science and medicine. Indeed, without a satisfactory answer to these problems, the interpretation of subcellular structures as revealed by three-dimensional reconstruction lacks the rigor found in other fields of three-dimensional imaging, such as radiological imaging, which have the luxury of being able to work with complete data sets.

After years of experimentation, very powerful approaches to 3D restoration have been found, and first results on "real data" begin to emerge which indicate that the solutions found are viable approximations to the undegraded object sought. A similar breakthrough occurred in the mid-1980s in the field of astronomy when two-dimensional maximum entropy restorations of stellar maps, now commonplace, were first presented in the scientific literature. The contribution by Jose-María Carazo gives a unified view of several restoration methods from the standpoint afforded by the theory of measurement spaces. Two methods, namely the maximum entropy method and the method of projection onto convex sets, receive the most comprehensive treatment because of their relative importance as compared to other approaches.

Part III, Methods, proceeds in the sequence in which the data are processed: the electron microscopic experiment, alignment of the scanned data, and the various tools of visualization and interpretation. No details on the computational steps and the software used in the 3D reconstruction step are provided, because the actual algorithmic implementation of formulas given in the theoretical chapters is currently probably the fastest changing aspect of the subject matter, owing to the rapid development of computer architecture.

The data collection in the electron microscope is covered by James Turner and Ugo Valdré, who have designed numerous specialized goniometers for conventional- and high-voltage electron microscopes. Their chapter is followed by a brief chapter in which Michael Lawrence describes the method of least-squares alignment of projections using colloidal gold markers, and by an equally brief account, jointly given by Joachim Frank and Bruce McEwen, on the use of the cross-correlation function to refine translational alignment.

One of the greatest challenges posed by electron tomography lies in the interpretation of the resulting 3D data. As of this writing, there exists no satisfactory tool that allows a biologist to cut and paste a 3D data set interactively and, thus, to study various selected components of the resulting volume from different aspects.

Because of the importance of this issue, three chapters are devoted to visualization of reconstructed data: ArDean Leith reviews the current techniques of surface and volume rendering, while Zvi Kam *et al.* describe the approach that David Agard's group at the University of California at San Francisco have taken to track and visualize structural components such as fibers in three dimensions. Finally, Kaveh Bazargan presents a novel solution to 3D display in the form of a holographic multiplexing technique.

In Part IV, Results, Bruce McEwen gives an account of results obtained to date by HVEM tomography of thick sections. Essentially, his contribution gives an idea about the broad range of biological problems that can be tackled by 3D imaging despite the present imperfections of this technique. Chris Woodcock reviews tomographic results that have provided insight into the organization of chromosomes and chromatin. Lastly, Terence Wagenknecht summarizes the impressive results obtained to date in the attempts to visualize macromolecular assemblies using noncrystallographic imaging techniques, notably those developed by Radermacher *et al.*

It is hoped that this book, in combining and systematically presenting the experience of several experts in this field, will become a resource for many investigators who are eager and ready to use the new tool of electron tomography in their research.

ACKNOWLEDGMENT

This work was supported, in part, by NIH grants GM 29169 and GM 40165.

REFERENCES

Amos, L. A., Henderson, R., and Unwin, P. N. T. (1982). Three-dimensional structure determination by electron microscopy of two-dimensional crystals. Prog. Biophys. Mol. Biol. **39**:183–231.

Andrews, H. C. (1970). *Computer Techniques in Image Processing.* Academic Press, New York.

Bracewell, R. N. and Riddle, A. C. (1967). Inversion of fan-beam scans in radio astronomy. *Astrophys. J.* **150**:427–434.

Chalcroft, J. P. and Davey, C. L. (1984). A simply constructed extreme-tilt holder for the Philips eucentric goniometer stage. *J. Microsc.* **134**:41–48.

Colsher, J. G. (1976). Iterative three-dimensional image reconstruction from tomographic projections. *Comput. Gr. Image Process* **6**:513–537.

Cormack, A. M. (1964). Representation of a function by its line integrals, with some radiological applications. I. *J. Appl. Phys.* **35**:2908–2912.

Crowther, R. A., Amos, L. A., Finch, J. T., and Klug, A. (1970a). Three-dimensional reconstruction of spherical viruses by Fourier synthesis from electron micrographs. *Nature (London)* **226**:421–425.

Crowther, R. A., DeRosier, D. J., and Klug, A. (1970b). The reconstruction of a three-dimensional structure from its projections and its application to electron microscopy. *Proc. R. Soc. London A* **317**:319–340.

DeRosier, D. and Klug, A. (1968). Reconstruction of three-dimensional structures from electron micrographs. *Nature (London)* **217**:130–134.

Frank, J. (1989). Three-dimensional imaging techniques in electron microscopy. *BioTechniques* **7**:164–173.

Frank, J. and Radermacher, M. (1986). Three-dimensional reconstruction of nonperiodic macromolecular assemblies from electron micrographs. In: Advanced Techniques in Biological Electron Microscopy. J. Koehler, ed. Springer-Verlag, Berlin, pp. 1–72.

Frank, J., Goldfarb, W., Eisenberg, D. and Baker, T. S. (1978). Reconstruction of glutamine synthetase using computer averaging. *Ultramicroscopy* **3**:283–290.

Gilbert, P. F. C. (1972). Iterative methods for the three-dimensional reconstruction of an object from projections. *J. Theor. Biol.* **36**:105–117.

Hegerl, R. and Altbauer, A. (1982). The "EM" program system. *Ultramicroscopy* **9**:109–116.

Henderson, R. and Unwin, P. N. T. (1975). Three-dimensional model of purple membrane obtained by electron microscopy. *Nature (London)* **257**:28–32.

Herman, G. T., ed. (1979). *Image Reconstruction from Projections.* Springer-Verlag, Berlin.

Herman, G. T. and Lewitt, R. M. (1979). Overview of image reconstruction from projections, in *Image Reconstruction from Projections* (G. T. Herman, ed.), pp. 1–7, Springer-Verlag, Berlin.

Hoppe, W. (1972). Dreidimensional abbildende Elektronenmikroskope. *Z. Naturforsch.* **27a**:919–929.

Hoppe, W. (1981). Three-dimensional electron microscopy. *Ann. Rev. Biophys. Bioeng.* **10**:563–592.

Hoppe, W. (1983). Elektronenbeugung mit dem Transmissions-Elektronenmikroskop als phasenbestimmendem Diffraktometer—von der Ortsfrequenzfilterung zur dreidimensionalen Strukturanalyse an Ribosomen. *Angew. Chem.* **95**:465–494.

Hoppe, W., Gassmann, J., Hunsmann, N., Schramm, H. J., and Sturm, M. (1974). Three-dimensional reconstruction of individual negatively stained yeast fatty-acid synthetase molecules from tilt series in the electron microscope. *Hoppe-Seyler's Z. Physiol. Chem.* **355**:1483–1487.

Hoppe, W., Langer, R., Knesch, G., and Poppe, Ch. (1968). Protein-Kristallstrukturanalyse mit Elektronenstrahlen. *Naturwissenschaften* **55**:333–336.

Klug, A. (1983). From macromolecules to biological assemblies. *Angew. Chem.* **22**:565–582.

Lewitt, R. M. and Bates, R. H. T. (1978a). Image reconstruction from projections I: General theoretical considerations. *Optik (Stuttgart)* **50**:19–33.

Lewitt, R. M. and Bates, R. H. T. (1978b). Image reconstruction from projections. III: Projection completion methods (theory). *Optik (Stuttgart)* **50**:189–204.

Lewitt, R. M., Bates, R. H. T., and Peters, T. M. (1978). Image reconstruction from projections. II: Modified back-projection methods. *Optik (Stuttgart)* **50**:85–109.

McEwen, B. F. and Frank, J. (1990). Application of tomographic 3D reconstruction to a diverse range of biological preparations, in *Proc. XII Int. Congr. Electron Microscopy* (L. D. Peachey and D. B. Williams, eds.), Vol. I, pp. 516–517, San Francisco Press, San Francisco.

OE Reports (1990). The development of computerized axial tomography. No. 79 (July 1990), p. 1.

Radermacher, M. (1980). *Dreidimensionale Rekonstruktion bei kegelförmiger Kippung im Elektronenmikroskop.* Thesis, Technical University, Munich.

Radermacher, M. (1988). Three-dimensional reconstruction of single particles from random and nonrandom tilt series. *J. Electron. Microsc. Tech.* **9**:359–394.

Radermacher, M. and Hoppe, W. (1980). Properties of 3D reconstruction from projections by conical tilting compared to single axis tilting, in *Proc. 7th European Congr. Electron Microscopy*, Den Haag, Vol. I, pp. 132–133.

Radermacher, M., Wagenknecht, T., Verschoor, A., and Frank, J. (1987a). Three-dimensional structure of the large subunit from *Escherichia coli. EMBO J.* **6**:1107–1114.

Radermacher, M., Wagenknecht, T., Verschoor, A., and Frank, J. (1987b). Three-dimensional reconstruction from a single-exposure random conical tilt series applied to the 50S ribosomal subunit of *Escherichia coli. J. Microsc.* **146**:113–136.

Radon, J. (1917). Über die Bestimmung von Funktionen durch ihre Integralwerte längs gewisser Mannigfaltigkeiten. Berichte über die Verhandlungen der Königlich Sächsischen Gesellschaft der Wissenschaften zu Leipzig. *Math. Phys. Klasse* **69**:262–277.

Smith, P. R., Peters, T. M., and Bates, R. H. T. (1973). Image reconstruction from a finite number of projections. *J. Phys. A* **6**:361–382.

Typke, D., Hoppe, W., Sessler, W., and Burger, M. (1976). Conception of a 3-D imaging electron microscope, in *Proc. Sixth European Congr. Electron Microscopy* (D. G. Brandon, ed.), Vol. I, pp. 334–335, Tal International, Israel.

Unwin, P. N. T. and Henderson, R. (1975). Molecular structure determination by electron microscopy of unstained crystalline specimens. *J. Mol. Biol.* **94**:425–440.

Vogel, R. H. and Provencher, S. W. (1988). Three-dimensional reconstruction from electron micrographs of disordered specimens. *Ultramicroscopy* **25**:223–240.

Zwick, M. and Zeitler, E. (1973). Image reconstruction from projections. *Optik* **38**:550–565.

I

Imaging in the Electron Microscope

2

The Electron Microscope as a Structure Projector

Peter W. Hawkes

1. INTRODUCTION

The intuitive understanding of the process of three-dimensional reconstruction is based on a number of assumptions, which are easily made unconsciously; the most crucial is the belief that what is detected is some kind of projection through the structure. This "projection" need not necessarily be a (weighted) sum or integral through the structure of some physical property of the latter; in principle, a monotonically varying function would be acceptable, although solving the corresponding inverse problem might not be easy. In practice, however, the usual interpretation of "projection" is overwhelmingly adopted, and it was for this definition that Radon

Peter W. Hawkes • Laboratoire d'Optique Electronique/CEMES du CNRS, 31055 Toulouse Cedex, France

Electron Tomography: Three-Dimensional Imaging with the Transmission Electron Microscope, edited by Joachim Frank, Plenum Press, New York, 1992.

(1917) first proposed a solution. In the case of light, shone through a translucent structure of varying opacity, a three-dimensional transparency as it were, the validity of this projection assumption seems too obvious to need discussion. We know enough about the behavior of x-rays in matter to establish the conditions in which it is valid in radiography. In this chapter, we enquire whether it is valid in electron microscopy, where intuition might well lead us to suspect that it is not. Electron–specimen interactions are very different from those encountered in x-ray tomography, the specimens are themselves very different in nature, creating phase rather than amplitude contrast, and an optical system is needed to transform the information about the specimen that the electrons have acquired into a visible image. If the electrons encounter more than one structural feature in their passage through the specimen, the overall effect is far from easy to guess, whereas in the case of light shone through a transparent structure, it is precisely the variety of such overlaps or superpositions that we use to effect the reconstruction. If intuition were our only guide, we might easily doubt whether three-dimensional reconstruction from electron micrographs is possible: there is no useful projection approximation for the balls on a pin-table! Why then has it been so successful? To understand this, we must examine in detail the nature of the interactions between the electrons and the specimen and the characteristics of the image-forming process in the electron microscope. Does the information about the specimen imprinted on the electron beam as it emerges from the latter represent a projection through the structure? How faithfully is this information conveyed to the recorded image? These are the questions that we shall be exploring in the following sections.

2. ELECTRON–SPECIMEN INTERACTIONS

2.1. Generalities

The electron microscope specimen is an ordered array of atoms—highly ordered if the object is essentially crystalline or has been organized during the preparation process into some kind of array, and organized, or ordered more locally, in the case of isolated particles. It may be stained, in which case the light atoms of which organic matter is mainly composed will be selectively bound to much heavier atoms, but if very high resolution is desired it will probably be unstained. Its thickness will depend on its nature and on the resolution to be achieved, for reasons that will gradually become clear. The electric potential within the specimen is not uniform, since in the simplest picture, each atomic nucleus creates a Coulomb potential, screened to an extent that depends on the nature of the atom in question by its electrons. In reality, the situation is complicated by the bonding between the atoms, but the potential distribution will be dominated by the pattern of screened Coulomb potentials associated with the atoms of specimen or stain.

The first step in electron image formation in a conventional (fixed-beam) transmission electron microscope (TEM) consists in irradiating the specimen with a beam of electrons, which may suffer various fates. Many will pass through the specimen, unaffected by its presence in their path. These are said to be unscattered.

In the atomic picture, they do not pass close enough to any of the specimen atoms to be deviated significantly from their course; recalling that, from the electrons' point of view, the interior of the specimen is a potential distribution, the unscattered electrons pass through a zone in which the electric transverse field is everywhere weak. Since the mean potential inside the specimen is different from that of the surroundings, the energy of the beam electrons will be altered slightly as they enter the specimen, reverting to its original value as they emerge.

Electrons that pass close to a specimen atom experience a strong transverse electric field and will hence be deflected laterally, or *scattered*; this scattering may or may not be accompanied by a transfer of energy from the beam electron to the specimen. If the energy transferred is negligible, the electron is said to be scattered *elastically*; if, however, there is an appreciable recoil on the part of the specimen atoms, the scattering is *inelastic*. The likelihood of these various events is measured by their scattering cross sections, which tell us not only the probability of elastic or inelastic scattering as a function of incident beam energy and of the atomic number of the scattering atom but also the angular scattering probability. The mean scattering angle, for example, can thus be determined.

Scattered electrons may then emerge from the specimen or may be scattered afresh, elastically or inelastically; the likelihood of a second scattering event increases with specimen thickness, as we should expect.

This qualitative picture of the electron–specimen interaction goes some way to dispel any doubts that the transmitted beam carries enough information about the specimen for reconstruction to be possible. It also shows us that the simple picture of intensity attenuation must be replaced by the very different idea of a scattering pattern directly related to the electric potential distribution within the specimen, itself determined by the atomic arrangement and dominated by the fields created by heavy atoms, if any are present. In the case of unstained specimens, therefore, we can hope to recover information from all the specimen atoms, but if these are light the signal will be weak or, in other words, only a small fraction of the incident beam electrons will be carrying useful information when they leave the specimen. In the case of stained specimens, whether the stain is negative or positive, the recorded image will be dominated by contrast from the stain.

In order to translate the foregoing discussion into more quantitative language, we now analyze the electron–specimen interaction, especially the elastic case, with the aid of quantum mechanics. We only reproduce here the main steps of the argument, a full account of which may be found in texts on electron microscopy (Reimer, 1989).

It is sufficient for our purposes to represent the incident beam by an extended plane wave

$$\psi = \psi_0 \exp(2\pi i k z) \tag{1}$$

traveling along the microscope axis, which coincides with the z coordinate axis. The effect of imperfect beam coherence (a spread in the incident beam direction and energy) will be considered in Section 3.2. Elastic scattering by a specimen atom generates a spherical wave, so the wave function becomes

$$\psi = \psi_0 \exp(2\pi i k z) + \psi_{sc} \tag{2}$$

where

$$\psi_{sc} = i\psi_0 f(\theta) \frac{\exp(2\pi ikr)}{r} \tag{3}$$

The function $f(\theta)$ is the complex scattering amplitude, $f(\theta) = |f(\theta)| \exp i\eta(\theta)$, and k is the wave number:*

$$k = \frac{1}{\lambda} \tag{4}$$

We can use this basic formalism in different ways, and it is important to distinguish between these, since they provide an understanding of the two important contrast mechanisms in the electron microscope. The contrast is always phase contrast, in the sense that virtually all the electrons incident on the specimen emerge beyond it and are, hence, available to participate in image formation. Two very different mechanisms operate to convert the phase information into a visible image and, hence, into intensity variations in some plane downstream from the specimen. We return to these in Section 3, but they inevitably influence the present discussion.

2.2. Amplitude Contrast

We first use Eq. (3) to calculate the number of electrons scattered through an angle greater than some threshold value α, treating the scattering medium as a film of given mass thickness μ (product of density ρ and thickness t). It is easy to show that the decrease in the number of electrons in the beam scattered through angles smaller than α is given by

$$\frac{dn}{n} = -N\sigma(\alpha)\, d\mu \tag{5}$$

where $N = $ (Avogadro's number)/(atomic weight) and $\sigma(\alpha)$ is the scattering cross section:

$$\sigma(\alpha) = \int_\alpha^\pi |f(\theta)|^2\, 2\pi \sin\theta\, d\theta \tag{6}$$

From (5) we see that

$$\begin{aligned} n &= n_0 \exp\{-N\sigma(\alpha)\mu\} \\ &= n_0 \exp\left(-\frac{\mu}{\mu_t}\right), \qquad \mu_t = \frac{1}{N\sigma(\alpha)} \end{aligned} \tag{7}$$

* Note that two conventions are equally common in the literature: $k = 1/\lambda$ and $k = 2\pi/\lambda$.

If, therefore, the angle α corresponds to the angle of acceptance of the objective aperture, n/n_0 tells us what fraction of the incident beam will pass through the aperture and reach the image. This is a most important result, for two reasons. First, it tells us that the contrast variations at the final image are directly related to the projection of the density through the specimen along the path of the electron. Second, it reassures us that this relation is single-valued since the exponential function varies monotonically.

The fraction intercepted varies with the nature and position of the scattering atoms and hence generates *scattering contrast*. The phase shifts are closely related to the angular deflections of the electrons in the specimen and depend on the nature of the scattering atoms encountered. They are converted into amplitude contrast by the objective aperture, which intercepts electrons that have been scattered through larger angles and have passed close to the heavier atoms in the specimen. Virtually all the medium-resolution contrast (greater than 2–3 nm, say) seen in electron micrographs is generated by this mechanism, and this therefore includes contrast from negatively stained specimens, for which the resolution is not better than this figure.

Equation (7) is valid only if multiple scattering is rare, though an expression of the same form is obtained if elastic and inelastic single scattering are allowed (see section 6.1 of Reimer, 1989); the total cross-section is the sum of the elastic and inelastic cross-sections. An estimate of the maximum acceptable specimen thickness can therefore be based on the mean free path between scattering events, which is denoted by μ_t in Eq. (7). (Published values of this path length frequently represent the product of density and path length, just as the mass–thickness is the product of density and thickness.) Figure 1 shows how the total elastic and inelastic cross-sections vary in the range from 20 to 500 kV for elements between carbon (Z = 6, atomic weight = 12, density = 2.2 g cm^{-3}) and platinum (Z = 78, atomic weight = 195, density = 21.5 g cm^{-3}). Table 1 shows the true elastic mean free path for a light element (carbon) and a heavy element (platinum) as a function of accelerating voltage. For 100 kV, for example, the elastic mean free path is about 200 nm for pure carbon (graphite) but falls to 40 nm for germanium (Z = 32) and below 10 nm for platinum (Z = 78), which is comparable with the common staining elements osmium (Z = 76) and uranium (Z = 92). For further data, see Reimer and Sommer (1968), Arnal *et al.* (1977), Reichelt and Engel (1984), and, for guidance through the extensive literature, Chapters 5 and 6 of Reimer (1989).

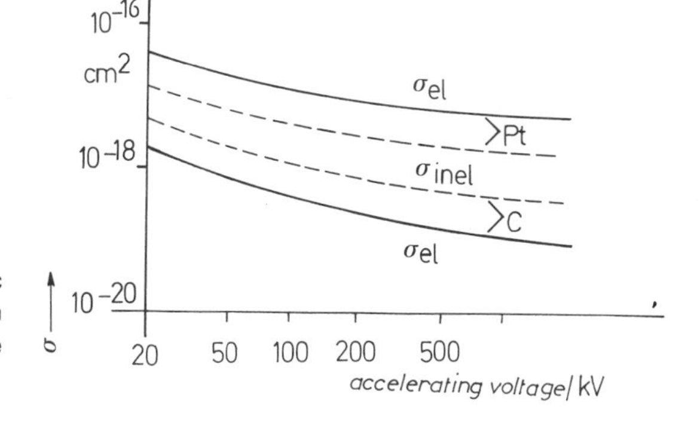

FIGURE 1. The elastic and inelastic cross-sections for carbon and platinum for accelerating voltages in the range 20–500 kV. (After Reimer, 1989.)

TABLE 1. Elastic Mean Free Paths for Carbon and Platinum (in nanometers) as a Function of Accelerating Voltage[a]

Accelerating voltage (kV)	Carbon	Platinum
17.3	45.9	3.03
25.2	65.5	3.78
41.5	102	5.41
62.1	145	6.57
81.8	181	7.83
102.2	216	8.95
150	321	10.9
300	518	14.7
750	632	23.6

[a] After Reimer (1989).

The simple expression for n can be improved by including inelastic scattering, but is unreliable if the specimen thickness is increased to the point beyond which multiple scattering is common. Nevertheless, in normal circumstances, the current density distribution emerging from the objective aperture does reflect faithfully the pattern of scattering atoms in the specimen, especially the heavy atoms which will probably be those of any stain. Even though the model of the specimen that enabled us to reach this conclusion was crude, we can conclude that, so far as this contrast mechanism is considered, the images used for three-dimensional reconstruction are essentially projections through the specimen. All the early, successful attempts at three-dimensional reconstruction relied on contrast of this kind (DeRosier and Klug, 1968; DeRosier, 1971), and the reasoning employed to justify the reconstruction procedure was considerably less laborious than that given here. The specimen is there represented by a mass–density distribution, $\rho_t(x, y, z)$, the Fourier transform of which is, of course,

$$F(X, Y, Z) = \iiint \rho_t(x, y, z) \exp\{-2\pi i(Xx + Yy + Zz)\}\, dx\, dy\, dz \tag{8}$$

A central section through this three-dimensional function yields a two-dimensional distribution, and, in particular, for $Z = 0$, we have

$$\begin{aligned} F(X, Y, 0) &= \iint \left\{ \int \rho_t(x, y, z)\, dz \right\} \exp\{-2\pi i(Xx + Yy)\}\, dx\, dy \\ &= \iint \mu(x, y) \exp\{-2\pi i(Xx + Yy)\}\, dx\, dy \end{aligned} \tag{9}$$

The quantity μ is the projection in the z direction of the mass–density distribution, and, as we have seen, the intensity distribution in the image is given by $n/n_0 = \exp(-\mu/\mu_t)$. The foregoing analysis is, however, necessary to understand the assumptions underlying this reasoning.

2.3. Phase Contrast

At higher resolution, a very different contrast mechanism comes into play, and we must reexamine the expression for the wave function emerging from the specimen. The latter is calculated by dividing the specimen into a large number of very thin slices, each of which is represented by a two-dimensional multiplicative, complex "transparency." The incident wave is thus modified by the first slice, allowed to propagate to the following slice, is again modified, propagates further, and so on, until the far side of the specimen is reached. This technique has been studied in very great detail, in connection with crystal structure determination in particular, and is widely used for image simulation. The aspect of this theory of interest here is the *phase-grating approximation.* In certain circumstances, the outgoing wave from the specimen can be represented by a multiplicative specimen transparency function:

$$S(x, y) = |S(x, y)| \exp\{i\varphi(x, y)\} \tag{10}$$

in which the phase shift φ and the amplitude term $|S|$ are projections of the potential and absorption of the specimen. In this approximation, multiple scattering is permitted. A careful examination of the derivation of this result shows that it is essentially a high-energy approximation, strictly valid for vanishing wavelength, when the Ewald sphere* becomes indistinguishable from a plane. The phase shift φ is given by

$$\begin{aligned} \varphi(x, y) &= -\frac{\pi\gamma}{\lambda e\hat{U}} \int V(x, y, z)\, dz \\ &\approx -\frac{\pi}{\lambda e U} \int V(x, y, z)\, dz \end{aligned} \tag{11}$$

in which λ is the electron wavelength:

$$\lambda = \frac{h}{(2m_0 e\hat{U})^{1/2}} \approx \frac{1.2}{\hat{U}^{1/2}} \text{ nm} \tag{12}$$

e is the absolute value of the charge on the electron, m_0 is its rest mass, U is the accelerating voltage, and $\hat{U}$ is its relativistically corrected value:

$$\hat{U} = U\left(1 + \frac{eU}{2m_0c^2}\right) \approx U(1 + 10^{-6}U); \qquad \gamma = 1 + \frac{eU}{m_0c^2} \approx 1 + 2 \times 10^{-6}U$$

(U in volts). The integral is taken through the specimen in which the potential distribution is $V(x, y, z)$.

* The Ewald sphere is a geometrical construct widely used in the study of crystalline specimens. The diffraction pattern of such specimens can be predicted by identifying the points of intersection of this sphere with the nodes of the reciprocal lattice, which is determined directly by the crystal lattice of the specimen. The radius of the Ewald sphere is $1/\lambda$, so that for short wavelengths its curvature can often be neglected.

Although it is not necessary to adopt the approximations that lead to this phase-grating approximation when calculating the image of a given structure, it is difficult to see how we could proceed in the opposite direction without it. If the specimen cannot be represented by a multiplicative specimen transparency function, or projection, the interpretation of any attempted three-dimensional reconstruction will be obscure. Indeed, attempts to perform such reconstructions would be likely to fail or, at least, to prove incapable of furnishing high-resolution detail owing to the inconsistency of the data in the different views through the specimen when these views are not strictly projections.

In conclusion, therefore, we may say that the information conveyed by the emergent electron wave is essentially projection information provided that the conditions of the phase-grating approximation are satisfied. The most important of these is that the wave function does not "spread" laterally as it propagates through the specimen ("column approximation"). This is stated with great clarity by Spence (1988):

> This important qualitative result [that the broadening is typically less than 0.4 nm for thicknesses below 50 nm] indicates that the axial, dynamical image wavefunction on the exit-face of a crystal is locally determined by the crystal potential within a small cylinder whose axis forms the beam direction and whose diameter is always less than a few ångströms for typical HREM conditions. This is essentially a consequence of the forward scattering nature of high-energy electron diffraction.

Thus, the thinner the specimen, the higher the accelerating voltage (with little further gain beyond about 600 kV), and the lighter the atoms involved, the better will be the phase grating and hence the projection approximation.

This qualitative observation is useful but how can it be rendered quantitative? A valuable rule-of-thumb is given by Spence (1988), whose chapter on nonperiodic specimens is recommended as background reading. He notes that the effects of Fresnel diffraction in the specimen are neglected in the phase-grating approximation, which is tantamount to assuming that the curvature of the Ewald sphere is negligible. The properties of Fresnel wave propagation can therefore be used to estimate the maximum tolerable thickness (t) for which the phase-grating approximation is acceptable. Spence finds

$$t \leqslant \frac{d^2}{\lambda}$$

where d is effectively the desired resolution. Figure 2 shows t as a function of λ (or accelerating voltage) for $d = 0.5$ nm, 1 nm, and 2 nm.

It is easier to establish the reliability of the phase-grating approximation in the case of crystalline materials. Indeed, any attempt to simulate the image of a structure by the multislice method effectively imposes a periodic character on the latter for, if the structure has no natural periodicity, then for the purposes of the calculation it is regarded as one member of an array of replicas of itself. The amount of calculation becomes proportionately larger, which explains why accurate image simulations for complex, nonperiodic specimens are rare. Among the few such calculations are those by Grinton and Cowley (1971), for a negatively stained

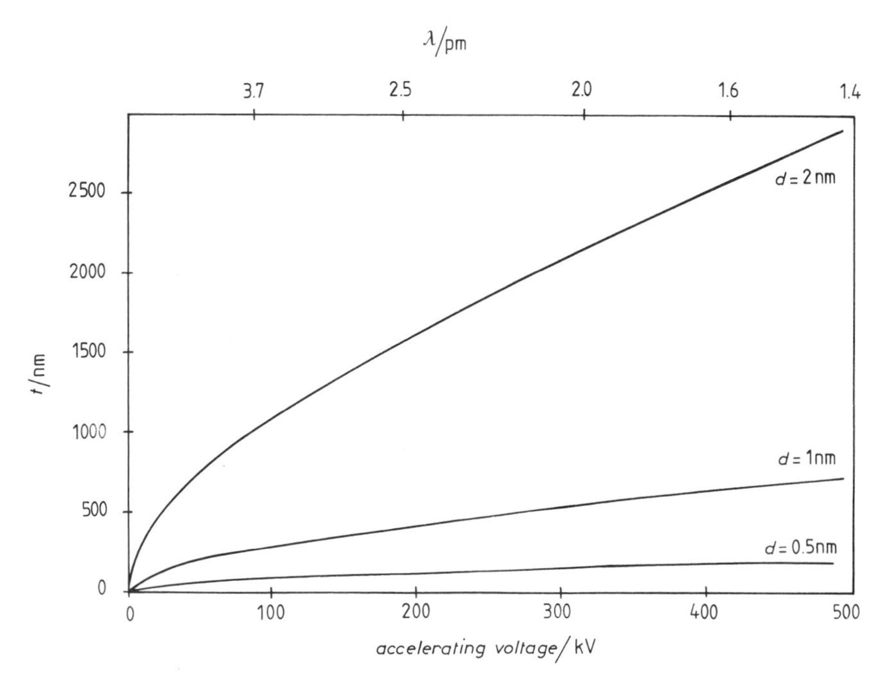

FIGURE 2. The maximum thickness t as a function of accelerating voltage (U) or wavelength (λ) for a resolution of 0.5 nm, 1 nm, and 2 nm.

rod-shaped structure (schematically representing a virus), who find that the approximation breaks down at about 10 nm for a resolution of 0.3 nm at about 60 kV ($\lambda = 5$ pm); this is more pessimistic than Spence's estimate (18 nm).

Although it is dangerous to extrapolate their results beyond the rather limited conditions in which the latter were obtained, we draw attention to the careful study by Jap and Glaeser (1980) of the domains of validity of two single-scattering approximations (the kinematic or first Born approximation and the weak phase object approximation) for two organic crystalline materials: anhydrous cytosine and DISOPS (disodium-4-oxypyrimidine-2-sulfinatehexahydrate). Since these are crystalline, Jap and Glaeser consider both individual diffraction spots and the projected potential. Their findings cannot be encapsulated in a single conclusion, but their curves for the reliability of the projected potential given by the two single-scattering approximations considered do suggest some general guidelines. For resolutions in the 3–5 Å range, the single-scattering approximations are "quite reliable" for thicknesses of 100–150 Å at 100 kV and of 200–250 Å at 500 kV. (The dissimilarity factor, in terms of which reliability is defined by Jap and Glaeser, is a well-defined quantity for crystalline specimens; the ranges given above correspond to different values of this factor.) We must, however, insist that these results, though indicative, must be treated with caution, even though they agree reasonably with the estimates given earlier. Jap and Glaeser concluded:

> It seems likely that the single-scattering approximations will tend to have a greater validity for large, complex structures than for small, simple structures because it is more likely in the case of small structures that atom centers can overlap in projection. This effect gives

> rise to large values of the Fourier coefficients of the crystal potential, and a non-linear dependence of the transmitted electron wave function upon the crystal potential. On the other hand, organic structures with much larger unit-cell dimensions will produce a situation in which very many more beams are simultaneously excited than is the case for cytosine or DISOPS. Thus the domain of validity of the single-scattering approximations for larger structures again cannot be accurately predicted from the present results.

Our present concern being with specimens possessing little or no periodicity, these warnings are particularly apposite. It is clear that there is an urgent need for much more numerical study of the validity of the various approximations for such specimens. The increasing availability of very fast multiprocessor computers has now made it much more reasonable for electron microscopists to embark on such calculations.

2.4. *Amplitude Contrast Reexamined*

It is convenient, particularly on a first encounter, to discuss the scattering contrast generated by the objective aperture, which intercepts some of the scattered electrons, and the higher resolution phase contrast as though they were separate and disjoint mechanisms. This is, however, strictly incorrect and, more seriously, leaves a twilight zone somewhere between the high- and medium-resolution regimes unaccounted for. In fact, we could abandon the notion of scattering contrast altogether and include the role of the aperture in the discussion of the mechanisms that convert phase variations in the wave function into amplitude variations in the image. This is indeed done in image simulation. Nevertheless, a great many electron images can be explained satisfactorily by the arguments of Section 2.2, and there is no reason to adopt a complicated train of thought when a simpler one is not misleading. The fact that the two approaches give identical results for single atoms can be demonstrated explicitly (see Section 6.3.3 of Reimer, 1989, for a proof of this).

2.5. *Radiation Damage*

The problem of the damage that the electron beam may cause as it passes through the specimen goes rather beyond the subject of this chapter and is dealt with more fully by Luther in Chapter 3. Indeed, one of the conditions to be satisfied if the electron microscope is to furnish true projections through the structure is that the passage of the electrons leaves the latter unscathed. Fortunately, many of the requirements for the preservation of high-resolution information that we have already encountered are the same as those that protect the specimen from radiation damage, since the aim is to keep inelastic scattering as low as possible. The accelerating voltage should not be too low, and the specimen should be thin and consist of light atoms. For stained specimens, the remedy is less obvious, for heavy atoms are numerous and the beam may cause loss of mass and stain migration. Some idea of the magnitude of the effects may be gained by noting that both the elastic and inelastic cross sections are roughly halved by an increase in accelerating voltage from 100 to 300 kV. The reduction in inelastic scattering in a given specimen should be accompanied by diminution of radiation damage, though other factors are also involved. We refer to Chapter 3 of this volume and to Chapter 10 of Reimer (1989), which has an extensive bibliography, and to Zeitler (1982, 1984) for discussion of the problem and of the various ways of avoiding it. The use of cryotechniques is described in detail by Robards and Sleytr (1985) and by Steinbrecht and Zierold (1987).

3. *IMAGE FORMATION IN THE TRANSMISSION ELECTRON MICROSCOPE*

3.1. *The Coherent Limit*

The illumination in an electron microscope is partially coherent. The effective source is small but not pointlike, and the energy spread is narrow but still appreciable. It is, however, convenient to analyze image formation on the assumption that the source is perfectly coherent, that is, has vanishingly small dimensions and negligible energy spread, and introduce the effects of partial coherence by a perturbation procedure. As before, we represent the wave emerging from the specimen in the form

$$\psi = S(x_0, y_0)\psi_s \tag{13}$$

and we allow this wave to propagate through the objective lens field to the objective aperture and thence to the image plane of the microscope. Each of these steps is represented by a Fourier transform, an immediate consequence of the form of the Schrödinger equation (for details, see Glaser, 1952, Hawkes, 1980a, or Hawkes and Kasper, forthcoming). The effect of the objective lens aberrations may be represented by a multiplicative function in the objective aperture plane (strictly, the plane conjugate to the source, but by considering a plane incident wave, the source is automatically conjugate to the back focal plane of the objective, close to which the aperture is in principle located); see Born and Wolf (1980) for a proof of this. With no new approximations, therefore, and the sole assumption that the illumination is coherent, we may write down the wave function (ψ) and hence the current density ($\propto |\psi\psi^*|$) in the image plane of the microscope. In practice, however, the spherical aberration of the objective lens is the only geometrical aberration that need concern us, and this renders the expression for the image intensity substantially more simple.

A straightforward calculation, set out in detail in most modern texts on electron image formation (Reimer, 1989; Spence, 1988; Hawkes, 1980a), tells us that the wave function in the image plane of the microscope, $\psi(x_i, y_i)$ is related to that emerging from the object, $\psi(x_0, y_0)$, by a relation of the form

$$E_i\psi(x_i, y_i, z_i) = \frac{1}{M}\iint K\left(\frac{x_i}{M} - x_0, \frac{y_i}{M} - y_0\right)\psi(x_0\, y_0 z_0)E_0\, dx_0\, dy_0 \tag{14}$$

in which E_i and E_0 are quadratic phase factors and K is a function characterizing the transfer of information along the microscope:

$$K(x_i, y_i; x_0, y_0) = \frac{1}{\lambda^2 f^2} \iint a(x_a, y_a) \exp[-i\gamma(x_a, y_a)] \times \exp\left[-\frac{2\pi i}{\lambda f}\left\{\left(x_0 - \frac{x_i}{M}\right) x_a + \left(y_0 - \frac{y_i}{M}\right) y_a\right\}\right] dx_a \, dy_a \tag{15}$$

and x_a, y_a are position coordinates in the aperture plane.

The function $a(x_a, y_a)$ is equal to unity in the objective aperture and zero outside. The phase shift $\gamma(x_a, y_a)$ is determined by the spherical aberration coefficient of the objective lens C_s and the defocus Δ measured in the object plane (that is, the distance from the specimen to the plane optically conjugate to the image plane). An additional term can be included in γ to describe any residual astigmatism, but we shall neglect this since it is not directly relevant to the theme of this chapter. Thus,

$$\gamma(x_a, y_a) = \frac{2\pi}{\lambda}\left\{\frac{1}{4} C_s \frac{(x_a^2 + y_a^2)^2}{f^4} - \frac{1}{2} \Delta \frac{x_a^2 + y_a^2}{f^2}\right\} \tag{16}$$

Since (14) has the form of a convolution, its Fourier transform will be the direct product of the transforms of the member functions. Writing

$$\tilde{\psi}_0(p_x, p_y) = \iint E_0 \psi(x_0, y_0) \exp\{-2\pi i(p_x x_0 + p_y y_0)\} \, dx_0 \, dy_0$$

$$\tilde{\psi}_i(p_x, p_y) = \iint E_i \psi(Mx_i, My_i) \exp\{-2\pi i(p_x x_i + p_y y_i)\} \, dx_i \, dy_i$$

$$\tilde{K}(p_x, p_y) = \iint K(x, y) \exp\{-2\pi i(p_x x + p_y y)\} \, dx \, dy \tag{17}$$

Eq. (14) becomes

$$\tilde{\psi}_i = \frac{1}{M} \tilde{K} \tilde{\psi}_0 \tag{18}$$

This extremely important exact equation tells us that, provided we can neglect geometrical aberrations other than spherical aberration, the electron microscope behaves as a linear scalar filter acting on the complex wave function ψ. If we could record the image wave function, the object wave function could be deduced from it straightforwardly. Unfortunately, only the amplitude of the wave function, and not its phase can generally be recorded unless we are prepared to abandon the traditional imaging modes and invoke holography or accept the heavy and difficult computing needed to solve the "phase problem." This difficulty is readily understood in simple physical terms. We have seen in the discussion of scattering that deflection of an electron by the electric field close to an atom is equivalent to introducing a phase variation in the wave function. The phase of the image wave is likewise intimately associated with the direction in which the electrons are traveling at the image plane. We can record the points of arrival of electrons, but we do not know where they were coming from. This is the essence of the phase problem, and shows why it can be solved, in theory at least, by taking two micrographs at different defocus values (or, more generally, in different imaging conditions).

In order to proceed further, we replace the wave function ψ_0 at the exit surface of the specimen by the product of the specimen transparency and the (plane) wave incident on the object:

$$\psi_0 = S(x_0, y_0)\psi_s \tag{19}$$

Since the modulus of S is frequently close to unity, it is convenient to write

$$S = (1 - s)\exp(i\varphi) \tag{20}$$

In the case of a weakly scattering specimen, both s and φ are small; we recall that the presence of s allows for inelastic scattering. The exponential may then be expanded, giving

$$S \approx (1 - s)(1 + i\varphi - \cdots) \approx 1 - s + i\varphi \tag{21}$$

if quadratic and higher-order terms are neglected. Substituting the approximation (21) into (18), we find

$$\tilde{\psi}_i = \frac{1}{M}\tilde{K}\{\delta(p_x, p_y) - \tilde{s} + i\tilde{\varphi}\} \tag{22}$$

in which δ is the Dirac delta function and $\tilde{s}$, $\tilde{\varphi}$ are the Fourier transforms of s and φ. From the definitions of K and $\tilde{K}$, Eqs. (15), (17), we know that

$$\tilde{K}(p_x, p_y) = A(\lambda f p_x, \lambda f p_y) \tag{23}$$

where $A = ae^{-i\gamma}$. The current density in the image j_i is proportional to $\psi_i\psi_i^*$. After some straightforward calculation, again neglecting quadratic terms in s and φ, we find that

$$\begin{aligned} j_i(Mx_i, My_i) &\equiv M^2\,|\psi_i(Mx_i, My_i)|^2 \\ &= 1 - 2\iint a\tilde{s}\cos\gamma\exp\{2\pi i(p_x x_i + p_y y_i)\}\,dp_x\,dp_y \\ &\quad + 2\iint a\tilde{\varphi}\sin\gamma\exp\{2\pi i(p_x x_i + p_y y_i)\}\,dp_x\,dp_y \end{aligned} \tag{24}$$

Writing $j_i = 1 + C$, we see immediately that

$$\tilde{C} = -2a\tilde{s}\cos\gamma + 2a\tilde{\varphi}\sin\gamma \tag{25}$$

where $\tilde{C}$ is the Fourier transform or spatial frequency spectrum of C, itself a measure of the image contrast.

This result tells us how the phase and amplitude of a weakly scattering specimen are represented in the image. Each is modulated by a sinusoidally varying function, the behavior of which is governed by the spherical aberration coefficient of the objective lens, C_s, and the choice of the defocus value. Since the microscopist has no control over C_s, the value of the defocus is chosen in such a way that as wide a range of values as possible of the spatial frequencies of the specimen phase reach the image undistorted by the function $\sin\gamma$. This in turn implies that a range of sizes of specimen detail will be present in the image and can be interpreted directly.

However, Eq. (25) is a linear equation. Thus, even if all the information needed for subsequent three-dimensional reconstruction is not present in a single micrograph, it is, in principle, possible to form a linear superposition of several micrographs, taken at different defocus values and suitably weighted, to create an image less disturbed by the transfer function $\sin\gamma$. We shall not give details of this technique (see Schiske, 1968, 1973; Saxton, 1986; and the survey in Hawkes, 1980a), but it may be important to ensure that information about the specimen that is carried by the electron beam, but is destroyed or distorted by the passage through the microscope, does appear correctly in the image used for three-dimensional reconstruction.

The form of the phase contrast transfer function, $\propto \sin\gamma$, changes rapidly as the defocus is varied. Before discussing this behavior, we scale, for convenience, the defocus Δ and the coordinates (x_a, y_a) in the aperture plane in such a way as to suppress the dependence of p_x, p_y, and Δ on λ and C_s. We write

$$\begin{aligned} \bar{\Delta} &= \frac{\Delta}{(C_s\lambda)^{1/2}}, \quad \bar{f} = \frac{f}{(C_s\lambda)^{1/2}}, \\ \bar{p}_x &= (C_s\lambda^3)^{1/4}\, p_x = \frac{\bar{x}_a}{\bar{f}}, \quad \bar{p}_y = (C_s\lambda^3)^{1/4}\, p_y = \frac{\bar{y}_a}{\bar{f}} \end{aligned} \tag{26}$$

where $\bar{x}_a = x_a/(C_s\lambda^3)^{1/4}$ and likewise for $\bar{y}_a$. It has been suggested (Hawkes, 1980b) that the scaling factors be given names. A defocus of $(C_s\lambda)^{1/2}$ is thus said to be 1 Sch (one scherzer), and a spatial frequency of $(C_s\lambda^3)^{1/4}$ is said to be 1 Gl (one glaser). With this new scaling, we have

$$\gamma = \frac{\pi}{2}\bar{p}^4 - \pi\bar{\Delta}\bar{p}^2, \qquad \bar{p}^2 = \bar{p}_x^2 + \bar{p}_y^2 \tag{27}$$

Figure 3 shows $\sin\gamma$ as a function of $\bar{p}$ for various values of $\bar{\Delta}$ in the vicinity of $\bar{\Delta} = 1$, $\sqrt{3}$, and $\sqrt{5}$, where the function has wide fairly constant regions and varies least rapidly with $\bar{\Delta}$. These are therefore the values of defocus to be preferred for

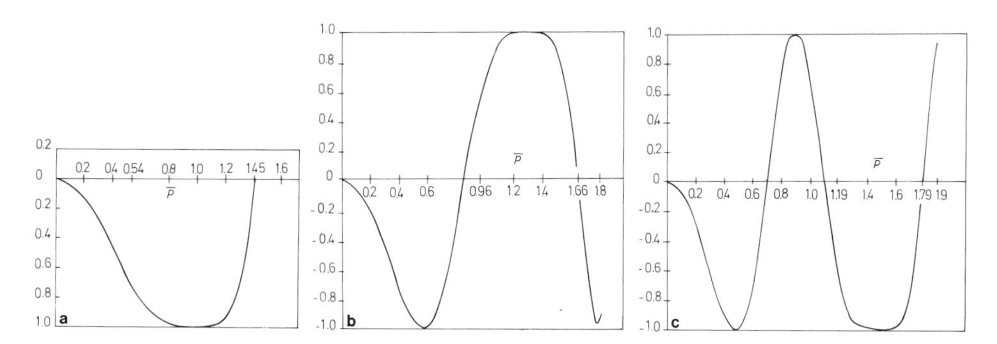

FIGURE 3. The function $\sin\gamma$ as a function of reduced spatial frequency $\bar{p}$ for reduced defocus $\bar{\Delta}=1$ (a), $\sqrt{3}$ (b), and $\sqrt{5}$ (c).

high-resolution imaging. In practice, the so-called Scherzer focus, corresponding to $\bar{\Delta}=1$, $\Delta=(C_s\lambda)^{1/2}$, is the most convenient for direct interpretation of the recorded image.

3.2. *Real Imaging Conditions: Partial Coherence*

In practice, almost all of the assumptions that we have been making are unjustified. The illuminating beam is not perfectly coherent, and the specimen rarely scatters so weakly that φ^2 can be truly neglected. In this section, we briefly consider the relatively innocuous effects of partial coherence. We return to the problems associated with the weak-scattering approximation in Section 4.

We have assumed that the illuminating beam has no energy spread (perfect temporal coherence) and is emitted from a vanishingly small source (perfect spatial coherence). In practice, the energy spread may reach a few electron volts, the exact value depending on the nature of the emitter and the gun optics. The finite source size has the effect of spreading the directions of arrival of the electrons at the specimen. Instead of all arriving parallel to the microscope axis (plane wave), they will be incident over a narrow range of angles centred on this same axis.

Each of these effects can be represented to a good approximation by an *envelope function*, in the sense that the coherent transfer function $\sin\gamma$ is to be multiplied by a function representing the effect of finite energy spread and another representing that of finite source size. These envelopes attenuate the transfer of information, and it is therefore important to ensure that, if high resolution is to be achieved, the microscope source is sufficiently small and has a narrow enough energy spread for this information to survive in the image. Figure 4 shows a typical situation.

3.3. *Real Imaging Conditions: Specimen Tilt*

In practice, the incident beam, however close to perfection the spatial coherence may be, is not incident normal to the specimen surface for the simple reason that a three-dimensional reconstruction is based on a set of closely spaced views through the specimen from different directions. These are commonly obtained

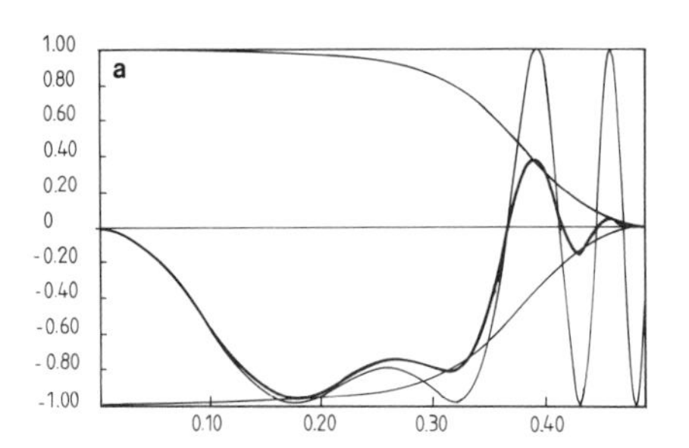

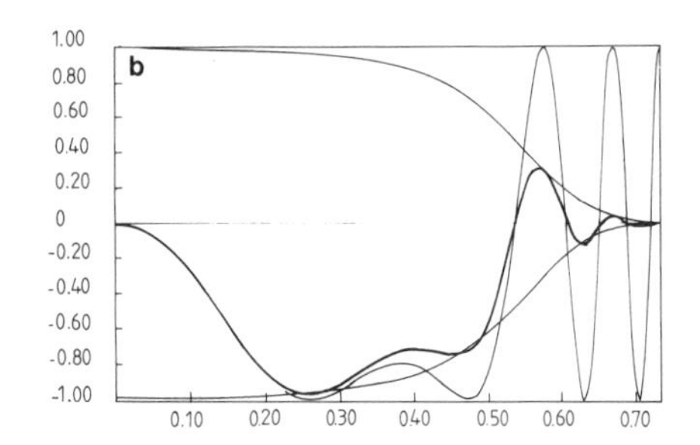

FIGURE 4. The attenuation of the transfer function ($\sin \gamma$) due to partial coherence for the following conditions: (a) $U = 200$ kV; $C_s = 2$ mm; defocus $= -841$ Å; beam divergence (partial spatial coherence) semiangle $= 0.6$ mrad; defocus spread (partial temporal coherence) $= 69$ Å. (b) $U = 400$ kV; $C_s = 0.9$ nm; defocus $= -457$ Å; semiangle $= 0.64$ mrad; spread $= 56$ Å.

by tilting the specimen, and any known symmetry properties of the structure may be used to supplement the information obtained from the tilted images. Alternatively, a single tilted image may be sufficient if the specimen consists of particles that tend to adopt a particular orientation on the specimen support but have random orientations in the plane of the latter. In all cases, however, the problem remains: the microscope transfer function is highly defocus dependent, and we have stated without comment that the microscopist must choose the defocus value with care and forethought. How can we reconcile this with the fact that, for specimen tilts that may reach 45° or more, the defocus difference between different sides of the image may well be of the order of micrometers? This problem has other ramifications of considerable importance. For example, averaging to enhance the signal-to-noise ratio for low-dose images of fragile specimens is not possible if the defocus varies across the object; the contrast transfer equations themselves cannot be used in the form given above, and the necessary corrections introduce artefacts. Various proposals for overcoming the problem by digital processing of the recorded tilt images before three-dimensional reconstruction have been made over the years (e.g., Amos *et al.*, 1982; Schiske, 1982; Henderson and Baldwin, 1986; Henderson *et al.*, 1986), but the most satisfactory solution is surely that developed by Zemlin (1989). This requires modification of the microscope to permit the image to be formed as a mosaic of small subimages, each created by illuminating a small zone of the specimen; the illumination spot is then moved to the adjoining zone until the whole specimen area has been covered. In this way, the defocus spread need never exceed that corresponding to the width of an illuminated zone since the microscope defocus can be adjusted as the illuminating spot climbs up the slope of the specimen.

This technique is not a panacea; there are so many conflicting requirements in high-resolution electron microscopy that any perturbation of the delicate balance, however advantageous, is almost certain to have drawbacks as well. In the case of this dynamic focusing idea, care must be taken to ensure that the inevitable loss of spatial coherence due to the decrease in spot size does not cause an unacceptable attenuation of the transfer function. Zemlin estimates that this diminution of the coherence can be kept within acceptable bounds, however. Furthermore, the situation that we have been considering is the most unfavorable; when the structure of isolated particles is to be established, the defocus difference across an individual particle will be much smaller, and the problem consequently less serious. For discussion of the choice of defocus for such specimens, see Radermacher (1988).

3.4. Real Imaging Conditions: Thick Specimens

For the vast majority of the specimens employed in three-dimensional reconstruction, the image is the result of scattering contrast, as described in Section 2.2. Electrons scattered through large enough angles by the specimen atoms are intercepted by the objective aperture and, hence, are absent from the image. Their absence is indicated by dark contrast against the bright background and, as Eq. (7) indicates, the number of electrons per unit area at an image pixel is proportional to $\exp(-\mu/\mu_t)$, where μ is the mass thickness through the corresponding object area.

In an ideal recording system, therefore, in which the number of electrons per pixel could be measured, the logarithm of this quantity would represent a projection through the object directly. Cameras that do indeed yield such measurements accurately over a wide range of values of the number of incident electrons are becoming available. If the images are recorded on film, the situation is a little more complicated. At electron microscope energies, every electron causes sufficient ionization of the grains of the photographic emulsion to render at least one crystal developable. If E electrons are incident per unit area of an emulsion consisting of silver halide crystals of mean cross-sectional area A, there will be on average EA collisions per crystal, so the fraction F of all crystals hit at least once becomes

$$F = 1 - \exp(-EA) \qquad \text{(Poisson distribution)} \tag{28}$$

After development, the emulsion is characterized by its optical density D, which is defined in terms of the fraction of light transmitted through the emulsion T by

$$D = \log_{10}\left(\frac{1}{T}\right) \tag{29}$$

This may be written as

$$D = D_{\text{sat}}\{1 - \exp(-EA)\} \tag{30}$$

where D_{sat} is the saturation density, the value reached after an exposure long enough for every grain to be rendered developable. Very often, it is sufficient to expand the exponential and neglect quadratic and higher-order terms, whereupon Eq. (30) becomes

$$D = D_{\text{sat}} AE \propto E \tag{31}$$

(For extensive discussion, see Valentine, 1966, and also Section 4.6.2 of Reimer, 1989.)

To summarize the result of this section, the natural logarithm of the optical density gives a direct measurement of the projected density through the structure, apart from a scaling factor.

4. THE MICROSCOPE AS PROJECTOR: ASSESSMENT

The object of this chapter is to discuss the extent to which the electron microscope provides a projection of some aspect of the specimen structure that can be used to form a reliable three-dimensional reconstruction of the latter. There is always a danger that a critical examination of a complicated situation will give the reader the impression that the situation is, if not hopeless, certainly far from rosy. The list of things that can go wrong seems interminable: insufficient source coherence, a strongly scattering specimen, appreciable inelastic scattering, structural dimensions that cross the vague frontier between scattering contrast and phase contrast, loss of information thanks to the form of the contrast transfer function, inadequacy of the column approximation. And yet, there are plenty of successful three-dimensional reconstructions in the literature, and the structures obtained seem in harmony with structural information obtained by other means and enable the molecular biologist, for example, to confirm or invalidate earlier hypotheses about structure and function. In this last section, we attempt to draw together the strands of the interwoven arguments of the preceding sections, and we conclude that, provided that certain cautionary remarks are borne in mind, the electron microscope image does contain the projection information on which a three-dimensional reconstruction can be safely based.

First, we must distinguish between stained and unstained specimens (for more technical details, see Chapters 13–15). If a heavy-atom stain is employed, then for several reasons the highest resolutions cannot be expected. The stain will itself not cling to all the fine details of the object and is often incapable of penetrating into all the narrow interstices of the structure. Furthermore, the presence of heavy atoms makes the weak scattering approximation untenable, so all the reasoning based on this becomes invalid. Conversely, the scattering properties of this heavy stain are excellent for creating scattering contrast, and such contrast does indeed represent a projection through the object. We recall that it is important to use not the optical density but its logarithm. A set of views through a stained specimen may be expected to yield a three-dimensional reconstruction that is reliable down to the resolution of the specimen preparation and staining procedures, at best around 2 nm but sometimes worse than this. For a much fuller discussion of all practical aspects of this branch of three-dimensional reconstruction, see the long review by Amos *et al.* (1982) and the book edited by Turner (1981), especially the chapter by Steven (1981), as well as Chapters 13 and 14 in this book.

When we wish to establish the three-dimensional structure of a specimen at higher resolution, better than 1.5 or 2 nm, say, the situation changes radically. Very different techniques of specimen preservation are brought into play, the beam electrons convey information about the positions of light atoms, and the mechanism of image formation is that appropriate to a phase object.

If the information recorded in the image is to be used for reconstruction, and hence represents a projection through the specimen, certain precautions must imperatively be taken. If not, the different views used for the reconstruction will at best be found to be incompatible, and the reconstruction methods described in the remainder of Part II will fail to generate a high-resolution model. At worst, the views will be compatible but wrong, yielding a reconstructed structure that is false.

These precautions are of two kinds. First, we must be sure that the phase shift of the wave function leaving the specimen is essentially a projection through the latter, and, second, we must arrange that this phase information is converted faithfully into amplitude variations in the recorded image, at least in the size range of interest. In other words, certain rules must be respected during the preparation of the specimen, and the mode of observation in the microscope must likewise be rigorously controlled. So far as the specimen is concerned, it is most important that it can be represented by a multiplicative transparency [Eq. (13)], with the implications that we have already mentioned: the beam accelerating voltage should not be too low, nor the specimen thickness too great. It is difficult to give figures, since the limiting thickness is a function of beam voltage and of the atomic number of the heaviest atoms present, but for unstained biological material observed between 100 and 200 kV the thickness should probably not exceed 10 or 20 nm.

For transfer along the microscope, we are obliged to use linear transfer theory, since no useful conclusions can be drawn from the more complicated nonlinear theory that we obtain if we do not assume that the specimen scatters weakly. The beam convergence must be small, the energy spread small, and the defocus value suitably chosen. Some way of circumventing the problem of the variation in the defocus of a steeply tilted specimen must be found, such as that described in Section 3.3.

For the very highest resolution, all these precautions and many others relating to the actual reconstruction process and described in subsequent chapters are mandatory. For more modest, but still high, resolution, however, we may enquire what is likely to happen if we relax the requirements a little. If the specimen is thicker than ideally we should like, the columns of the "column approximation" will broaden and the resolution of the phase-grating approximation will deteriorate slowly. Inelastic scattering will increase and add a lower-resolution background to the image, which will in consequence display a lower signal-to-noise ratio. The curvature of the Ewald sphere will have a small effect on the fidelity of the reconstruction (Cohen *et al.*, 1984).

The theoretical assumption that is the worst respected in practice is, however, that the specimen is a weak scatterer and, hence, that the phase shifts introduced are small. Even if we relax this by regarding the specimen phase variation as a function with only small variations about its mean value, many specimens that preserve very fine detail will not be truly weak scatterers. It is difficult to discuss the consequences of semi-weak scattering, but we can anticipate that the effects will be, if a little misleading, not catastrophic. If we continue the series expansion of $\exp(i\varphi)$, we obtain an alternating sequence of real and imaginary terms, of which the former will join the amplitude contrast term [s in Eq. (21)] and the latter will alter the absolute value of the phase contrast term. Modest violation of the weak-scattering requirement should therefore not be too serious.

There are hazards in three-dimensional reconstruction of many kinds. Here we have concentrated on a particular kind and have deliberately refrained from comment on those dealt with elsewhere in the book, notably radiation damage, which is extensively discussed by Luther, and the practical problems of data acquisition, which fill much of Part III. We have shown that at both high and modest resolution projection information can be recorded. How this information is used to generate three-dimensional structures, such as those discussed in Part IV, is the subject of the remainder of this book.

REFERENCES

Almost all the reviews on the various aspects of three-dimensional reconstruction neglect the question examined in this chapter, taking it for granted that the electron microscope does project the specimen structure onto the image, or at least onto the exit surface of the object. The surveys by Frank and Radermacher (1986) and by Glaeser (1985) are rare exceptions. The list of further reading is concerned with three-dimensional reconstruction in general, therefore, and hardly at all with the specific topic of projection.

Arnal, F., Balladore, J. L., Soum, G., and Verdier, P. (1977). Calculation of the cross-sections of electron interaction with matter. *Ultramicroscopy* **2**:305–310.

Amos, L. A., Henderson, R., and Unwin, P. N. T. (1982). Three-dimensional structure determination by electron microscopy of two-dimensional crystals. *Prog. Biophys. Mol. Biol.* **39**:183–231.

Born, M. and Wolf, E. (1980). *Principles of Optics.* Pergamon, New York.

Cohen, H. A., Schmid, M. F., and Chiu, W. (1984). Estimates of validity of projection approximation for three-dimensional reconstructions at high resolution. *Ultramicroscopy* **14**:219–226.

Deans, S. R. (1983). *The Radon Transform and Some of Its Applications.* Wiley-Interscience, New York.

DeRosier, D. J. (1971). The reconstruction of three-dimensional images from electron micrographs. *Contemp. Phys.* **12**:437–452.

DeRosier, D. J. and Klug, A. (1968). Reconstruction of three-dimensional structures from electron micrographs. *Nature* **217**:130–134.

Frank, J. and Radermacher, M. (1986). Three-dimensional reconstruction of nonperiodic macromolecular assemblies from electron micrographs, in *Advanced Techniques in Biological Electron Microscopy III* (J. K. Koehler, ed.), pp. 1–72. Springer, Berlin and New York.

Glaeser, R. M. (1982). Electron microscopy, in *Methods of Experimental Physics* (G. Ehrenstein and H. Lecar, eds.), Vol. 20, pp. 391–444. Academic Press, New York.

Glaeser, R. M. (1985). Electron crystallography of biological macromolecules. *Ann. Rev. Phys. Chem.* **36**:243–275.

Glaser, W. (1952). *Grundlagen der Elektronenoptik.* Springer, Vienna.

Grinton, G. R. and Cowley, J. M. (1971). Phase and amplitude contrast in electron micrographs of biological material. *Optik* **34**:221–233.

Hawkes, P. W. (1980). Image processing based on the linear theory of image formation, in *Computer Processing of Electron Microscope Images* (P. W. Hawkes, ed.), pp. 1–33. Springer, Berlin and New York.

Hawkes, P. W. (1980b). Units and conventions in electron microscopy, for use in ultramicroscopy. *Ultramicroscopy* **5**:67–70.

Hawkes, P. W. and Kasper, E. (forthcoming). *Principles of Electron Optics*, Vol. 3. Academic Press, London.

Henderson, R. and Baldwin, J. M. (1986). Treatment of the gradient of defocus in images of tilted, thin crystals, in *Proc. 44th Ann. Meetg. EMSA* (G. W. Bailey, ed.), pp. 6–9. San Francisco Press, San Francisco.

Henderson, R., Baldwin, J. M., Downing, K. H., Lepault, J., and Zemlin, F. (1986). Structure of purple membrane from *Halobacterium halobium*: Recording, measurement and evaluation of electron micrographs at 3.5 Å resolution. *Ultramicroscopy* **19**:147–178.

Jap, B. K. and Glaeser, R. M. (1980). The scattering of high-energy electrons. II: Quantitative validity domains of the single-scattering approximations for organic crystals. *Acta Crystallogr. A* **36**:57–67.

Radermacher, M. (1988). Three-dimensional reconstruction of single particles from random and non-random tilt series. *J. Electron Microsc. Technique* **9**:359–394.

Radon, J. (1917). Über die Bestimmung von Funktionen durch ihre Integralwerte längs gewisser Mannigfaltigkeiten. *Ber. Verh. K. Sächs. Ges. Wiss. Leipzig, Math.-Phys. Kl.* **69**:262–277. For an English translation, see Deans (1983).

Reichelt, R. and Engel, A. (1984). Monte Carlo calculations of elastic and inelastic electron scattering in biological and plastic materials. *Ultramicroscopy* **13**:279–294.

Reimer, L. (1989). *Transmission Electron Microscopy*, 2nd ed. Springer, Berlin and New York.

Reimer, L. and Sommer, K. H. (1968). Messungen und Berechnungen zum elektronenmikroskopischen Streukontrast für 17 bis 1200 keV-Elektronen. *Z. Naturforsch.* **23a**:1569–1582.

Robards, A. W. and Sleytr, U. B. (1985). *Low Temperature Methods in Biological Electron Microscopy*. Elsevier, New York.

Saxton, W. O. (1986). Focal series restoration in HREM, in *Proc. XIth Int. Cong. Electron Microscopy*, suppl. to *J. Electron Microsc.* **35**, Post-deadline paper 1.

Schiske, P. (1968). Zur Frage der Bildrekonstruktion durch Fokusreihen, in *Electron Microscopy 1968* (D. S. Bocciarelli, ed.), Vol. I, pp. 147–148. Tipografia Poliglotta Vaticana, Rome.

Schiske, P. (1973). Image processing using additional statistical information about the object, in *Image Processing and Computer-Aided Design in Electron Optics* (P. W. Hawkes, ed.), pp. 82–90. Academic Press, New York.

Schiske, P. (1982). A posteriori correction of object tilt for the CTEM. *Ultramicroscopy* **9**:17–26.

Spence, J. C. H. (1988). *Experimental High-resolution Electron Microscopy*, 2nd ed. Oxford University Press, New York.

Steinbrecht, R. A. and Zierold, K., eds. (1987). *Cryotechniques in Biological Electron Microscopy*. Springer, Berlin and New York.

Steven, A. C. (1981). Visualization of virus structure in three dimensions, in *Methods in Cell Biology*, Vol. 22, *Three-Dimensional Ultrastructure in Biology* (J. N. Turner, ed.), pp. 297–323. Academic Press, New York.

Turner, J. N., ed. (1981). *Methods in Cell Biology*, Vol. 22, *Three-Dimensional Ultrastructure in Biology*. Academic Press, New York.

Valentine, R. C. (1966). The response of photographic emulsions to electrons. *Adv. Opt. Electron Microsc.* **1**:180–203.

Zeitler, E., ed. (1982). Cryomicroscopy and radiation damage. *Ultramicroscopy* **10**:1–178.

Zeitler, E., ed. (1984). Cryomicroscopy and radiation damage. II. *Ultramicroscopy* **14**:161–316.

Zemlin, F. (1989). Dynamic focusing for recording images from tilted samples in small-spot scanning with a transmission electron microscope. *J. Electron Microsc. Technique* **11**:251–257.

FURTHER READING

Frank, J. (1973). Computer processing of electron micrographs, in *Advanced Techniques in Biological Electron Microscopy* (J. K. Koehler, ed.), pp. 215–274. Springer, Berlin and New York.

Frank, J. (1980). The role of correlation techniques in computer image processing, in Hawkes (1980), pp. 187–222.

Frank, J. (1981). Introduction *and* Three-dimensional reconstruction of single molecules, in *Methods in Cell Biology*, Vol. 22, *Three-Dimensional Ultrastructure in Biology* (J. N. Turner, ed.), pp. 119–213, 325–344. Academic Press, New York.

Frank, J. (1989). Image analysis of single macromolecules. *Electron Microsc. Rev.* **2**:53–74.

Hawkes, P. W., ed. (1980). *Computer Processing of Electron Microscope Images*. Springer, Berlin and New York.

Henderson, R. and Glaeser, R. M. (1985). Quantitative analysis of image contrast in electron micrographs of beam-sensitive crystals. *Ultramicroscopy* **16**:139–150.

Hoppe, W. and Hegerl, R. (1980). Three-dimensional structure determination by electron microscopy (nonperiodic specimens), in Hawkes (1980), pp. 127–185.

Hoppe, W. and Typke, D. (1979). Three-dimensional reconstruction of aperiodic objects in electron microscopy, in *Advances in Structure Research by Diffraction Methods* (W. Hoppe and R. Mason, eds.), Vol. 7, pp. 137–190. Vieweg, Braunschweig.

Lewitt, R. M. (1983). Reconstruction algorithms: transform methods. *Proc. IEEE* **71**:390–408.

Mellema, J. E. (1980). Computer reconstruction of regular biological objects, in Hawkes (1980), pp. 89–126.

Moody, M. F. (1990). Image analysis of electron micrographs, in *Biophysical Electron Microscopy* (P. W. Hawkes and U. Valdrè, eds.), Academic Press, New York.

Robinson, D. G., Ehlers, U., Herken, R., Herrmann, B., Mayer, F., and Schürmann, F.-W. (1987). *Methods of Preparation for Electron Microscopy, An Introduction for the Biomedical Sciences.* Springer, Berlin and New York.

Sommerville, J. and Scheer, U., eds. (1987). *Electron Microscopy in Molecular Biology, A Practical Approach.* IRL Press, Oxford and Washington.

Stewart, M. (1988). Introduction to the computer image processing of electron micrographs of two-dimensionally ordered biological structures. *J. Electron Microsc. Technique* **9**:301–324.

Stewart, M. (1988). Computer image processing of electron micrographs of biological structures with helical symmetry. *J. Electron Microsc. Technique* **9**:325–358.

3

Sample Shrinkage and Radiation Damage

Pradeep K. Luther

1. INTRODUCTION

For successful electron microscope tomography, an understanding of the effects of the electron beam on the specimen is important. Radiation damage in the electron microscope manifests itself in several ways: loss of high-resolution structure, mass loss, and specimen shrinkage. These effects must be considered for biological materials of all types in different preparations. The methods employed to minimize the effects of electron irradiation include low-dose imaging techniques and cooling the sample to liquid nitrogen or liquid helium temperatures.

Pradeep K. Luther • The Blackett Laboratory, Imperial College, London SW7 2BZ, England

Electron Tomography: Three-Dimensional Imaging with the Transmission Electron Microscope, edited by Joachim Frank, Plenum Press, New York, 1992.

For a study of high-resolution structures with better than 10 Å detail, the material must be unstained. Unstained samples are highly radiation sensitive, as observed by the rapid fading of the electron diffraction pattern. Hence, very low doses (<1 e/Å^2) must be used for recording the images, which are consequently very noisy (Amos *et al.*, 1982). To improve the signal-to-noise ratio and show the underlying structure, one needs to use highly ordered 2D material so that averaging can be done. Some aperiodic objects can be averaged albeit with greater difficulty (see Frank, 1990). In the case of aperiodic objects that cannot be averaged, staining is a necessity. Staining with heavy-metal salts enhances the contrast and makes the specimen less sensitive to radiation, but the resolution is limited to $\sim$20 Å. One of the main effects of the electron beam at this lower resolution is specimen shrinkage normal to the plane of the grid. Shrinkage of this type (also called collapse or flattening) does not affect the image of the untilted sample since the image is formed as a projection along the same direction. However, if different views of a sample are recorded for carrying out 3D reconstruction, by tilting the sample through various angles, then shrinkage causes dimensional changes of structure normal to the sample plane. Any shrinkage that occurs during the period when tilt views are being recorded blurs the 3D reconstruction. Since the thickness found from the 3D reconstruction is the final reduced value, an independent estimate of the thickness is required. Past studies have shown that the shrinkage is uniform through the depth of the material (Bennett, 1974; Luther *et al.*, 1988; Jesior and Wade, 1987). The uniform shrinkage allows a simple geometrical factor to be applied on the final reconstruction in order to scale it to the independently determined starting thickness.

In this chapter, the effects of electron irradiation on biological samples are described, especially those related to shrinkage. The types of sample discussed include bulk material which has been cryosectioned or resin embedded and sectioned and particulate material which is crystalline or noncrystalline. The effects of radiation appear to be similar for these different preparations. Although the emphasis here will be on resin-embedded and sectioned material, there are more quantitative results about radiation damage effects on crystalline material. Methods to measure sample thickness prior to and during electron microscopy will be described. Finally, we suggest strategies to minimize the effects of radiation damage.

2. THEORETICAL ASPECTS OF RADIATION DAMAGE

Several researchers have written reviews on the effects of the electron beam on biological samples (Stenn and Bahr, 1971; Grubb, 1974; Lamvik, 1991). Electron microscope radiation has the primary effect of producing intense ionization in organic materials, which results in the formation of free radicals and ions. This causes bond scission and molecular fragments are formed. These primary effects appear to occur at all temperatures (Siegel, 1972). At room temperature, the free radicals and molecular fragments can undergo diffusion and produce cross-linking or further chain scission. Damage to the secondary structure occurs at an electron dose of less than 1 e/Å^2. Further exposure causes the tertiary structure to undergo dramatic reorganization following the loss of specific groups and altered structural composition. The dominant effect finally is that of mass loss from the sample, which preferentially involves H and O in comparison with C and N. The mass loss is accompanied or followed closely by shrinkage of the sample normal to the beam.

For samples cooled to liquid helium temperatures, Siegel (1972) has proposed that the primary damaging effects of electron irradiation appear to be unchanged, but the resultant molecular fragments remain trapped. The preservation of structure is shown by the persistence of the diffraction intensities (Siegel, 1972; Wade, 1984). Siegel (quoted in Wade, 1984) showed very elegantly that a sample exposed to half the latent dose (the latent dose is defined as the dose which causes the spot intensity to reduce to 1/2 of the initial value) at low temperatures displayed no noticeable change, but when warmed up, showed the same crystalline disorder as a sample exposed to the same dose at room temperature.

As mentioned in the introduction, one of the main effects of electron irradiation at conventional illumination levels is to cause specimen shrinkage normal to the plane of the sample. The effect of the shrinkage in reciprocal space is illustrated in Fig. 1. The example is based on a structure with cubic symmetry (a), which collapses in thickness by 50% (c). With a conventional tilt holder, a series of views about a single-tilt axis are recorded in the range $-60°$ to $+60°$. The 3D transform in (b), obtained by combining the transforms of the individual views, has missing

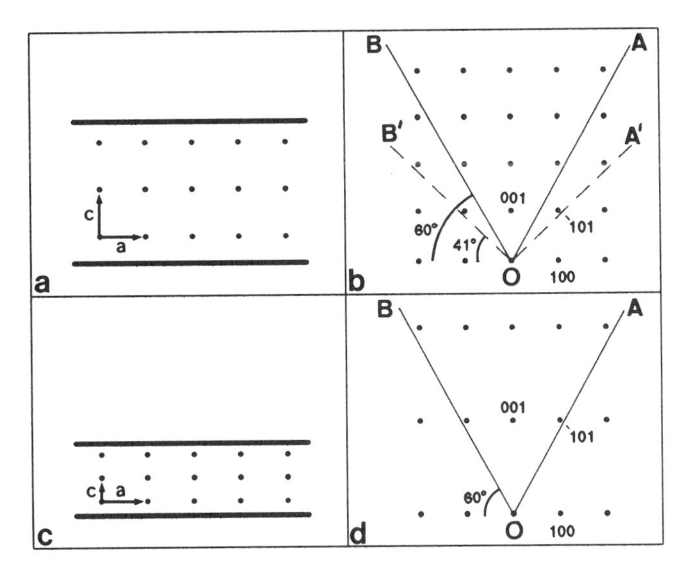

FIGURE 1. Illustration of the worsened "missing cone" (in this case, "missing wedge") problem due to specimen shrinkage in the electron microscope. (a) Projection of a cubic crystal (for example) viewed edge-on with the *c* axis parallel to the electron beam. (b) Reciprocal lattice for the projection in (a). A conventional tilt holder in the electron microscope can be tilted in the range $-60°$ to $+60°$. When a tilt series is recorded about a single-tilt axis and the 3D transform is calculated by combining the individual transforms, data will be missing from the 3D transform inside the "wedge" *AOB* of angle 60°. If the sample thickness reduces by 50% as in (c), then the corresponding reciprocal lattice shown in (d) is stretched 100% along c^*. The projection for the 101 diffraction spot, comfortably included in the tilt series for the unshrunk sample (a, b) now lies within the missing wedge. In relation to the original reciprocal lattice, the missing wedge is now described in (a) by *A′OB′* and has an angle of 98°. For 50% shrinkage, the tilt holder effectively only covers the range $-41°$ to $+41°$ in relation to the original sample.

from it data in the 60° wedge *AOB*. In the case of the 50% collapsed sample (c), the reciprocal space is stretched 100% in the corresponding direction (d). The volume of the 3D transform that can be sampled is now much reduced: e.g., the spot 101 present in (b) is missing from the transform (d). In relation to the original sample, the missing wedge (*A'OB'*) in the 3D transform has an angle of 98°. The effective tilt range is now only ±41°. Sample shrinkage therefore directly reduces the resolution normal to the sample plane, and efforts must be made to curtail the shrinkage.

It is appropriate to describe the terminology of Amos *et al.* (1982) for the various imaging modes in the electron microscope and the electron dose involved in each case. A total dose on the sample of 0.5 to 4 e/Å^2 is considered as a very low dose, which is appropriate for very high resolution studies of unstained crystalline specimens. A total dose around 10–20 e/Å^2 is termed a minimal dose and is used for stained or noncrystalline specimens. Conventional microscopy involves doses of ~50–500 e/Å^2 due to the time involved in searching and focusing. Minimal- and low-dose methods require a search of suitable areas to be done at very low magnification, about ×2000, during which the dose should be extremely low (~0.02 e/Å^2). The focusing is done on an adjacent area at high magnification. The image is then recorded at the desired magnification. Most modern electron microscopes provide low-dose imaging modes.

3. *METHODS OF MEASURING SAMPLE THICKNESS*

3.1. *Introduction*

Knowledge of the starting thickness of a sample is necessary for carrying out 3D reconstructions and their interpretation. Knowledge of the changes that occur to the sample dimensions during electron microscopy is also important. Methods to measure the initial thickness of a sample and changes that occur following the microscopy are described in this section.

3.2. *Methods Not Involving the Electron Microscope*

3.2.1. Ultramicrotome Advance Setting

The advance setting on an ultramicrotome is potentially a good estimate of the thickness of a section. The microtome setting may apply especially to sections belonging to a well-cut ribbon in which the sections have uniform interference colour. However, past research has shown that there is considerable variation in section thickness within a ribbon. Gunning and Hardham (1977) used a Reichert OMU3 ultramicrotome to cut ribbons of sections from a Spurr's resin block. By monitoring the thickness of the individual sections by interference microscopy (next section), they found a 33% variation in the thickness of the sections about the mean value. Ohno (1980) evaluated the section thickness in relation to the microtome setting. He used sections of Epon-embedded rat kidney. By resectioning the sections transversely (see later), he found a variation of 20% for sections of thickness less than 100 nm. For thicker sections ranging from 0.2 to 0.9 μm, the thickness was found to be closer to the microtome setting with a relative variation of about 7%.

3.2.2. Interference Color Scale

The actual thickness of sections cut at a particular advance setting on a microtome can be judged from the interference color of the sections as they float in the knife trough. The thickness scale of interference colors proposed by Peachey in 1958 is probably the most widely used. These measurements were done on methacrylate sections with an ellipsometer. Several groups have reported sightly differing scales (e.g., Porter and Blum, 1953; Williams and Meek, 1966). The author has found the thickness scale of Yang and Shea (1975) based on Epon-Araldite sections and measured by the resectioning technique (see later) to be more useful, especially for the important range of less than 100 nm (Table 1).

In general, the interference color of a section is only used as a rough guide to its thickness. This is probably because, with current methods, thickness estimation from the observed color is subjective. It is the author's view that more quantitative estimates can be obtained from the interference colors. For this to be possible, two provisions are essential: (i) an accurately reproduced color scale and (ii) a standard light source. Since different light sources have different spectral properties, thus giving different color biases, a standard light source used to illuminate the sections and the color scale simultaneously will provide a standard scale. Perhaps a single manufacturer could provide these items worldwide.

The interference color of a section becomes even more valuable when it is realized that the color of a section mounted on an uncoated grid viewed subsequent to the microtomy is the same color as that observed for the section when it was floating on the trough during the microtomy. The physical basis for the color to be the same in the two situations is as follows. The color observed is due to the interference of the light reflected off the top and bottom surfaces of the section. When light reflects from a surface of higher refractive index, there is a phase change of $\pi/2$. This change occurs only for the reflection off the top surface of a section. Reflection off the bottom surface does not produce a phase change, since the refractive index of a resin section (~ 1.54) is greater than that of water (1.3) when

TABLE 1

	Thickness (nm)	
Interference color	Peachey (1958)	Yang and Shea (1975)
Gray	<60	<40
Silver	60–90	40–60
Yellowish silver		60–67
Pale gold		78–90
Gold	90–150	90–100
Dark gold		100–110

the section is on the trough, and of air (1.0) when a section is viewed after the microtomy. There appears to be little effect of any absorbed water or satin.

Well-cut cryosections mounted on coated grids, negatively stained and dried, also show interference colors when viewed in white light (Morris and Luther, unpublished results). By taking into account the thickness of the support film judged from its interference color, or by using holey-support films, one could estimate the thickness of the cryosections.

3.2.3. Interference Microscopy

An important method of determining the thickness of thin transparent samples uses the transmitted light interference microscope. The technique has been used in

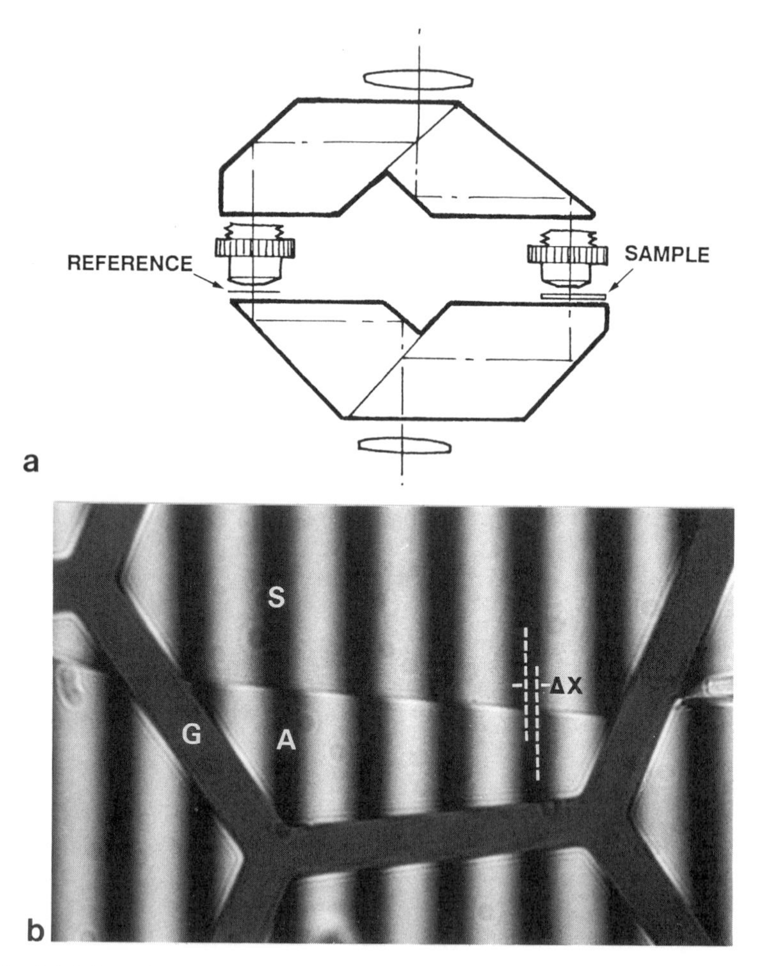

FIGURE 2. (a) Schematic representation of the Mach–Zehnder interference microscope. In this microscope, the incident light is divided into two beams which pass through the test sample and the reference. (b) Fringe pattern across a section and grid in the interference microscope. A section (S) mounted on a grid (G) causes a shift Δx in the fringe pattern relative to that in air (A).

several studies on the measurement of section thickness (Cosslett, 1960; Gillis and Wibo, 1971; Gunning and Hardman, 1977). Unfortunately, transmitted light interference microscopes do not appear to be manufactured commercially any more. However, they are much cherished and can be tracked down in a few microscopy or physics research laboratories.

The use of an interference microscope fitted with a Jamin–Lebedeff (Mach–Zehnder principle) interference system has been described by Gillis and Wibo (1971) and Spencer (1982). In this microscope (Fig. 2*a*), a polarized beam of light is divided in two. The measuring beam passes through the transparent sample, and the reference beam passes through the reference medium. A small path difference in the measuring beam is introduced by the sample, and it can be measured from the shift in the interference fringes obtained on combining the two beams (Fig. 2*b*). The thickness t is calculated from the fractional shift $\Delta x/x$ of the fringes, fringe separation x, and the wavelength of the light λ employed:

$$(\mu - n)\, t = \frac{\Delta x}{x} \cdot \lambda$$

where μ is the refractive index of the section, about 1.54 for epoxy resins, and n is the refractive index of the reference medium ($=1$ for air).

Fig. 2*b* shows an example of the fringe pattern as viewed in a Mach–Zehnder interference microscope and the shift in the pattern caused by a section mounted on a grid. The relative shift in the fringe pattern needs to be determined as accurately as possible. The method used by Gornall and Luther (unpublished) for measuring the shift in the fringe pattern involves photographing the section and its surroundings through the microscope and scanning the negative with a microdensitometer. On the computer, we integrate separately the fringe patterns over the section and over air, and find the shift by correlating the two arrays. Cosslett (1960) and Gillis and Wibo (1971) estimate that the error involved is about 2% when 10 or more measurements are done.

3.3. Methods of Thickness Measurement by Electron Microscopy

3.3.1. Folds

One of the simplest methods of estimating the thickness of a section during electron microscopy is by measurement of the width of any folds that may occur in the section (Small, 1968). Folds or crimps sometimes occur during electron microscopy on sections that are not well stuck to the grid coating. The minimum width of the fold is equal to twice the section thickness. Recording of such views must be done at low or minimal electron doses in order to avoid errors due to shrinkage.

3.3.2. Thickness Measurement from Cross Sections

Probably the most direct method of measuring the sample thickness is by embedding the sample, cutting cross sections, and viewing the cross sections in the electron microscope. The main drawback of the method is the loss of the sample. There is the dilemma of cutting cross sections from unviewed samples or from radiation-damaged samples after electron microscopy. If the samples used are resin sections, Bedi (1987) has suggested the following remedy. After a block face is trimmed a small score is applied. Upon sectioning, two ribbons are obtained of which one can be used for experimental viewing in the electron microscope and the other for measuring thickness by the above method.

Cutting cross sections at any random position in a sample is straightforward. However, if one wants to cut cross sections across a region precisely identified by electron microscopy, for example, across the region where tilt views were obtained, then the meticulous methods developed by Jesior (1982) are recommended. Although most of his work was done on negatively stained crystalline material, the method applies just as well to positively stained resin sections. Jesior's method requires application of latex particles along with the test material onto a coated finder grid. The size of the latex particles (1.3 μ) is chosen to enable viewing at electron and light microscopy levels. The method is illustrated in Fig. 3. In the electron microscope the grid is viewed and the path of the cross section required is noted relative to the grid holes. At the same time the position of the required object is measured carefully in terms of its distance and angle relative to the edges of the grid hole, and the latex particles in the vicinity are mapped. The whole grid is reembedded in

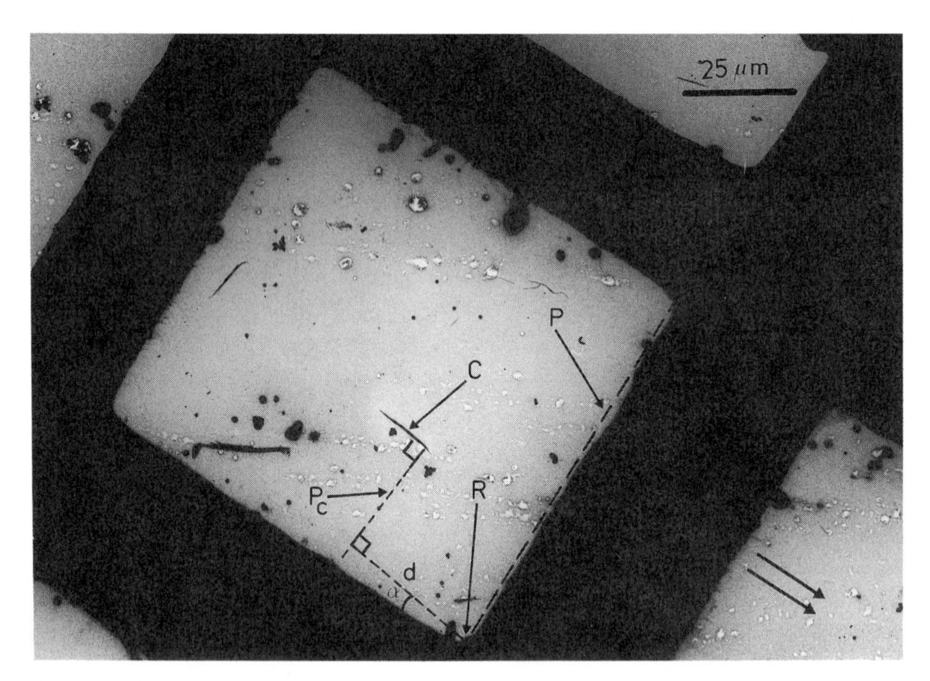

FIGURE 3. Method for cutting cross sections of a specimen along a path precisely defined by electron microscopy (From Jesior, 1982). For the chosen sample, in this case a fibrinogen crystal (labeled C), the surrounding latex particles are mapped and the reference plane *P* and distance *d* of the reference plane to the crystal are noted. After embedding, the block is trimmed and then oriented in the microtome so that the principal plane is parallel to the knife. The required area is approached by cutting several thick sections. (From Jesior, 1982; courtesy of J-C. Jesior.)

epoxy resin, and the block is then trimmed with a glass knife. Once the grid bars surrounding the required object are exposed, the block face and the sloping sides are cut. The required region is then approached by first cutting on the microtome a precise number of thick sections. A diamond knife can then be safely used for cutting the required sections. Sections from material stained prior to embedding do not require restaining.

3.3.3. Thickness Measurement from Internal Periodicity—Bennett Method

An elegant method devised by Bennett (1974) exploits the internal periodicity of a sample to measure the effect of electron irradiation on the section thickness. Bennett used a pellet of paracrystals of light meromyosin (LMM), which is the α-helical rod part of a myosin molecule, fixed and embedded in Araldite. The LMM paracrystals viewed directly in negative stain show a strong banding pattern with a period of 43 nm along their length. In the embedded pellet, the paracrystals have a variety of sizes and orientations. Consequently, in a section cut from the block, the majority of the paracrystals do not show any banding pattern, since they do not lie in the plane of the section. Tilting the section perpendicular to the length of a selected paracrystal brings the banding pattern into view. Some paracrystals fortuitously oriented in the plane of the section show a sharp banding pattern in the untilted section. Bennett found that the periodicity measured from the in-plane paracrystals was different from that measured in paracrystals which required tilting of the section. The paracrystals which required tilting had a smaller periodicity than that predicted by the geometry. By assuming a shrinkage to a fraction f of the original thickness, she derived the relationship between the tilt angle ϕ, required to view the banding pattern sharply, the true periodicity p, and the measured periodicity q (Fig. 4):

$$q^2 = p^2 - p^2(1 - f^2)\sin^2\phi$$

Plotting q^2 versus $\sin^2\phi$ allows the relative thickness and periodicity to be found. Luther *et al.* (1988) modified the above equation by introducing a term to account for the in-plane shrinkage.

A serious limitation of the method is that a variable amount of time is required in selecting suitable paracrystals and in tilting the section until the banding pattern is sharp. The high accumulated dose means that this method is more suited for the measurement of the final thickness in conventional-dose electron microscopy. Bennett noted a 50% reduction in the thickness of Araldite sections.

This method gives changes in thickness of a section, not the absolute value. But the principle involved, that of measuring the relative thickness of a section from a knowledge of the periodicity, is widely applicable, as is shown later for cryosections (Sjostrom *et al.*, 1991).

3.3.4. Electron Diffraction

Berriman and Leonard (1986) have used a method first proposed by Dorset and Parsons (1975) for using the electron diffraction pattern to measure the

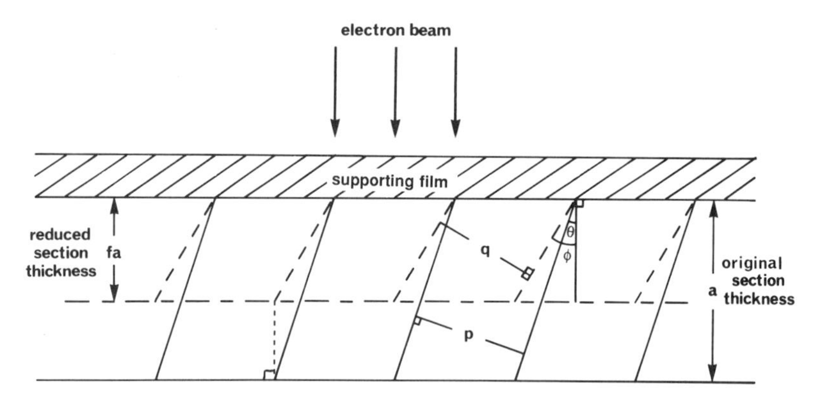

FIGURE 4. Measurement of the relative shrinkage in periodic objects (Bennett, 1974). This method is suitable for samples which have 3D order or 2D order out of the plane of the sample; i.e., the sample has to be tilted in the electron microscope to visualize the repeating pattern. For a sample with original periodicity p, the electron beam causes a reduction in thickness by a factor f and a new smaller periodicity q. Consequently, the sample has to be tilted to a larger angle ϕ to visualize the banding pattern. (From Bennett, 1974.) The fact that the banding pattern for paramyosin crystals, as used by Bennett (1974), and tropomyosin paracrystals, as used by Luther *et al.* (1988), can be visualized sharply at the reduced thickness state shows that the shrinkage occurs uniformly through the depth of the sample. The reduction in thickness does not occur due to, for example, etching from each of the surfaces because that would require that the periodicity remain unchanged.

thickness of crystals several unit cells thick. To understand this method, note that the 3D Fourier transform of a crystal one unit cell thick consists of a reciprocal lattice of spikes perpendicular to the a^*b^* plane. The diffraction pattern of such a crystal tilted to $\theta°$ samples the 3D transform along the plane tilted to the same angle (Fig. 5*a*). By tilting the crystal at various angles, we sample the 3D transform at different planes and information is built up for 3D reconstruction. For a crystal more than one unit cell thick, the spikes in the 3D transform are broken up into layers or zones (Fig. 5*b*). The width of these Laue zones is related to the thickness of the crystal, and the separation of the zones is related to the unit cell spacing c. From the tilt angle, c can be calculated geometrically from the separation of the zones. The advantage of this method is that it does not require any specific orientation of the crystal relative to the beam, only a knowledge of the tilt angle. Hence, it allows low-dose experiments to be performed.

3.3.5. Use of Gold Particles

The application of colloidal gold particles to a sample for determining its thickness has been investigated by Berriman *et al.* (1984) and Luther *et al.* (1988). Since the mathematics relating to the latter method is described in Chapter 8, only the basic geometry involved will be described here.

The stereo pair in Fig. 6 shows a section labeled with gold particles on both surfaces. Using gold particles of different diameters (5 and 15 nm, respectively) to mark each surface allows areas labeled simultaneously on the top and bottom surfaces to be identified while being viewed in the electron microscope. The

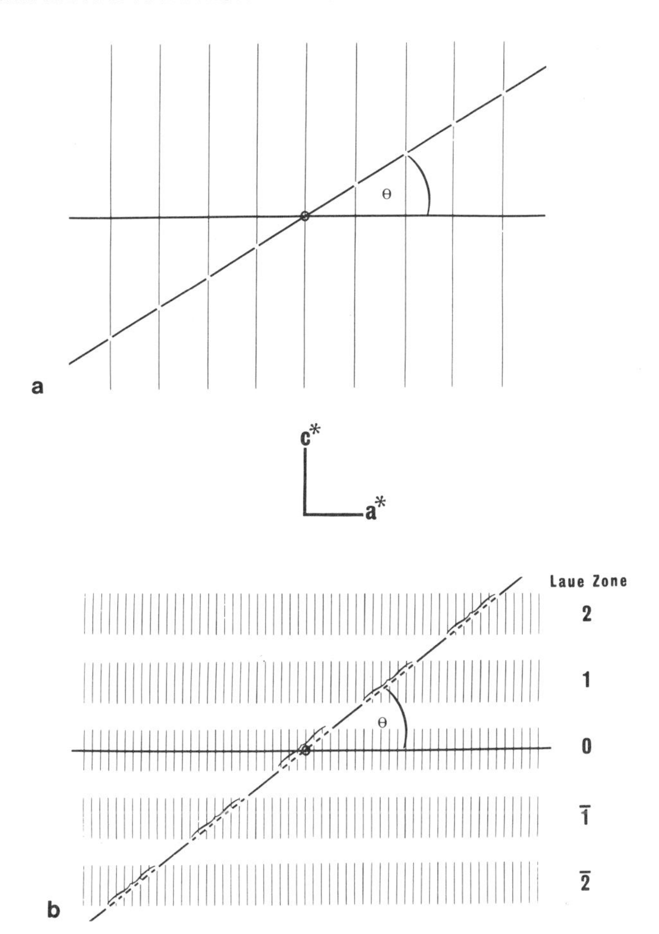

FIGURE 5. Electron diffraction method for measuring lattice spacing and crystal thickness for a thin crystal (Berriman and Leonard, 1986; Dorset and Parsons, 1975). The c axis is normal to the plane of the crystal. (a) For a crystal 1 unit cell thick, the diffraction intensities are continuous spikes which can be sampled by tilting at any angle. (b) For a crystal a few unit cells thick, the diffraction intensities lie along bands or zones with a periodicity equal to c^*. The thickness of the crystal can be calculated from the width of the bands (Dorset and Parsons, 1975). This method allows relative thickness measurements (from c) to be found using low-dose conditions. (From Berriman and Leonard, 1986).

geometry of the method, illustrated in Fig. 7, is derived from the edge-on view of a section tilted to an angle β. Particles 1 and 2 mark one surface, and particle 3 marks the other surface. They are imaged on the micrograph as 1′, 2′, and 3′. The coordinate system for the section has the y axis parallel to the tilt axis, the x axis in the plane of the section, and the z axis normal to the plane. Coordinates u, v are measured on the micrograph, with the u axis parallel to the x axis. If the shrinkage in the plane is m_x and the shrinkage normal to the plane is m_z, then

$$u = m_x x \cos \beta + m_z z \sin \beta \tag{1}$$

$$v = m_x y \tag{2}$$

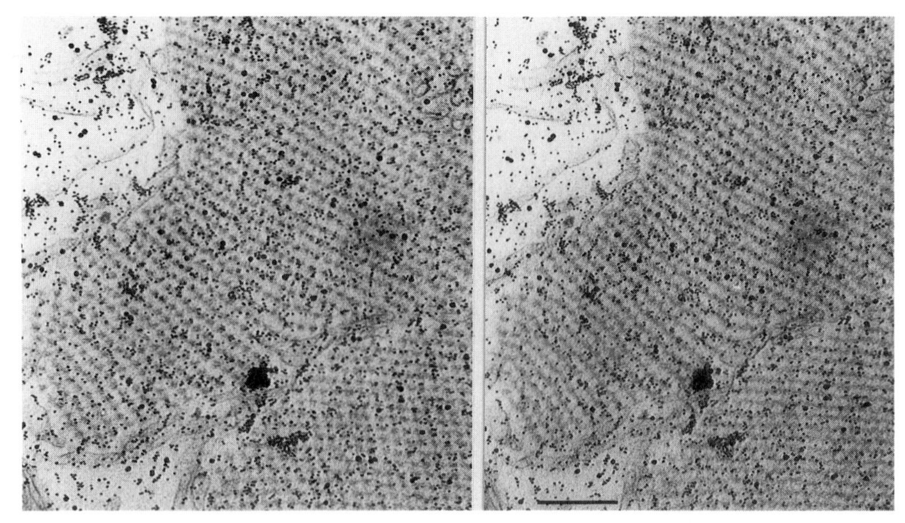

FIGURE 6. Thickness determination by labeling with gold particles. Stereo view of a section marked with gold particles. The two surfaces of the section are labeled with gold particles of different sizes to enable identification in the electron microscope of areas labeled on both sides. Scale bar 200 nm.

Shrinkage in the plane of the section m_x can be found from (2), and then the shrinkage m_z normal to the plane can be found.

At the start of the experiment, the sample is tilted to 45° in the electron microscope, a suitable region is found at very low magnification, and the microscope is focused at an adjacent area. The effects of very low doses can be investigated since the minimum potential dose is the amount required to record one micrograph. Note that the applied gold particles can obscure valuable structural details. Therefore, the particles must be applied carefully to ensure suitable distribution on each surface. The method used by the author is to float a grid onto a drop of the colloidal gold

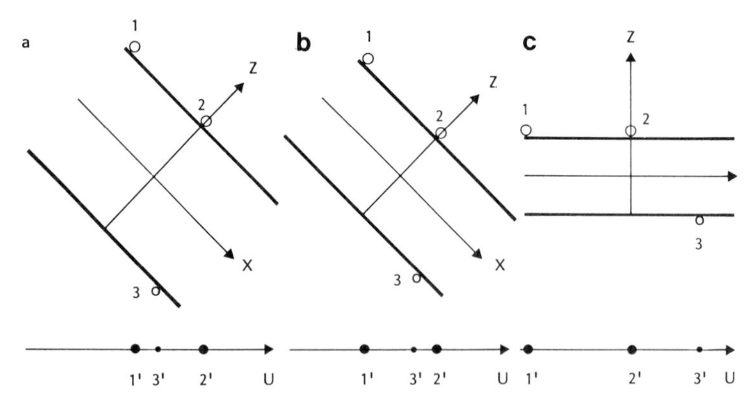

FIGURE 7. Geometry of a section labeled with gold-particles viewed edge-on to measure section thickness and collapse. Gold particles 1 and 2 mark one of the surfaces and particle 3 (smaller) marks the other. (a) Section tilted to 45°. In the micrograph the images of the three particles occur at 1′, 2′, and 3′. (b) Subsequent state after section collapse. In the micrograph, 3′ has moved relative to 1′ and 2′. (c) Final image taken at 0° tilt. (From Luther *et al.*, 1988.)

solution for about 30 s and then blot the grid lightly and allow it to air-dry. This is repeated for the other grid surface. The grid is finally rinsed gently or poststained (with uranyl acetate and Reynolds lead citrate). An excellent method of preparing colloidal gold particles of different sizes is described by Slot and Geuze (1985).

4. EFFECTS OF IRRADIATION ON STAINED SAMPLES AT ROOM TEMPERATURES

4.1. "Collapse" of Resin Sections

In her pioneering work on the effect of the electron beam on sections, Cosslett (1960) used the interference microscope to measure the section thickness before and after the electron microscopy. She used two media for the reference beam, which allowed her to measure changes in thickness and refractive index of the section. The resins examined were methacrylate, Araldite, Vestopal, and Aequon (water-miscible medium). For methacrylate the effect of the beam was the greatest: the sections reduced in thickness by 50% of the starting value. Sections of the the other resins reduced in thickness by 20–30%, in some cases by 50%. In each case Cosslett noted that the refractive index increased from 1.54 to 1.9. The increase in refractive index was attributed to changes in internal structure of the resins. The work of Bennett (1974), using sections of LMM paracrystals embedded in araldite, showed a reduction in section thickness by 50% following electron microscopy at conventional doses.

A detailed study of the variation in thickness with electron dose was carried out by Luther *et al.* (1988), using the gold particles method. Sections of tropomyosin paracrystals embedded in araldite were used. Their results are shown in Fig. 8. In each panel, the upper trace shows the shrinkage measured in the plane of the section, and the lower trace shows the shrinkage measured in the depth. There are two parts to each curve. First low dose rates were used, ~ 0.5 (e/Å^2)/s. Then this was increased to 4 (e/Å^2)/s. Despite the low dose rates used at a magnification of 20,000×, "strict" minimal dose conditons could not be used to record the image of a pristine section because the beam, although weak, caused sudden large planar movements in the section and, occasionally, breakage of the section. Therefore the sample was first viewed at very low magnification, and then the selected area was translated into the beam by hand. There were still gross movements within the irradiated region, and when the first image was recorded after 15 s of irradiation (allowing the section to stabilize), the section had already experienced a dose of 7 e/Å^2. The results in Fig. 8 show that there is a rapid collapse to 70% within 3 min (an accumulated dose of 90 e/Å^2) and then the shrinkage levels off. The collapse occurred in a similar fashion in areas with resin only (Fig. 8*a–c*) and in areas with embedded tissue regions (Fig. 8*d*). Increasing the dose rate from 0.5 to 5 (e/Å^2)/s (after 20 min in Part *a* and 15 min in Parts *b*, *c*, *d*) resulted in a new shrinkage curve, which leveled off at a final thickness of 60% of the original. Since the curves level off after each set dose rate level, this indicates that the changes that occur in a sample allow it to dissipate the energy at the set dose rate. This implies that the

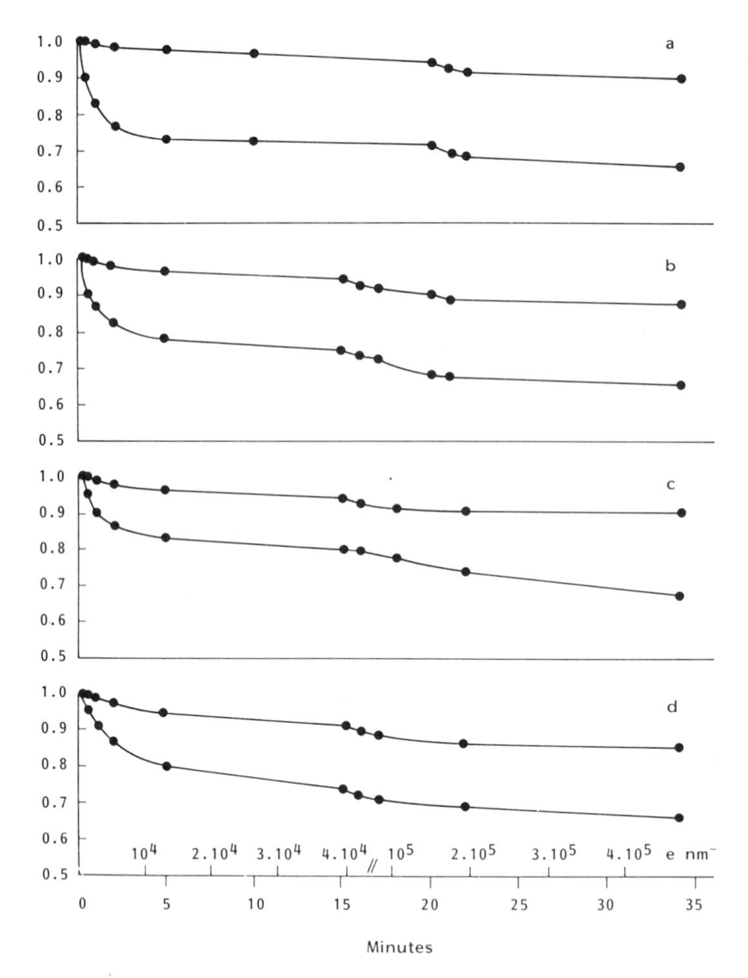

FIGURE 8. Measurement of section collapse (lower trace) and planar shrinkage (upper trace) as a function of electron dose. Panels (a, b, c) are from gold particles over resin only (Araldite), and panel (d) is from gold particles over a paracrystal. The dose rate, initially 0.5 (e/Å^2)/s, was increased to 4 (e/Å^2)/s after 20 min for (a) and after 15 min for (b, c, d). (From Luther *et al.*, 1988.)

electron microscopy should be carried out at the lowest dose rate possible for a sample. The shrinkage in the plane of the section following irradiation is much lower. There is a small shrinkage of 5% at the lower dose level, and a further 5% at the higher dose level.

Luther *et al.* (1988) also discussed the "clearing" observed with a stained section after very low doses (Fig. 9). The effect seen is that of enhanced contrast of the stained material due to the increased transparency of the surrounding resin. To demonstrate the effect, a small disk within the field of view was irradiated with a dose of 0.01 e/Å^2 at low magnification ($\times$3000). Then the beam was spread to cover a larger region. With the resulting reduced illumination, the exposed disk and the immediate surround were photographed. The process of photography causes the surround to be slightly irradiated as well. In the micrographs the main exposed

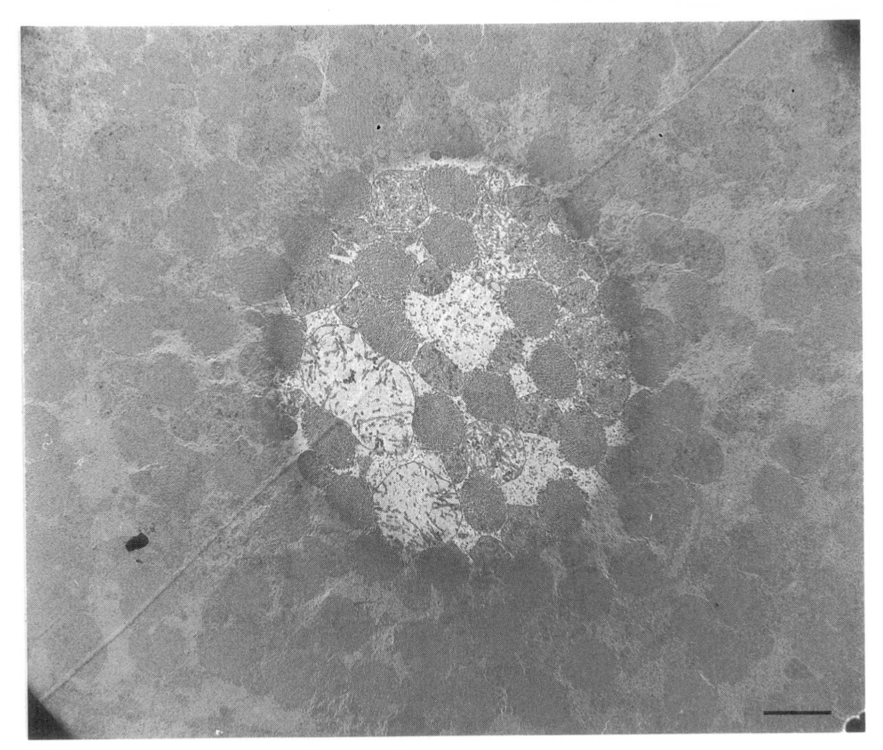

FIGURE 9. The "clearing" effect as observed in an Araldite section of *Drosophila* flight muscle. The electron dose on the "cleared" central patch is 0.1 e/Å^2, compared with the much lower dose (0.02 e/Å^2) for photography for the surrounds (Luther *et al.*, 1988). Scale bar = 2 μm.

region showed a marked improvement in contrast of the stained material compared to the slightly exposed surround. For the total dose received, ~0.15 e/Å^2, the shrinkage curves in Fig. 8 imply that no significant shrinkage occurred. Hence, Luther *et al.* concluded that clearing could be due to mass loss or some change in chemical structure which causes less electron scattering and that the effect must precede the depth shrinkage.

Some reduction in radiation sensitivity may be provided by subjecting resin blocks to microwave radiation from a domestic microwave oven (Kinnamon and Young, 1989). This suggestion has been tested by Hillenbrand and Luther (unpublished results). Blocks of araldite-embedded muscle were placed for 2 min in a domestic microwave oven operating at full power. Sections were cut and labeled with gold particles, and the response to the accumulated electron dose was measured. Although the collapse response followed a similar course to samples not treated with microwave irradiation (Fig. 8), the sections appeared more stable and were less prone to movement at initial low-dose viewing conditions.

4.2. "Flattening" of Negatively Stained Samples

Jesior and Wade (1987) examined the shrinkage in negatively stained 2D crystalline arrays of the bladder membrane, following "high-dose" electron microscopy with a total dose of 1000 e/Å^2. By embedding the sample and cutting cross sections, they found that the membrane had "flattened" to 60% of the native thickness. Berriman and Leonard (1986) used electron diffraction to study the variation in thickness of negatively stained crystals as a function of electron dose. Thin crystals of catalase were used. From the electron diffraction pattern the separation of the Laue zones; and hence the *c* spacing of the unit cell, was measured. As shown in trace 5 of Fig. 10, the *c* spacing reduced as the logarithm of the accumulated electron dose. After a dose of 1000 e/Å^2, the crystals had reduced in thickness to 60% of the starting value. Berriman and Leonard also studied how samples that are cooled to liquid nitrogen temperatures are affected in the electron microscope, and these results will be described in a later section.

The effect of electron irradiation on negatively stained cryosections has recently been investigated by Sjostrom *et al.* (1991). The material used was longitudinal cryosections of fish muscle, negatively stained in ammonium molybdate. In the ideal case, tilting such sections about the myofibril axis to view in turn the [10] and [01] projections of the myosin filament hexagonal array would require a total tilt of 60°. However, Sjostrom *et al.* found that the angle of tilt required to observe the two views was about 120°. From this they concluded that the sample had collapsed to nearly 33% of the original thickness. This value includes any shrinkage that may have occurred prior to the electron microscopy.

4.3. Shrinkage Occurs along Directions Lacking Restraint and Is Uniform

Bennett (1974) first suggested that reduction in section thickness occurs in the electron microscope because this is the only dimension of the section which is not supported on each side. In the plane of the section, shrinkage is restrained by attachment of the section to the grid bars or support film. Berriman and Leonard

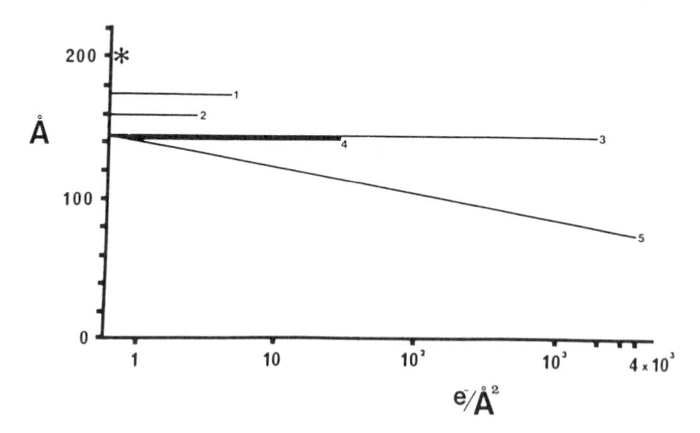

FIGURE 10. Investigation of the *c* unit cell size of thin catalase crystals (hence, crystal thickness) with accumulated dose under different experimental conditions. The * indicates the value (206 Å) found by x-ray diffraction. (1) Frozen hydrated crystals at 120 K. (2) Glucose embedded at room temperature. (3) Dried negatively stained crystals in uranyl acetate at 120 K. (4) Negatively stained at 120 K with a contaminating layer of ice or deposited carbon layer. (5) Negatively stained at room temperature. (From Berriman and Leonard, 1986.)

(1986) have provided strong evidence for this idea (Fig. 11). They applied catalase crystals to a holey carbon-coated grid and viewed carefully those crystals that partly covered the holes. When viewed at very low illumination levels (1 e/Å^2 accumulated dose), crystal edges werc continuous across the hole and the support film. With increased accumulated dose (100 e/Å^2), the part of the crystal across the hole contracted in the plane of the grid such that the edge of the crystal was drawn toward the rest of the crystal. The amount of contraction was estimated by viewing the lattice spacings.

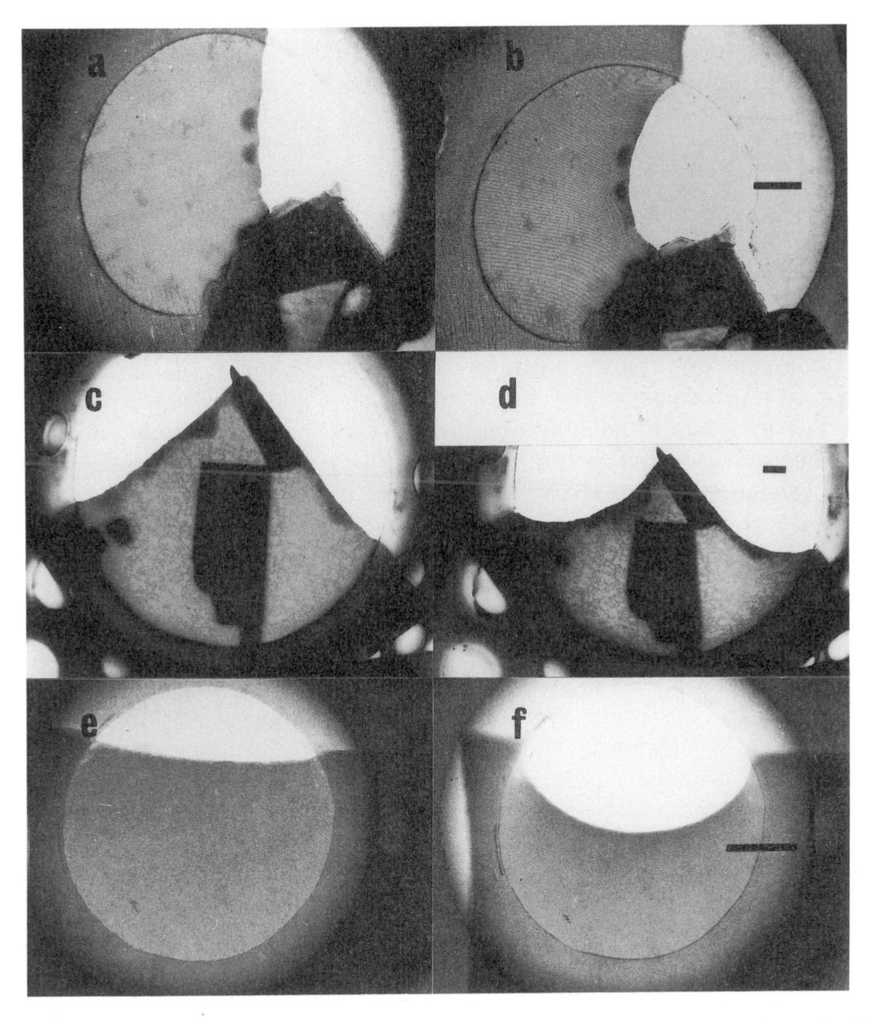

FIGURE 11. Demonstration of sample shrinkage in the electron microscope occurring only along directions lacking constraint. The figure shows images of catalase crystals spanning holes in the supporting film, viewed under low dose, 1 e/Å^2 (left panel), and high dose, 100 e/Å^2 (right panel). In each case the straight edge of the catalase crystals imaged in low-dose conditions gets rounded with accumulated dose. With the points of attachment remaining unaffected, the roundedness is due to shrinkage of the crystal sheets perpendicular to the edges. Scale bars represent 0.1 μm. (From Berriman and Leonard, 1986; courtesy of J. Berriman.)

Concomitant with the dimensional changes, Unwin (1974) has shown that there are changes in the images and diffraction patterns of stained samples, indicating that radiation causes movement in the stain. Berriman and Leonard (1986) suggest that this redistribution of stain may be due to the stain moving along with the parts of the sample which are free to contract in the electron beam.

A consistent result in past studies on sample shrinkage is that the shrinkage is uniform through the depth of a sample. It does not, for example, occur due to etching from the surfaces of a sample, an effect which would cause overall shrinkage without the reduction in periodicity observed in paracrystals (Bennett, 1974; Luther *et al.*, 1988). Jesior and Wade (1987) have also argued that there is good correlation between the projection maps calculated from x-ray data and the negatively stained (Steven and Navia, 1980) and frozen-hydrated projected structure obtained by electron microscopy (Lepault and Pitt, 1984).

5. LOW TEMPERATURES REDUCE RADIATION DAMAGE

5.1. Thin Crystals

It is now well established that radiation-damage effects are greatly reduced by electron microscopy at low temperatures. For example, studies on purple membrane (Hayward and Glaeser, 1979) have shown that at liquid nitrogen temperatures, there is a reduction in radiation sensitivity by a factor of at least 4, as judged by the fading of the diffraction spots.

A careful and systematic study of catalase thickness under various conditions has been carried out by Berriman and Leonard (1986). They monitored the effects of radiation dose by measuring the separation of the Laue zones in the electron diffraction pattern. They also measured the effect of the preparation on starting thickness prior to any electron microscopic irradiation (Fig. 10). From x-ray diffraction of crystals in the hydrated state, the cell c dimension was measured to be 206 Å (* in Fig. 10). With negatively stained crystals, they noted that this dimension was reduced to 70% of the hydrated value at the start of the electron microscope investigation. As mentioned earlier, for the sample kept at room temperature, trace 5 shows a logarithmic shrinkage of a further 50% following a dose of 10^4 e/Å^2. Maintaining the negatively stained crystals at 120 K, they found no change in c for high levels of dose $\sim 2 \times 10^3$ e/Å^2 (trace 3). Negatively stained crystals with a contaminating layer of ice or a layer of carbon deposited, kept at 120 K, follow the same course (trace 4) as crystals without contamination (trace 3), but they are more radiation sensitive. Berriman and Leonard (1986) also investigated the effect of "embedding" the crystals in glucose and viewed them at room temperature with low dose. Trace 2 shows that the starting thickness was 80% of the native value. Although the sample did not shrink, it was highly radiation sensitive, and the pattern disappeared after a dose of only 3 e/Å^2. Viewing frozen hydrated crystals gave the value of c closest to the native value (90%, trace 1). There was no shrinkage, and the sample, although radiation sensitive, is more resistant than the glucose-embedded sample at room temperature; here the pattern is destroyed after a dose of 8 e/Å^2.

5.2. *Resin Sections*

Recent results of the author (unpublished) on the effect of cooling araldite-embedded sections to liquid nitrogen temperatures ($\sim -170\,^{\circ}\mathrm{C}$) have shown greatly reduced shrinkage when the sections were viewed at low illumination levels [~ 2 (e/Å^2)/s]. The material used was longitudinal sections of insect flight muscle coated with gold particles.

The liquid-nitrogen-cooled sections were observed to be stable with very little of the planar movements that occur in ambient temperature sections viewed at very low dose rates [~ 0.5 (e/Å^2)/s; see Luther *et al.*, 1988].

5.3. *Mass-Thickness Studies*

When x-ray microanalysis of a sample is carried out in the electron microscope, it is the preservation of the original mass rather than its thickness which is paramount. The relative mass or mass thickness can be measured by electron energy loss spectroscopy (EELS; Leapman *et al.*, 1984) or from energy dispersive spectroscopy (EDS; Hall and Gupta, 1974; Cantino *et al.*, 1986) systems fitted in a scanning transmission electron microscope. The physical thickness of a sample is related to the mass thickness measured by these methods. The important quantity required for estimating the mass thickness of a sample is the inelastic cross section per unit mass. The great merit of determining the mass thickness by these methods is that exceedingly low doses, ~ 0.1 e/Å^2 (Cantino *et al.*, 1986), can be used to measure the initial mass thickness before any observable effects of radiation damage can occur.

Cantino *et al.* (1986) used EDS to measure the effect of electron dose on the mass thickness for three different specimens: muscle homogenate, salivary gland sections, and albumin. The samples were examined at room temperature (296 K), and at low temperature (93 K), when liquid nitrogen was used for cooling. To quantify the rate of mass loss, it is common to measure the electron dose, $D_{1/e}$, required to reduce the mass thickness to $1/e$ of the initial value; i.e., 37% of the original value remains. The value of $D_{1/e}$ for these samples at room temperature varied from 40 to 190 e/Å^2. When the samples were cooled to 93 K, the $D_{1/e}$ values ranged from 310 to 3750 e/Å^2. This represents a 10-fold improvement in reduction of radiation sensitivity of the samples at the lower temperature.

Lamvik *et al.* (1987) investigated the mass loss in a thin film of collodion using EELS at three different temperatures: room temperature (290 K), liquid nitrogen cooled (130 K), and liquid helium cooled (<10 K). The values of $D_{1/e}$ measured were 7, 30, and 1200 e/Å^2, respectively. In this remarkable result, the improvement in radiation sensitivity over the room-temperature maintained sample is fourfold at 130 K but nearly 100-fold at <10 K. Later work by Lamvik *et al.* (1989) with collodion placed on the better-conducting titanium substrate, instead of carbon as in the above work, showed a smaller, but still very significant, improvement ($D_{1/e} = 400$ e/Å^2) in radiation sensitivity at <10 K.

The reduction in radiation sensitivity when a sample is further cooled from liquid nitrogen to liquid helium is variable. Wade (1984) noted little difference in the degree of structural preservation between the two temperatures for crystalline l-valine, polyethylene, and paraffin. Lamvik (1991) reports a 70-fold improvement for Epon 812 between room temperature and liquid helium temperature, an important result which implies that a complete tilt series can be obtained with the same radiation damage as would occur with a single image recorded at room temperature.

6. CONCLUSIONS

The overwhelming conclusion from research during the last decade must be that to limit the effects of electron irradiation a sample must be cooled to low temperatures (about $-150\,^{\circ}\mathrm{C}$ or less) and images must be recorded with minimal-electron-dose methods. For stained material, the use of low temperatures means that there is virtually no specimen shrinkage (Berriman and Leonard, 1986). Hence, tilt series views can be obtained over the full effective range of the microscope tilt holder (Fig. 1). Because of reduced beam sensitivity, minimal-dose microscopy is much easier to carry out, and several tilt views can be obtained with little radiation damage.

It is now common practice to preirradiate a section at very low illumination levels until it appears to "clear" and become stable. Preirradiation allows the rapid collapse phase shown in Fig. 8 for Araldite sections to take place. For a particular electron dose rate, the thickness then reduces slowly. A complete tilt series can be obtained if care is taken that the accumulated dose, in relation to the slow shrinkage phase (Fig. 8), produces little further shrinkage. The lowest electron dose rate must be used in order that the sample become stable at the least "shrunk" state.

The above procedure of preirradiation, prior to recording tilt data, is equally applicable to negatively stained materials since these materials have a similar rapid collapse phase followed by a period of reduced thinning.

Reduction in radiation sensitivity may be provided in certain samples by subjecting the resin blocks to microwave irradiation prior to sectioning (Kinnamon and Young, 1989; Hillenbrand and Luther, unpublished results). To reduce the effect of specimen movement due to the electron beam, the method of spot scanning (Bullough and Henderson, 1987) can be used, in which a narrow 50-nm-diameter beam is scanned systematically over a sample to build up the complete image.

ACKNOWLEDGMENTS

I am indebted to Drs. J. M. Squire, E. P. Morris, J. J. Harford, P. Bennett, R. A. Crowther, and J. Berriman for many helpful discussions and for critically reading the chapter. This work was supported by grants to Dr. J. M. Squire from the Leverhulume Trust and the British Medical Research Council.

REFERENCES

Amos, L. A., Henderson, R., and Unwin, P. N. T. (1982). Three-dimensional structure determination by electron microscopy of two-dimensional crystals. *Prog. Biophys. Mol. Biol.* **39**:183–231.

Bedi, K. S. (1987). A simple method of measuring the thickness of semi-thin and ultra-thin sections. *J. Microsc.* **148**:107–111.

Bennett, P. M. (1974). Decrease in section thickness on exposure to the electron beam; the use of tilted sections in estimating the amount of shrinkage. *J. Cell Sci.* **15**:693–701.

Berriman, J., Bryan, R. K., Freeman, R., and Leonard, K. R. (1984). Methods for specimen thickness determination in electron microscopy. *Ultramicroscopy* **13**:351–364.

Berriman, J. and Leonard, K. R. (1986). Methods for specimen thickness determination in electron microscopy. II: Changes in thickness with dose. *Ultramicroscopy* **19**:349–366.

Bullough, P. and Henderson, R. (1987). Use of spot-scan procedure for recording low-dose micrographs of beam-sensitive specimens. *Ultramicroscopy* **21**:223–230.

Cantino, M. E., Wilkinson, L. E., Goddard, M. K., and Johnson, D. E. (1986). Beam induced mass loss in high resolution biological microanalysis. *J. Microsc.* **144**:317–327.

Cosslett, A. (1960). The effect of the electron beam on thin sections. In *Proc. 1st European Conf. Electron Micros.*, Delft, Vol. II, pp. 678–681.

Dorset, D. L. and Parsons, D. F. (1975). Thickness determination of wet protein microcrystals: Use of Laue zones in cross-grating electron diffraction patterns. *J. Appl. Phys.* **46**:938–940.

Frank, J. (1990). Classification of macromolecular assemblies studied as "single particles". *Q. Rev. Biophys.* **23**:281–329.

Gillis, J-M. and Wibo, M. (1971). Accurate measurement of the thickness of ultrathin sections by interference microscopy. *J. Cell Biol.* **49**:947–949.

Grubb, D. T. (1974). Radiation damage and electron microscopy of organic polymers. *J. Mater. Sci.* **9**:1715–1736.

Gunning, B. E. S. and Hardham, A. R. (1977). Estimation of the average section thickness in ribbons of ultra-thin sections. *J. Microsc.* **109**:337–340.

Hall, T. A. and Gupta, B. L. (1974). Beam induced loss of organic mass under electron microprobe conditions. *J. Microsc.* **100**:177–188.

Hayward, S. B. and Glaeser, R. M. (1979). Radiation damage of purple membrane at low temperature. *Ultramicroscopy* **4**:201–210.

Jesior, J-C. (1982). A new approach for the visualization of molecular arrangement in biological microcrystals. *Ultramicroscopy* **8**:379–384.

Jesior, J-C. and Wade, R. H. (1987). Electron-irradiation-induced flattening of negatively stained 2D protein crystals. *Ultramicroscopy* **21**:313–320.

Kinnamon, J. C. and Young, S. S. (1989). Three-dimensional reconstructions from serial sections using the IBM PC. *Eur. J. Cell Biol.* **48**:Suppl. 25, 65–68.

Lamvik, M. K. (1991). Radiation damage in dry and frozen hydrated organic material. *J. Microsc.* **161**:171–181.

Lamvik, M. K., Davilla, S. D., and Klatt, L. L. (1989). Substrate properties affect the mass loss rate in collodion at liquid helium temperature. *Ultramicroscopy* **27**:241–250.

Lamvik, M. K., Kopf, D. A., and Davilla, S. D. (1987). Mass loss rate in collodion is greatly reduced at liquid helium temperature. *J. Microsc.* **148**:211–217.

Leapman, R. D., Fiori, C. E., and Swyt, C. R. (1984). Mass thickness determination by electron energy loss for quantitative x-ray microanalysis in biology. *J. Microscopy* **133**:239–253.

Lepault, J. and Pitt, T. (1984). Projected structure of an unstained, frozen-hydrated *T*-layer of *Bacillus brevis*. *EMBO J.* **3**:101–105.

Luther, P. K., Lawrence, M. C., and Crowther, R. A. (1988). A method for monitoring the collapse of plastic sections as a function of electron dose. *Ultramicroscopy* **24**:7–18.

Ohno, S. (1980). Morphometry for determination of size distribution of peroxisomes in thick sections by high-voltage electron microscopy. I: Studies on section thickness. *J. Electron Microsc.* **29**:230–235.

Peachey, L. D. (1958). A study of section thickness and physical distortion produced during microtomy. *J. Biophys. Biochem. Cytol.* **4**:233–242.

Porter, K. R. and Blum, J. (1953). A study in microtomy for electron microscopy. *Anat. Rec.* **117**:685.

Siegel, G. (1972). Der Einflus tiefer Temperaturen auf die Strahlenschädigung von organischen Kristallen durch 100 keV-Electronen. *Z. Naturforsch* **27a**:325–332.

Slot, J. W. and Geuze, H. J. (1985). A new method of preparing gold probes for multiple-labeling cytochemistry. *Eur. J. Cell Biol.* **38**:87–93.

Sjostrom, M., Squire, J. M., Luther, P., Morris, E., and Edman, A-C. (1991). Cryoultramicrotomy of muscle; improved preservation and resolution of muscle ultrastructure using negatively stained ultrathin cryosections. *J. Microscopy* **163**:29–42.

Small, J. V. (1968). Measurements of section thickness, in *Proc. 4th European Congr. on Electron Microscopy* (S. Bocciareli, ed.), Vol. 1, pp. 609–610. Tipografia, Poliglotta Vaticana, Rome.

Spencer, M. (1982). *Fundamentals of Light Microscopy*. IUPAB Biophysics Series.

Stenn, K. and Bahr, G. F. (1971). Specimen damage caused by the beam of the transmission electron microscope, a correlative reconsideration. *J. Ultrastructure Res.* **31**:526–550.

Steven, A. C. and Navia, M. A. (1980). Fidelity of structure representation in electromicrographs of negatively stained protein molecules. *Proc. Nat. Acad. Sci. USA* **77**:4721–4725.

Unwin, P. N. T. (1974). Electron microscopy of the stacked disk aggregate of tobacco mosaic virus proteins. II: The influence of electron irradiation on the stain distribution. *J. Mol. Biol.* **87**:657–670.

Wade, R. H. (1984). The temperature dependence of radiation damage in organic and biological materials. *Ultramicroscopy* **14**:265–270.

Williams, M. A. and Meek, G. A. (1966). Studies on thickness variation in ultrathin sections for electron microscopy. *J. Roy. Microsc. Soc.* **85**:337–352.

Yang, G. C. H. and Shea, S. M. (1975). The precise measurement of the thickness of ultrathin sections by a "re-sectioned section" technique. *J. Microsc.* **103**:385–392.

II

Mathematics of Reconstruction

4

Reconstruction with Orthogonal Functions

Elmar Zeitler

1. INTRODUCTION

In 1917 Johann Radon posed the question of whether the integral over a function with two variables along an arbitrary line can uniquely define that function such that this functional transformation can be inverted. He also solved this problem as a purely mathematical one, although he mentions some relations to the physical potential theory in the plane. Forty-six years later, A. M. Cormack published a paper with a title very similar to that by Radon, yet still not very informative to the general reader, namely "Representation of a Function by Its Line Integrals"—but now comes the point—"with Some Radiological Applications." And another point is that the paper appeared in a journal devoted to applied physics. Says Cormack, "A method is given of finding a real function in a finite region of a plane given by its line integrals along all lines intersecting the region. The solution found is applicable to three problems of interest for precise radiology and radiotherapy."

Elmar Zeitler • Fritz-Haber-Institut der Max-Planck-Gesellschaft, W-1000 Berlin 33, Germany

Electron Tomography: Three-Dimensional Imaging with the Transmission Electron Microscope, edited by Joachim Frank, Plenum Press, New York, 1992.

Today we know that the method is useful and applicable to the solution of many more problems, including that which won a Nobel Prize in medicine, awarded to A. M. Cormack and G. N. Hounsfield in 1979. Radon's pioneering paper (1917) initiated an entire mathematical field of integral geometry. Yet it remained unknown to the physicists (also to Cormack, whose paper shared the very same fate for a long time). But the problem of projection and reconstruction, the problem of tomography as we call it today, is so general and ubiquitous that scientists from a variety of fields stumbled on it and looked for a solution, without, however, looking back or looking to other fields. Today there is a vast literature which cannot comprehensively be appreciated in this short contribution. It was Cormack (1963, 1964) who first made use of orthogonal functions for the solution of Radon's problem. Not only is their application elegant, but it also provides a good understanding about the intrinsic relations of a structure to its projections. The goal of this contribution is to demonstrate these relations.

2. ORTHOGONAL POLYNOMIALS

Because we do not assume the reader has any knowledge of orthogonal polynomials, we compile some results, which are derived from knowledge available to anybody engaged in advanced electron microscopy, diffraction theory, and Fourier techniques. We are concerned with the properties of orthogonal polynomials, but abstain from general derivations. Rather, we demonstrate them for the special functions that are needed in tomography and point out that their properties are generally valid for all orthogonal polynomials.

Without calculation we all know that the integral over a periodic function vanishes when a number of plus and minus cycles proper for a complete cancellation fall within the range of integration; for instance,

$$\int_0^\pi \cos n\phi \, d\phi = 0 \tag{1}$$

Only if we change n, the number of nodes (zeros) within the interval, to zero does the integration result in a finite value equal to π. But actually no node means also that we violate the original agreement of periodic functions. One can retain the periodicity and yet avoid the cancellations by squaring the integrand; thus, again without calculation and just by reasoning, one can arrive at

$$\int_0^\pi \cos^2 n\phi \, d\phi = \frac{\pi}{2} \qquad n \neq 0$$

Multiplication of two such harmonic functions leads to

$$\cos n\phi \cos m\phi = \tfrac{1}{2}\cos(n+m)\phi + \tfrac{1}{2}\cos(n-m)\phi$$

i.e., to an additive mixture of a sum tone and a difference tone whose integral amounts to zero except when the number of nodes m and n are the same.

The product of two vectors can be zero, although their lengths are not. This happens when their directions are orthogonal to each other. In analogy, the harmonic cosine functions are called orthogonal because of the properties we found.

Orthogonality:

$$\frac{2}{\pi}\int_0^\pi \cos n\phi \cos m\phi \, d\phi = \delta_{n,m}(1+\delta_{n,0}) \tag{2}$$

where

$$\delta_{n,m} = \begin{cases} 1 & n=m \\ 0 & \text{otherwise} \end{cases}$$

A typical feature of these orthogonal functions is the existence of a recurrence relation which connects three members of the set.

Recurrence:

$$\cos n\phi = 2\cos\phi\cos(n-1)\phi - \cos(n-2)\phi \tag{3}$$

This relation, which follows from the simple rules of compounded angles, enables one to set up a complete system:

$$\begin{aligned}
\cos 0\phi &= 1\\
\cos 1\phi &= \cos\phi\\
\cos 2\phi &= 2\cos\phi\cos\phi - 1 = 2\cos^2\phi - 1\\
\cos 3\phi &= 2\cos\phi\cos 2\phi - \cos\phi = 4\cos^3\phi - 3\cos\phi\\
\cos 4\phi &= 2\cos\phi\cos 3\phi - \cos 2\phi = 8\cos^4\phi - 8\cos^2\phi + 1
\end{aligned} \tag{4}$$

Two remarkable facts are contained here. First, solving the last equation for $\cos^4\phi$, we obtain

$$\cos^4\phi = \tfrac{1}{8}\cos 4\phi + \tfrac{1}{2}\cos 2\phi + \tfrac{3}{8}$$

which amounts to the Fourier decomposition of $\cos^4\phi$ into its harmonics; in other words [see Eq. (2)],

$$\frac{1}{8} = \frac{2}{\pi}\int_0^\pi \cos^4\phi\cos 4\phi\, d\phi$$

is the corresponding Fourier coefficient usually found by means of orthogonality as shown in the preceding.

The second fact is that the harmonic functions $\cos n\phi$ with n nodes can be expressed by polynomials in powers of $\cos\phi$ whose degree equals n. The right side of Eq. (4) shows polynomials, while the left side shows orthogonal harmonics; hence, we are very close to orthogonal polynomials.

We introduce

$$\cos\phi = x \qquad \text{or} \qquad \phi = \cos^{-1} x$$

Then

$$\cos 3\phi = \cos 3(\cos^{-1} x) = 4x^3 - 3x = T_3(x)$$

This polynomial in x is called the Chebyshev polynomial $T_3(x)$ of third degree. With this variable change from ϕ to x, the results from the above equations also lead to a new formulation of orthogonality.

Orthogonality:

$$\frac{2}{\pi}\int_{-1}^{+1} T_n(x)\, T_m(x) \frac{dx}{\sqrt{1-x^2}} = \delta_{n,m}(1 + \delta_{n,0}) \tag{5}$$

The integrand contains $(1-x^2)^{1/2}$ as a so-called weight or weighting factor which, besides the range of integration, is typical for the particular set of polynomials.

Recurrence:

$$T_{n+2}(x) = 2xT_{n+1}(x) - T_n(x) \tag{6}$$

So far we have introduced only the $T_n(x)$, the first kind of Chebyshev polynomials. But after this lengthy preparation, we may apply more formal and quicker steps to arrive at those polynomials of the second kind. In the complex plane, $z_1 = e^{i\phi}$ represents a point on the unit circle with coordinates x and $y = (1-x^2)^{1/2}$. Instead of dealing only with the real part we take the complex function

$$(\cos n\phi + i \sin n\phi) = e^{in\phi} = (\cos\phi + i\sin\phi)^n = (x + i\sqrt{1-x^2})^n = z_1^n$$

$$e^{i0\phi} = 1 = T_0(x)$$

$$e^{i1\phi} = x + i\sqrt{1-x^2} = T_1(x) + i\sqrt{1-x^2}\, U_0(x)$$

$$e^{i2\phi} = x^2 - (1-x^2) + 2ix\sqrt{1-x^2} = T_2(x) + i\sqrt{1-x^2}\, U_1(x)$$

$$e^{i3\phi} = T_3(x) + i\sqrt{1-x^2}\, U_2(x)$$

In order to avoid the unappealing square root, the modern literature designates as the Chebyshev polynomials of second kind:

$$\Im \frac{z_1^{n+1}}{\sqrt{1-x^2}} = \Im \frac{e^{i(n+1)\phi}}{\sin\phi} = \frac{\sin(n+1)\phi}{\sin\phi} = U_n(x); \qquad x = \cos\phi$$

$$\sin 0\phi = 0$$

$$\sin 1\phi = \sqrt{1-x^2} = \sqrt{1-x^2}\, U_0(x)$$

$$\sin 2\phi = \sqrt{1-x^2}\, 2x = \sqrt{1-x^2}\, U_1(x)$$

$$\sin 3\phi = \sqrt{1-x^2}\,(4x^2 - 1) = \sqrt{1-x^2}\, U_2(x).$$

From

$$e^{i(n+2)\phi} = (e^{i\phi} + e^{-i\phi})e^{i(n+1)\phi} - e^{in\phi} = 2 \cos \phi e^{i(n+1)\phi} - e^{in\phi}$$

follows the recurrence relation

$$U_{n+2}(x) = 2xU_{n+1}(x) - U_n(x) \tag{7}$$

and the orthogonality expression

$$\frac{2}{\pi}\int_{-1}^{+1} U_n(x)\, U_m(x) \sqrt{1-x^2}\, dx = \delta_{n,m} \tag{8}$$

Sometimes the so-called generating function can be helpful. When expanded as a power series, it has as coefficients the polynomials to be generated. It can readily be derived that

$$U(t, x) = \frac{1}{1 - 2xt + t^2} = \sum_0^{\infty} t^n U_n(x) \tag{9}$$

For the many connections of these polynomials to other classical orthogonal polynomials like Gegenbauer's or Jacobi's, the literature must be consulted. Here we shall try to involve only those types which occur in tomography, like the Zernike polynomials.

We conclude this section with the reminder that any function can be approximated by orthogonal polynomials as long as the range of its independent variable is the same:

$$f(x) = \sum_0^n a_n T_n(x) = \frac{2}{\pi} \sum_0^n T_n(x) \int_{-1}^{+1} f(\xi) \frac{T_n(\xi)}{\sqrt{1-\xi^2}}\, d\xi \tag{10}$$

The expansion coefficients (Fourier coefficients) a_n are readily calculated. The error of the approximation is always the minimum of the mean-square deviation. If further improvement is required, more terms must be included. The advantage is that the coefficients from the previous approximations remain valid for the refined approximation; new ones merely are added. This is a very economical feature—coefficients calculated once keep their value. In the next section we demonstrate the usefulness of these properties for solving the problem of interpolation.

3. INTERPOLATION AND QUADRATURE

Measurements of a physical property always render discrete quantitative values for a discrete set of independent parameters, freely chosen and typical for that experiment. On the other hand, the theoretical treatment of a measured sequence is often facilitated when an analytical expression for the data can be established in order to perform the required mathematical operations effectively. In other words, one needs values not only at the discrete spots but elsewhere as well. This problem is solved by interpolation. Therefore this section is included to show how orthogonal polynomials are most suited for this purpose. With an interpolating polynomial at hand, the problem of determining the integral over an interpolated function is readily achieved. As this problem of quadrature occurs in our reconstruction procedures as well, its solution is also included in this section. Although all these matters go all the way back to Lagrange and Gauss, a brief review is still desirable.

At every "abscissa point" x_j we have performed a measurement and found the outcome $M(x_j)$. We have obtained a table $(x_j, M(x_j))$ or a curve with m points, $j = 1, 2, 3,..., m$. In accord with Lagrange we seek an analytic expression

$$\begin{aligned} F(x) &= \sum_1^m L(x, x_j)\, M(x_j) \\ F(x_j) &= M(x_j) \end{aligned} \tag{11}$$

This equation is a discrete version of an integral representation

$$F(x) = \int L(x, y)\, M(y)\, dy$$

Instead of a continuous kernel we have m kernel functions $L(x, x_j)$. (For convenience we suppress the indicator m; it should be either L_m instead of L or $x_{m,j}$ instead of x_j.) The problem is solved if

$$L(x_i, x_j) = \delta_{ij} \tag{12}$$

The polynomial of mth degree,

$$p(x) = \prod (x - x_j), \qquad j = 1, 2,..., m \tag{13}$$

has m zeros at $x = x_j$; its first-order Taylor expansion about $x = x_i$ is

$$\begin{aligned} p(x - x_i) &= p(x_i) + (x - x_i)\, p'(x_i) \\ &= (x - x_i)\, p'(x_i) \end{aligned} \tag{14}$$

so that $p(x)/[(x - x_i)\, p'(x_i)] = L(x, x_i)$ has the required property. Remember that the locations of the zeros x_j have not been committed. Instead of opting for the usually equidistant division, we select the locations such that the sampling of the outcome of our measurements, namely the sampling of the projections, becomes most profitable as input into the structure synthesis (reconstruction). This point will be considered more extensively later on.

First of all, the zeros, $|x_j|$, fall into the range 0 to 1. Second, as the projection length through the unit circle shortens toward the rim away from the center, we like to compensate for that fact by sampling more densely there. If $x = \cos\phi$, the projection length is $\sin\phi$, so we should avoid sampling points at $x_j = \pm 1$ or $\phi_j = 0, \pi$. All those requirements are fulfilled when we sample in an equiangular division in ϕ rather than in an equidistant division in x, by means of the function $\sin(m+1)\phi/\sin\phi$. This function has m zeros at $\phi_j = (\pi/(m+1))j$ and $j = 1, \ldots, m$ with the corresponding zeros $x_j = \cos\phi_j$ ($\phi_0 = 0$ and $\phi_{m+1} = \pi$ are not zeros!).

We recognize

$$p(x) = \frac{\sin(m+1)\phi}{\sin\phi} = U_m(\cos\phi) = U_m(x) \tag{15}$$

as the mth Chebyshev polynomial of the second kind, of which x_j is one of its m zeros (see Fig. 1). The Lagrange interpolation function is then

$$L(x, x_j) = \frac{U_m(x)}{(x - x_j)\, U'_m(x_j)} \tag{16}$$

There are now two properties: one generally valid for orthogonal polynomials, the other especially for the derivatives of $U_m(x)$, which permit simplification of $L(x, x_j)$.

a. The Christoffel-Darboux relation

$$U_m(x)\, U_{m-1}(y) - U_{m-1}(x)\, U_m(y) = 2(x - y) \sum_0^{m-1} U_k(x)\, U_k(y)$$

becomes, upon identifying y with a zero x_j on account of

$$U_m(x_j) = 0 \qquad \text{and of} \qquad U_{m-1}(x_j) = (-1)^{j+1}$$

identical to

$$\frac{U_m(x)}{x - x_j} = 2(-1)^{j+1} \sum_0^{m-1} U_k(x)\, U_k(x_j) \tag{17}$$

a formula that is reminiscent of the unity operator made up of eigenfunctions.

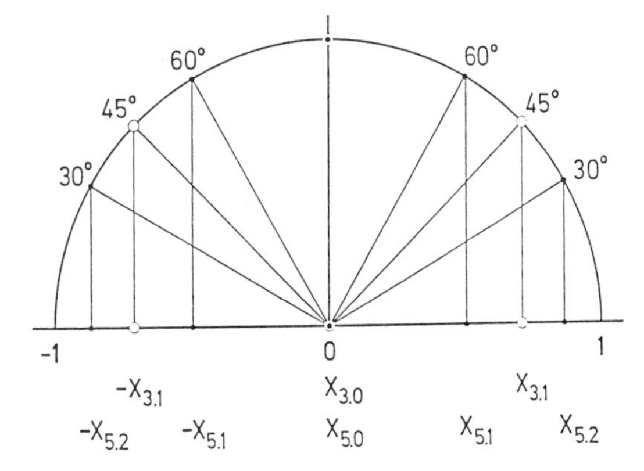

FIGURE 1. Two sets of zeros $x_{3,j}$ and $x_{5,j}$ of the Chebyshev polynomials $U_3(x)$ and $U_5(x)$. Note the interlacing of the sets and the higher density toward the limits ± 1.

b. The expression for the derivative of the Chebyshev polynomial at the general coordinate x simplifies considerably at the location of a zero

$$U'_m(x_j) = (-1)^{j+1} \frac{m+1}{1-x_j^2} \tag{18}$$

With these results we obtain the Lagrange polynomials suited for our problem:

$$L(x, x_j) = \frac{2}{m+1} \sum_0^{m-1} U_k(x_j)(1-x_j^2)\, U_k(x) \tag{19}$$

The representation found for $F(x)$ can be seen as an expansion in m terms of orthogonal polynomials, namely

$$F(x) = \sum_0^{m-1} B_k U_k(x)$$

The Fourier coefficients are

$$B_k = \frac{2}{m+1} \sum_1^m (1-x_j^2)\, U_k(x_j)\, F(x_j) \tag{20}$$

whereby $F(x_j) = M(x_j)$ has been utilized. As mentioned in Section 2, the usual way to obtain the Fourier coefficients makes use of the orthogonality, which here gives

$$B_k = \frac{2}{\pi} \int_{-1}^{1} \sqrt{1-x^2}\, F(x)\, U_k(x)\, dx \tag{21}$$

Equating the two expressions is the essence of a Gaussian quadrature: Integrals of the type

$$I = \int_{-1}^{1} \sqrt{1-x^2}\, G(x)\, dx \tag{22}$$

can be approximated by weighted sums

$$I = \sum_1^m w_j G(x_j)$$

whose error vanishes if $G(x)$ is a polynomial of power less than $2m$ because m weights w_j and m sampling points x_j are available. Indeed, it can be shown that the error in equating both results is less than

$$R_m = \frac{\pi}{2} \frac{2^{-2m}}{(2m)!} \frac{d^{2m}}{d\xi^{2m}} G(\xi), \qquad |\xi| < 1$$

Depending on the weight under the integral and on the range of the integration, different sets of orthogonal polynomials are invoked for the Gaussian quadrature.

From the "derivation" given we can understand that for integrals of the type

$$I=\int_{-1}^{1} G(x)\frac{dx}{\sqrt{1-x^2}}$$

the approximation

$$I=\sum_{1}^{m} w_j G(x_j) \tag{23}$$

is valid, whereby the sampling points coincide with the zeros of the Chebyshev polynomials $T_n(x)$ of the first kind:

$$\phi_j=\frac{2j-1}{m}\frac{\pi}{2}, \qquad w_j=\frac{\pi}{m} \tag{24}$$

and the weights are conveniently constant; on the other hand, the remainder R_m is four times larger than the former one.

Incidentally, integrals over the area of the unit circle which are of the type, say, $\int_0^1 r^k G(r) r\, dr$ are approximated by sums over $G(r_j)$ whose sampling points are the zeros of certain Zernike polynomials, which are of course orthogonal. In the general literature they are known as shifted Jacobi polynomials; i.e., their argument is $2r^2-1$. Weights, zeros, and remainder are rather complicated and, hence, are omitted here.

Equally important as the radial integration over the unit circle is the azimuthal integration over 2π-periodic functions as, for example, over a two-dimensional function at a fixed radius. The theory based on the orthogonal functions $f_k(\theta)=e^{ik\theta}$ gives as a formula for a Gaussian quadrature

$$\int_0^{2\pi} f(\theta)\, d\theta=\frac{2\pi}{m}\sum_{1}^{m} f(\theta_j)+R_m(f) \tag{25}$$

The angular sectors are of equal size around the circumference

$$\theta_j=\frac{2\pi}{m}(j-1)+\alpha, \qquad w_j=\frac{2\pi}{m}, \qquad j=1, 2,\dots, m \tag{26}$$

with an arbitrary start angle α and weights w_j, which are constant. The remainder $R_m(f_k)$ is zero as long as k is less than m.

4. A PRIORI *KNOWLEDGE AND ASSUMPTIONS ABOUT THE OBJECT*

The 3D object is composed of thin 2D slices. Within each slice, the mass density ρ varies from point to point (x, y), but does not vary along the third dimension. This requirement translates into a limitation of the slice's real thickness. The mass density is always positive and real; it differs from zero within a finite region.

1. We assume this region to be the unit disk

$$x^2 + y^2 = r^2 \leqslant 1 \tag{27}$$

2. We assume that the mass density is periodic with respect to rotations about an axis normal to the unit disk and through its center. Introducing cylindrical coordinates r and ϕ, we have

$$\rho(r, \phi) = \rho(r, \phi + 2\pi) \tag{28}$$

The mathematical consequence of these realistic assumptions is that the mass density can be written as a Fourier series

$$\rho(r, \phi) = \sum_{-\infty}^{\infty} \rho_n(r) e^{in\phi} \tag{29}$$

whose coefficients determine the radial dependence. Symmetry considerations, which are always quite powerful, restrict the mathematical form of the radial dependence.

3. We assume for the sake of argument that the highest occurring periodicity is of order m. We know from the previous section that

$$\begin{aligned} \Re e^{im\phi} &= \cos m\phi = T_m(\cos \phi) \\ &= a \cos^m \phi - b \cos^{m-2} \phi + \cdots \end{aligned} \tag{30}$$

Hence, the highest exponent which $x = r \cos \phi$ as such can assume in the power expansion of the regular mass density function is also m. However, in terms of $r^2 = x^2 + y^2$, which are rotationally invariant, higher exponents may occur. It follows that the coefficient ρ_m must be of the form

$$\rho_m(r) = r^m(a_0 + a_1 r^2 + a_2 r^4 + \cdots + a_k r^{2k} + \cdots) \tag{31}$$

The Bessel function of order m, which plays an important role in the theory of diffraction and of reconstruction, fulfills the above requirement.

4. We assume that the total mass of the disk is unity (normalized); that means

$$\int_0^{2\pi} \int_0^1 \rho(r, \phi) r \, dr \, d\phi = 1 = 2\pi \int_0^1 \rho_0(r) \, dr \tag{32}$$

The total mass is contained in the "dc" coefficient ρ_0, whereas the "ac" coefficents ρ_j effect "decorative" yet massless changes of the distribution.

5. Finally, we assume that the physical response of the electron microscope is proportional to the optical pathlength or to the mass thickness traversed by the electron, which we write as

$$\sigma[L] = \int_L \rho(x, y)\, ds \tag{33}$$

We call $\sigma[L]$ the projection of the mass distribution ρ along the path L.

The problem at hand is to record many projections for various paths L and derive from the results the mass distribution inside the disk. With the assumptions made, the problem can be formulated in a clear-cut way and thus solved. We prefer to present the attainment of this goal in an inductive manner rather than by rigorous mathematical deduction.

5. THE PROJECTION OPERATION

Consider a beer coaster, in scientific language a homogeneous disk of uniform density. The projection operation consists of performing the integration

$$\sigma(x) = \int_{-\sqrt{1-x^2}}^{\sqrt{1-x^2}} 1\, dy = 2\sqrt{1-x^2} \tag{34}$$

i.e., summing up the density values 1 along the projection axis (in this case along the y axis from rim to rim). The result is just the length of the secant of the line within the disk.

Since we need to perform this operation quite often, let us introduce a symbol, an operator, for it. We write

$$PY(1) = 2\sqrt{1-x^2}, \qquad \text{with} \quad PY(\ ) = \int_{-\sqrt{1-x^2}}^{\sqrt{1-x^2}} (\ )\, dy \tag{35}$$

Trying a few more projections,

$$PY(x) = x2\sqrt{1-x^2} \qquad PY(y) = 0$$

$$PY(x^2) = x^2 2\sqrt{1-x^2} \qquad PY(y^2) = \frac{1}{3}(1-x^2)2\sqrt{1-x^2}$$

$$PY(x^{2n}) = x^{2n} 2\sqrt{1-x^2} \qquad PY(y^{2n}) = \frac{1}{2n+1}(1-x^2)^n\, 2\sqrt{1-x^2} \tag{36}$$

we can draw the following conclusions:

a. Functions of x are merely multiplied by the path length through the disk.
b. Odd functions of y yield zero projections; even functions project into even functions in x multiplied by the path length. We realize that path length

is the same square root that we encountered in the Chebyshev polynomials of the second kind. Normalizing the projections by this (trivial) length does remove the unappealing root. It also makes sense since it reflects only the a priori knowledge of the object's spatial limits.

Recall that density functions with arbitrary powers in r are not allowed; rather, their lowest term must be of the form $(re^{i\phi})^n$. The projection of this expression leads to a surprise:

$$PY(r^n e^{in\phi}) = PY((x+iy)^n) = \frac{2}{n+1}\Im(x+i\sqrt{1-x^2})^{n+1}$$

$$= \frac{2}{n+1}\sqrt{1-x^2}\,U_n(x)$$

$$PR(r^n e^{in\phi}) = 2\int_{|x|}^{1} r^n \frac{T_n(x/r)}{\sqrt{1-(x/r)^2}}\,dr \tag{37}$$

the last step showing that the projection operation can also be performed as an integration over the radius r instead of over y. In either case our reasonable assumptions for the projection operation introduce Chebyshev polynomials of either kind and point out their interrelation. The reason is that the projection operation is a simple line integral between two conjugated points on the perimeter of the unit circle in the complex plane:

$$PY(\rho(x, y)) = -i\int_{\bar{z}_1}^{z_1} f(z)\,dz = 2\Im F(z_1)$$
$$z_1 = x + \sqrt{1-x^2} = e^{i\phi}; \qquad x = \cos\phi \tag{38}$$

Also for the allowed combinations $z^n r^{2k}$ we can retain the convenience of this "analytical" projection in writing for r^2,

$$r^2 = z\bar{z} = 2xz - z^2 \tag{39}$$

Let us compile a short table of permitted powers and their normalized projections PYN, as it can be done with ease and without computer.

$$PYN(z^n) = \frac{1}{2\sqrt{1-x^2}}PY(z^n) = \frac{1}{n+1}U_n(x)$$

$$PYN(z^n r^2) = \frac{2x}{n+2}U_{n+1} - \frac{1}{n+3}U_{n+2}$$

$$PYN(z^n r^4) = \frac{4x^2}{n+3}U_{n+2} - \frac{4x}{n+4}U_{n+3} + \frac{1}{n+5}U_{n+4} \tag{40}$$

Again the special suitability of the Chebyshev polynomials is pointed out by the readily applicable recurrence formula, which enables us to project terms of $z^n r^{2k}$ into pure sums of Chebyshev polynomials:

$$PYN(z^n) = \frac{1}{n+1} U_n(x)$$

$$PYN(z^n r^2) = \frac{1}{n+2} U_n(x) + \frac{1}{(n+2)(n+3)} U_{n+2}(x)$$

$$PYN(z^n r^4) = \frac{1}{n+3} U_n(x) + \frac{2}{(n+3)(n+4)} U_{n+2}(x) + \frac{2}{(n+3)(n+4)(n+5)} U_{n+4}(x)$$

$$\binom{n+2k}{k} PNY(z^n r^{2k}) = \sum_{0}^{k} \binom{n+2k+1}{k-s} \frac{U_{n+2s}}{n+2k+1}(x) \tag{41}$$

As a hint of things to come, observe the possibility of creating, by proper choice of the factors $a_{n,j}$, expressions

$$e^{in\phi} R_{n,k}(r) = z^n \sum_{0}^{k} a_{n,j} r^{2j} \tag{42}$$

which project into a single Chebyshev polynomial U_{n+2k}.

The problem consists of finding the reverse of the matrix $A(a_{n,j})$. This, however, is easy to do since A is already triangular. The polynomials $R_{n,k}(r)$ turn out to be orthogonal as well. They are the famous Zernike polynomials. Before we conclude this section, we mention one more point about the projection operator PR, which integrates over those values of the radius r that keep the product $r \cos \phi = x$ constant.

$$\begin{aligned} PY(r^n e^{in\phi} g(r)) &= \int_{-\sqrt{1-x^2}}^{\sqrt{1-x^2}} r^n e^{in\phi} g(r)\, dy \\ &= 2\Re \int_{|x|}^{1} r^n e^{in\phi} g(r) \frac{r\, dr}{\sqrt{r^2 - x^2}} \\ &= 2 \int_{|x|}^{1} r^n g(r) \frac{T_n(x/r)}{\sqrt{1-(x/r)^2}}\, dr \end{aligned} \tag{43}$$

It is certainly not accidental that the projection operation invokes the Chebyshev polynomial of the first kind as well since the cosine is typical for projections. However, the operation reveals also the relation between the two kinds of polynomials, namely

$$\frac{2}{n+1} \sqrt{1-x^2}\, U_n(x) = 2 \int_{|x|}^{1} r^n \frac{T_n(x/r)}{\sqrt{1-(x/r)^2}}\, dr \qquad \text{when} \quad g(r) = 1 \tag{44}$$

that is, as a projection $PR(z^n)$, as shown earlier.

The same effect of an r projection could be achieved by a delta function $\delta(x - r\cos\phi)$ and an integration over the entire domain r and ϕ, that is, by a 2D integration over the unit circle. How does this projection kernel relate to the previous one? The delta function is an even periodic function in ϕ, and hence it can be represented by a Fourier cosine series. Since expansions are the topic of this chapter, we ask for the expansion coefficients $G_n(x, r)$ of this series. They are found directly as

$$G_n(x, r) = \frac{1}{\pi}\int_0^{\pi} \delta(x - r\cos\phi)\cos n\phi\, d\phi$$

$$= \frac{1}{\pi}\int_{-r}^{+r} \delta(x - u)\, T_n\left(\frac{u}{r}\right)\frac{du}{\sqrt{r^2 - u^2}}$$

$$= \frac{1}{\pi}\frac{T_n(x/r)}{\sqrt{r^2 - x^2}}, \qquad |x| < r$$

$$= 0, \qquad |x| > r \tag{45}$$

In answering the above questions we see that both kernels render the same projection result. In the second case the azimuthal integration will first filter out the n-fold symmetric structure, while the consecutive r integration will add the seemingly missing r factor.

With these findings the various aspects of the projection operation are adequately presented.

6. FOURIER TRANSFORMATION AND PROJECTION

Confirmed by the many contributions to this volume, Fourier transformations, decompositions, and syntheses, whether continuous or discrete, play a major role in the processing of information. There is a simple computational aspect to it; that is, very fast and economical implementations exist, and, furthermore, many other mathematical procedures like convolution and correlation can be effected with great advantage by advanced Fourier algorithms. But there is also a philosophical side. Questions as to the extension and the contrast of an image detail are recast in Fourier language into questions of how many lines per unit length, of which intensity, resemble best the image detail under investigation. The physical connection between these two reciprocal quantities of length and of lines per length is established by the theory of diffraction or experimentally by the interaction of waves with matter. We like to capitalize on this knowledge.

We continue with our very simple 2D object, the homogeneous disk of radius a. Its 2D Fourier transformation leads to the famous Fraunhofer diffraction pattern

$$\oint\int_0^a e^{i\omega r\cos\phi} r\, dr\, d\phi = 2\pi a\frac{J_1(\omega a)}{\omega} \tag{46}$$

where $J_1(x)$ is the Bessel function of first order. If we normalize the integral by the area of the disk πa^2, we obtain the well-known Airy function:

$$A_{0,0}(\omega a) = \frac{2J_1(\omega a)}{\omega a} \tag{47}$$

which behaves very much like its one-dimensional pendant

$$\int_{-a}^{+a} e^{i\eta y}\, dy = \frac{2\sin(\eta a)}{\eta} \tag{48}$$

which, after normalization (division by $2a$), turns into the sinc function

$$\operatorname{sinc}(\eta a) = \frac{\sin(\eta a)}{\eta a} \tag{49}$$

Both functions have very similar properties; they are even, they are unity for zero argument, and they decay in an oscillatory fashion. Later, we encounter more general Airy functions

$$A_{n,k}(x) = \varepsilon_{n,k} \frac{J_{n+2k+1}(x)}{x} \tag{47a}$$

with

$$\varepsilon_{n,k} = (1 + \delta_{n,0}\delta_{k,0})$$

hence the special notation.

When the spatial frequency η is chosen to be zero, the Fourier transform, Eq. (48), becomes tantamount to an integration or projection, and the result is just a path length of the integral. A projection of the unit disk parallel to the y axis at a distance x off center gives the length $2(1-x^2)^{1/2}$ and hence a Fourier transform

$$\frac{2\sin(\eta\sqrt{1-x^2})}{\eta} \tag{50}$$

We can readily remove the square root by expressing the "impact parameter x," the normal distance from the center, by $x = \cos\phi$.

The sine function thus becomes periodic in ϕ and can be represented by a Fourier series. The result can be found in pertinent tables:

$$2\sin(\eta\sin\phi) = 4\sum_{0}^{\infty} J_{2k+1}(\eta)\sin(2k+1)\phi \tag{51}$$

the left side, representing an FM sine wave, is decomposed in a multitude of AM sidebands whose amplitudes (Fourier coefficients) are Bessel functions of odd order. If we go back from ϕ to x and apply the functions already introduced, we obtain quite naturally

$$\frac{2\sin(\eta\sqrt{1-x^2})}{\eta} = \sqrt{1-x^2}\, 4\sum_{0}^{\infty} \varepsilon_{0,k} A_{0,k}(\eta)\, U_{2k}(x) \tag{52}$$

as a generating function for either the Airy functions or the Chebyshev polynomials. Making use of their orthogonality, we arrive at their respective integral representations.

In order to repeat in Cartesian coordinates the 2D Fourier transform of the disk, whose result in polar coordinates we already know, we perform or look up the following integral:

$$\int_{-1}^{+1} e^{i\xi x} \frac{2\sin(\eta\sqrt{1-x^2})}{\eta} dx = \pi A_{0,0}(\sqrt{\xi^2+\eta^2}) \tag{53}$$

We recover the previous Airy disk and recognize the rotational symmetry on account of the quadratic addition of the Cartesian frequency components

$$\omega^2 = \xi^2 + \eta^2$$

Every central section through reciprocal Fourier space of the homogeneous disk looks the same. One projection suffices for reconstructing the disk. This situation will quickly change when more interesting structures arise.

If we perform the same transformation on the single members of the right side in Eq. (52) we again obtain an Airy function:

$$\int_{-1}^{+1} e^{i\xi x} U_{2k}(x)\sqrt{1-x^2}\, dx = \pi(-1)^k (2k+1) A_{0,k}(\xi)\varepsilon_{0,k}^{-1} \tag{54}$$

and the previous 2D transform, the diffraction pattern of the disk, appears as a completely symmetric expression

$$A_{0,0}(\omega) = 4\sum_{0}^{\infty} (-1)^k (2k+1) A_{0,k}(\xi) A_{0,k}(\eta)\varepsilon_{0,k}^{-1} \tag{55}$$

The most important message of Eq. (54), however, is that the one-dimensional FT of the Chebyshev polynomials which, as we saw in Section 3, results from a (not normalized) y projection, is related to the Airy function, which itself is a product of the 2D Fourier transformation in polar coordinates. To clarify this important connection, we investigate the 2D transformations of the allowed powers $z^n r^{2k}$ introduced earlier:

$$\frac{1}{2\pi}\oint\int_0^1 r^{n+2k} e^{in\phi} e^{i\omega r\cos\phi} r\, dr\, d\phi = i^n \int_0^1 r^{n+2k} J_n(\omega r) r\, dr \tag{56}$$

With $J_n(\omega r)$ as kernel the last expression appears as a special transformation in the radial domain of our 2D problem. This transformation is referred to as a general Hankel transformation (HT, general because n is not zero). Since $(x^{n+1}J_{n+1}(x))' = x^{n+1}J_n(x)$, the Hankel transforms of $r^n r^{2k}$ can be readily evaluated. After having the recurrence formula applied to the expressions $J_m(x)/x$, let us compile a few such transformations:

$$\mathrm{HT}(r^n) = A_{n,0}(\omega)\varepsilon_{k,0}^{-1}$$

$$\mathrm{HT}(r^{n+2}) = \frac{n+1}{n+2} A_{n,0}(\omega)\varepsilon_{k,1}^{-1} - \frac{1}{n+2} A_{n,1}(\omega)$$

$$\mathrm{HT}(r^{n+4}) = \frac{n+1}{n+3} A_{n,0}(\omega)\varepsilon_{k,1}^{-1} - \frac{2}{n+4} A_{n,1}(\omega) + \frac{2}{(n+3)(n+4)} A_{n,2}(\omega)\varepsilon_{k,2}^{-1}$$

$$\binom{n+2k}{k} \mathrm{HT}(r^{n+2k}) = \sum_0^k (-1)^s \binom{n+2k+1}{k-s} \frac{n+2s+1}{n+2k+1} A_{n,s}(\omega)\varepsilon_{n,s}^{-1} \tag{57}$$

The last equation gives an answer for the allowed powers of r a sum of Airy functions $A_{n,s}(\omega)$ already introduced. Here we clearly recognize them as diffraction patterns of disk-like structures which have an n-fold azimuthal symmetry. The first index n refers to that symmetry, whereas the second index refers to the highest power of (the rotation invariant) r^2 which occurs in the series describing the radial dependence of the mass density in this n-fold symmetric structure.

These generalized Airy functions have for a fixed n-fold symmetry the welcome property of orthogonality; that is,

$$\int_0^\infty A_{n,s}(\omega)\, A_{n,t}(\omega)\omega\, d\omega = \frac{1}{2(n+2s+1)} \delta_{s,t}\varepsilon_{n,s}^2 \tag{58}$$

so that in reciprocal space frequency functions with that symmetry can readily be expanded into sums over generalized Airy functions.

Now we look for such combinations of allowed powers in r that Hankel-transform into a single "clear-cut" Airy function. The table shows that, for example,

$$\mathrm{HT}\{(n+1)r^n - (n+2)r^{n+2}\} = A_{n,1}(\omega)$$

$$\mathrm{HT}\{(n+1)(n+2)r^n - 2(n+2)(n+3)r^{n+2} + (n+3)(n+4)r^{n+4}\} = 2A_{n,2}(\omega) \tag{59}$$

We could go on finding more expressions by inverting the triangular equation system (matrix), but there is a simpler way of generating these interesting polynomials, which are the Zernike polynomials $R^n_{n+2k}(r)$. As their definition we require that their Hankel transform yield an Airy function with identical indices:

$$(-1)^k \int_0^1 R^n_{n+2k}(r)\, J_n(\omega r) r\, dr = A_{n,k}(\omega)\varepsilon_{nk}^{-1} \tag{60}$$

The straightforward yet slightly cumbersome stepwise inversion described above can be eased by applying directly the inverse Hankel transform (which, by the way, is self-reciprocal) to Eq. (60).

We obtain the mass-density function that diffracts into an Airy function by transforming the pattern back into real space:

$$(-1)^k \varepsilon_{n,k} R^n_{n+2k}(r) = \int_0^\infty A_{n,k}(\omega) J_n(r\omega)\omega \, d\omega \tag{61}$$

When you look up these integrals in a table, you will find them written in the form of the more general Jacobi polynomials; hence, more information is found by checking under those (e.g., in the well-known handbook by Abramowitz and Stegun, 1965). We could have synthesized the Zernike polynomials also by orthogonalizing the powers of r^{2k} with a weighting function r^{2n+1} over an integration range from 0 to 1.

We summarize briefly the results obtained thus far. We have dealt with three operations: (a) the projection along a line (PY), which could be seen as a special case of (b) the second transformation, namely, the one-dimensional Fourier transform, and (c) the 2DFT. In Cartesian coordinates the 2DFT is but a duplication of the FT. In polar coordinates, however, the azimuthal transform leaves as the second one-dimensional transform over the radius a special transform known as the Hankel transform.

Throughout our problem all three transforms are intimately connected, as demonstrated by our finding of special functions and polynomials that transform to each other. As a special bonus, they are orthogonal and remain so after transformation.

7. ZERNIKE POLYNOMIALS AND SELECTION RULE

Up to now we have always encountered the Zernike polynomials in connection with some other orthogonal function. In this section we introduce them more formally in their own right.

The Zernike polynomials are chosen such that the functions

$$\psi_{n,s}(r,\phi) = e^{in\phi} R^{|n|}_{|n|+2s}(r) \tag{62}$$

form a complete orthogonal system over the unit circle

$$\oint\int_0^1 \psi_{n,s}(r,\phi)\,\bar{\psi}_{n',s'}(r,\phi) r \, dr \, d\phi = \frac{\pi}{\lambda_{n,s}} \delta_{n,n'} \delta_{s,s'} \qquad \lambda_{n,s} = |n| + 2s + 1 \tag{63}$$

The polynomials $R^n_{n+2s}(r)$ are then orthogonal over the interval (0, 1) with a weighting function r and normalized such that

$$R^n_{n+2s}(1) = 1 \qquad \int_0^1 R^n_{n+2s}(r)\, R^n_{n+2t}(r) r \, dr = (2\lambda_{n,s})^{-1} \delta_{s,t} \tag{64}$$

Most important for our purpose is the generating function

$$g_n(t, r) = \sum_{s=0}^{\infty} t^s R_{n+2s}^n(r) = \frac{1}{R(t, r)} \left[\frac{1 + t - R(t, r)}{2tr} \right]^n \tag{65}$$

whereby

$$R(t, r) = \sqrt{1 - 2t(2r^2 - 1) + t^2} \tag{66}$$

If we identify t with $e^{i2\theta}$, we obtain very simple expressions, namely

$$R(e^{i2\theta}, r) = 2e^{i\theta} \sqrt{\cos^2\theta - r^2}$$
$$1 + t = 2e^{i\theta} \cos\theta$$

and with $x = \cos\theta$,

$$g_n(x, r) = \left(\frac{x}{r} - \sqrt{\left(\frac{x}{r}\right)^2 - 1} \right)^2 \frac{e^{-i(n+1)\theta}}{2\sqrt{x^2 - r^2}} \tag{67}$$

Observing that the imaginary part

$$\mathfrak{I}\left(\frac{(x/r - \sqrt{(x/r)^2 - 1})^n}{\sqrt{x^2 - r^2}} \right) = \frac{T_n(x/r)}{\sqrt{r^2 - x^2}} \tag{68}$$

is just the kernel for the projection operator, the generating function assumes a very simple meaning:

$$2g_n(x, r)e^{i(n+1)\theta} = \sum_0^{\infty} R_{n+2s}^n(r) e^{i(n+2s+1)\theta} \tag{69a}$$

or

$$\frac{2T_n(x/r)}{\sqrt{r^2 - x^2}} = \sqrt{1 - x^2} \sum_0^{\infty} R_{n+2s}^n(r)\, U_{n+2s}(x) \tag{69b}$$

This formula encompasses all the relations between the set of orthogonal functions in real space involved in projection and reconstruction. A Fourier transform of the formula translates the relations into the reciprocal space.

Multiplying both sides of the formula by a Zernike polynomial $R_{n+2k}^n(r)r$ and integrating over r leads, on the left side, to a projection, whereas on the right side orthogonality singles out one Chebyshev polynomial $U_{n+2k}(x)$ with the same indices as the projected Zernike polynomial. Playing this trick again, yet multiplying both sides now with a Chebyshev polynomial $U_{n+2k}(x)$, leads to a reconstruction of a Zernike polynomial. In other words, the projection operator has reconstructive properties as well; this time, however, in the x or angular domain. And indeed, this is what tomography is all about: to obtain a structure from many projections of a tilted object. Therefore, it pays to rewrite this reconstruction operation in angles:

$$\frac{x}{r} = \cos \alpha, \qquad T_n\left(\frac{x}{r}\right) = \cos n\alpha$$

Then

$$R_{m+2s}^{m}(r) = \frac{2}{\pi} \int_{0}^{\pi/2} U_{m+2s}(r \cos \alpha) \cos m\alpha \, d\alpha \tag{70}$$

Forgetting the weight $\cos m\alpha$ for a moment, this integral expressing the structure R_{m+2s}^{m} is the sum of projections taken under various tilts. This operation is often referred to as back-projection. However, the analysis shows that back-projection can be correct only if the cosine weighting factor is included. It is remarkable that due to the symmetries an angular range from zero to $\pi/2$ suffices. This goes beyond the fact that a projection $\sigma(x, 0)$ changes to $\sigma(-x, \pi)$ when flipped over. That is a feature of the function system employed.

We conclude this section with some remarks about the selection rule. The Zernike polynomials have two distinct indices, n and s, while the Chebyshev polynomials and Airy functions have only one, say m, which is the sum $m = n + 2s$. For a given m this diophantine equation has many solutions (n, s), the parity of m and n being, however, always the same. How can we understand this ambiguity?

The "quantum number" n was already introduced as the number of spokes or, better, zero lines running through the n-fold symmetric structure. The "quantum number" $2s$ is the number of zeros (nodes) which occur in the radial extension of one particular Zernike polynomial. These are $2s$ concentric grooves running around the center of the "eigenstructure" (n, s). The spokes and the grooves add up to $m = n + 2s$ zeros. In the projection only the total number of zeros counts and not their type, azimuthal or radial, as in the eigenstructure. This phenomenon, which is typical for a reduction of dimensions as it occurs in the process of projection, for example, is well known in other fields. In crystallography the term *aliasing* has been coined.

8. THE NUMERICAL SOLUTION TO THE RADON PROBLEM

In Fig. 2 we present the coordinate systems of the experimental setup. The recording "plane" is the x axis. The projection is performed parallel to the optical y axis. The object with its own coordinate system (ξ, η) or (r, ϕ) centered on the tilt axis of the stage is tilted by an angle θ with respect to the x axis. The measured results are a set of projections σ obtained at discrete tilts θ.

The problem is to find from these results the density distribution $\rho(r, \phi)$ within the object (we consider the angle still as a continuous variable). As pointed out, it

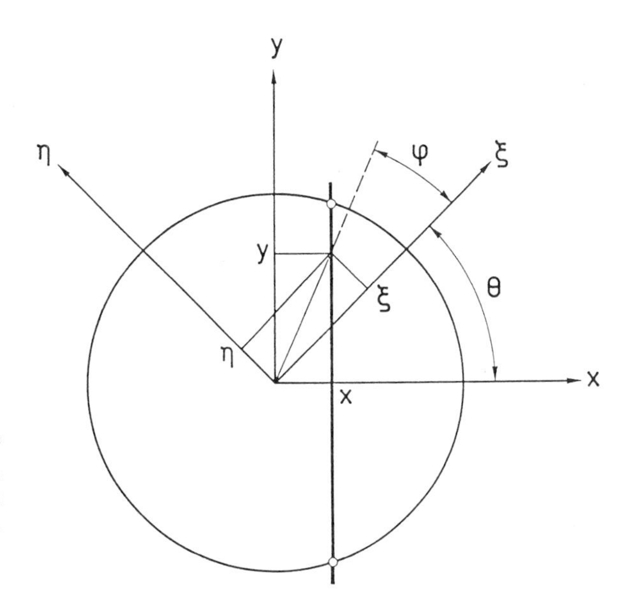

FIGURE 2. Coordinate systems of the projection experiment. System (ξ, η) is fixed within the tilted object, and θ is the tilt angle. System (x, y) is fixed in the microscope (y is the optical axis).

is sensible to normalize the projections by the length $2(1-x^2)^{1/2}$, which contributes within the object to this projection. The Radon integral expresses the relation between the measured results and the mass density to be deduced from them:

$$\tau(x, \theta) = \frac{\sigma(x, \theta)}{2\sqrt{1-x^2}} = \oint\int_0^1 G(x, \theta; r, \phi)\, \rho(r, \phi) r\, dr\, d\phi \tag{71}$$

The Radon kernel

$$G(x, \theta; r, \phi) = \frac{\delta(x - r\cos(\theta + \phi))}{2\sqrt{1-x^2}} \tag{72}$$

is a delta function whose Fourier components we could already identify as the kernel of the projection operator for the n-fold symmetric mode. The general solution must include all modes and all tilt angles. Therefore, we introduce in analogy to the $\psi_{n,s}$ of the previous section the two-dimensional projection polynomials $\pi_{n,s}$

$$\pi_{n,s}(x, \theta) = U_{|n|+2s}(x) e^{-in\theta} \tag{73}$$

which also are orthogonal over the domain $[-1, 1]$ and $[-\pi, \pi]$:

$$\oint\int_{-1}^{+1} \pi_{n,s}(x, \theta)\, \pi_{m,t}(x, \theta)\, dx\, d\theta = \pi^2 \delta_{n,m} \delta_{s,t} \tag{74}$$

and the Radon kernel then becomes

$$G(x, \theta; r, \phi) = \frac{1}{\pi} \sum_{n=-\infty}^{\infty} \sum_{s=0}^{\infty} \pi_{n,s}(x, \theta)\, \bar{\psi}_{n,s}(r, \phi) \tag{75}$$

a product kernel each of whose factors belongs to one of the two orthogonal spaces spanned by the functions $\pi_{n,s}$ and $\psi_{n,s}$.

The density function expanded in the appropriate space is represented by the series

$$\rho(r, \phi) = \sum_{n=-\infty}^{\infty} \sum_{s=0}^{\infty} \rho_{n,s} \psi_{n,s}(r, \phi) \tag{76}$$

whose Fourier coefficients are

$$\rho_{n,s} = \frac{\lambda_{n,s}}{\pi} \oint \int_0^1 \rho(r, \phi)\, \bar{\psi}_{n,s}(r, \phi) r \, dr \, d\phi \tag{77}$$

And, similarly, in the dual space the normalized projected mass density is represented by

$$\tau(x, \theta) = \sum_{n=-\infty}^{\infty} \sum_{s=0}^{\infty} \sigma_{n,s} \pi_{n,s}(x, \theta) \tag{78}$$

whose coefficients are

$$\sigma_{n,s} = \frac{1}{\pi^2} \oint \int_{-1}^{+1} \tau(x, \theta)\, \pi_{n,s}(x, \theta) \sqrt{1 - x^2} \, dx \, d\theta \tag{79}$$

Inserting the expansion of $\rho(r, \phi)$ into the Radon integral with its kernel also expanded and performing the integrals over the orthogonal system's domain, we obtain the very simple relation between the expansion coefficients of the density function and the projected density function. They are simply proportional to each other:

$$\sigma_{n,s} = \frac{\rho_{n,s}}{\lambda_{n,s}} \tag{80}$$

Therefore, the solution of the integral equation, that is, the inversion of the experimental data into the sought-for information, is achieved by solving this relation for $\rho_{n,s}$:

$$\rho_{n,s} = \lambda_{n,s} \sigma_{n,s} \tag{81}$$

Herewith the problem is reduced to one of preparing the $\sigma_{n,s}$ values from the experimental raw material. Fortunately, the necessary integrals are all of the type ideally suited for Gaussian quadrature. If the experimentalist interpolates discrete measurements onto abscissa points demanded by the Gaussian quadrature, a reconstruction can be performed with ease and high accuracy and, more importantly, with a reliable estimate of the errors.

The theory relates the data $\tau(x, \theta)$ in the double integral to

$$\sigma_{n,s} = \frac{1}{\pi^2} \int_0^{2\pi} e^{in\theta} \, d\theta \int_{-1}^{1} \tau(x, \theta) \, U_{n+2s}(x) \sqrt{1-x^2} \, dx \tag{82}$$

In Gaussian quadrature this double integral becomes a sum over the integrand at special "spokes" θ_j and at special coordinates x_k, namely the zeros of the Chebyshev polynomial, multiplied by the appropriate weights v_j and w_k. Assume the option of $2l$ spokes and m zeros. Then

$$\theta_j = \frac{\pi}{l}(j-1), \qquad j = 1, 2, 3, \ldots, 2l$$

$$v_j = \frac{\pi}{l}$$

$$x_k = \cos \alpha_k, \quad \alpha_k = \frac{\pi}{m+1} k, \qquad k = 1, 2, 3, \ldots, m$$

$$w_k = \frac{\pi}{m+1}(1 - x_k^2)$$

$$\sigma_{n,s} = \frac{\pi}{l} \frac{\pi}{m+1} \sum_{j=1}^{2l} e^{in\theta_j} \sum_{k=1}^{m} \tau(x_k, \theta_j) \times U_{n+2s}(x_k)(1 - x_k^2) + R \tag{83}$$

Two simplifications are possible. One is merely a cancellation of the weights w_k by those occurring in τ and U_{n+2s}, which gives

$$\tau(x_k, \theta_j) \, U_{n+2s}(x_k)(1 - x_k^2) = \sin(\lambda_{n,s} \alpha_k) \tag{84}$$

The second simplification brings an experimental advantage. It makes use of the fact that

$$\tau(x_k, \theta_j + \pi) = \tau(-x_k, \theta_j)$$

$$e^{in(\theta_j + \pi)} = (-1)^n e^{in\theta_j}$$

so that, for the same tilt, measurements are paired in a consistent fashion and thus increase the accuracy of the procedure (this is why an even number of tilt angles should enter the analysis).

$$\sigma_{n,s} = \frac{\pi^2}{l(m+1)} \sum_{j=1}^{l} e^{in\theta_j} \sum_{k=1}^{m} (\sigma(x_k, \theta_j) + (-1)^n \sigma(-x_k, \theta_j)) \sin(\lambda_{n,s} \alpha_k) + R$$

Having obtained these values, it is a simple task to obtain the mass density, for we have

$$\rho(r, \phi) = \sum_{n=-m}^{+m} \sum_{s=0}^{k} \lambda_{n,s}\sigma_{n,s}R_{|n|+2s}^{|n|}(r)e^{in\phi}$$

The problem of finding a judicious truncation was pointed out and solved by P. R. Smith (1978), by resorting to the selection rules. (See the following section.)

9. REFERENCES TO THE RADON PROBLEM

This section is primarily intended as a guide through the electron microscopy literature, which was already rather complete by the mid-1970s. But first we say a word about the general development.

Mathematical research has continued to develop along the lines typical for this field—that is, towards generalization. Whereas the method presented operates in a 2D space, mathematicians study the general n-dimensional problem. Deans (1979) describes a Radon inversion formula which holds in spaces of even or odd dimensions. As transformation pairs he finds Gegenbauer polynomials of which Chebyshev polynomials are a special type. Mathematically one can construct various function pairs which erect dual orthogonal spaces (e.g., Lerche and Zeitler, 1976). As we have seen, the orthogonal functions are typified by the weight functions. Therefore, weight functions most suited for the particular Radon problem must be sought. In our case we used a sharp cutoff in the weighting function without investigating its consequences. One knows, however, from optics or signal theory that such measures will lead to ringing—to Gibbs phenomena—which in optics are avoided by masks with soft transitions (apodization). Mathematical investigations in this direction would be welcomed by experimentalists. A fairly modern account of the mathematical status of the Radon problem is given by Helgason (1980).

Among the practical papers we can only mention the large array from electrical engineers and computer specialists, who are concerned with improving medical tomography. In this field many restrictions typical for the electron microscopic tomography are not operative. Furthermore, the projection schemes used in medicine differ from those in electron microscopy, the latter relying on a strictly parallel projection. In astronomy, image reconstruction had begun earlier than in electron microscopy. There the line integration (projection) is taken perpendicular rather than parallel to the propagation of the radiation, a fact that confirms the breadth of applicability of the Radon problem. An overview of the implementation and application of image reconstruction methods outside the field of electron microscopy was compiled and edited by Herman (1979).

We continue now with tomography for electron microscopy. This field was opened very suddenly by DeRosier and Klug in 1968. They approached the problem as crystallographers accustomed to reconstructing regular structures from diffraction patterns and to Fourier-transforming the information desired from reciprocal space into real space. Since the electron microscope forms an image of the object, this step should be unnecessary. Where is the error? The crux is that the electron microscopic image is a projection and its Fourier transform is just a central section through the reciprocal space of the object. The idea of DeRosier and Klug was then to fill the entire reciprocal space by Fourier transforms of many tilted exposures and perform the reconstruction just as crystallographers would do. This approach is called the Fourier approach.

Very early in the game, a group around R. H. T. Bates in New Zealand concerned itself with image reconstruction from finite numbers of projections (Smith *et al.*, 1973). They were concerned with the fact that interpolation between data is required when the Fourier approach is to be applied. So they were looking for an appropriate and reliable interpolation scheme which also permitted economical usage of the computer. As a result of their research, they found polynomials which are identical to those of Cormack, identical to the orthogonal Chebyshev and Zernike polynomials, including the selection rule with which this chapter deals. They showed how one can use this rule as a consistency condition to estimate a posteriori the quality of the input data and the validity of the information derived from them, including statements about the attained resolution. The most recent paper along the same line of reliability is by Howard (1988); it neither mentions the paper by Smith *et al.* (1973) nor does it surpass their results.

A number of years ago a computer program based on these ideas was implemented and run by P. R. Smith at the Biozentrum in Basel, Switzerland, and it remains in use there and in several other labs. A comprehensive description of this program system for processing micrographs of biological structures MDPP, including a comparison with other systems, was prepared by Smith (1978). An important trilogy on image reconstruction from projections by Lewitt and Bates (1978a, b) and Lewitt *et al.* (1978) addresses the question of error sensitivity toward truncation and the influence of sampling. The basic difference between the Fourier approach and the orthogonal function method can be pinpointed to the difference of the expansion of the 2D Fourier kernel in cylindrical coordinates. The Fourier approach chooses Bessel functions and harmonic functions for its decomposition, of which only the latter are orthogonal, whereas the Cormack system of orthogonal polynomials is readily introduced by expanding the kernel right from the beginning into orthogonal functions, namely into Airy functions and Chebyshev polynomials (e.g., Zeitler, 1974). Also the convolution method of Ramachandran and Lakshiminarayanan (1971) becomes unnecessary in this system, since the Fourier transform of a projection leads to the same Airy function as does the 2D Fourier transformation form of the structure.

The most obvious extention of the expansion method into three dimensions has been implemented by Provencher and Vogel, 1988; Vogel and Provencher, 1988. They describe the 3D structure of single particles by the orthogonal system of spherical harmonics which are generally known from quantum mechanics through their application in calculating the electronic states of atoms. Instead of tilting a single object, they start out with an array of statistically oriented particles and study in an iterative fashion how the projections obtained fit the proposed structure. They obtained most impressive results of a globular virus (Semliki Forest virus) suspended in vitrified water. It must be pointed out, however, that the functions of the projections are not orthogonal.

10. SUMMARY

The premise of the chapter is that the experimental results are given as numerical data. The mathematical solution of reconstructing a two-dimensional object from its one-dimensional projections is based on the orthogonal function expansion method such that the experimental data can be utilized with ease and least error.

The necessary steps toward this goal are summarized in Scheme 1. Arrows pointing from structure function ρ to projection functions σ indicate the projection operation, whereas arrows of opposite sense denote the inverse operation, i.e., reconstruction, and hence the solution of the problem. Hankel transforms applied to the structure functions and Fourier transforms applied to the projection functions lead into frequency space. Here the most suitable representation of the Fourier components is found as an expansion in orthogonal Airy functions $A_{n,s}(\omega)$. In the real space of the structure, the Airy functions correspond to the orthogonal Zernike polynomials $R^n_{n+2s}(r)$, and in the space of the projections they correspond to the orthogonal Chebyshev polynomials $U_{n+2s}(x)$. When the Fourier components ρ_n and σ_n are expanded in terms of the pertinent orthogonal sets, the expansion coefficients $\rho_{n,s}$ and $\sigma_{n,s}$ are proportional to each other. Hence, if the $\sigma_{n,s}$ are determined from the experimental data, the reconstruction is effected by merely inserting $\sigma_{n,s}$ for $\rho_{n,s}$ into the structure expansion.

The relation among the three orthogonal sets of functions allows the decomposition of diffraction patterns into elemental eigenmodes $A_{n,s}(\omega)$. The corresponding

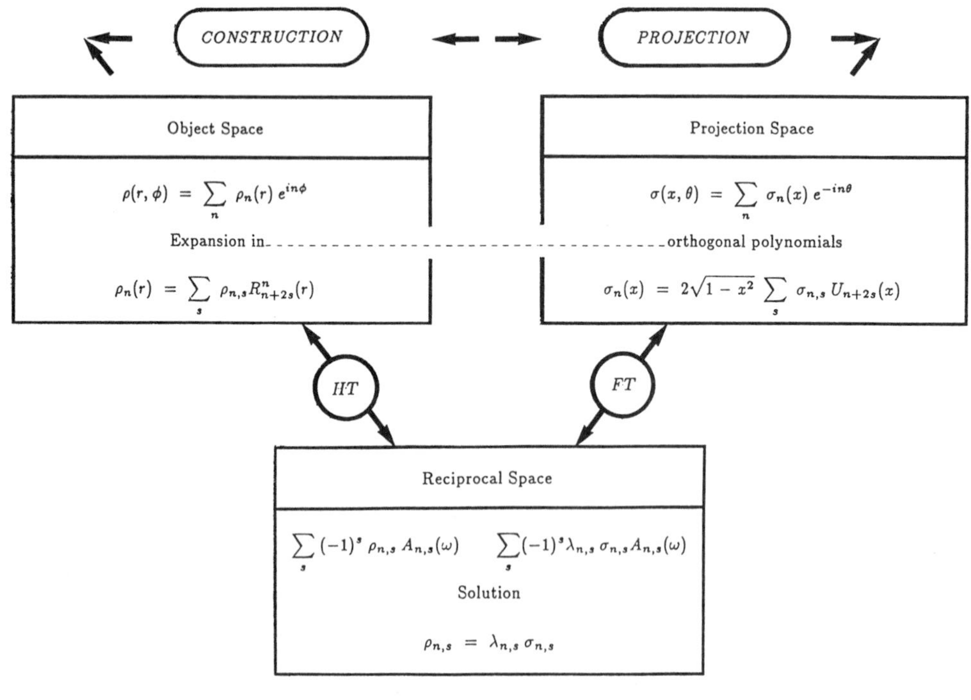

SCHEME 1

structure element which gives rise to a single mode is a Zernike polynomial $R^n_{n+2s}(r)$. Projecting a structure element leads to a projection element—that is, to a Chebyshev polynomial $U_{n+2s}(x)$ whose diffraction pattern is again a simple mode $A_{n,s}(\omega)$.

This triangular relation is the essence of the projection theorem on which tomography is based. The essence of this expansion approach is the additional condition that all the operations involved conserve the orthogonality of the various functions.

REFERENCES

Abramowitz, M. and Stegun, I. A. (1965). *Handbook of Mathematical Functions.* Dover, New York.

Cormack, A. M. (1963). Representation of a function by its line integrals, with some radiological applications. *J. Appl. Phys.* **34**:2722–2727.

Cormack, A. M. (1964). Representation of a function by its line integrals, with some radiological applications. II. *J. Appl. Phys.* **35**:2908–2912.

Deans, S. R. (1979). Gegenbauer transforms via Radon transforms. *SIAM J. Math. Am.* **10**:577–585.

DeRosier, D. J. and Klug, A. (1968). Reconstruction of three-dimensional structures from electron micrographs. *Nature* **217**:130–134.

Helgason, S. (1980). *The Radon Transform.* Birkhäuser, Boston.

Herman, G. T. (1979). *Image Reconstruction from Projections.* Springer-Verlag, Berlin.

Howard, J. (1988). Tomography and reliable information. *J. Opt. Soc. Am.* **5**:999–1014.

Lerche, I. and Zeitler, E. (1976). Projections, reconstructions and orthogonal functions. *J. Math. Anal. Appl.* **56**(3):634–649.

Lewitt, R. M. and Bates, R. H. T. (1978a). Image reconstruction from projections. I: General theoretical considerations. *Optik* **50**(1):19–33.

Lewitt, R. M. and Bates, R. H. T. (1978b). Image reconstruction from projections. III: Projection completion methods (theory). *Optik* **50**(3):189–204.

Lewitt, R. M., Bates, R. H. T., and Peters, T. M. (1978). Image reconstruction from projections. II: Modified back-projection methods. *Optik* **50**(2):85–109.

Provencher, S. W. and Vogel, R. H. (1988). Three-dimensional reconstruction from electron micrographs of disordered specimens. I: Method. *Ultramicroscopy* **25**:209–222.

Radon, J. (1917). Über die Bestimmung von Funktionen durch ihre Integralwerte längs gewisser Mannigfaltigkeiten. *Ber. Verh. Sächs. Akad.* **69**:262–277.

Ramachandran, G. N. and Lakshiminarayanan, A. V. (1971). Three-dimensional reconstruction from radiographs and electron micrographs: Application of convolutions instead of Fourier transforms. *Proc. Nat. Acad. Sci. USA* **68**(9):2236–2240.

Smith, P. R. (1978). An integrated set of computer programs for processing electron micrographs of biological structures. *Ultramicroscopy* **3**:153–160.

Smith, P. R., Peters, T. M., and Bates, R. H. T. (1973). Image reconstruction from finite numbers of projections, *J. Phys.* **6**:319–381.

Vogel, R. H. and Provencher, S. W. (1988). Three-dimensional reconstruction from electron micrographs of disordered specimens. II: Implementation and results. *Ultramicroscopy* **25**:223–240.

Zeitler, E. (1974). The reconstruction of objects from their projections. *Optik* **39**:396–415.

5

Weighted Back-Projection Methods

Michael Radermacher

1. INTRODUCTION

Traditionally, three-dimensional reconstruction methods have been classified into two major groups, *Fourier reconstruction methods* and *direct methods* (e.g., Crowther *et al.*, 1970; Gilbert, 1972). Fourier methods are defined as algorithms

Michael Radermacher • Wadsworth Center for Laboratories and Research, New York State Department of Health, Albany, New York 12201-0509

Electron Tomography: Three-Dimensional Imaging with the Transmission Electron Microscope, edited by Joachim Frank, Plenum Press, New York, 1992.

that restore the Fourier transform of the object from the Fourier transforms of the projections and then obtain the real-space distribution of the object by inverse Fourier transformation. Included in this group are also equivalent reconstruction schemes that use expansions of object and projections into orthogonal function systems (e.g., Cormack, 1963, 1964; Smith *et al.*, 1973; Zeitler, Chapter 4). In contrast, direct methods are defined as those that carry out all calculations in real space. These include the convolution back-projection algorithms (Bracewell and Riddle, 1967; Ramachandran and Lakshminarayanan, 1971; Gilbert, 1972) and iterative algorithms (Gordon *et al.*, 1970; Colsher, 1977). Weighted back-projection methods are difficult to classify in this scheme, since they are equivalent to convolution back-projection algorithms, but work on the real-space data as well as the Fourier transform data of either the object or the projections. Both convolution back-projection and weighted back-projection algorithms are based on the same theory as Fourier reconstruction methods, whereas iterative methods normally do not take into account the Fourier relations between object transform and projection transforms. Thus, it seems justified to classify the reconstruction algorithms into three groups: *Fourier reconstruction methods, modified back-projection methods, and iterative direct space methods*, where the second group includes convolution back-projection as well as weighted back-projection methods.

Each reconstruction algorithm requires a set of projections of the object, recorded under different projecting directions. While the choice of the data collection geometry is determined by the properties of the specimen, e.g., by its radiation sensitivity, by the degree of variation among particles in the preparation, and by orientational preferences of the particles, this geometry in turn determines which algorithm is most efficient in calculating the three-dimensional structure.

As explained in more detail in Chapter 1, four kinds of data collection schemes are used in electron microscopy: single-axis tilting, conical tilting, random-conical tilt, and general random tilt. For the collection of a single-axis tilt series, the specimen is tilted in the microscope in a range of typically $-60°$ to $+60°$ in small increments, e.g., $1°$ to $5°$, and an image of the same particle is recorded for each specimen position. For a conical tilt series the specimen is tilted by one fixed angle in the range of $45°$ to $60°$ and then rotated within this plane by small angular increments. Again an image of the same particle is recorded for each specimen position. Both data collection schemes are mainly used for preparations that are radiation resistant or contain particles that individually have different shapes; i.e., averaging over different particles is either not required or not possible. Reconstruction schemes based on a random or random-conical tilt geometry intrinsically average over a large number of identical particles, and the electron exposure of each single particle can be kept very low. The random-conical data collection scheme (Frank and Goldfarb, 1980; Radermacher *et al.*, 1986, 1987) makes use of the fact that many particles tend to exhibit a preferred orientation with respect to the specimen support plane, leaving only one rotational degree of freedom, which is a rotation around an axis perpendicular to this plane. The set of images from many such particles in a micrograph of a tilted specimen form a conical tilt series with random azimuthal angles. To achieve the maximum amount of three-dimensional information, a large tilt angle is required, typically between $45°$ and $65°$. If a preferred orientation is not present, then any micrograph of the specimen showing particles in random orientation provides a random tilt series (e.g., Vogel and Provencher, 1988). Intermediate to the latter situation is a specimen of particles with more than one preferred orientation. Each such orientation can be used as a random-conical tilt series, or the sets of particles with different preferred orientations can be combined into a single random-tilt series (Frank *et al.*, 1988; Carazo *et al.*, 1989).

Among all types of reconstruction algorithms, most work has been devoted to the reconstruction of objects from a single-axis tilt series with equal angular increments. The Fourier algorithms have been extensively optimized, and most of them perform very efficiently. Many variations of the convolution and weighted back-projection algorithms can be found that are designated for single-axis tilt geometry (e.g., Ramachandran and Lakshminarayanan, 1971; Gilbert, 1972; Kwok *et al.*, 1977).

For the regular conical-tilt geometry, the choice of reconstruction algorithms is much smaller. Although many iterative direct space algorithms can be used that do not rely on a specific geometry, their performance can be very slow. In many Fourier reconstruction methods the problem of Fourier interpolation and the associated matrix inversion becomes too large to be feasible on most computers. Methods that use orthogonal series expansions still need more development to be efficient in applications to regular conical tilt geometries. Weighted back-projection methods, on the other hand, have been extended to the reconstruction from conical projection series (Radermacher and Hoppe, 1978; Radermacher, 1980) and offer a very efficient solution.

A similar situation can be found for the reconstruction from arbitrary non-regular geometries with randomly distributed tilt angles. As in the case of regular conical tilt geometries, the iterative methods are directly applicable. However, as the number of input images in reconstructions from randomly distributed projections is necessarily much larger than in reconstructions from regular tilt geometries, the problem of computing time requirements increases even further. Again the weighted back-projection method in the form recently developed to deal with arbitrary geometries (Radermacher *et al.*, 1986, 1987; Harauz and van Heel, 1986) has become the most efficient reconstruction algorithm. In this context we must also mention the series expansion method developed for arbitrary geometries (Provencher and Vogel, 1988; Vogel and Provencher, 1988), yet for asymmetrical particles this procedure still requires multiple tilts and is limited, on current computer systems, in the number of projections that can be handled.

For the reasons stated, weighted back-projection methods are currently the most widely used algorithms for the reconstruction of single, asymmetrical particles in electron microscopy. The version developed for single-axis tilt geometry is mostly used for the reconstruction of structures embedded in thick sections from micrographs recorded in the high-voltage electron microscope (e. g., McEwen *et al.*, 1986; Frank *et al.*, 1987), whereas the weighted back-projection algorithm for arbitrary geometry is used for the reconstruction of macromolecular structures from single-exposure random-conical tilt series (Radermacher *et al.*, 1986), for which this algorithm was specifically developed.

Besides performing faster on large data sets, weighted back-projection algorithms have the important advantage over most iterative algorithms that all operations involved are linear and the outcome of the reconstruction is entirely determined by the experimental input data. This linearity facilitates a description of weighted back-projection methods using the concept of *point spread functions* and *transfer functions.*

The chapter starts with a short introduction to the concept of linear systems. Following this, the transfer function of a simple back-projection is determined, which yields the weighting functions for the weighted back-projection. First, the most general weighting function, for a reconstruction from projection with arbitrary angular distribution, is derived. Afterward specific analytical forms of the weighting functions are shown that can be derived for data sets that consist of projections obtained with equal angular increments.

Many other weighting functions have been proposed which differ from the functions shown here mainly by the incorporation of low-pass filters. The functions derived here do not contain implicit low-pass filters; instead, the formulae will be given that describe the dependence of the resolution on the number of projections available. With these formulae available the reconstructions can be low-pass-filtered independently of the reconstruction algorithm used. This approach allows for flexibility in the choice of the low-pass filter function and ensures that maximum use is made of the available data. At the end of this section some examples of computer implementations of the reconstruction algorithms will be given.

2. *THE CONCEPT OF POINT-SPREAD FUNCTIONS AND TRANSFER FUNCTIONS*

The point-spread function of an imaging system describes the image of a single point as it results after using a perfect point as input to the system. Such a system can be any linear optical system or an algorithm that "images" an input distribution onto an output distribution, or even the combination of optical and digital systems. The complete system involved in three-dimensional electron microscopy is the microscope that images the object onto a photographic plate, the digitizer that transfers the image to the computer, the algorithms used to process the images, and the camera and film used to record the final result.

For a more detailed analysis of point-spread and transfer functions, we refer to Goodman (1968), whose terminology will be closely followed, and to Papoulis (1968).

A system is defined as a "black box" that maps a set of input functions onto a set of output functions. The system will be described by the operator S. If $g_1(x', y')$ represents a two-dimensional input and $g_2(x, y)$ is the output of the system, then we use the notation

$$g_2(x, y) = S\{g_1(x', y')\} \tag{1}$$

Because the extension to three- and higher-dimensional systems is straightforward, only two-dimensional systems will be regarded here. If the operator S (or the system it is associated with) is linear, then

$$S\{a \cdot f(x', y') + b \cdot g(x', y')\} = a \cdot S\{f(x', y')\} + b \cdot S\{g(x', y')\} \tag{2}$$

The response of a linear system can be described as the superposition of responses to elementary functions into which the input can be decomposed. One possible decomposition of the function $g_1(x', y')$ would be

$$g_1(x', y') = \iint g_1(\xi, \eta)\, \delta(x' - \xi, y' - \eta)\, d\xi\, d\eta \tag{3}$$

where δ is the two-dimensional delta function. The function g_1 passing the system results in

$$g_2(x, y) = S\left\{\iint g_1(\xi, \eta)\, \delta(x' - \xi, y' - \eta)\, d\xi\, d\eta\right\} \tag{4}$$

and from the linearity of S it follows that

$$g_2(x, y) = \iint g_1(\xi, \eta)\, S\{\delta(x' - \xi, y' - \eta)\}\, d\xi\, d\eta \tag{5}$$

Here the input function $g_1(\xi, \eta)$ simply appears as a coefficient of the elementary function $\delta(x' - \xi, y' - \eta)$. To describe the mapping of g_1, only the system's response h to the delta-function, called the *impulse response* or (in optics) the *point-spread function*, needs to be known for every location (x, y):

$$h(x, y; \xi, \eta) = S\{\delta(x' - \xi, y' - \eta)\} \tag{6}$$

Note, however, that the system's response still can depend on the location of the point.

If the system is *shift-invariant*, a property also called *isoplanatic* in optics, then the system response is independent of the absolute coordinates and depends only on the difference vector $(x - \xi, y - \eta)$. The point-spread function then can be written as

$$h(x, y; \xi, \eta) = h(x - \xi, y - \eta) \tag{7}$$

Combining Eqs. (5)–(7), we see that the mapping by a linear shift-invariant system can be described as the convolution of the input function with the point-spread function:

$$g_2(x, y) = \iint g_1(\xi, \eta) \cdot h(x - \xi, y - \eta)\, d\xi\, d\eta \tag{8}$$

In connection with linear shift-invariant systems, the convolution theorem of Fourier transforms is of importance:

$$g * f = \mathbf{F}^{-1}\{\mathbf{F}\{g\} \cdot \mathbf{F}\{f\}\} \tag{9}$$

The asterisk is used to indicate convolution, and **F** indicates the Fourier transform. The convolution of a function g with a function f is the inverse Fourier transform of the product between the Fourier transform of g and the Fourier transform of f. This relationship will be extensively used. Let the function $f(x, y)$ be mapped onto the function $g(x, y)$:

$$g(x, y) = f(x, y) * h(x, y) \tag{10}$$

where $h(x, y)$ is the point-spread function of the system. The Fourier transform of $g(x, y)$ is

$$G(X, Y) = F(X, Y) \cdot H(X, Y) \tag{11}$$

where $H(X, Y)$, the Fourier transform of $h(x, y)$, is the transfer function of the system, and $F(X, Y)$ is the Fourier transform of $f(x, y)$. Throughout this chapter capital letters are used for coordinates in Fourier space. The original function $f(x, y)$ can be recovered from $g(x, y)$ by dividing the Fourier transform of $g(x, y)$ by the Fourier transform of $h(x, y)$ followed by an inverse Fourier transform:

$$f(x, y) = \mathbf{F}^{-1}\left\{\frac{G(X, Y)}{H(X, Y)}\right\} \quad \text{for} \quad H \neq 0 \tag{12}$$

The division of the Fourier transform $G(X, Y)$ by $H(X, Y)$ corresponds to a deconvolution of $g(x, y)$ by the function $h(x, y)$.

3. THREE-DIMENSIONAL RECONSTRUCTION

The input data to any three-dimensional reconstruction are projections of a three-dimensional distribution. Bright-field images in electron microscopy are in approximation projections of the potential distribution of the object (see Chapter 2). Use will be made of the projection theorem, which states that the Fourier transform of a projection is a central section through the Fourier transform of the three-dimensional distribution (see also Chapter 4).

For understanding weighted back-projection methods, it is necessary to understand the simple back-projection or summation technique, which will be explained first.

A weighted back-projection is a simple back-projection followed by a deconvolution with the point spread function of the simple back-projection algorithm. This deconvolution is done by a division of the Fourier transform of the back-projection by its transfer function. One divided by the transfer function is called the *weighting function*, a term from which the term *weighted back-projection* originates. Convolution back-projections are algorithms that perform the equivalent convolution in real space. The convolution kernel for the real-space convolution is the inverse Fourier transform of the weighting function.

The weighting functions derived in the following sections is correct only for the transfer function of the simple back-projection algorithm. Specifically they do not take into account the limits to the resolution caused by the limited number of projections available. Thus, to obtain a faithful reconstruction of the object, the three-dimensional reconstruction must be low-pass-filtered to the resolution that is determined, by virtue of Shannon's sampling theorem, by the number and spacing of the available projections. The resolution formulae will be given in Section 4.

In the derivation of the analytical form of the weighting function for single-axis tilting, it will be assumed that data are available over the full tilt range from $-\pi/2$ to $+\pi/2$. Distortions of the image point that are caused by the missing wedge in single-axis tilt geometries or the missing cone in conical-tilt geometries cannot be corrected by a weighting function or deconvolution, because the Fourier transform of the point-spread function $H(X, Y, Z)$ becomes 0 in these regions and thus violates the condition stated in Eq. (12).

Distortions caused by missing data can be reduced, however, using a priori knowledge, such as the possible density range or the overall size of the reconstructed object. The technique of projection onto convex sets (Carazo, Chapter 6) allows this information to be combined with the reconstruction. Essentially, data consistent with the a priori information are inserted into the missing range of the Fourier transform of the reconstructed object, while data that are determined by the experimental projections are left unchanged.

3.1. The Simple Back-Projection Algorithm

An intuitively simple reconstruction method is the back-projection or summation technique. The technique can most easily be explained using the example of a simple two-dimensional binary object consisting, e.g., of three disk-shaped regions with value 1 inside and value 0 outside. This object is to be reconstructed from its one-dimensional projections.

The object O (Fig. 1*a*) appears in the projections P_1, P_2, P_3, P_4, and P_5 at angles θ_1, θ_2, θ_3, θ_4, and θ_5. As a first step in the reconstruction process, the projections are smeared out to form so called back-projection bodies (Hoppe *et al.*, 1976); (Fig. 1*b*). To reconstruct the object, one has to sum all the back-projection bodies (Fig. 1*c*).

The simple back-projection calculates only an approximation of the object. In the example above, the original three disks are identifiable in the back-projection because of their higher intensity relative to the surrounding (Fig. 1*d*). If the object were not binary, the ambiguity in the back-projection would be even greater. The quality, however, improves when more projections are used. It can be shown that a simple back-projection reconstructs the object with a point-spread function that enhances the low-spatial-frequency components.

Let $f(x, y, z)$ be a three-dimensional distribution, which is projected under the angles θ_j, ϕ_j to form a series of projections $p_j(x^j, y^j)$. Let $\mathbf{r}^j = (x^j, y^j, z^j)$ be the coordinates in the coordinate system of the projection p_j, which forms the (x^j, y^j) plane. The geometrical relation between the object coordinates $\mathbf{r} = (x, y, z)$ and $\mathbf{r}^j = (x^j, y^j, z^j)$ can be described using the rotation matrices D_{θ_j}, D_{ϕ_j}:

$$\mathbf{r}^j = D_{\theta_j} \cdot D_{\phi_j} \cdot \mathbf{r} \tag{13}$$

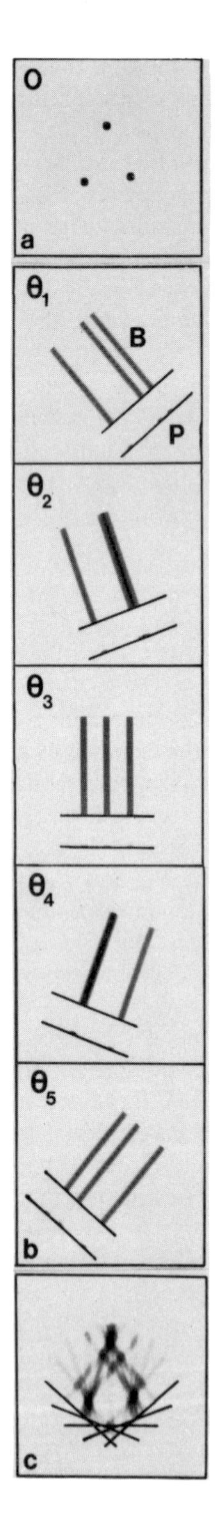

FIGURE 1. Principle of a simple back-projection. A binary object consisting of three points (a) is projected onto five projections P at angles θ_1, θ_2, θ_3, θ_4, θ_5. (b) From these projections back-projection bodies B are created and the object is reconstructed by addition of these back-projection bodies (c).

with

$$D_{\theta_j} = \begin{pmatrix} \cos\theta_j & 0 & -\sin\theta_j \\ 0 & 1 & 0 \\ \sin\theta_j & 0 & \cos\theta_j \end{pmatrix} \tag{14}$$

and

$$D_{\phi_j} = \begin{pmatrix} \cos\phi_j & \sin\phi_j & 0 \\ -\sin\phi_j & \cos\phi_j & 0 \\ 0 & 0 & 1 \end{pmatrix} \tag{15}$$

The object is first rotated by ϕ_j around its z axis, then tilted by $-\theta_j$ around the new y axis, and then projected onto the (x^j, y^j) plane. Only two angles are needed to describe all projecting directions. A third angle would just describe a rotation in the plane of the projection, which will be disregarded here.

A projection along the direction z^j with angles θ_j, ϕ_j can be written as

$$p_j = \int f(x^j, y^j, z^j)\, dz^j \tag{16}$$

A back-projection body is formed by convolution of p_j in the (x^j, y^j) plane with the two-dimensional spread function

$$l_j = \delta(x^j, y^j)\, c(z^j) \tag{17}$$

with

$$c(z^j) = \begin{cases} 1 & \text{for} \quad -a \leqslant z^j \leqslant +a \\ 0 & \text{otherwise} \end{cases} \tag{18}$$

In practical applications $2a$ is chosen such that it is larger than the diameter of the object in all projecting directions. The back-projection body can thus be written as

$$\begin{aligned} p_j^b(x^j, y^j, z^j) &= \iint p_j(x^j, y^j) \cdot l(x^j - x'^j, y^j - y'^j, z^j)\, dx'^j\, dy'^j \\ &= \iiint f(x'^j, y'^j, z'^j)\, dz'^j \cdot l(x^j - x'^j, y^j - y'^j, z^j)\, dx'^j\, dy'^j \end{aligned} \tag{19}$$

and the back-projection algorithm becomes

$$b(x, y, z) = \sum_j p_j^b(x^j, y^j, z^j) \tag{20}$$

Simple back-projection is a linear and shift-invariant algorithm. Thus, its performance can be described as the convolution of the original object with the point-spread function of the algorithm. To obtain a faithful reconstruction, the back-projected density has to be corrected. This correction can be achieved by the convolution with a function that acts as the inverse of the point-spread function. In weighted back-projection methods this deconvolution is achieved by a division of the Fourier transform of the back-projected density by the transfer function of the reconstruction algorithm.

3.2. *Weighting Function for Arbitrary Tilt Geometry*

To find the weighting function for a weighted back-projection for arbitrary geometry, we must first analyze the point-spread function of a simple back-projection in more detail. The three-dimensional object can be rewritten as a superposition of δ functions, using the three-dimensional equivalent to Eq. (3):

$$f(x, y, z) = \iiint f(\xi, \eta, \zeta)\, \delta(x-\xi, y-\eta, z-\zeta)\, d\xi\, d\eta\, d\zeta \tag{21}$$

The point-spread function then can be found by analyzing how the back-projection algorithm affects the function

$$q = \delta(x, y, z) \tag{22}$$

The projection of q at angles θ_j, ϕ_j is

$$p_j(x^j, y^j) = \delta(x^j, y^j) \tag{23}$$

The back-projection body becomes, according to Eqs. (17) and (19),

$$p_j^b(x^j, y^j, z^j) = \delta(x^j, y^j)\, c(z^j) \tag{24}$$

and the point back-projected in three dimensions is found by summation over θ_j, ϕ_j.

$$b(x, y, x) = \sum_j \delta(x^j, y^j)\, c(z^j) \tag{25}$$

Thus, $b(x, y, z)$ is the point-spread function of a back-projection calculated from a set of projections taken with arbitrary angles θ_j, ϕ_j. The transfer function is the Fourier transform of (25):

$$\begin{aligned} H(X, Y, Z) &= \mathbf{F}\{b(x, y, z)\} \\ &= \mathbf{F}\left\{\sum_j \delta(x^j, y^j)\, c(z^j)\right\} \\ &= \sum_j \mathbf{F}\{\delta(x^j, y^j)\, c(z^j)\} \end{aligned} \tag{26}$$

and

$$\mathbf{F}\{\delta(x^j, y^j)\, c(z^j)\} = \int_{-\infty}^{+\infty} \iint \delta(x^j, y^j)\, c(z^j)\, e^{-2\pi i(x^j X^j + y^j Y^j + z^j Z^j)}\, dx_j\, dy_j\, dz_j$$

$$= 1 \int_{-a}^{+a} e^{-2\pi i z^j Z^j}\, dz^j = \frac{\sin(2\pi a Z^j)}{\pi Z^j}$$

$$= 2a \operatorname{sinc}(2a\pi Z^j) \tag{27}$$

where $\operatorname{sinc}(x) = \sin(x)/x$.

With the rotation matrices given in Eqs. (14) and (15), Z^j can be expressed in the coordinate system (X, Y, Z) of the object:

$$Z^j = X \sin\theta_j \cos\phi_j + Y \sin\theta_j \sin\phi_j + Z\cos\theta_j \tag{28}$$

and the transfer function becomes (Radermacher *et al.*, 1986)

$$H(X, Y, Z) = \sum_j 2a \operatorname{sinc}[2a\pi(X \sin\theta_j \cos\phi_j + Y\sin\theta_j \sin\phi_j + Z\cos\theta_j)] \tag{29}$$

Hence, the corresponding weighting function for arbitrary geometry is

$$W_a(X, Y, Z) = \frac{1}{H(X, Y, Z)}$$

$$= \left\{\sum_j 2a \operatorname{sinc}[2a\pi(X\sin\theta_j\cos\phi_j + Y\sin\theta_j\sin\phi_j + Z\cos\theta_j)]\right\}^{-1} \tag{30}$$

Equation (30) is valid only for $H \neq 0$. In the implementation of the algorithm a lower threshold for H is set to avoid division by 0 and to limit the enhancement of noise. The original three-dimensional distribution $o(x, y, z)$ can be recovered from the back-projection $b(x, y, z)$ by multiplication of its Fourier transform $B(X, Y, Z)$ with $W_a(X, Y, Z)$ followed by an inverse Fourier transform:

$$o(x, y, z) = \mathbf{F}^{-1}\{O(X, Y, Z)\} = \mathbf{F}^{-1}\{B(X, Y, Z)\, W_a(X, Y, Z)\} \tag{31}$$

The weighting function (30) is three-dimensional and is applied to the three-dimensional Fourier transform of an object reconstructed by simple back-projection. Since the Fourier transforms of projections are central sections through the three-dimensional Fourier transform of the object and since simple back-projection is essentially a summation which commutes with the Fourier transform operation, the weighting function can also be applied to the Fourier transform of the projections before the back-projection step. However, in an arbitrary tilt geometry the location of a point in the central section relative to all other central sections is not independent of an inplane rotation of the projection, and therefore a third Euler angle has to be included into the transformation (13). Only in a random conical tilt series and simpler geometries can this third angle be kept 0 by aligning the tilt axis parallel to the same axis in each projection. Thus, for the application to the Fourier transforms of the projections, transformation (13) becomes

$$\mathbf{r}^j = D_{\psi_j} D_{\theta_j} D_{\phi_j} \mathbf{r} \tag{32}$$

Z^j in Eqs. (28) and (29) becomes the Z component of the vector

$$\mathbf{R}^j = D_{\psi_j} D_{\theta_j} D_{\phi_j} D^{-1}_{\phi_k} D^{-1}_{\theta_k} D^{-1}_{\psi_k} \mathbf{R}^k \tag{33}$$

where $\mathbf{R}^k$ is the vector $(X^k, Y^k, 0)$ in the section in Fourier space at angles ψ_k, θ_k, ϕ_k to which the weighting function is being applied.

The advantage of applying the weighting function to the projection is that in most cases H is nonzero along the corresponding sections through the three-dimensional Fourier transform, because each section is also the origin of one of the superimposed sinc functions in H. The disadvantage of this mode of application is that the number of calculations needed is usually larger. Assuming that the dimensions of the Fourier transforms of projections and object are equal to their real-space dimensions, a common situation is that the xy dimensions of the projections are larger than the xy dimensions of the object and, second, that the number of projections is larger than the number of z slices in the reconstructed object. The most appropriate implementation of the algorithm is one that allows a choice of the mode of application.

In practical applications the weighting function (Eq. 30) performs well if the number of projections per angular interval varies smoothly. If, however, large gaps exist in the angular sampling and the number of projections per angular interval varies rapidly, then fringes may occur in the reconstruction. These artifacts can be eliminated by the use of a weighting function that is the Fourier transform of an apodized point-spread function. The discontinuous step function $c(z^j)$ in Eq. (18), which describes the length of the rays in the backprojection step, is replaced by a continuous step function with Gaussian falloff. This function can be created by a convolution of the discontinuous step function with a Gaussian function. Because of the convolution theorem, its Fourier transform is the product of the sinc function (Eq. 18) and the Fourier transform of the Gaussian, which also is a Gaussian function. For the calculation of the weighting function based on the apodized point-spread function the elements in the sum in Eq. (29) are replaced by products of sinc functions and Gaussians.

3.3. Weighting Function for Single-Axis Tilt Geometry with Equal Angular Increments

The weighting function derived in the previous section is applicable to any tilt geometry. However, for a single-axis tilt geometry with equal angular increments, an analytical form of the weighting function can be found. Although the weighting function for single-axis tilting can be derived from the inverse Radon transform or a comparison between a simple back-projection and a Fourier inversion method (e.g., Cormack, 1963, 1964; Ramachandran and Lakshminarayanan, 1971;

Vainshtein, 1971; Vainshtein and Orlov 1972; Gilbert, 1972, and Section 5 of this chapter), it will be derived here as a special case of the weighting function for arbitrary geometry.

In a single-axis tilt geometry the object is tilted around a single axis in equal angular increments, and a projection is recorded for each orientation. Without loss of generality the tilt axis is assumed to be the y axis (Fig. 2); thus, θ in Eq. (14) becomes the tilt angle and ϕ in Eq. (15) is kept at 0. Now we make the following approximations: (i) the variable a in Eq. (18) is assumed to be infinite, which corresponds to back-projection bodies that are infinitely extended in the z^j direction; (ii) a series of projections is available over a continuous range of θ from $-90°$ to $+90°$. With these approximations and with

$$\lim_{a \to \infty} a \operatorname{sinc}(a \cdot x) = \delta(x) \tag{34}$$

Eq. (29) becomes

$$H(X, Y, Z) = \sum_j \delta(X \sin \theta_j + Z \cos \theta_j) \tag{35}$$

From condition (ii) the sum in Eq. (35) can be replaced by an integral:

$$H(X, Y, Z) = \int_{-\pi/2}^{+\pi/2} \delta(X \sin \theta + Z \cos \theta)\, d\theta \tag{36}$$

Since the tilt axis is assumed to be the y axis, $H(X, Y, Z)$ is independent of Y. The transfer function is constant in the direction of the tilt axis and varies only in the planes perpendicular to Y. If the coordinates X and Z are replaced by the cylindrical coordinates R and Γ within the XZ plane

$$X = R \cos \Gamma, \qquad Z = R \sin \Gamma, \qquad Y = Y \tag{37}$$

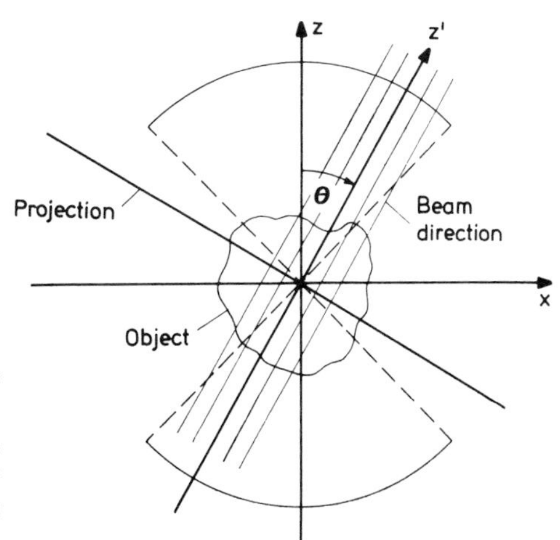

FIGURE 2. Geometry of single-axis tilting. The tilt axis y is perpendicular to the plane of the page, θ is the tilt angle, x, y, z the coordinate system fixed to the object, x', y', z' the coordinate system, with x', y' being the plane of the projection. (From Radermacher, 1988, with permission of Alan R. Liss, Inc.)

Eq. (36) becomes

$$H(R, Y, \Gamma) = \int_{-\pi/2}^{+\pi/2} \delta(R\cos(\Gamma - \Theta))\, d\Theta \tag{38}$$

The solution to this integral can be found using the identity

$$\delta(f(x)) = \sum_n \frac{1}{|f'(x_n)|}\, \delta(x - x_n), \qquad \text{with} \quad f'(x_n) = \left.\frac{df(x)}{dx}\right|_{x = x_n} \tag{39}$$

for $f'(x_n) \neq 0$. The x_n are the zeros of $f(x)$ within the integration range. There is only one zero of f within the integration interval $\Theta \varepsilon (-\pi/2, \pi/2]$; $f(\Theta) = 0$ for $\Theta = \Gamma - \pi/2$, and the solution of (38) becomes

$$H(R, Y, \Gamma) = \frac{1}{R} \tag{40}$$

We thus obtain the result that, for the deconvolution of a simple back-projection from a single-axis tilt series with equal angular increments, the Fourier transform of the back-projections has to be multiplied by the weighting function

$$W_s(R, Y, \Gamma) = R \tag{41}$$

where R is the radius in Fourier space perpendicular to the tilt axis Y. Again the multiplication with R can be applied either directly to the three-dimensional Fourier transform of the back-projection is calculated. If applied to the projection, the application of the weighting function corresponds to a multiplication of the Fourier transform of the projection by X.

3.4. *Weighting Function for Conical Tilt Geometry with Equal Angular Increments*

For collection of a conical tilt series in the microscope the specimen is tilted by a fixed angle θ_0 and then rotated in this inclined plane by equal angular increments $\Delta\phi$. In each position a projection is recorded. The geometry is equivalent to moving the beam direction along the surface of a cone (Fig. 3).

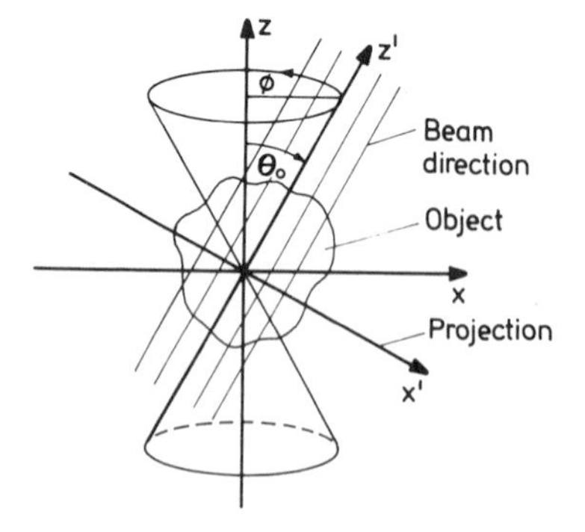

FIGURE 3. Geometry of conical tilting. As in Fig. 2, y axis and y' axis are perpendicular to the plane ($\phi = 0$), x, y, z are object coordinates, x', y', z' the projection coordinates, θ_0 the fixed tilt angle around y, and ϕ the azimuthal angle around z. (From Radermacher 1988, with permission of Alan R. Liss, Inc.)

To calculate the geometrical relation between the coordinate system in the projections and the coordinate system fixed to the object, again the rotation matrices (14) and (15) are employed with $\theta_j = \theta_0$ constant, and $\phi_j = j\,\Delta\phi$. To find the analytical form of the weighting function, the same assumptions are made as in the calculation of the analytical weighting for single-axis tilting; i.e., the back-projection bodies are infinitely extended in the z^j direction, and the azimuthal angle ϕ is assumed to be continuous. As before, this results in a replacement of the sinc function in Eq. (29) by a δ function and a replacement of the sum by an integral. The transfer function of a simple back-projection from a regular conical tilt series thereby becomes

$$H(X, Y, Z) = \int_{-\pi}^{\pi} \delta(Z_\phi)\, d\phi$$

$$= \int_{-\pi}^{\pi} \delta[X \sin\theta_0 \cos\phi + Y \sin\theta_0 \sin\phi + Z \cos\theta_0]\, d\phi \tag{42}$$

By expressing X, Y, Z in cylindrical coordinates R, Γ, Z and making use of Eq. (39), we finally obtain (Radermacher and Hoppe, 1978; Radermacher, 1980)

$$H(R, \Gamma, Z) = \frac{2}{|R \sin\theta_0 \sin \arccos[-Z \cot(\theta_0)/R])|} \tag{43}$$

Equation (43) is valid in the region $R > Z \cot\theta_0$, i.e., only in the region of the Fourier transform where measured data are available. It is not valid within the missing cone. Because $0 \leqslant \theta_0 \leqslant \pi/2$,

$$\arccos\left(\frac{-Z \cot\theta_0}{R}\right) = \arcsin\left(\sqrt{1 - \frac{Z^2 \cot^2\theta_0}{R^2}}\right) \tag{44}$$

and Eq. (43) becomes

$$H(R, \Gamma, Z) = \frac{2}{\sqrt{R^2 \sin^2\theta_0 - Z^2 \cos^2\theta_0}} \tag{45}$$

The weighting function that needs to be applied to the three-dimensional Fourier transform of the back-projection from a regular conical tilt series then becomes

$$W_c(R, \Gamma, Z) = \tfrac{1}{2}\sqrt{R^2 \sin^2\theta_0 - Z^2 \cos^2\theta_0} \tag{46}$$

Here, as in the previous cases, the weighting function may also be applied to the projections before the back-projection is carried out. In the central section in Fourier space that corresponds to the projection at the angles θ_0 and ϕ_j, Eq. (47) becomes

$$W_c(X^j, Y^j) = Y^j \sin\theta_0 \tag{47}$$

Equation (47) was derived by reverting to Cartesian coordinates in Eq. (46) and transforming to the coordinate system of the projection at angles θ_0, ϕ_j, using

$$\mathbf{r}^j = D_{\theta_0} D_{\phi_j} \mathbf{r} \tag{48}$$

with D_ϕ and D_θ as previously defined in Eqs. (14) and (15).

For a reconstruction from a conical tilt series with equal angular increments, the weighting corresponds to a multiplication of the Fourier transforms of the projections by $Y \sin \theta_0$, where y is the direction of the tilt axis belonging to the angle θ_0. In contrast to the weighting function for single-axis tilt geometry, which only affects the projection perpendicular to the tilt axis, here the weighting function increases in the direction of the tilt axis and is constant perpendicular to it.

3.5. Other Forms of Weighting Functions

For reconstruction from projections with arbitrary projecting angles, a weighting function different from the one derived in Eq. (29) has been proposed (Harauz and van Heel, 1986). The filter, which the authors term "exact filter," is an approximation to the function shown here. The essential difference is that the sinc function is replaced by the simpler triangular function. The use of the triangular function in the summation of the single projection's contribution to the weighting function poses smaller computational demands and thus results in a faster performance of the algorithm. However, if the sine function that appears in Eq. (29) as part of the sinc function calculation is tabulated, the difference in computational effort between the two approaches should be minimal.

For single-axis tilting, a large variety of weighting functions can be found that combine the R weighting with a window of the Fourier transform that corresponds to a low-pass filter (e.g., Ramachandran and Lakshminarayanan, 1971; Kwok *et al.*, 1977; Suzuki, 1983). The inclusion of a window function $w(R)$, however, becomes necessary if the deconvolution is to be carried out in real space. The real-space convolution kernel is calculated as the inverse Fourier transform of $R \cdot w(R)$. As R is not square integrable and does not have a compact support, its inverse Fourier transform does not exist. The window function is needed to limit the weighting function to a compact support, make it square integrable, and, possibly, to carry out this inverse Fourier transform.

4. BANDLIMIT OF A RECONSTRUCTION FROM A LIMITED NUMBER OF PROJECTIONS

For all reconstructions only a discrete, limited number of projections is available. If no bandlimit is imposed on the reconstruction, then there exists an infinite number of possible reconstructions that are all consistent with the available measurements, and the result obtained would be just one of these possible solutions. However, by using Shannon's sampling theorem (Shannon, 1949) a bandlimit can be determined up to which the reconstruction is unique, presenting a faithful image of the original object (see also Zeitler, Chapter 4). The bandlimitation is not taken into account by the weighting functions derived above and thus needs to be imposed in a separate step. Like the weighting function, the low-pass filter can be applied either to the projections before the reconstruction or to the reconstructed object. The application to the projections has the advantage that it avoids multiple Fourier transformations, since it can be done simultaneously with the application of the weighting function. On the other hand, the application of the low-pass filter to the reconstruction allows for more flexibility.

4.1. Resolution for Single-Axis Tilt Geometry

The form in which the resolution depends on the number of projections in a single-axis tilt geometry with equal angular increments has been determined by Bracewell and Riddle (1967) and Crowther *et al.* (1970). Let d be the resolution of the reconstruction and D the diameter of the reconstruction volume, which is assumed to be cylindrical, the axis of the cylinder coinciding with the tilt axis. The reconstruction volume is defined as the volume large enough to include the complete object. Let $\Delta\theta$ be the angular increment, and let $N = \pi/\Delta\theta$. Then the best resolution achievable is

$$d = \pi \frac{D}{N} \qquad \text{for a tilt range from } -90^\circ \text{ to } +90^\circ \tag{49}$$

or

$$d = D\,\Delta\theta \tag{50}$$

4.2. Resolution for Conical Tilt Geometry

Let θ_0 be the fixed tilt angle, d the resolution, D the object diameter, and N the number of projections evenly spaced in ϕ. The resolution of the reconstruction then depends on the number of projections in the following way (Radermacher, 1980, 1988; Radermacher and Hoppe, 1980):

$$d = 2\pi \frac{D}{N} \sin\theta_0 \qquad \text{for } N \text{ even} \tag{51}$$

and

$$d = \frac{2\pi D \sin\theta_0 \cos\theta_0}{N\sqrt{\sin^2\theta_0(\pi/2N)^2 + \cos^2\theta_0}} \qquad \text{for } N \text{ odd} \tag{52}$$

Inversion of (52) yields

$$N = \frac{\pi D}{2d} \tan\theta_0 \sqrt{16\cos^2\theta_0 - (d/D)^2} \tag{53}$$

The resolution formulae above do not take into account the missing regions in the Fourier transform of the object caused by the limited tilt range for single-axis tilting or by $\theta_0 < 90°$ for conical tilting. In a single-axis tilt series with a limited tilt range around the y axis from $-\alpha_{max}$ to $+\alpha_{max}$ in increments of $\Delta\theta$, Eq. (49) gives the resolution in the x direction. The resolution in the z direction depends on the maximum tilt angle, and the resolution parallel to the tilt axis y is equal to the resolution of the input projections. Similarly, for a conical tilt series, where the fixed tilt by the angle θ_0 is around y, Eqs. (51) and (52) give the resolution in the directions parallel to the xy plane, whereas the resolution in the z direction depends on the size of the missing cone (Radermacher, 1980, 1988).

4.3. Resolution in a Flat Extended Reconstruction Volume

For the derivation of Eqs. (49)–(52) a spherical shape of the reconstruction volume has been assumed, which is appropriate for most reconstructions of macromolecular assemblies. However, especially in reconstructions of subcellular components in thick sections, the assumption of a flat extended slab for the shape of the reconstruction volume is more appropriate. Equations (49)–(52) could be applied by using for D the diameter of a sphere that circumscribes the complete slab. However, the resulting number of projections necessary to obtain a reconstruction with a reasonable resolution becomes prohibitive if calculated this way. Indeed, this approach is unnecessarily conservative. If the image point is analyzed more closely, it can be seen that, for a given resolution d, D in any Eq. (49) to (52) is the radius of the volume surrounding a single image point where artifacts are minimal. If the object is considered as being built up by a superposition of image points, each of which carries a halo of rays in the directions of the contributing projections (Fig. 4), starting at a distance D from the center of the image point, then a faithful reconstruction can be calculated up to a section thickness of

$$T = D \cos \gamma \tag{54}$$

where γ is the maximum tilt angle α_{max} for single-axis tilting or the fixed angle θ_0 for conical tilting. Substitution of

$$D = \frac{T}{\cos \gamma} \tag{55}$$

in Eqs. (49) to (52) will therefore give the resolution of a reconstruction of an extended slab with thickness T.

4.4. Resolution for Random and Random-Conical Geometry

For reconstructions from randomly distributed projections, no formulae are available that predict the theoretical value of the final resolution. For reconstructions from a random-conical tilt series, however, the angular coverage can be analyzed and a conservative estimate of the resolution can be made by using the largest angular step between two adjacent projections as the value for $\Delta\theta = 2\pi/N$ in Eq. (52).

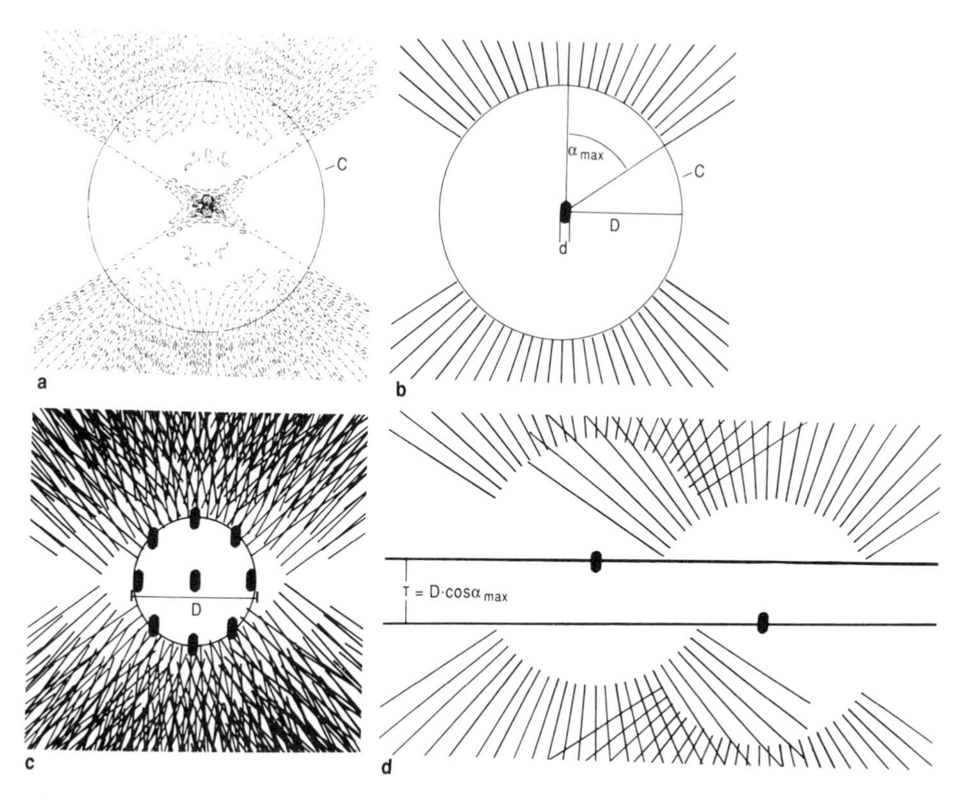

FIGURE 4. Reconstruction of an object within a spherical reconstruction volume and within an extended slab. (a) Contour plot of a cross section through a three-dimensional image point reconstructed from a single-axis tilt series containing 25 projections in the range of $-60°$ to $+60°$, using R-weighted back-projection. Contour levels: dashed lines 2% increments <0 to 8%, –·–·– 0-line, solid lines at 10%, 20%, 40% to 100%. Percentages relative to image maximum. (b) Sketch of the image point of Fig. 1*a*. C = circle with radius D (= diameter of reconstruction volume in Eq.), d = resolution. (c) Superposition of image points on the surface of a sphere with diameter D. The inside of the sphere is virtually artifact-free. (d) Superposition of two image points at the surface of an extended slab. Within the thickness T of the slab no overlap of the artifacts surrounding the image points occurs.

5. RELATION BETWEEN THE INVERSE RADON TRANSFORM, FOURIER INVERSION, AND WEIGHTED BACK-PROJECTION METHODS

The inverse Radon transform (Radon, 1917; Cormack, 1963, 1964), Fourier inversion methods, and weighted back-projection are equivalent formulations of the reconstruction problem in a single-axis tilt geometry. This is true for a single-axis tilt geometry only, where the problem can be reduced to the problem of reconstructing a two-dimensional image from one-dimensional projection.

The n-dimensional Radon transform is obtained by integration over $(n-1)$-dimensional hyperplanes which for $n=1$ results in line integrals. For $n=3$ the integration has to be carried out over planes. Projections obtained in the electron microscope correspond to line integrals of the potential distribution of the object. In a single-axis tilt series with tilt axis y, the set of lines taken from each projection at the same coordinate y_0 form the two-dimensional Radon transform of the xz section through the object at the location y_0. In a conical or arbitrary tilt geometry the reconstruction problem can no longer be reduced to a set of planar reconstructions from a set of lines extracted from the projections. The set of two-dimensional projections in a conical tilt geometry does not constitute the three-dimensional Radon transform of the object. The latter would require a set of integrals over all planes and not the set of line integrals as is found in two-dimensional projections. A method for reconstruction from a set of one-dimensional projections that form a three-dimensional Radon transform, however, can be found in earlier three-dimensional imaging techniques applied in nuclear magnetic resonance imaging (Shepp, 1980).

To show the equivalence between Radon inversion, Fourier inversion, and weighted back-projection methods, we follow the book of Deans (1983, p. 96ff). (See also, e.g., Crowther *et al.*, 1970; Ramachandran and Lakshminarayanan, 1971; Gilbert, 1972; Vainshtein and Orlov, 1972; Smith *et al.*, 1973; Zwick and Zeitler, 1973; Zeitler, Chapter 4.)

The two-dimensional Radon transform of an object $f(\mathbf{r})$, $\mathbf{r} \in \mathbf{R}^2$, is the function

$$\check{f}(p, \xi) = \int f(\mathbf{r})\, \delta(p - \xi \cdot \mathbf{r})\, d\mathbf{r} \tag{56}$$

which is equivalent to a line integral along the line $p = \xi \cdot \mathbf{r}$, with ξ being the unit vector in the direction of the line, $\xi = (\cos\theta, \sin\theta)$, and the integration is carried out over the plane in $\mathbf{R}^2$. The Fourier transform of $f(\mathbf{r})$ is

$$\begin{aligned} F(\mathbf{R}) = \mathbf{F}_2\{f(\mathbf{r})\} &= \int f(\mathbf{r})\, e^{-2\pi i \mathbf{r} \cdot \mathbf{R}}\, d\mathbf{r} \\ &= \int f(\mathbf{r})\, e^{-2\pi i t}\, \delta(t - \mathbf{R} \cdot \mathbf{r})\, dt\, d\mathbf{r} \end{aligned} \tag{57}$$

where $\mathbf{F}_2$ indicates the two-dimensional Fourier transform. The Fourier transform along sections in the direction of the unit vector ξ can be found by replacing $\mathbf{R}$ with $s\xi$. Substituting sp for t then yields

$$\begin{aligned} F(s \cdot \xi) &= |s| \int_{-\infty}^{+\infty} \int f(\mathbf{r})\, e^{-2\pi i s p}\, \delta(sp - s\xi \cdot \mathbf{r})\, d\mathbf{r}\, dp \\ &= \int_{-\infty}^{+\infty} \int f(\mathbf{r})\, \delta(p - \xi \cdot \mathbf{r})\, d\mathbf{r}\, e^{-2\pi i s p}\, dp \end{aligned} \tag{58}$$

Here, $g(p, \xi) = \int f(\mathbf{r})\, \delta(p - \xi \cdot \mathbf{r})\, d\mathbf{r}$ is the Radon transform of $f(\mathbf{r})$ or the one-dimensional projection of f along ξ. Equation (58) essentially is the projection theorem, which states that the two-dimensional Fourier transform of the function $f(x, y)$ is the one-dimensional Fourier transform along p of the two-dimensional

Radon transform of f. The function $f(x, y)$ can be recovered by application of an inverse two-dimensional Fourier transform to the one-dimensional Fourier transform along p of the Radon transform. This is the basis for all Fourier reconstruction methods (see Chapter 4).

The equivalence of the weighted back-projection method and Fourier reconstruction methods can be seen from the following. Let $p_\phi(q)$ be a one-dimensional projection of a two-dimensional object $o(x, y)$ at the angle ϕ. In the coordinate system of the object we have $q = x\cos\phi + y\sin\phi$. Let $f(q, \phi)$ be the function describing the set of projections for a continuous angle ϕ. The back-projection in its continuous form then is

$$b(x, y) = \int_0^\pi f(x\cos\phi + y\sin\phi, \phi)\, d\phi = \int f(q(x, y, \phi), \phi)\, d\phi \tag{59}$$

Let $F(\mathbf{R})$ be the one-dimensional Fourier transform of $f(q, \phi)$ along q:

$$F(R, \phi) = \int_{-\infty}^{+\infty} f(q, \phi)\, e^{-2\pi i q R}\, dq \tag{60}$$

Replacing f in Eq. (59) by the inverse Fourier transform of F yields

$$\begin{aligned} b(x, y) &= \int_0^\pi \int_{-\infty}^{+\infty} F(R, \phi)\, e^{2\pi i R q(x, y, \phi)}\, dR\, d\phi \\ &= \int_0^\pi \int_{-\infty}^{+\infty} F(R, \phi)\, e^{2\pi i R(x\cos\phi + y\sin\phi)}\, dR\, d\phi \end{aligned} \tag{61}$$

and after introduction of polar coordinates r, γ with $x = r\cos\gamma$, $y = r\sin\gamma$, Eq. (61) becomes

$$\begin{aligned} b(r, \gamma) &= \int_0^\pi \int_{-\infty}^{+\infty} F(R, \phi)\, e^{2\pi i R r\cos(\gamma - \phi)}\, dR\, d\phi \\ &= \int_0^{2\pi} \int_0^\infty \frac{1}{R} F(R, \phi)\, e^{2\pi i R r\cos(\gamma - \phi)} R\, dR\, d\phi \end{aligned} \tag{62}$$

which is the two-dimensional inverse Fourier transform of $F(R, \phi)$ multiplied by $1/R$. The weighting function for single-axis tilting, R, compensates for this factor. Thus, a weighted back-projection is equivalent to a reconstruction by Fourier inversion methods.

ACKNOWLEDGMENTS

This work was supported by NIH grant #1R01 GM29169 and 1S10 rr03998 and NSF grant #8313405. I wish to thank Joachim Frank for many discussions and critically reading this manuscript.

A. APPENDIX: NOTES ON THE COMPUTER IMPLEMENTATION OF THE ALGORITHMS

Many schemes of computer implementation of algorithms can be found in the literature on numerical methods, and the following does not claim to be the most efficient way of programming the algorithms outlined in the foregoing chapter, but might be helpful in translating the equations into program code.

A.1. Implementation of the Simple Back-Projection Algorithm

For the calculation of a simple back-projection, the projection bodies are summed into the three-dimensional volume. Both the volume and the projections are available on a discrete sampling grid, and an interpolation is necessary to sum the rays of the back-projection body into the volume. One possible solution uses the following scheme. In a loop over the volume elements the coordinates of each point in the volume are calculated in the projection P_j at the angles θ_j, ϕ_j. The value of the projection at this point then is interpolated from the values of the surrounding points in the projection and added to the volume element. For a bilinear interpolation in the back-projection the formula is

$$O(m, n, o) = \sum_i P_i(p, q)(1 - \Delta x)(1 - \Delta y) + P_i(p+1, q)\, \Delta x(1 - \Delta y) + P_i(p, q+1)(1 - \Delta x)\, \Delta y + P_i(p+1, q+1)\, \Delta x\, \Delta y \tag{63}$$

where $O(m, n, o)$ is the volume element at x, y, z indices m, n, o with $-M \leqslant m \leqslant +M$, $-N \leqslant n \leqslant +N$, and $-O \leqslant o \leqslant +O$. Here, p and q are the x and y indices of pixels in the projections and are calculated using the matrices (14) and (15):

$$\begin{aligned} x &= m \cos\theta_i \cos\phi_i + n \cos\theta_i \sin\phi_i - o \sin\theta_i \\ p &= \text{integer truncation of } x, \qquad \Delta x = x - p \end{aligned} \tag{64}$$

and

$$\begin{aligned} y &= -m \sin\phi_i + n \cos\phi_i \\ q &= \text{integer truncation of } y, \qquad \Delta y = y - q \end{aligned} \tag{65}$$

Of course, other interpolation schemes can be used, the fastest but crudest being a nearest-neighbor selection, and a much more elaborate one is a spline interpolation.

A.2. Implementation of the Weighting Scheme for Arbitrary Geometry

The following contains two sketches for possible layouts of a program that calculates and applies the weighting function for arbitrary geometry. One shows the application of the weighting function to the projections, and the second shows the application of the weighting function to the volume after simple back-projection (Eqs. 29, 28, 31).

A.2.1. Application to the Fourier Transform of the Projections

Read the angles ψ_j, θ_j, ϕ_j for all projections
Loop 1 over all projections p_k, $1 \leqslant k \leqslant J$.
calculate the transfer function $H_k(X^k, Y^k)$ for projection p_k at angles θ_k, ϕ_k.
Loop 2 over all Fourier coordinates X^k, Y^k in the Fourier plane corresponding to the projection p_k.
Loop 3 over all projection angles ψ_j, θ_j, ϕ_j; $1 \leqslant j \leqslant J$.
calculate the Z_j coordinate of point (X^k, Y^k), i.e., the Z coordinate of point $\mathbf{R}^k$ in the coordinate system of projection p_j at angles ψ_j, θ_j, ϕ_j using
$\mathbf{R}^j = D_{\psi_j} D_{\theta_j} D_{\phi_j} D_{\phi_k}^{-1} D_{\theta_k}^{-1} D_{\psi_k}^{-1} \mathbf{R}^k$
calculate $H(X^j, Y^j) = H(X^j, Y^j) + 2a\,\mathrm{sinc}(2\pi a Z^j)$
end of loop 3 over projection angles
If $H(X^j, Y^j) <$ threshold, then $H(X^j, Y^j) =$ threshold
end of loop 2 over Fourier plane coordinates
Fourier-transform projection $P_k(X^k, Y^k) = \mathbf{F}(p_k(x^k, y^k))$
compute $P_k(X^k, Y^k)/H_k(X^k, Y^k) := P_k^w(X^k, Y^k)$.
Inverse Fourier-transform: $\mathbf{F}^{-1}\{P_k^w(X^k, Y^k)\} := p_k^w(x, y)$.
end of loop 1 over projections.

A.2.2. Application to the Three-Dimensional Fourier Transform

Read the angles ψ_j, θ_j, ϕ_j for all J projections
Loop 1 over all Fourier coordinates X, Y, Z in the three-dimensional Fourier transform.
Loop 2 over all J projection angles ψ_j, θ_j, ϕ_j.
calculate the Z^j coordinate of point $\mathbf{R} = (X, Y, Z)$, i.e., the z coordinate of point $\mathbf{R}$ in the coordinate system of projection p_j at angles θ_j, ϕ_j (ψ can be neglected) using $\mathbf{R}^j = D_{\theta_j} D_{\phi_j} \mathbf{R}$
calculate $H(X, Y, Z) = H(X, Y, Z) + 2a\,\mathrm{sinc}(2\pi a Z^j)$
end of loop 2 over projection angles.
If $H(X, Y, Z) <$ threshold, then $H(X, Y, Z) =$ threshold
end of loop 1 over Fourier coordinates
Fourier transform back-projected volume: $\mathbf{F}\{o^b(\mathbf{r})\} := O^b(\mathbf{R})$
Divide: $O^b(\mathbf{R})/H(\mathbf{R}) := O(\mathbf{R})$
Inverse Fourier transform: $\mathbf{F}^{-1}\{O(\mathbf{R})\} := o(\mathbf{r})$

REFERENCES

Bracewell, R. N. and Riddle, A. C. (1967). Inversion of fan-beam scans in radio astronomy. *Astrophys. J.* **150**:427–434.

Carazo, J-M., Wagenknecht, T., and Frank, J. (1989). Variations of the three-dimensional structure of the *Escherichia coli* ribosome in the range of overlap views. *Biophys. J.* **55**:465–477.

Colsher, J. G. (1977). Iterative three-dimensional image reconstruction from tomographic projections. *Comput. Graph. Image Proc.* **6**:513–537.

Cormack, A. M. (1963). Representation of a function by its line integrals, with some radiological applications. *J. Appl. Phys.* **34**:2722–2727.

Cormack, A. M. (1964). Representation of a function by its line integrals, with some radiological applications. II. *J. Appl. Phys.* **35**:2908–2913.

Crowther, R. A., DeRosier, D. J., and Klug, A. (1970). The reconstruction of a three-dimensional structure from projections and its application to electron microscopy. *Proc. R. Soc. London A* **317**:319–340.

Deans, S. R. (1983). *The Radon Transform and Some of Its Applications.* Wiley, New York.

Frank, J. and Goldfarb, W. (1980). Methods for averaging of single molecules and lattice fragments, in *Electton Microscopy at Molecular Dimensions* (W. Baumeister and W. Vogell, eds.), pp. 261–269. Springer-Verlag, Berlin.

Frank, J., McEwen, B. F., Radermacher, M., Turner, J. N., and Rieder, C. L. (1987). Three-dimensional tomographic reconstruction in high-voltage electron microscopy. *J. Electron Microsc. Technique* **6**:193–205.

Frank, J., Carazo, J-M., and Radermacher, M. (1988). Refinement of the random conical reconstruction technique using multivariate statistical analysis and classification. *Eur. J. Cell Biol. Suppl. 25* **48**:143–146.

Gilbert, P. F. C. (1972). The reconstruction of a three-dimensional structure from projections and its application to electron microscopy. II:Direct methods. *Proc. R. Soc. London B* **182**:89–102.

Goodman, J. W. (1968). *Introduction to Fourier Optics.* McGraw-Hill, New York.

Gordon, R., Bender, R., and Herman, G. T. (1970). Algebraic reconstruction techniques (ART) for three-dimensional electron microscopy and x-ray photography. *J. Theor. Biol.* **29**:471–481.

Harauz, G. and van Heel, M. (1986). Exact filters for general three-dimensional reconstruction. *Optik* **73**:146–156.

Hoppe, W., Schramm, H. J., Sturm, M., Hunsmann, N., and Gaßmann, J. (1976). Three-dimensional electron microscopy of individual biological objects. I:Methods. *Z. Naturforsch.* **31a**:645–655.

Kwok, Y. S., Reed, I. S., and Truong, T. K. (1977). A generalized $|\omega|$-filter for 3D-reconstruction. *IEEE Trans. Nucl. Sci.* **NS24**:1990–2005.

McEwen, B. F., Radermacher, M., Rieder, C. L., and Frank, J. (1986). Tomographic three-dimensional reconstruction of cilia ultrastructure from thick sections. *Proc. Nat. Acad. Sci. USA* **83**:9040–9044.

Papoulis, A. (1968). *Systems and Transforms with Applications in Optics.* McGraw–Hill, New York; reprint, Robert E. Krieger, Florida, 1986.

Provencher, S. W. and Vogel, R. H. (1988). Three-dimensional reconstruction from electron micrographs of disordered specimes. I:Method. *Ultramicroscopy* **25**:209–222.

Radermacher, M. (1980). *Dreidimensionale Rekonstruktion bei kegelförmiger Kippung im Elektronenmikroskop.* Ph.D. thesis, Technische Universität München, Germany.

Radermacher, M. (1988). Three-dimensional reconstruction of single particles from random and non-random tilt series. *J. Electron. Microsc. Technique* **9**:359–394.

Radermacher, M. and Hoppe, W. (1978). 3-D reconstruction from conically tilted projections, in *Proc. 9th Int. Congr. Electron Microscopy*, Vol. 1, pp. 218–219.

Radermacher, M. and Hoppe, W. (1980). Properties of 3-D reconstructions from projections by conical tilting compared to single axis tilting, in *Proc. 7th Europ. Congr. Electron Microscopy*, Vol. 1, pp. 132–133.

Radermacher, M., Wagenknecht, T., Verschoor, A., and Frank, J. (1986). A new 3-D reconstruction scheme applied to the 50*S* ribosomal subunit of *E. coli. J. Microsc.* **141**:RP1–RP2.

Radermacher, M., Wagenknecht, T., Verschoor, A., and Frank, J. (1987). Three-dimensional reconstruction from a single-exposure random conical tilt series applied to the 50*S* ribosomal of *Escherichia coli. J. Microsc.* **146**:113–136.

Radon, J. (1917). Über die Bestimmung von Funktionen durch ihre Integralwerte längs gewisser Mannigfaltigkeiten. *Ber. Verh. König. Sächs. Ges. Wiss. Leipzig, Math. Phys. Kl.* **69**:262–277.

Ramachandran, G. N. and Lakshminarayanan, A. V. (1971). Three-dimensional reconstruction from radiographs and electron micrographs: Application of convolution instead of Fourier transforms. *Proc. Nat. Acad. Sci. USA* **68**:2236–2240.

Shannon, C. E. (1949). Communication in the presence of noise. *Proc. IRE* **37**:10–22.

Shepp, L. A. (1980). Computerized tomography and nuclear magnetic resonance zeugmatography. *J. Comput. Assist. Tomogr.* **4**:94–107.

Smith, P. R., Peter, T. M., and Bates, R. H. T. (1973). Image reconstruction from a finite number of projections. *J. Phys. A* **6**:361–382.

Suzuki, S. (1983). A study on the resemblance between a computed tomographic image and the original object, and the relationship to the filterfunction used in image reconstruction. *Optik* **66**:61–71.

Vainshtein, B. K. (1971). Finding the structure of objects from projections. *Sov. Phys. Crystallogr.* **15**:781–787.

Vainshtein, B. K. and Orlov, S. S. (1972). Theory of the recovery of functions from their projections. *Sov. Phys. Crystallogr.* **17**:253–257.

Vogel, R. W. and Provencher, S. W. (1988). Three-dimensional reconstruction from electron micrographs of disordered specimes. II:Implementation and results. *Ultramicroscopy* **25**:223–240.

Zwick, M. and Zeitler, E. (1973). Image reconstruction from projections. *Optik* **38**:550–565.

6

The Fidelity of 3D Reconstructions from Incomplete Data and the Use of Restoration Methods

José-María Carazo

José-María Carazo • Centro Nacional de Biotecnología and Centro de Biología Molecular, Universidad Autonoma, 28049 Madrid, Spain

Electron Tomography: Three-Dimensional Imaging with the Transmission Electron Microscope, edited by Joachim Frank, Plenum Press, New York, 1992.

1. INTRODUCTION

During the last two decades it has become increasingly evident that electron microscopy images of typical thin biological specimens carry a large amount of information on the three-dimensional (3D) structure of the object. It has been shown many times how the information contained in a set of images (2D signals) can determine a useful estimate of the 3D structure of the specimen under study. Naturally, the mathematical methods used in these studies have become more and more elaborate as the complexity of the structural problems increased.

Stated in its most general form, the 3D reconstruction problem is defined by the following statement: Given a collection of images (2D data) g, determine the 3D structure f that produced the images g. We are interested in knowing under which conditions g is adequate for determining f or at least some $\hat{f}$ that is close to f in some sense. We are aiming to solve practical cases, so we must be aware that our image data set g will always be corrupted by some noise n. Our questions are then: How large is the effect of the noise in obscuring f? Can we do something to ameliorate the situation? Do small changes in g produce radical changes in f? Formulated in this way it becomes clear that the topics covered under electron tomography belong to the broad class of signal recovery problems.

For practical reasons we always have two basic limitations in the collection of the image data set g: one is the noise, and the other is the restriction to a finite number of images. We will see how these physical limits directly affect the fidelity of the estimated 3D structures by providing a framework in which the specific problems encountered in electron tomography can be formulated and the different approaches discussed.

Unfortunately there is very little practical work done in this area, and experimental results are still scarce. The work in this chapter is therefore of a more general nature than most of the other chapters in this book.

2. DATA COLLECTION AND MISSING INFORMATION

2.1. General Considerations

To perform the 3D reconstruction of a macromolecular structure from electron microscopy images, it is necessary to collect a set of views of the specimen from different directions. Depending on the symmetry of the specimen and the complex into which it may aggregate, the way these different views are obtained varies in practice.

The simplest case is presented by specimens that are aggregated into a helical suprastructure (for the purpose of this presentation it does not matter whether these aggregates are natural or artificially induced). In this case it is clear that a single image of the helix already contains views from different directions of the individual objects; this is so because the relative orientation between such an object and the electron beam changes along the helix in a defined way. This property makes it possible to perform a complete 3D reconstruction of the helix from one image of the helical aggregate (at least up to some aliasing-limited rsolution)

(DeRosier and Moore, 1970). There are other types of aggregations that are also highly symmetrical; this is the situation, for instance, for icosahedral viruses, where a general view of the specimen already provides 59 other symmetry-related views (Crowther *et al.*, 1970). For a general case, however, we cannot count on symmetries, since we may well have an isolated and asymmetric particle object.

In this latter situation the different views are obtained by tilting the specimen grid. The same situation occurs for specimens arranged in a 2D crystal; since most of the symmetry elements are not orientation dependent in the direction perpendicular to the specimen grid plane (except for a possible screw axis), the different views must also be obtained by tilting.

In all cases there is an obvious practical limitation to the total number of different views that can actually be taken. Furthermore, both in the general case in which no symmetries are assumed and in the case of specimens arranged in a 2D crystal, the different views are obtained by tilting the specimen grid within the microscope with the help of a goniometer. There are technical limits to the maximum tilt angle that commercial goniometers can achieve, usually around $\pm 60°$. Even if specially designed goniometers are constructed providing, in principle, unlimited tilting (Chalcroft and Davey, 1984), there are physical limits to the maximum tilt angle that can be reached while still obtaining useful images. These limits arise from the increase in the effective specimen thickness according to 1/cos(tilt angle), which makes multiple scattering events more probable.

The general problem in electron tomography is, thus, the 3D reconstruction of a structure from a finite set of noisy 2D projection images over a restricted angular range (usually $\pm 60°$).

2.2. Fourier Space Formulation

As mentioned before, tilting by more than $\pm 60°$ is often impossible. This fact introduces a restriction to the total data angular range available for 3D reconstruction. Much insight into how this limitation may affect a reconstruction is gained by introducing the Fourier transform (FT) operator. The Fourier transform of a function $f(\mathbf{r}) \in L^2$ (the Hilbert space $\mathscr{H}$ of all square-integrable functions in R^3) is defined as follows:

$$F(\mathbf{R}) = \int_{-\infty}^{\infty} \int_{-\infty}^{\infty} \int_{-\infty}^{\infty} f(\mathbf{r}) \exp(-2\pi i \mathbf{r} \cdot \mathbf{R})\, d(\mathbf{r}) \tag{1}$$

with $\cdot$ denoting the inner product.

In Cartesian coordinates with the Euclidean distance, this definition becomes

$$F(X, Y, Z) = \int_{-\infty}^{\infty} \int_{-\infty}^{\infty} \int_{-\infty}^{\infty} f(x, y, z) \exp(-2\pi i (xX + yY + zZ)\, dx\, dy\, dz \tag{2}$$

In turn, we define the inverse Fourier transform of F as

$$f(\mathbf{r}) = \int_{-\infty}^{\infty} \int_{-\infty}^{\infty} \int_{-\infty}^{\infty} F(\mathbf{R}) \exp(2\pi i \mathbf{r} \cdot \mathbf{R})\, d(\mathbf{R}) \tag{3}$$

We thus obtain pairs of functions $(f, F) \in \mathscr{H} \times \mathscr{H}$ where F is the Fourier transform of f. The convention of denoting these pairs by lowercase and uppercase letters (lowercase letters associated with the real-space quantity and uppercase ones with the quantity in Fourier space) will be used throughout this chapter, both relating to functions and to variables. After applying the FT operator to the original function f, the coordinate system associated with the variable $\mathbf{R}$ of the new function F is different from the coordinate system of the variable $\mathbf{r}$ of the original function f. In the following we will use the term *reciprocal*, or *Fourier*, *space* if we are based in the coordinate system of F, and the term *real space* if we are based in the original coordinate system of f.

One of the most important results relating f and F to each other is the so-called central section theorem, which states that the Fourier transform of a 2D projection image is equal to a central section of the 3D Fourier transform of the object (e.g., Crowther *et al.*, 1970). Figure 1 shows this relationship. Let us consider the z axis pointing in the direction of the electron beam. Without loss of generality, let us place the tilt axis along the x axis. If the projection image g is obtained by tilting the object by θ, then its Fourier transform $G = \mathrm{FT}(g)$ will be a central section of the 3D Fourier transform of the object tilted around the X axis by Θ. It is usual to measure tilt angles between $\pm 90°$ with reference to the z (or Z) axis; so, if θ is, say, $+60°$ (with the usual convention that clockwise rotations are negative), then G is a central section that includes the X axis and makes an angle of $\Theta' = -30°$ with the Z axis.

The central section theorem will be fundamental in the following discussion,

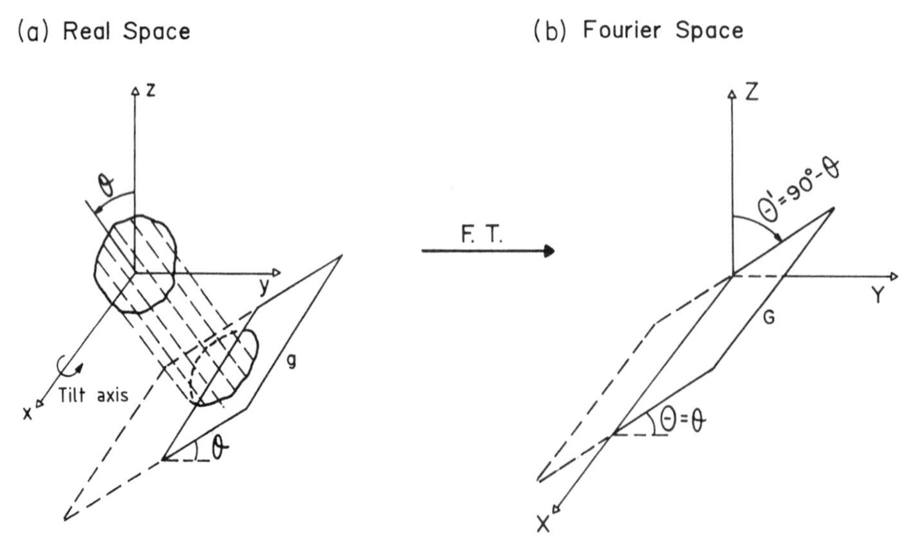

FIGURE 1. The central section theorem. Real space coordinate axes are labeled (x, y, z), while corresponding ones in Fourier space are labeled (X, Y, Z). (a) An object is shown in real space placed near the origin, and one of its projection images is presented on plane g. The projection image is formed by parallel beams that are tilted with respect to the z axis by θ. (b) The Fourier transform of the object shown before extends over the entire Fourier space. The Fourier transform of the projection plane g is a section G through the origin of the Fourier space (a central section) tilted by Θ' with respect to the Z axis.

helping us to visualize how the different practical data collection procedures necessarily result in the partial absence of the information content of the object. Because of its importance in electron tomography, two types of data collection procedures will be studied in detail; these are the *single-axis* tilt and *conical tilt* series.

2.3. Single-Axis Tilt Series

In this scheme of data collection the object is tilted by small increments in the microscope around a fixed axis, and a micrograph of the specimen is obtained in each orientation (Fig. 2*a*).

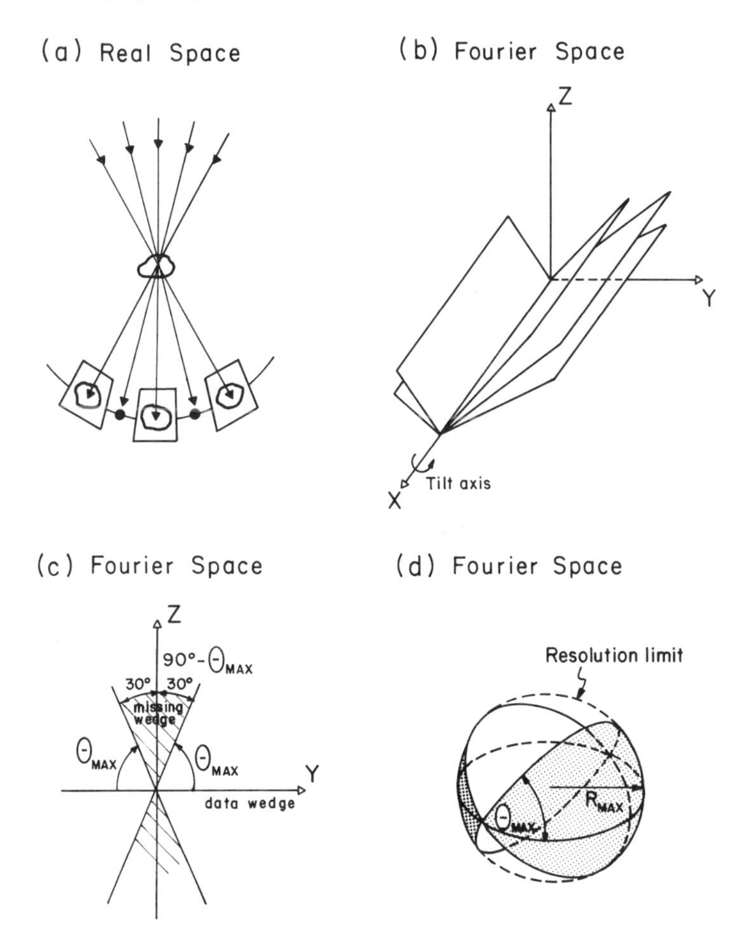

FIGURE 2. Single-axis tilt data collection geometry. (a) Tilting in real space around an axis perpendicular to the plane of this page (which is assumed to be the *x* axis). (b) Fourier space representation of the information presented in (a), where the different projection views are central planes through the three-dimensional Fourier transform of the object with the *X* axis in common. (c) Close-up view of (b) from the *X* axis; a missing wedge of 30° zenithal angle around the *Z* axis is shown in dashed lines. (d) Spatial representation in Fourier space in a practical situation where Θ_{max} is the maximum tilt angle and R_{max} is the maximum frequency. [(a) and (d) from Radermacher, 1980, reproduced with permission.]

By calculating the Fourier transform of each of these projections and using the central section theorem, we obtain a set of tilted planes in reciprocal space that have the tilt axis in common (Fig. 2*b*, *c*, *d*). Note that for tilting around a single axis, the three-dimensional problem can be divided into a set of two-dimensional problems, each being the reconstruction of a plane perpendicular to the tilt axis with a thickness equal to the expected resolution in the direction of the tilt axis.

As we discussed before, the maximum tilt angle achievable in the microscope is limited, usually to $\pm 60°$, and this limit defines the maximum tilt angle of the corresponding sections in reciprocal space (Fig. 2*c*, *d*). It is then clear that there exists a wedge-shaped region in reciprocal space where no data can be measured (Fig. 2*c*, *d*), and it is consequently termed the *missing wedge* region. Correspondingly, the measurable area is also wedge shaped, and it is usually termed the *data wedge*.

The missing wedge is centered on the Z axis in Fourier space and, assuming a maximum tilt angle of $\theta = \pm 60°$, has a width of 60° (from $-30°$ to $+30°$ off the Z axis), while the data wedge is centered around the tilt axis in the XY plane and has a width of 120°. Being more precise, and following the usual notation of negative and positive frequencies, there are two missing wedges and two data wedges (Fig. 2*c*, *d*); however, these are symmetrically placed with respect to the origin and, therefore, are Friedel related: $F(-\mathbf{R}) = F^*(\mathbf{R})$.

2.4. Conical Tilt Series

A conical tilt series can be collected by tilting the specimen in the microscope by the maximum attainable tilt angle, θ_{max}, and then rotating it in the tilted plane by small angular increments (Fig. 3*a*). As before, a micrograph is recorded at each position. This data collection geometry has been introduced in practice only in recent years, but its impact in many areas is becoming very large (see Radermacher, 1988, for a review).

Following the same line of reasoning that was used in the previous study of the single-axis tilt series data collection process, it is clear that there exists a portion of the Fourier transform of the object that remains undetermined. The shape of this *missing region* in Fourier space is that of a cone centered on the Z axis and with a halfwidth of $90° - |\phi_{max}|$ (Fig. 3*b*), which is consequently termed the *missing cone* region. This 3D geometry cannot be analyzed by reduction to a set of identical 2D problems, but it has to be treated explicitly in three dimensions.

When we calculate the volume of this missing cone in Fourier space (Radermacher, 1980), and assume a spherical limitation in Fourier space with a maximum radius (resolution) of R_{max}, we obtain

$$V = \frac{8}{3} R_{max}^3 \left(\frac{\pi}{2} - \theta_{max} \right) \tag{4}$$

if θ_{max} is expressed in radians. A similar calculation for the missing wedge results in

$$V = \tfrac{4}{3} \pi R_{max}^3 (1 - \sin \theta_{max}) \tag{5}$$

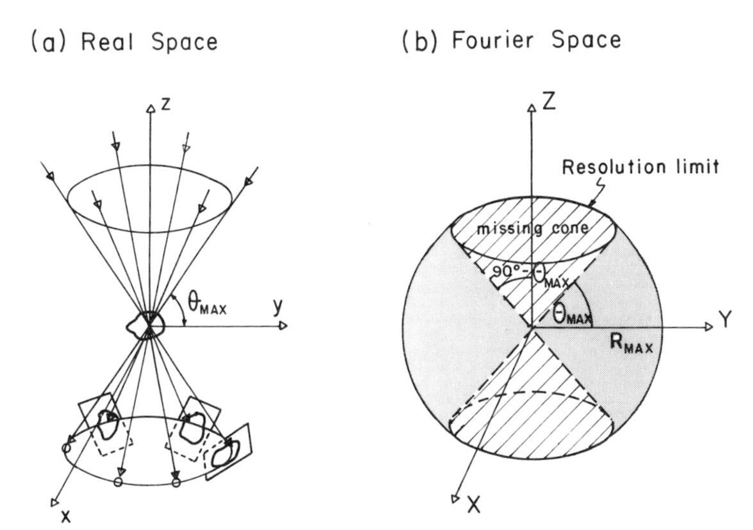

FIGURE 3. Conical tilt data collection geometry. (a) Tilting in real space by θ_{max} around multiple tilt axis in the *xy* plane. (b) Fourier space representation of the information collected in (a); no experimental data are available in a "missing cone" around *Z*. (From Radermacher, 1980, reproduced with permission.)

Considering that the total volume of the sphere is given by $V = \frac{4}{3}\pi R^3_{\text{max}}$, the relative percentage of missing versus total volume (considering $\theta_{\text{max}} = |60|°$) is 33% for the missing wedge case and 13.4% for the missing cone case. It then follows that much less data are missing from a conical tilt series than from a single-axis tilt series. In both data collection procedures, however, all frequencies along the reciprocal axis $(0, 0, Z)$ are missing (Figs. 2*c*, *d* and 3*b*). These frequencies determine the mean density values on planes perpendicular to the electron beam (the z axis); their absence results in all planes perpendicular to z having the same mean density value, whether they contain any portion of the specimen structure or not.

In the next section we study in detail the effects that the missing cone or wedge has on the final 3D reconstructions. Here we just wish to stress that a significant amount of information is not experimentally available and that any reconstruction method applied to these data, as well as any interpretation of the results, has to take this fact into consideration.

3. *MISSING INFORMATION AND STRUCTURAL DEGRADATION, A PSF AND OTF APPROACH*

3.1. Linear Systems and Point-Spread Functions

In this section we intend to provide the reader with some insights into the effects that the absence of data have on the results of a reconstruction process. We will study how a mathematically perfect point would be reconstructed if only measurements within the "data region" were available. Such a model reconstruction is usually termed the point-spread function (PSF), and, in general, it depends on the spatial location of the point being reconstructed.

The PSF approach is especially useful in those cases where the signal formation system is linear, since then any output signal (i.e., a 3D reconstruction) can be interpreted in terms of a linear combination of suitably placed PSFs. Furthermore, if the PSF does not depend on the spatial coordinates of the point, then the system is said to be shift invariant. The type of situations covered in this chapter are characterized by shift-invariant PSFs. In this case the final 3D reconstruction can be conceptualized by considering each point of the object to be replaced by the system PSF multiplied by the value of the object point. In any case, and even for nonlinear formulations, the knowledge of how the simplest object—just a single point—is finally rendered by the system is a very useful piece of information that may enable us to "graphically" understand the effects of the missing information.

The Fourier transform of the PSF is termed the *optical transfer function* (OTF). The OTF and the PSF both carry the same information; however, sometimes it is easier to work in Fourier space with the OTF, and sometimes in real space with the PSF. Obviously, if the system is shift invariant, it will be characterized by one OTF.

In the following we study in some detail both the process of image formation in electron microscopy and the problems originated by the restricted angular range. In the first case we use the instrumental imaging system OTF, in the second the PSF associated with the 3D reconstruction process.

3.2. Image Formation in Electron Microscopy

Not surprisingly, the electron microscope cannot be considered a perfect imaging system. It is well known that it introduces some degradations into the images it renders. Erickson and Klug (1971) and Hawkes, in Chapter 2, present an analysis on how these instrumental degradations can be modeled, within certain limitations, in terms of a single-system OTF [in this field, the OTF is usually called the *contrast transfer function* (CTF)].

With respect to the contents of this chapter, knowledge of the CTF is important because it tells us that the relationship between the 3D object and the 2D image obtained from it is not that the image is a simple projection of the object, but a projection filtered by the CTF. This fact should be kept in mind when modeling the image formation system in any recovery approach.

3.3. General Calculation of the 3D Point-Spread Function

A point object is described by a Dirac delta distribution centered at the point's coordinates. Let us assume that the point object is located at the origin. Then its mathematical representation is $\delta(x, y, z)$. The Fourier transform of a delta distribution is a constant; that is, all Fourier space should be filled with a uniform value. Let us assume that the only data degradation is the absence of data in the missing region. Since we do not know the actual values within the missing region, we may as well fill it with zeros. Mathematically, this situation can be formulated in Fourier space in the following way,

$$\text{FT(reconstructed point)} = C \cdot X_\theta \tag{6}$$

where $C = \mathrm{FT}(\delta(x, y, z))$ is the Fourier transform of the Dirac delta distribution, and X_θ is a function, usually termed the missing region's characteristic function, that has a value of 1 within the data region and 0 outside. Because of the duality $\cdot/\circ$, multiplication/convolution, between Fourier and real space, each reconstructed point in real space will be the convolution of the ideal point object with the inverse Fourier transform of X_θ, which is precisely the system PSF. In the following the system PSF will be denoted by h.

Within this formulation, the 3D reconstruction of the object obtained directly from information within the data region, $\hat{f}$, is related to the unknown 3D structure, f, by a convolution with the system PSF, h; that is,

$$\hat{f}(x, y, z) = \int_{-\infty}^{\infty} \int_{-\infty}^{\infty} \int_{-\infty}^{\infty} f(\alpha, \beta, \gamma)\, h(x - \alpha, y - \beta, z - \gamma)\, d\alpha\, d\beta\, d\gamma \qquad (7)$$

where (x, y, z) and (α, β, γ) refer to the coordinates of the output and input signals, respectively. In passing, we note that Eq. (7) is a Fredholm integral equation of the first kind. This remark points to some interesting approaches to this problem (Rushforth, 1987), which, however, cannot be treated here for reasons of space.

3.4. Point-Spread Function for Single-Axis Tilting

As discussed earlier, there exists a missing wedge of information associated with the single-axis tilt data collection geometry. Because of this absence of data the reconstruction of a point will be distorted. This distorted point is in fact the PSF of the system. Expressed in Cartesian coordinates, with h denoting the system PSF and $\theta_{\max}$ the maximum tilt angle around the x axis (recall that the missing wedge is centered on the Z axis and that the problem can be reduced to a set of 2D problems), we have (Tam and Perez-Mendez, 1981a, b)

$$h(y, z) = \frac{1}{\pi^2(z^2 \tan \theta_{\max}) - y^2/(\tan \theta_{\max})} \qquad (8)$$

for $(y, z) \neq (0, 0)$.

The distortion of a point source caused by the missing wedge is illustrated in Fig. 4, where a maximum tilt angle of $\pm 60°$ has been considered. From Eq. (8) and, more intuitively, from Fig. 4*a*, it is apparent that the three main sources of distortion are (i) the elongation of the PSF along the z axis (Figs. 4*b*, *c*), (ii) the appearance of two pronounced side minima along the y axis (Fig. 4*d*), and (iii) the presence of four ridges originating from the point source and decaying with distance: two positive ones and two negative ones bordering the lines $y = \pm z \tan(\theta_{\max})$ (Fig. 4*a*).

The effect of each of these distortions on the final 3D reconstruction is different. The elongation of the PSF along z (which is $\simeq 1.8$ for the case $\theta_{\max} = \pm 60°$; see Radermacher, 1980; Radermacher and Hoppe, 1980; Frank and Radermacher, 1986; Glaeser *et al.*, 1990) makes the reconstruction anisotropic, in particular by blurring (superimposing) features parallel to the z axis. The two side minima result in the possible destruction of structural information along the y axis. The presence of the other four ridges, together with additional ridges along the z and y axis if there are sharp frequency cutoffs, is of much less importance.

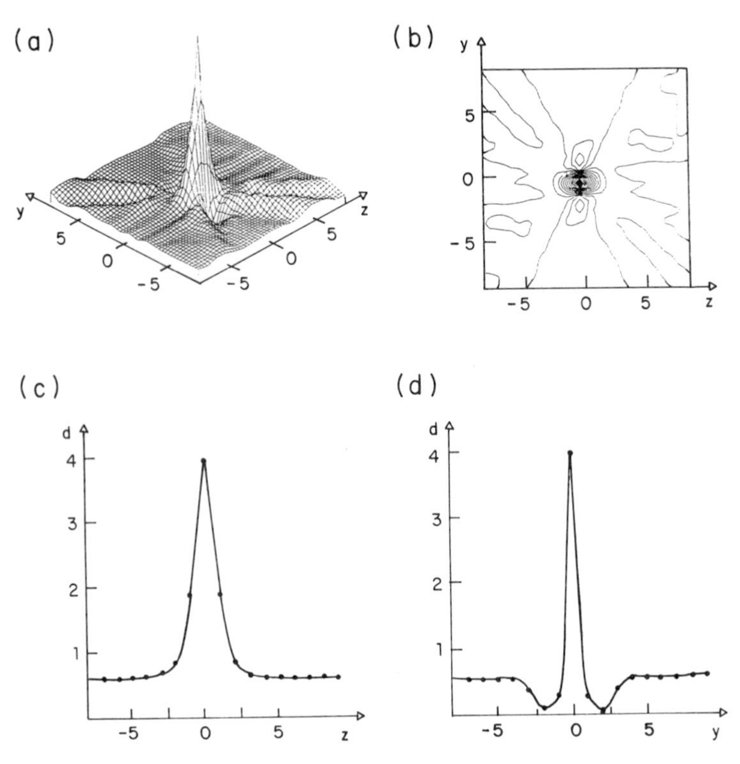

FIGURE 4. Point-spread function for single-axis tilting. (a) Pictorial representation of the PSF where the height at each point (y_p, z_p) corresponds to the value of $h(y_p, z_p)$. (b) Contour level representation of h. (c) Plot of $h(y, z) = d$ along the z axis $(y = 0)$. (d) Plot of $h(y, z) = d$ along the y axis $(z = 0)$.

3.5. Point-Spread Function for Conical Tilting

Associated with this data collection geometry is a missing cone of information where data cannot be measured. Many studies have dealt with the characterization of the PSF of this system, aiming, ideally, to obtain a closed form for it (Chiu *et al.*, 1979; Radermacher and Hoppe, 1980; Radermacher, 1988). Although no such closed form has yet been found, the PSF can be expressed in cylindrical coordinates (Chiu *et al.*, 1979) as

$$h(r, z) = 2\,\frac{R_{\max}^2}{z} \int_0^1 \varepsilon J_0(\alpha\varepsilon r) \sin(\beta\varepsilon z)\, d\varepsilon \tag{9}$$

where $\varepsilon = R/R_{\max}$, $\alpha = 2\pi R_{\max}$, $\beta = 2\pi R_{\max}/(\tan\theta_{\max})$, and J_0 is the Bessel function of zeroth order, $(r, z)(R, Z)$ refer to the real and Fourier space coordinate systems, respectively (the angular symmetry in the xy plane has already been taken into account), and $R_{\max}$ is the maximum reciprocal resolution assuming a spherical (isotropic) frequency limitation.

Figure 5*a* represents the PSF for $|\theta_{\max}| = 60°$. Although the overall shape of the PSFs associated with single-axis tilt and conical tilt is similar, there are very

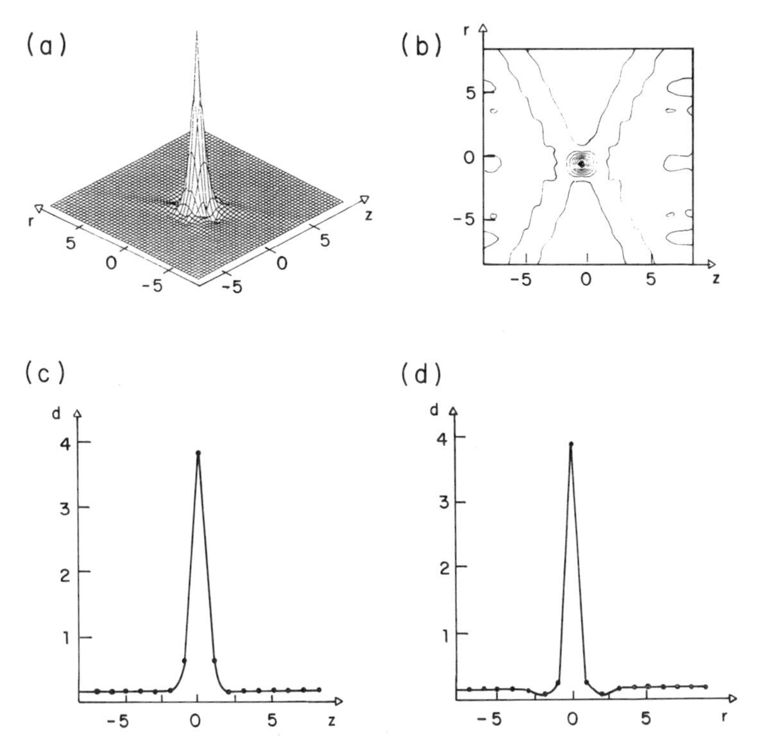

FIGURE 5. Point-spread function for conical tilting. (a) Pictorial representation of the PSF as in Fig. 4. (b) Contour level representation of h. (c) Plot of $h(r, z) = d$ along the z axis ($r = 0$). (d) Plot of $h(r, z) = d$ along the r axis ($z = 0$).

important differences in the point elongation along the z axis (which is now reduced to $\simeq 1.3$ for $\theta_{max} = \pm 60°$; see Radermacher and Hoppe, 1980) (Figs. 5*b*, *c*) and in the reduced side minima along r (Fig. 5*d*).

It is now clear that the outcome of a 3D reconstruction from data obtained either by a single-tilt or a conical tilt series procedure is a structure whose features are necessarily degraded. The aim of a restoration process is to minimize these degradations in some way.

4. GENERAL MATHEMATICAL FRAMEWORK

4.1. The Concepts of Measurement and Null Spaces

Section 2 showed that there is an essential lack of information in the available measurements about the unknown 3D structure of the specimen. The effects of this situation were first explored in Section 3 with the help of the system PSF. In this section we introduce another way to present this limitation mathematically, based on the concept of the null space of functions associated with the measurement geometry (Hanson and Wecksung, 1983; Hanson, 1987). This new approach may be less intuitive than the study of the PSF, but it will provide a powerful and simple mathematical framework where both the missing data problem and the efforts to compensate for it by the use of restoration methods can be clearly expressed. The actual close relationship between these two approaches will be treated in the next section.

With the approximation that EM images are true projection images of the 3D object density distribution (see Chapter 2), each individual measurement g_i (the density at a given point of a given image) may be written as a weighted three-dimensional integral of the unknown function $f(x, y, z)$, representing the specimen's 3D structure, in the following way:

$$g_i = \int_{-\infty}^{\infty} \int_{-\infty}^{\infty} \int_{-\infty}^{\infty} l_i(x, y, z) f(x, y, z)\, dx\, dy\, dz = l_i f \qquad (10)$$

where l_i are the weighting functions and $i = 1, ..., j, j+1, ..., 2j, 2j+1, ..., k \cdot j$, for $k \cdot j$ individual measurements grouped into k images of j pixels. Each l_i has nonzero real values only within that region of the object that makes a contribution to the ith measurement, typically a narrow strip through the object. For the ideal case of zero strip width and a perfect imaging system, Eq. (10) becomes a line integral.

Much insight into the basics of the 3D reconstruction process can be gained by a geometrical interpretation of Eq. (10). The integral in (10) can be understood as the inner product of two functions, l_i and f ($l_i, f \in \mathscr{H}$). In general, the inner product of two vectors (a, b) may be regarded as the projection of vector b onto vector a. Equation (10) may then be understood as saying that each measurement g_i is the projection of function f onto the weighting vector l_i. It is then clear that the available measurements carry information only on those components of f that lie in the subspace M spanned by the set of all l_i's. Subspace M is closed linear (Oskoui-Fard and Stark, 1988), so $M^\perp$, the orthogonal complement of M is also a closed linear subspace, and every $u \in \mathscr{H}$ has a unique representation of the form $u = s + t$, with $s \in M$ and $t \in M^\perp$. In order to stress these concepts, M is usually termed the *measurement space* and $M^\perp$ the *null space*. The missing information problem can then be formulated as the absence of those components of f that lie in $M^\perp$. Note that M and $M^\perp$ are solely determined by the l_i's, that is, by the data collection strategy, and not by f, the unknown specimen structure.

That the null space is, in fact, a very tangible entity can be shown by decomposing a computer simulated image (Fig. 6*a*) into its components in M (the measurement space) and $M^\perp$ (the null space) for a given data collection strategy. A set of 11 projections spanning 90° in view angle were calculated. A reconstruction was performed with the information contained in this data set only; the result is shown in Fig. 6*b*. As we discussed before, such a reconstruction can only render those components of the original signal (Fig. 6*a*) that lie in the measurement space M. The null space components of the image could be easily calculated in this test case by subtraction. However, in a real-data case such a subtraction would obviously be impossible because the original signal is unknown. In fact, Fig. 6*c* is one of the many possible images that lie solely in $M^\perp$. The information content of Fig. 6*c* is quite high, showing graphically the important amount of information that was lost in the reconstruction from this limited set of measurements.

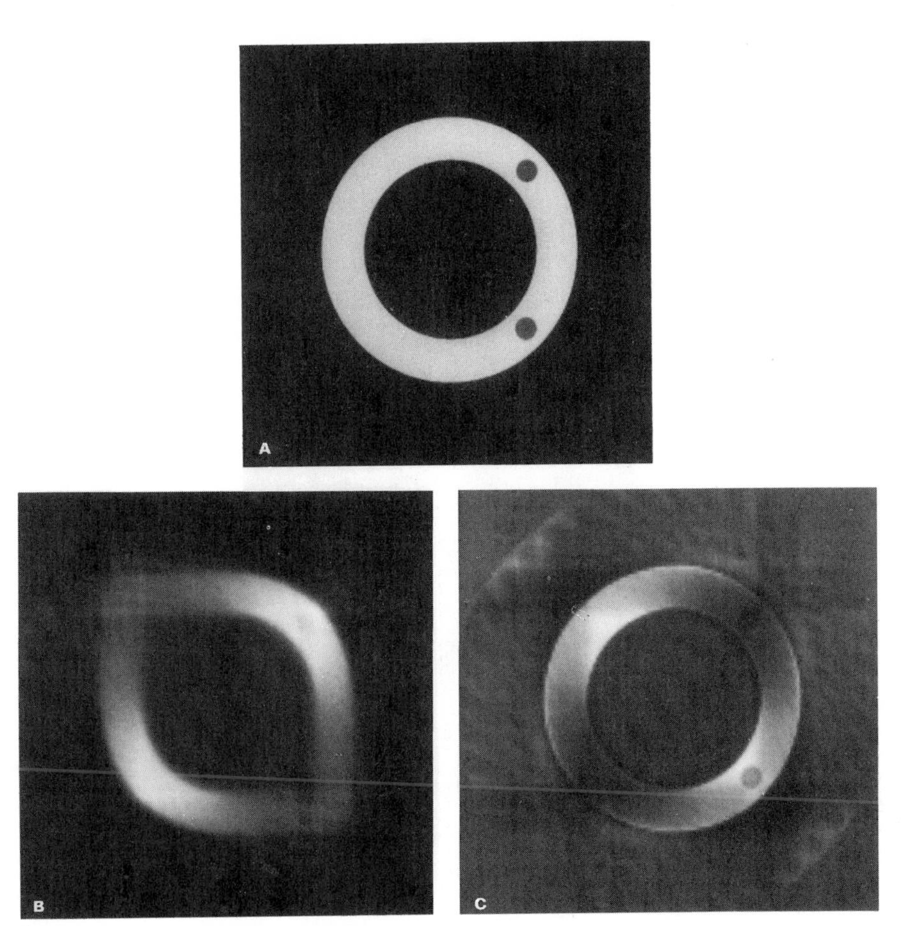

FIGURE 6. Decomposition of an object consisting of an annulus with two half-density holes (A) into the measurement space M (B) and the null space $M^{\perp}$ (C) contributions, corresponding to the experimental situation of having 11 measured projections covering 90°. Note that the projections of (C) along each of the measured directions are zero! This example clearly illustrates that for any given measurement scheme, every possible solution is a sum of a component that is measured and a component that is not. (From Hanson, 1987, © Academic Press, reproduced with permission.)

4.2. Digital Considerations, a Discrete Approach

In the course of introducing the concepts of measurement and null space, we have adopted a very general position, namely, that the unknown 3D structure f was a continuous function of (x, y, z) with, in principle, an infinite number of degrees of freedom, and that it was possible to actually work with infinite sets of numbers in calculating integrals. However, in the vast majority of cases, calculations are done with the help of a digital computer, implying that it is possible to process a finite set of numbers only. Furthermore, imaging systems are not ideal either, and usually they render an approximation to a low-pass-filtered version of the original image, as we discussed in Section 3.2. Fortunately, it is well known (Whittaker, 1914) that a bandlimited signal has a finite number of degrees of freedom, say P, implying that, as far as sampling is concerned, it is indeed possible to work, in a mathematically well-defined way, with a "real image" in a "real computer."

A point that needs clarification is how equations derived in a continuous case, like Eq. (7) can be reformulated in a discrete way. Assuming that the object function f is sampled satisfying the Nyquist criterion, that is, $\Delta x \leqslant 1/2X_{\max}$, $\Delta y \leqslant 1/2Y_{\max}$, $\Delta z \leqslant 1/2Z_{\max}$, with $X_{\max}$, $Y_{\max}$, $Z_{\max}$ being the maximum frequencies in either direction along X, Y, Z, then the object function f can be correctly represented by its samples over an array of equally spaced voxels, provided the distance between voxels is smaller than, or equal to, the minimum sampling distance along x, y, and z. Since data can only be known for a finite number of points, it is necessary to deal with finite arrays. This limitation (truncation) will result in the introduction of edge artifacts (see Trussell, 1984).

Let us assume that these arrays are cubic of dimensions $N \times N \times N$. By further assuming that the PSF is periodic with a period of N, we can explicitly consider the discrete nature of the data in Eq. (7) by changing integrals to finite sums; that is,

$$\hat{f}(m, n, l) = \sum_{i=0}^{N-1} \sum_{j=0}^{N-1} \sum_{k=0}^{N-1} f(i, j, k)\, h_p(m-i, n-j, l-k) \tag{11}$$

where $\hat{f}$ is the 3D reconstruction obtained directly from information within the data region, f is the real 3D structure of the object, and $h_p(m, n, l)$ is the periodic extension of $h(m, n, l)$; that is,

$$h_p(m+iN, n+jN, l+kN) = h(m, n, l) \tag{12}$$

Equation (11) is most conveniently written in matrix notation. This can be done by using stacked notation (Andrews and Hunt, 1977) to describe the three-dimensional volumes, that is, by lexicographically concatenating the voxel values. It is then possible to define a stacked vector for the volume $f(i, j, k)$ as

$$f(i, j, k) = \begin{bmatrix} f(0,0,0) \\ f(1,0,0) \\ \vdots \\ f(N-1,0,0) \\ f(0,1,0) \\ f(1,1,0) \\ \vdots \\ f(N-1,1,0) \\ \vdots \\ f(N-1,N-1,0) \\ f(0,0,1) \\ \vdots \\ f(N-1,N-1,N-1) \end{bmatrix} \tag{13}$$

We are thus stacking the planes of the volume $f(i, j, k)$ on top of each other and, within each plane, stacking columns on top of columns. In this way vector f is represented as an $N^3 \times 1$ matrix.

Similar processing can be done with vector $\hat{f}$. Equation (11) can then be written as

$$\hat{f} = Hf \tag{14}$$

where f and $\hat{f}$ are $N^3 \times 1$ and H is $N^3 \times N^3$. It is easy to show that matrix H is block circulant, and that each block H_i is also an $N^2 \times N^2$ block-circulant matrix with $N \times N$ circulants. It is known that any circulant matrix is diagonalized by the DFT (discreate Fourier transform) operator. We may therefore use the FFT (fast Fourier transform) implementation of the DFT to diagonalize matrix H, thus allowing for faster matrix multiplications in Fourier space.

Similarly, Eq. (10) can also be changed into a matrix formulation. For each individual measurement g_i, expressing both f and l_i as one-dimensional vectors of dimensions $N^3 \times 1$, we have

$$g_i = l_i^{\mathrm{T}} f \tag{15}$$

where the superscript T denotes the transpose operator, which converts a column vector into a row vector.

All L equations corresponding to the L individual measurements may be summarized by

$$g_L = H_L f \tag{16}$$

For simplicity, we will assume that the images are square arrays of dimensions $N \times N$, and that k is the total number of collected images; therefore $L = k \times N^2$. In the present context, H_L is a linear operator that is entirely determined by the measurement process. It maps the function $f(x, y, z)$, which is defined on the 3D continuous domain of (x, y, z), onto the discrete set of measurements enumerated by the index i and represented as a whole by g_L.

This situation can be easily formulated within the framework of the measurement/null space. Since the only components of f that can be measured lie in the subspace spanned by the l_i's, M, it is natural to expand the measurement-determined estimate of $f(x, y, z)$, $\hat{f}$, in terms of the l_i's; that is,

$$\hat{f}(x, y, z) = \sum_{i=1}^{L} a_i l_i(x, y, z) \tag{17}$$

Formally, this equation is identical to the familiar back-projection process, where the value a_i is added to the volume along the strip function $l_i(x, y, z)$. It is then immediately clear that a 3D reconstruction obtained by back-projection from just the available measurements necessarily lies in M. In general, it is true that 3D reconstructions obtained by *any* method that uses only the measurement data will lie in M.

If all l_i's are linearly independent, then the dimensionality of M is L. It then follows that in the continuous case, where f may have an infinite number of degrees of freedom, the dimension of $M^{\perp}$ is also infinite, since the dimension of M is finite only. It is in this context that Theorem 4.2 of Smith *et al.* (1977) should be understood: "A finite set of radiographs [projection images] tells nothing at all [about the structure of the specimen]". (Some interesting considerations about uniqueness can also be found in Katz, 1978). The situation is very different if f itself has a maximum number of degrees of freedom P. The dimension of $M^{\perp}$ is then finite, namely $P-L$. We will further study the implications of this dimensionality reduction in the next section.

Equation (16) can be further extended by considering the image degradation introduced by the electron microscope. As we discussed in Section 3.2, this degradation can be well explained by a linear formulation, resulting in an electron-microscope-determined 2D PSF. The action of this 2D PSF can be expressed in matrix formulation by an $N^2 \times N^2$ matrix (assuming images of $N \times N$ pixels), which we will note by H_{EM}. In order to express H_{EM} in the same vector/matrix notation as Eq. (16) we have to define a new matrix H_{EM}^{D} of $L \times L$ elements containing k submatrices of dimension $N^2 \times N^2$ of the type H_{EM} placed along its diagonal, the rest being zeros. We then have

$$g_L = H_{\mathrm{EM}}^{D} H_L f = H_{\mathrm{Total}} f \tag{18}$$

Dropping all subscripts, we arrive at

$$g = Hf \tag{19}$$

This equation will be referred to several times in the next sections, since it is the basic relationship between the measurement data and the object structure.

One of the important points gained by this formulation is that it offers a particularly simple approach to the 3D reconstruction problem, since it allows us to think in terms of one-dimensional vectors and matrices, instead of general functions and operators. In the remainder of this work we will make extensive use of this algebraic representation.

4.3. Reconstruction and Restoration: Need for A Priori Knowledge

In all the analysis presented so far, it has been assumed that the only known structural information about the object was the one derived from the actual image set. In many cases, however, some additional pieces of information about the object are also known, such as, for instance, approximate boundaries in space, approximate intervals for the density fluctuations in the sample, similarities to already known structures, etc. Collectively, these pieces of information are termed *a priori* knowledge.

At this point we can give a precise definition of what we understand by "reconstruction" and by "restoration" in the context of this work. By reconstruction we mean the process of obtaining those components of f that are lying in M. In principle (assuming noiseless data), those components can be derived uniquely from the projection data alone. By restoration we mean the process of obtaining those components of f that lie in $M^{\perp}$. They can be derived solely from the a priori knowledge. Collectively, we will use the term *recovery* for the combined reconstruction and restoration procedures. Certainly, such a combination of approaches represents a step forward from the conventional processing of 3D data in electron tomography (Carazo and Carrascosa, 1986).

Since practical data collection strategies lead to a "missing data" problem, the first question to face is what would be "the best" way to use the available (albeit incomplete) information. Three elements are needed to provide a practical answer to this question: A mathematical model of the process of 3D reconstruction from incomplete data, a criterion as to what "the best" means, and a numerical algorithm. In the next section we will address the choice of the mathematical model.

4.3.1. Noise-Free Data Case

Our mathematical model will be the one provided by Eq. (19). The most straightforward way to solve this equation for f is by inverting H, that is, by performing

$$f = H^{-1}g \tag{20}$$

The problem with this approach is twofold: first, Eq. (19) does not take into account the always present experimental noise (as it will be discussed in the next section), and, second, H^{-1} seldom exists because H is usually not invertible. We may reformulate the problem as one of finding an estimate of f, $\hat{f}$ in M, such that $g = H\hat{f}$. The idea is thus to find an estimate which, when used as input to the model, gives the recorded data. There is no unique solution to this problem. An entire class of solutions is defined by

$$\hat{f} = H^{-}g + f_0 \tag{21}$$

where H^{-} is the pseudoinverse matrix (defined below) and f_0 is *any* vector in the null space of H, that is, with the property $Hf_0 = 0$. In particular, note that f_0 does not need to be zero. To set $f_0 = 0$ amounts to selecting that function f which satisfies Eq. (21) with the condition that its norm be a minimum. However, there is no conceptual base for imposing this condition and, therefore, such "unicity" is not justified. The pseudoinverse H^{-} used in Eq. (21) is, basically, the inverse of H involving only the measurement space. This concept is very well illustrated by introducing the notion of singular value decomposition (SVD). For a matrix $H(O \times O)$ let $U = \{u_1, ..., u_o\}$ and $V = \{v_1, ..., v_o\}$ be the eigenvectors of matrices HH^{T} and $H^{T}H$, respectively. The eigenvalues of these matrices, λ_j, represented by a diagonal matrix Λ, are the same (see Andrews and Hunt, 1977, p. 157). These eigenvalues are all nonnegative and ranked, such that $\lambda_1 \geqslant \lambda_2 \geqslant \cdots \geqslant 0$. A typical eigenvalue plot for the case of a reconstruction process from equally spaced projection images is shown in Fig. 7. The general form of the graphics indicates that eigenvalues can be roughly divided according to their values into two well-defined groups, those with a high value and those with an extremely low value, and that the separation

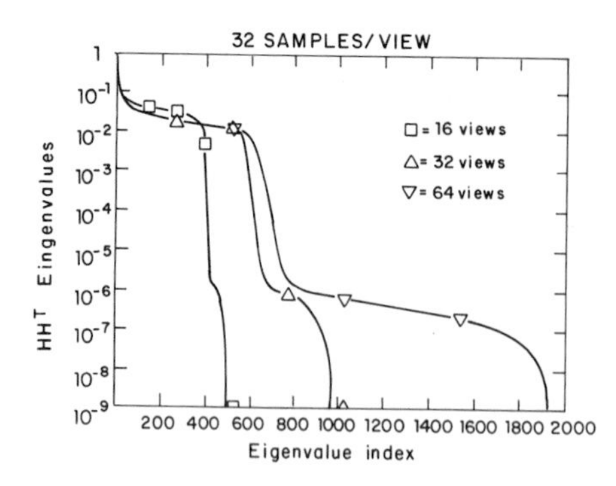

FIGURE 7. Eigenvalue map of HH^T corresponding to various numbers (16, 32, and 64) of equally spaced projections, each consisting of 32 samples. (From McCaughey and Andrews, 1977, © IEEE, reproduced with permission.)

between these two groups is rather sharp. Let us define the sequence $\mu_j = 1/\sqrt{\lambda_j}$. We may now write

$$H = U\Lambda^{1/2}V^{\mathrm{T}} \tag{22}$$

It is then clear that

$$H^{-1} = V\Lambda^{-1/2}U^{\mathrm{T}} \tag{23}$$

The elements of the diagonal of $\Lambda^{-1/2}$ are the sequence μ_j, which clearly goes to infinity if $\lambda_j = 0$. We may thus define a new matrix $\Lambda'^{-1/2}$ whose diagonal elements are the quantities γ_j such that

$$\gamma_j = \begin{cases} \dfrac{1}{\sqrt{\lambda_j}} & \text{if} \quad \lambda_j \neq 0 \\ 0 & \text{otherwise} \end{cases} \tag{24}$$

The pseudoinverse of H, H^-, is then defined as

$$H^- = V\Lambda'^{-1/2}U^{\mathrm{T}} \tag{25}$$

Therefore, any element $\hat{f} \in M$ can be expressed as

$$\hat{f} = \sum_{j=1}^{L} \gamma_j (g, u_j) v_j \tag{26}$$

This equation emphasizes that the recovery of $\hat{f}$ from the data depends on performing a division (either explicitely or implicitely) by the eigenvalues λ_j. Since the sequence $\mu_j = 1/\sqrt{\lambda_j}$ increases as j increases (assuming $\lambda_j \neq 0$), small perturbations in the eigenvectors corresponding to small eigenvalues are multiplied by successively larger factors. This feature can cause small changes in g to manifest themselves as very large changes in the solution, thus showing the ill-conditioned nature of the whole process.

Going back to Fig. 7, the plot shows that for the case of 32 samples per view the number of high-valued eigenvalues asymptotically approaches an upper limit of about 700 as the number of projection angles continually increases. From the previous considerations, it is clear that only those terms associated with large eigenvalues are the ones we probably want to keep in the eigenvector expansion given by Eq. (26). We may therefore interpret their number as the "number of degrees of freedom" associated with our measurement/reconstruction process. This indicates that the dimension of the measurement space M reaches an upper limit due to the finite number of measurements per view, and that more projections than roughly 40 do not add much information (in the absence of noise considerations).

From Eq. (21) it is clear that the general solution is not unique. There are, however, some interesting considerations on the uniqueness of the solutions. It has already been discussed that a bandlimited function has a finite number of degrees of freedom P. It has also been shown that for a given data collection strategy there is a maximum dimension L for the subspace of solutions determined by the measurements (the measurement space M). If $P > L$, then the true function f (unknown) cannot be uniquely determined by the measurements g_L. Because of the complementarity of M and $M^{\perp}$ ($H = M + M^{\perp}$, + denoting direct sum), it is possible to span any vector $f \in H$ in terms of the elements of any base of M plus the elements of any base of $M^{\perp}$. This principle leads to some very interesting developments.

It would be very practical to obtain a set of basis functions such that it is possible to regard Eq. (26) as a "structural" filter. The whole point here is to obtain a "physically meaningful" base of H, that is, one that would allow us to discern structural features arising from components in M from those derived from components in $M^{\perp}$.

Klug and Crowther (1972) studied this problem in Fourier space by performing an analysis in cylindrical coordinates (r, ϕ). They assumed that they had a set of angularly equally spaced projection images of a spatially limited object collected by a single-tilt axis procedure. An appropriate set of elementary functions in which to expand an object of radius a for data in this geometry is provided by

$$\Psi_{ns} = \begin{cases} J_n(2\pi R_{ns} r) \exp(in\phi), & r \leqslant a \\ 0 & r > a \end{cases} \tag{27}$$

where R_{ns} is the sth root of the equation

$$J_n(2\pi R_{ns} a) = 0 \tag{28}$$

The important result they derived from this approach amounted to the definition of a subset M_1 of independent harmonics Ψ_{ns} ($\mathrm{Dim}(M_1) = L_1$, $L_1 < L$) such that $\hat{f} \in M_1$ is an ideal low-pass-filtered rendition of $f \in \mathscr{H}$. The PSF for such a reconstruction is isotropic (the standard Airy disk) and space invariant. It was also shown that if the views are not equally spaced in angle the basic principles still hold, but that the dimensionality of M_1 is lower. Smith *et al.* (1973) addressed the same basic problem of obtaining a reconstruction free of artifacts, using a different mathematical approach but reaching basically the same conclusions.

It is therefore indeed possible to obtain an isotropic 3D reconstruction free of artifacts from only a limited number of views. However, this is true only if we restrict ourselves to work in the subspace M_1, which is formed by very smooth functions (L_1 is several times smaller than L, depending on the data collection geometry). Also, the Klug and Crowther (1972) formulation is only valid for continuous projection images. However, only a finite set of image samples can practically be measured and processed by digital means. An extension of their original formulation, taking into account the discrete nature of the data, is given by Howard (1988), who stressed the importance of the a priori information in addressing the discrete problem. In general, it is easy to show that the reduction of the information content of the images associated with the discrete sampling may lead to serious problems in the practical formulation of any reconstruction/restoration method. For instance, assuming that the solution is known to be bandlimited, it can be proved (Schafer *et al.*, 1981) that it is always possible to find a discrete bandlimited sequence which is identically zero over any finite interval of samples. Indeed, there is an infinite number of such sequences. Since any of these "artificial" sequences could obviously be added to the discrete solution f and still lead to the same measurements g, it is clear that the result in the discrete case cannot be unique. This added complication may be expected to lead to severe noise sensitivity problems in real-data cases, since images are always treated as discrete entities in actual computations.

In many practical cases it often happens that we are aiming to solve structural features that are beyond the resolution attainable by solutions lying entirely in M_1. We are then forced to use more terms in the expansions, opening the possibility of creating artifacts in the 3D reconstructions and, certainly, anisotropies. It is therefore always necessary to proceed with caution in the interpretation of the so-derived 3D reconstructions.

4.3.2. Noisy (Real-Data) Case

As mentioned, the mathematical model described by Eq. (19) is generally insufficient to describe a real-data situation. Noise, which is always present in the measurements, has not been considered on this derivation. It is indeed very difficult to model the influence of noise on the measurements in a general way. Noise can be either signal dependent or signal independent, multiplicative or additive, or it can occur intermixed with the signal in some more complicated form. At this point some explicit simplifications are needed. The validity of these simplifications will be application dependent.

A particularly simple way to extend Eq. (19) is to model the noise as signal independent and additive. In this way we have

$$g = Hf + n \tag{29}$$

The simplicity of this latter equation allows us to better realize the crucial role of the noise in the entire recovery process: the measurements g are no longer uniquely related to the unknown structure f by the operator H. That is, the measurements could not be regenerated even if f and H were perfectly known. Under these conditions, the ill-conditioned behavior associated with the recovery of $\hat{f}$ from g, discussed in the last section in connection with Eq. (26), become even more pronounced. In the first place, it is obvious that when working on digital computers there is no difference between $\lambda_i \equiv 0$ and λ_i being very small (within the machine's numerical precision); therefore the practically available subspace M' may be of a lower dimensionality than M ($\text{Dim}(M') = L'$, $L' < L$). This is coupled with the fact that, for very small λ_i, γ_i (the weighting factor in Eq. 26) becomes very large, something we want to prevent since the data do contain noise components that might dominate the expansion. The situation is actually even worse; since the measurements g are noisy, they will lie outside the subset M over which H^- is defined, that is, outside the domain of H^- $[g \notin D(H^-)]$. In this case it is possible to show (Rushforth, 1987) that those components of g outside $D(H^-)$ will be greatly amplified (in principle without bonds) during the recovery of $\hat{f}$.

The general way to proceed is to modify the original problem described by Eq. (29) in such a way that it is possible to obtain some sort of *approximate solution* which is not too sensitive to small perturbations in the data. On the other hand, we want to modify the original problem as little as possible. The process of achieving an acceptable compromise between these two conflicting goals is referred to as *regularization*. Clearly, from the very moment that a given regularization method is used, certain assumptions about g, H, f, and n are being made, which should be based on a broad understanding of both the reconstruction process and the nature of the data.

In the following we present some possible ways to approach the restoration process on this basis.

5. RESTORATION IN PRACTICE: METHODS

5.1. General Considerations

In this section we address the question of how different restoration methods can treat the "missing data" problem that is typical in electron tomography. As a general remark, it is important to stress that "the best" restoration method does not seem to exist. Each technique makes its own assumptions about either the possible solutions f, the noise n or the type and extent of the a priori knowledge. Depending on the specific problem at hand, these assumptions may or may not be the most appropriate ones.

There are two important points directly related to electron tomography applications that need to be addressed here. One is the characterization of the image degradations introduced by the electron microscope as an image-forming system, the other is the noise model to be used in the recovery process.

As previously noted, the electron microscope is not an ideal instrument. It renders images that are not perfect "projection images" (in the sense of Eq. 10) of the specimen 3D structure due to the instrumental degradations discussed in Section 3.2. In a first approximation to the recovery problem one might think of neglecting this instrumental degradation. However, it is becoming more and more evident that an explicit consideration of all instrumental degradations can lead to more improved results than was previously expected. As yet there are very few works addressing this point, but the recent works of Barth *et al.* (1989) and Pavlovic and Tekalp (1990) clearly point in this direction. Therefore H should be of the form described by Eq. (18) and (19) where the instrumental degradations are explicitely considered.

Regarding the noise model, it is clear that part of the noise is due to electron counting, and thus follows a Poisson process. It is also clear that there are other contributions from the photographic emulsion and processing, photographic grain, and inaccuracies in the scanning devices. However, experience in the general field of image restoration tells us that Eq. (29) is still a reasonable approximation to reality. The parameters for this general noise model could be derived (Lawrence *et al.*, 1989) from an experiment consisting in photographing the same area of the specimen a number of times under, in principle, identical conditions. There are, however, some sources of noise that are very particular to electron tomography, such as beam-induced movements in the sample, specimen heterogeneity, and inaccuracies in the determination of the data collection strategy. Although it is known how specimen heterogeneity can affect a 3D reconstruction in some cases (Carazo *et al.*, 1988, 1989), little is known in general about these effects. Noise models should therefore be used with some caution.

In this section we are aiming at providing an overview of several restoration/reconstruction methods. Unfortunately, at the present state of the field there is still a considerable uncertainty as to which approach is ultimately going to bear fruit in a routine way in electron tomography. It is partly because of this uncertainty, and partly because it is always good to look at a problem from a variety of perspectives, that we present different restoration methods. We start by presenting statistically based methods, which seek to overcome the inherent uncertainty in the data by formulating the problem in statistical terms. We present pseudoinverse filtering, Wiener filtering, and classical maximum a posteriori filtering. We then discuss constrained and optimization techniques, describing unconstrained optimization, constrained optimization, exemplified by the maximum entropy method (MEM), and constrained restoration, exemplified by the method of projections onto convex sets (POCS). We also discuss some new developments that may prove valuable for future applications.

5.2. *Statistical Methods*

5.2.1. Pseudoinverse Filtering

One of the simplest ways to solve Eq. (29) ($g = Hf + n$) is to introduce a new operator W (often referred to as a *filter*) and then find the estimate $\hat{f} = Wg$ that best fits the data in the mean-square sense (the noise is assumed to be zero mean). This can be written as

$$\text{minimize } E\{\|g - H\hat{f}\|^2\} \text{ over } \hat{f} \tag{30}$$

The filter W that produces such a function $\hat{f}$ from g and H is given by (Trussell, 1984)

$$H^{\mathrm{T}} R_{gg} = H^{\mathrm{T}} H W R_{gg} \tag{31}$$

R_{gg} being the autocorrelation matrix of vector g. A solution to Eq. (31) is obtained by making $W = H^{-}$, where H^{-} is the pseudoinverse matrix defined by Eq. (25). We then have

$$\hat{f} = H^{-} g \tag{32}$$

Obviously, since H^{-} is defined only on the measurement space M, no components of $M^{\perp}$ are present in $\hat{f}$.

This same solution is reached for the case of stationary and uncorrelated Gaussian noise if the problem is formulated in terms of "maximum likelihood," that is, maximizing the probability that a particular estimate would produce the recorded data. Formally, this is written as

$$\text{maximize } p(g \mid \hat{f}) \text{ over } \hat{f} \tag{33}$$

where $p(g \mid \hat{f})$ is the conditional probability of g occurring, given that the original signal was $\hat{f}$.

That the solution provided by the pseudoinverse method is ill-conditioned can also be shown in terms of the discrete Fourier transform for the case in which $W = H^{-}$ and H is circulant. Denoting $F = \text{FT}(f)$, $\hat{F} = \text{FT}(\hat{f})$, $N = \text{FT}(n)$, and $\mathscr{H} = \text{FT}(H)$, we have

$$\hat{F}(M, N, L) = \begin{cases} F(M, N, L) + \dfrac{N(M, N, L)}{\mathscr{H}(M, N, L)} & \text{if } \mathscr{H}(M, N, L) \neq 0 \\ 0 & \text{otherwise} \end{cases} \tag{34}$$

It is then clear that small values of $\mathscr{H}(M, N, L)$ will result in the amplification of the noise term $N(M, N, L)$. Therefore, the major problem with the pseudoinverse filtering method is that noise can easily dominate the restoration, limiting severely its practical use.

Recently, Zhang and Frank (N. Y. Zhang, personal communication), analyzed the behavior of a pseudoinverse reconstruction approach applied to a simplified version of the problem of reconstructing an object from data obtained by single-axis tilting (namely, the reconstruction of a 2D image from 1D projections). They directly computed H^{-} by Eq. (25) and then obtained $\hat{f}$ by Eq. (32). This approach may have serious problems in real-data cases both because of noise-related instabilities and because of numerical difficulties in diagonalizing H. However, the results they have obtained in noiseless conditions already provide some interesting information. In particular, Zhang and Frank were able to obtain a pseudoinverse solution $\hat{f}$ that was close to the projection of f onto M. They then compared this estimate, $\hat{f}$, with the one obtained by convolution back-projection f_{BP}. It turned out that $\hat{f}$ and f_{BP} were noticeably different. The conclusion that can be derived from this comparison is that, although f_{BP} must be within the measurement space M (Eq. 17), clearly f_{BP} is not the projection of f onto M; that is, f_{BP} is not that element within M closest in norm to f (at least for discrete data over a restricted angular range). The implications of these results for real-data applications still remain to be studied.

5.2.2. Wiener Filtering

The statistical approach which yielded the pseudoinverse solution was obtained by a best mean-square fit to the data. A more direct method would be to require the expected value of the mean-square error between the estimate $\hat{f}$ and the original signal f (unknown) to be a minimum. Formally this is written as

$$\text{minimize } E\{\|f-\hat{f}\|^2\} \text{ over } \hat{f} \tag{35}$$

As before, the estimate $\hat{f}$ is obtained from the data g by an operator W (which is also called a filter); that is, $\hat{f}=Wg$. Assuming that the noise n is zero mean and uncorrelated with the signal f, the filter W can be written in terms of the autocorrelation of f and n as (Trussel, 1984)

$$W=R_{ff}H^{\mathrm{T}}(HR_{ff}H^{\mathrm{T}}+R_{nn})^{-1} \tag{36}$$

where R_{ff} and R_{nn} are the autocorrelation matrices of the original signal f and the noise n.

Note that the amount of a priori knowledge required for this recovery method has increased significantly over that for the pseudoinverse method. The only information required for the pseudoinverse filter is the knowledge of H. The minimum mean-square error (MMSE) filter in Eq. (36) requires additional knowledge about the correlation structure of the signal and the noise.

This latter equation can be written in terms of the Fourier transform using the approximation that R_{ff} and R_{nn} are circulant matrices and that the image represents a stationary random process (Trussel, 1984). If we call P_{ff} and P_{nn} the power spectra of f and n, respectively, and $\omega=\mathrm{FT}(W)$, we obtain

$$\omega(M,N,L)=\frac{\mathscr{H}^*(M,N,L)}{|\mathscr{H}(M,N,L)|^2+P_{nn}(M,N,L)/P_{ff}(M,N,L)} \tag{37}$$

Equations (35) and, within its approximations, (36) and (37) represent a rather general form of the Wiener filter. Within this formulation a significant reduction of the dimensionality of $M^{\perp}$ can be obtained through *a priori* information coded in the correlation structure of the signal and the noise. However, this filter is often used with the further assumption that the noise and the signal covariance matrices are proportional to the identity matrix, reflecting a total lack of a priori information. With these additional assumptions it can be easily shown that P_{ff} and P_{nn} are both constants (termed, respectively, C_1 and C_2), obtaining

$$\omega(M,N,L)=\frac{\mathscr{H}^*(M,N,L)}{|\mathscr{H}(M,N,L)|^2+C_2/C_1} \tag{38}$$

With these approximations the Wiener filter is an improvement over the pseudoinverse filter in that it allows a better handling of the noise level through the constant C_2/C_1, preventing the denominator for reaching zero. But it must be noted that this does not have any impact on $M^{\perp}$.

This filter has been used in the recovery of electron microscopy images from the degradation introduced by the contrast transfer function of the microscope (see, for instance, Welton, 1979; Lepault and Pitt, 1984). It has not been used, however, for the recovery of 3D data.

There are a number of recent developments (Sezan and Tekalp, 1990a, b; Sezan and Trussell, 1991) that promise an interesting and simple new approach to the Wiener filter which show improved results. The basic idea is to use the Wiener solution only for those frequencies that lie outside the neighborhoods of zeros of the transfer function, leaving for other restoration techniques the recovery of the remaining spectral information. The key point of the estimation of the power spectrum of the image is solved with an autoregressive (AR) model. This development is particularly interesting in that it mixes several restoration methods. We will further exploit this proposal in subsequent sections.

5.2.3. Maximum a posteriori (MAP) Filter (Classical Bayesian Approach)

Since the "recovery" problem is stated as "given the measurement vector g, find the best estimate $\hat{f}$ of f," a seemingly obvious approach would be to maximize the probability of $\hat{f}$ being the original signal given the recorded data g, that is, the posterior probability density function $p(\hat{f} \mid g)$. This probability can be stated using Bayes's theorem as

$$\text{maximize } p(\hat{f} \mid g) = \frac{p(g \mid \hat{f})\, p(\hat{f})}{p(g)} \text{ over } \hat{f} \tag{39}$$

in terms of the conditional probability of g given $\hat{f}$, $p(g \mid \hat{f}) = p(n)$, and the a priori probability functions of $\hat{f}$ and g separately, $p(\hat{f})$ and $p(g)$, respectively.

Maximizing Eq. (39) we obtain

$$\frac{\partial}{\partial \hat{f}} p(\hat{f} \mid g) = \frac{1}{p(g)} \frac{\partial}{\partial \hat{f}} (p(g \mid \hat{f})\, p(\hat{f})) = 0 \tag{40}$$

where $p(g)$ is independent of $\hat{f}$ and, therefore, its derivative with respect to $\hat{f}$ is zero. Assuming that the noise n is a zero-mean Gaussian process and that $p(\hat{f})$ is a multivariate Gaussian distribution with mean value of $\hat{f}_{\text{mean}}$ and covariance matrix R_{ff}, we obtain

$$f = [H^{\mathrm{T}} R_{nn}^{-1} H + R_{ff}]^{-1} [H^{\mathrm{T}} R_{nn}^{-1} g + R_{ff}^{-1} \hat{f}_{\text{mean}}] \tag{41}$$

which is of the form

$$f = Wg + b \tag{42}$$

where the b term is directly related to $\hat{f}_{\text{mean}}$; that is, it is present because we are not considering the ensemble of images having zero mean. Information about the mean is, therefore, needed to solve the MAP problem.

The general form of the MAP filter that has been presented is very powerful, since it allows the introduction of explicit information related to the statistical distribution of the set of possible solutions through $p(\hat{f})$. Furthermore, it is the one method presented so far that is able to account for nonlinear sensor characteristics. This flexibility may be important for a detailed characterization of the process of image data collection in electron microscopy. In this way the MAP solution may have an important null space contribution. Also, the MAP approach may be shown to provide a good mathematical framework where many other recovery methods based on statistical knowledge can be formulated by appropriate choices of the probability density functions $p(\hat{f})$ and $p(n)$.

In particular, under certain simplifying assumptions, the MAP filter can be shown to reduce to the Wiener and the maximum likelihood filters. To obtain the Wiener filter we have to consider that $\hat{f}_{\text{mean}} = \text{constant}$, and the maximum likelihood solution is obtained by assuming $p(\hat{f}) = \text{constant}$, reflecting our complete lack of a priori information about the set of possible solutions. As we shall see later, some forms of the maximum entropy method can also be formulated within this framework.

5.3. *Constrained and Optimizational Methods*

5.3.1. General Formulation

There are many problems for which all we know about their solution is that it has to satisfy a given set of conditions, also called constraints, that can be expressed mathematically in many different forms. In our type of applications we could require, for instance, that the recovered 3D structure should reproduce the experimental 2D data, that the specimen dimensions should be finite or that it should be somehow similar to an already known structure. We then try to find a solution that simultaneously complies with all these constraints. We will refer to these recovery problems as "constrained restoration problems," and they will be treated in some depth in Section 5.3.4.

There are other cases in which we can define a given property that should be *optimized* (either maximized or minimized) while still be limited (constrained) by conditions expressed by equalities or inequalities. For instance, find the extremes of the function $a(x_1, x_2, ..., x_T)$ under the conditions that $b_1(x_1, x_2, ..., x_T) = 0$ and $b_2(x_1, x_2, ..., x_T) = 0$ (T is the total number of points; if a, b_1, and b_2 are defined on cubic arrays of dimensions $N \times N \times N$, then $T = N^3$). We will refer to these problems as *constrained optimization problems*. In practical situations, the function a could be, for instance, the entropy (to be maximized) or the norm (to be minimized), b_1 could be some form of data consistency, and b_2 could be the voxel value bound.

A conceptually simple and elegant way to approach this problem was developed by Lagrange. In essence, this method provides a *necessary* condition that any extreme should fulfill. It starts by forming the following linear combination

$$\phi(x_1, x_2, ..., x_T) = a(x_1, x_2, ..., x_T) + \lambda_1 b_1(x_1, x_2, ..., x_T) + \lambda_2 b_2(x_1, x_2, ..., x_T) + \cdots + \lambda_m b_m(x_1, x_2, ..., x_T) \tag{43}$$

where λ_1, λ_2, and λ_m are quantities associated with m different constraints. Differentiating ϕ with respect to each of the T variables, and considering the system of $T+m$ equations

$$\begin{aligned} \frac{\partial\phi(x_1, x_2,..., x_T)}{\partial x_r} &= 0, \qquad r=1, 2,..., T \\ b_k(x_1, x_2,..., x_T) &= 0, \qquad k=1, 2,..., m \end{aligned} \tag{44}$$

Lagrange found that if the point $(x_1, x_2,..., x_T)$ is a solution of the optimization problem, then it has to be a solution of this system of $T+m$ equations too. In practice, we try to solve this system of equations for the $T+m$ "unknowns" $\lambda_1,..., \lambda_m$ and $x_1,..., x_T$. The quantities $\lambda_1,..., \lambda_m$, which are introduced only to find the real unknowns [the extreme of $a(x_1,..., x_T)$], are called the *Lagrange multipliers.* In some cases the Lagrange multipliers can be obtained explicitly, but in many other they have to be found by numerical methods, which is certainly computational demanding and subject to errors.

In fact, other methods have been proposed in the literature in an attempt to solve problems conceptually similar to the ones presented above, but without making use of Lagrange multipliers. Among those other methods we will present here the "penalty method" (Polak, 1971). In essence, this method amounts to the definition of a function $p(x_1, x_2,..., x_T)$, called a *penalty function,* such that the constrained optimization problem is solved by constructing a sequence of points $(x_1, x_2,..., x_T)_i \in R^{\mathrm{T}}$ which are optimal for a sequence of unconstrained optimization problems of the form

$$\text{optimize } \{a(x_1, x_2,..., x_T) + p_i(x_1, x_2,..., x_T) \mid (x_1, x_2,..., x_T) \in R^{\mathrm{T}}\}, \qquad i=0, 1,... \tag{45}$$

This sequence is constructed in such a way that $(x_1, x_2,..., x_T) \to (\hat{x}_1, \hat{x}_2,..., \hat{x}_T)$ as $i \to \infty$, $(\hat{x}_1, \hat{x}_2,..., \hat{x}_T)$ being the optimal solution for some constrained optimization problem for $a(x_1, x_2,..., x_T)$. That is, penalty methods may be used to transform a constrained optimization problem into a sequence of unconstrained problems.

In the following we present some approaches that fall under the headings of "unconstrained optimization," "constrained optimization," and "constrained restoration," and that have proved to be useful in electron tomography applications.

5.3.2. Unconstrained Optimization

Recently, Penczek *et al.* (1990) and Penczek *et al.* (in press) have used a form of unconstrained optimization to solve the 3D structure of the 70*S* monosome from *Escherichia coli.* They started by defining a simple statistic (a chi-square) to measure the degree of misfit between the trial structure being recovered and the 2D data from which it was being derived. That is,

$$\chi^2(\hat{f}) = \sum_{j=1}^{L} \frac{1}{\sigma_j^2} \left(\sum_{i=1}^{T} H_{ji} \hat{f}_i - g_j \right)^2 = \sum_{j=1}^{L} \frac{1}{\sigma_j^2} n_j^2 \tag{46}$$

Their idea was then to minimize the norm of the solution $\hat{f}$ ($\|\hat{f}\|^2 = \sum_{i=1}^{T} \hat{f}_i^2$) while still keeping a small discrepancy with respect to the experimental data, as measured by χ^2 (which then acted as a constraint). Within the formalism of penalty methods, this approach amounts to solve the sequence of unconstrained problems

$$\text{minimize } \{\|\hat{f}\|^2 + p_i(\hat{f}) \mid \hat{f} \in R^{\mathrm{T}}\}, \qquad i = 0, 1, 2, \ldots \tag{47}$$

with $p_i(\hat{f}) = C\chi^2(\hat{f})$ (C a constant) for all i's. That is, there is no real dependence on i, and, therefore, the problem can be formulated as the following single unconstrained optimization problem:

$$\text{minimize } \{\|\hat{f}\|^2 + C\chi^2(\hat{f}) \mid \hat{f} \in R^{\mathrm{T}}\} \tag{48}$$

The value of the parameter C is derived by considering the expected value of $\chi^2(\hat{f})$ for an acceptable solution:

$$E(\chi_T^2(n^2)) = T \tag{49}$$

Therefore, C is chosen such that $\chi^2(\hat{f}) = T$ at the minimum.

The method is, therefore, conceptually simple but with a potentially large impact in the null space. The results obtained with the 3D structure of the *E. coli* 70*S* monosome in ice (Fig. 8) are indeed very encouraging.

5.3.3. Constrained Optimization: Maximum Entropy Method

Early Approaches. For many decades it has been conjectured that the notion of entropy somehow defines a kind of measure on the space of probability distributions, such that those of high entropy are in some sense favored over the others, all

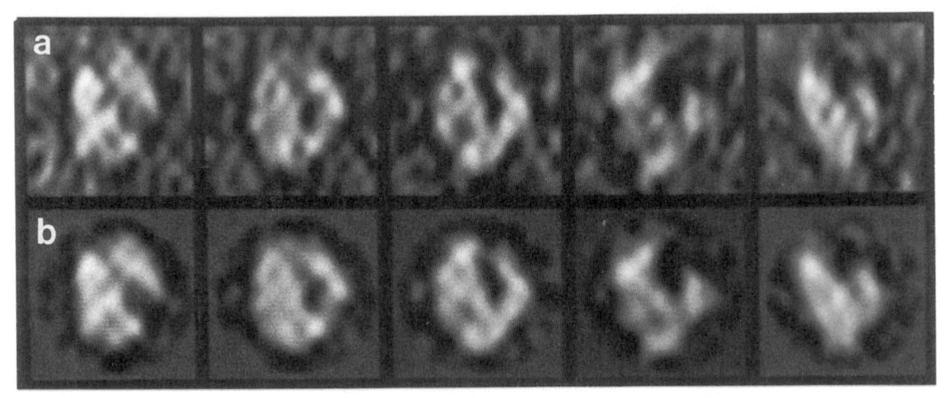

FIGURE 8. Comparison between the 3D structures of the *E. coli* 70S ribosome obtained with and without using an unconstrained optimization type of reconstruction. Selected *xz* slices are shown in each row (*z* is running vertical). The maximum tilt angle was 50°. (a) Reconstruction obtained by filtered back-projection. (b) Reconstruction obtained by unconstrained optimization. Note that the reconstruction shown in (b) is shorter in the direction of the missing cone than the one shown in (a); also, features seem to be better defined in (b) than in (a). (From Penczek and Frank, unpublished result, reproduced with permission.)

other things being equal. This notion, however, has been usually proposed on the basis of an intuitive interpretation of the missing information in terms of Shannon's information theory. Certainly, Shannon's axiomatic derivation is a formal description of what is required of an information measure, but to address the restoration problem we need a formal and direct description of what is required of a method for taking information into account. In fact, and as a result of this initial intuitive approach, the maximum entropy method has been derived in a number of different ways, which are not all mathematically equivalent. We will first discuss some of these earlier approaches because their simplicity allows us to understand more easily how the method works.

In the classical restoration work of Andrews and Hunt (1977) the maximum entropy formulation is described as

$$\begin{aligned} &\text{maximize } -\hat{f}^{\mathrm{T}} \ln \hat{f} \\ &\text{subject to } \|g - H\hat{f}\|^2 = \|n\|^2 \end{aligned} \tag{50}$$

That is, we desire to optimize a given property of the data (in this case to maximize the entropy), but still be consistent with the measured data. It is then a typical case of constrained optimization. Alternatively, and since maximization of the entropy has been formulated most often in statistical terms, it could also be considered a statistically based method. Note that the logarithm used in the definition of entropy in Eq. (50) forces the estimate $\hat{f}$ to be positive; that is, the maximum entropy method forces a "positivity constraint" on the solution. In fact, Hanson and Myers (1990) suggested that just this inherent positivity reinforcement may explain some of the good results obtained by this method.

This latter maximum entropy approach was analyzed by Trussell (1980), who found that the maximum entropy problem described by Eq. (50) satisfies the same equation as the MAP solution, described in the preceding section, if the a priori probability density function of $\hat{f}$ takes the form

$$p(\hat{f}) = k \exp(-\beta \hat{f}^{\mathrm{T}} \ln \hat{f}) \tag{51}$$

The general shape of $p(\hat{f})$ given by Eq. (51) is very much like a one-sided exponential. Thus, low values are common, whereas high values are rare. The parameter β in Eq. (51) determines the likelihood of encountering high and low values of the image. Furthermore, the points of $\hat{f}$ are independent of one another. That is, the method formulated in Eq. (50) will work well if the unknown function f indeed has the property of having relatively few high values with little correlation between them. A type of image which satisfies these descriptions is a star field from astronomical imagery, and this method has worked well on this type of data.

Another classical formulation of maximum entropy is due to Frieden (1972). His method starts by dividing the image space into n resolution cells (or n events). The frequency of occurrence of the ith event is identified as

$$p_i = \frac{\hat{f}_i}{\sum_{j=1}^{T} \hat{f}_j} \tag{52}$$

where $\hat{f}_i$ represents the gray tone intensity in the ith cell and T is the total number of cells (voxels). The entropy is then defined as

$$S(p_1,\ldots,p_T) = -\sum_{j=1}^{T} p_j \ln p_j \tag{53}$$

The sum of all intensities of $\hat{f}$ is constant, $\sum_{j=1}^{T} \hat{f}_j = \text{constant}$. Frieden (1972) proved that maximum entropy could be formulated in terms of $S(\hat{f}_1,\ldots,\hat{f}_T)$ as

$$\begin{aligned} &\text{maximize } S(\hat{f}_1,\ldots,\hat{f}_T) \\ &\text{subject to } \sum_{j=1}^{T} \hat{f}_j = C \qquad \text{(total intensity} = \text{constant)} \\ &\text{and } g_L = H_L \hat{f} + n_L \qquad \text{(data consistency constraint)} \end{aligned} \tag{54}$$

This problem can be solved by the method of Lagrange multipliers leading to a highly nonlinear system. Frieden proposed to solve this equation system by the Newton-Raphson method, a choice which, in practice, reduces the usefulness of this approach to problems with a small number of variables. Note that in the data consistency constraint we are using g_L and H_L directly (and, by extension, n_L), that is, the *actual* experimental measurements, meaning that no assumptions are being made about the "unmeasured" data and no interpolations are in principle required. Frieden also noted that maximization of $S(\hat{f}_1,\ldots,\hat{f}_T)$ did have a smoothing influence on $\hat{f}$, giving grounds for the intuitive interpretation of a maximum entropy solution as the smoothest possible solution compatible with the data. This notion of smoothness is also found in Trussell's (1980) MAP solution to the classical maximum entropy approach of Eq. (50). The probability density function of $\hat{f}$ giving by Eq. (51) can be regarded as a global measure of smoothness where neighboring points do not affect one another in the measurement. For instance, a typical star field image, consisting of a rather uniform background with peaked intensity maxima, will be "smooth" judged by the entropy measure, while any derivative-based smoothness criterion would not consider it "smooth."

Classic Axiomatic Derivation. A milestone in the development of maximum entropy methods was provided by the theoretical works of Shore and Johnson (1980), Johnson and Shore (1983), and Tikochinsky *et al.* (1984). They formally proved that under certain conditions, related to the reproducibility of results and explicitly listed below, the only consistent algorithm to take information in the form of constraints on expected values is the one leading to the distribution of maximum entropy subject to a set of given constraints. Furthermore, since the set of solutions $\hat{f}$ whose entropy is higher than a given constraint c, $\{\hat{f} \mid \hat{f} \ln \hat{f} \geqslant c\}$, is convex, there are no local extrema, and, therefore, the maximum entropy solution corresponds to a global maximum and is unique (i.e., the null space is empty). This result is fundamental because it provides firm formal grounds for the development of any particular maximum entropy method.

Basically, the four axioms (reproducibility conditions) required for this formulation are (Shore and Johnson, 1980) as follows:

1. Subset independence: it should not matter whether one treats an independent subset of system variables in terms of a separate conditional density or in terms of the full system density. This axiom explicitly rejects spatial correlations.
2. Invariance: The choice of coordinate system should not matter.
3. System independence: If a proportion q of a population has a certain property, then (in the absence of further data), the "best" choice for that proportion in any subpopulation is also q.
4. Uniqueness: The solution should be unique.

It is then proved that the "best" way to find the unknown structure f is by maximizing the functional

$$S(\hat{f}) = \int d\mathbf{r} \left\{ \hat{f}(\mathbf{r}) - m(\mathbf{r}) - \hat{f}(\mathbf{r}) \ln \left(\frac{\hat{f}(\mathbf{r})}{m(\mathbf{r})} \right) \right\} \tag{55}$$

where we have used Skilling's (1988) notation. Now $S(\hat{f})$ is defined as the entropy of $\hat{f}$, and m is the initial (structural) model used in the maximization problem. Of course, $S(\hat{f})$ has to be maximized subject to the constraint that $\hat{f}$ should be consistent with the measurements g_L, as indicated in Eq. (54).

One of the most fruitful methods to actually maximize $S(\hat{f})$ subject to $\sum_{i=1}^{T} \hat{f}_i = \text{constant}$ and $g_L = H_L \hat{f} + n_L$ is that of Skilling and Bryan (1984). In the following we present the main ideas used in that implementation. They start from the consideration that most attempts to satisfy the constraint $g_L = H_L \hat{f} + n_L$ based on a one-by-one voxel fitting will lead to the simultaneous consideration of a nonlinear system with at least the same number of unknowns as measurements. In the case of large images or 3D volumes the actual number of measurements may amount to millions, making this approach very difficult indeed. They therefore tried to use a single statistic to measure the misfit (usually a chi-squared). In fact, they used the same $\chi^2(\hat{f})$ that was used by Penczek *et al.* (1990), introduced in Section 5.3.2. The problem is therefore formulated as

$$\text{Maximize } S(\hat{f}) \quad \text{subject to } \chi^2(\hat{f}) = T \tag{56}$$

From this point on Skilling and Bryan (1984) developed a highly sophisticated method to maximize $S(\hat{f})$ within the set of possible solutions defined by the (hard) constraint $\chi^2(\hat{f}) = T$.

Electron Tomography Applications. Barth and co-workers (1988) used this approach to obtain the 3D reconstruction of the regular surface layer (HPI layer) of *Deinococcus radiodurans*. Barth *et al.* (1988) started collecting a very rich image data set, including projections up to $\pm 80°$ following a conical axis tilt–like data collection geometry. Using this data set, they calculated 3D reconstructions from projection data up to $\pm 60°$ and $\pm 80°$, with and without using the method of maximum entropy (performing a direct Fourier inversion in the latter case), and then compared the results. They concluded that their implementation of the maximum entropy method was able to reduce the point elongation due to the conical tilt–like PSF, but that some structural features which appeared when using higher tilt angle could not be retrieved from restricted angle data (Fig. 9). More recently, Barth *et al.* (1989) presented some interesting evidence suggesting that a substantial proportion of the retrieval limitations could be due to an incorrect modeling of the imaging process, more precisely, to neglecting the instrumental degradation introduced by the electron microscope (see Section 3).

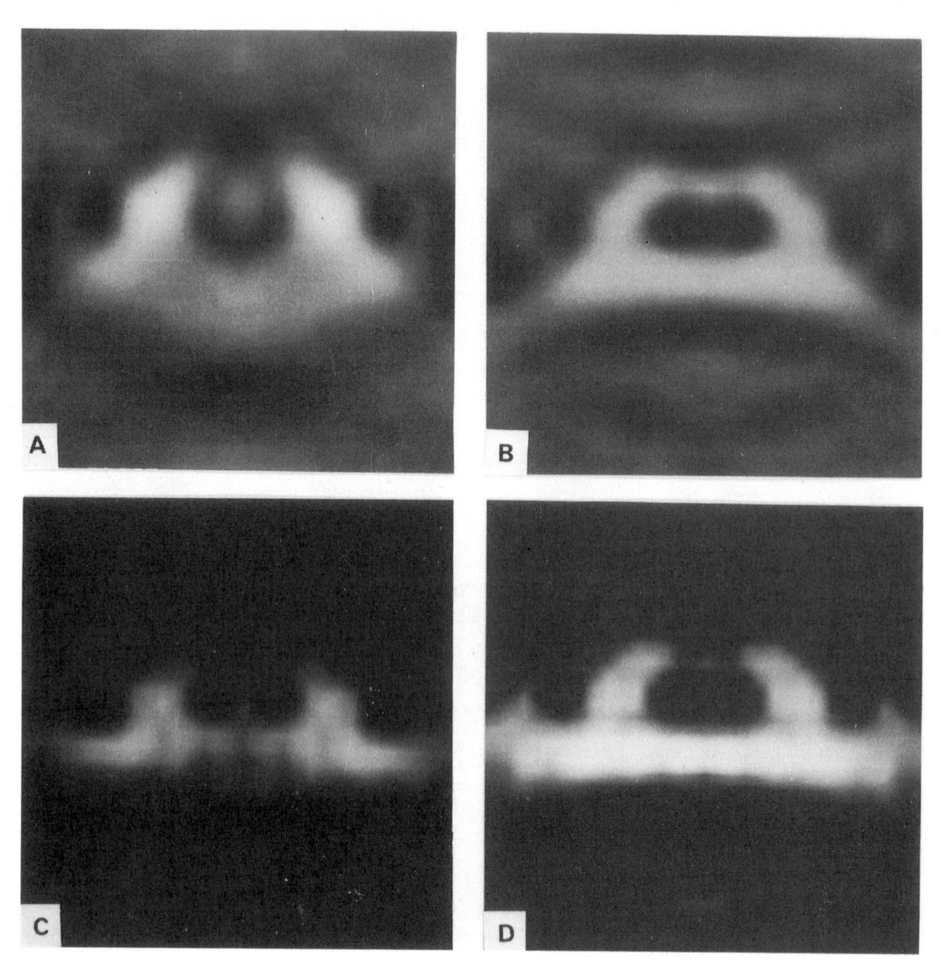

FIGURE 9. Comparison between the 3D structures of the regular surface layer (HPI layer) of *Deinococcus radiodurans* obtained with and without using maximum entropy-based methods. Central slices through the specimen are shown in each case; the *z* axis is running vertical and the *xy* plane is perpendicular to the figure. (a) and (b) reconstructions obtained by the direct Fourier method from projections up to 62° tilt (a) and 80° tilt (b). (c) and (d) reconstructions obtained by the maximum entropy method of Skilling and Bryan (1984) from data up to 62° tilt (c) and 80° tilt (d). Note that the protein on top of the central pore is very well defined in (b), while it disappears in the maximum entropy reconstructions (c) and (d). (From Barth *et al.*, 1988, © Scanning Microscopy International, reproduced with permission.)

A recent work using this same methodology is due to Lawrence and co-workers (Lawrence *et al.*, 1989). They used computer-generated and real data, comprising projection images in ranges up to $\pm 60°$ following a single-axis tilt data collection geometry. In their work they made important observations regarding the characterization of the reconstruction volume and the noise model. In particular, much work is devoted to the estimation of the standard deviation σ_j associated with each measurement value g_j. Some empirical tests on the effects of using incorrect specifications of σ_j in Eq. (46) were also carried out with computer-generated data. Using this latter data set, they found that maximum entropy always produced reconstructions closer to the (known) computer-generated object structure than weighted back-projection. Their results with real data could not be objectively assessed, however, both because the true object structure was obviously not known and because they used all their data in the maximum entropy reconstruction rather than just a subset, thus precluding part of the analysis of the results performed by Barth *et al.* (1988). Nevertheless, Lawrence *et al.* (1989) noticed that their real-data reconstructions were globally similar to those obtained by weighted back-projection (in the sense that the correlation coefficient between them was about 0.85), although it is very difficult to make meaningful global comparisons among reconstructions obtained by different approaches. In addition, and in view of the results by Barth *et al.* (1989), it is probable that these reconstructions could be improved when the instrumental transfer function introduced by the electron microscope is taken into account.

At this point it should be stressed that there are two different aspects that need to be considered when studying any given method. One is the theoretical foundation of the method, the other is how the theory is implemented in practice for real-data work. In the case of maximum entropy, the theoretical basis is linked to the four axioms given before. However, the first axiom (subset independence), which explicitly rejects any spatial correlation, is seldom met in electron tomography applications: pixels or voxels are very often correlated with their neighbors. Therefore some work needs to be done to relax this first axiom. Furthermore, Skilling and Bryan's (1984) method to implement maximum entropy has traditionally used the data consistency constraint in the form of $\chi^2(\hat{f}) = T$ (from Eq. 49). However, these authors have recently noted (Skilling, 1988; Gulf, 1988) that this constraint is not necessarily correct and that, in general, it underfits the data (i.e., the data consistency constraint is not sufficiently taken into account). Going back to some electron tomography results with these considerations in mind, it is probable that the apparent suppression of protein at the top of the *Deinococcus radiodurans* HPI layer when applying maximum entropy (Barth *et al.*, 1988; Fig. 9) was a result of an excessive regularization of the experimental data. Maximum entropy reconstruction/restoration would therefore show some beneficial features (like a reduction of the missing region PSF-induced elongation) and, at the same time, some negative ones that we have just mentioned. From this it is obvious that further theoretical and implementational developments in the maximum entropy field are needed.

New Developments. Recently, a number of authors (Skilling, 1988; Gulf, 1988; Bricogne, 1988) have proposed that the maximum entropy solution could be approached from the standpoint of Bayesian calculus. The basic formulation is then the one we presented in connection with the MAP filter. The maximum entropy problem is formulated in the context of Eq. (51), where we need to know $p(\hat{f})$ and $p(g \mid \hat{f})$. From Eq. (55) it is clear that $S(\hat{f})$ is a function of f and m, m being the initial model for f. In the context of maximum entropy it can be proved (Skilling, 1988) that $p(\hat{f})$ is a *monotonic function* of $S(\hat{f})$, which has a dimensional parameter, called α, that cannot be determined a priori. The solution to the maximum entropy problem depends therefore on the data g, the initial model m, and the parameter α. Consequently, it is the probability of $\hat{f}$ given g, m, and α, $p(\hat{f} \mid g, m, \alpha)$, that needs to be maximized. Note that α is found as part of the recovery process, and that it is a regularizing parameter which controls the relative influence of maximizing S while enforcing consistency with the measured data g. This formulation avoids the a priori setting of $\chi^2 = T$. Also, it shows that the practical results depend on the initial model m used in the maximization algorithm. It may therefore be beneficial not to start from a flat, uniform model (as it is the normal practice), but from a model with some structure in it. This approach to maximum entropy certainly leads to a very complex formulation, but it is free from the arbitrary specification of the set of possible solutions.

This approach has been used by Richard Bryan (personal communication) in problems connected with dynamic light scattering with much improved results. It seems to be an interesting research direction that could lead to valuable applications in the field of electron tomography, provided that the typically huge number of unknowns and the low signal-to-noise ratio of the projections can be accommodated in the practical implementation of the method.

A basic theoretical problem is still pending: the relaxation of axiom 1 of Shore and Johnson (1980). This important topic has recently led to the proposal of what has been called the "new maximum entropy model" (as opposed to the "classic model" presented above) (Gulf, 1988). Within this new approach it is postulated that the model m is produced by a set of "hidden variables" $\tilde{m}$ that are spatially uncorrelated. Model m is then produced by blurring $\tilde{m}$; that is, $m = G \circ \tilde{m}$ ($\circ$ denoting convolution). The hypothesis space associated with this new maximum entropy model is richer than before, because now G is also estimated. Some work using this approach has been performed so far with test images (Gulf, 1988), and it is very promising. However, the amount of information to derive from the data and the a priori information is very large indeed, involving the characterization of G, $\tilde{m}$, and α, and it is not yet clear how accurately this can be done. The future of maximum entropy methods in electron tomography depends on how these questions are answered in practice. The prospect is promising once a general formulation has been derived, although much development work remains.

5.3.4. Constrained Restoration: The Method of Projections onto Convex Sets

General Concepts. Many properties of the solution cannot be formulated in statistical terms but that can be easily described by mathematical functions. For instance, nonnegativity is a common property that is trivially expressed by the $\geqslant 0$ relation. Smoothness can be described by the spatial derivative or, in a more global sense, by entropy. Other common properties include space limitedness (finite support), frequency bandlimitedness, peakiness, and contrast. Constrained restoration seeks to incorporate as many general properties of the solution as possible in the restoration process. Each of these properties is formulated as a constraint on the possible solution, and hence the term *constrained restoration.*

Most of the methods that are based on satisfying a set of constraints are iterative; that is, they seek to obtain f (or an estimation $\hat{f}$) by calculating a sequence of successive approximations $\{\hat{f}_k\}$ such that $\hat{f}_{k+1} = F\hat{f}_k$. There are several important reasons to favor iterative methods. In the first place, they make unnecessary the calculation of H^{-1} (if it exist) or even H^{-} (which does exist but is very difficult to calculate). In the second place, F may be defined in such a way that it depends not only on H but also incoporates constraints based upon known properties of the desired solution. Furthermore, iterative methods can be very flexible, since F may also be defined as a function of the iteration k.

Ideally, we would like the sequence of approximations $\{\hat{f}_k\}$ to converge to a unique solution independent of the initial choice of f_0 (naturally, this solution should be f). Of course, convergence to the correct solution generally requires an infinite number of iterations; in practice, however, we are forced to stop the iteration somewhere, which necessarily leads to an approximation to f. Also, it often happens that the sequence $\{\hat{f}_k\}$ converges to a solution $\hat{f}$ that depends on the initial choice of f_0 (besides the choice of F needed to obtain $\{\hat{f}_k\}$).

Schafer *et al.* (1981) provided an in-depth review of constrained restoration algorithms. One of the main accomplishments of their work was to unify many apparently different constrained restoration procedures into a general scheme where F was designed considering both the signal formation process H and the known properties of the solution. These known properties were introduced in F by the definition of a constraint operator C, such that

$$f = Cf \tag{57}$$

if and only if f satisfies the constraint (that is, f is not modified when it meets the constraint). In general, for any f and C in Eq. (57), f is said to be a fixed point of C. Then F is constructed by a combination of H and C.

It is easily shown that for $\{\hat{f}_k\}$ to converge to a unique point independently of f_0, F must be a contraction mapping; that is, for any f_i and f_j obtained from the set of all possible structures (that is supposed to be closed) they must satisfy

$$\|Ff_i - Ff_j\| \leqslant r\,\|f_i - f_j\| \tag{58}$$

with $0 \leqslant r < 1$. If $r = 1$, the operator is said to be nonexpansive. Since the norm can be interpreted as the distance between vectors, we can say that contraction operators have the property that the distance between any two vectors tends to decrease as they are transformed by the operator.

On many occasions, however, the requirement that F must be a contraction is difficult to meet on the basis of our limited a priori knowledge. Instead, the much weaker requirement that F should be nonexpansive may be the most that can practically be sought. But, in general, neither uniqueness nor even existence of solutions is guaranteed in this case. Nevertheless, as mentioned, F depends on H (the reconstruction process) and also on the amount and type of a priori knowledge through the introduction of the constraint operator C. In fact, one of the major contributions to this field has been the identification of a broad class of applications where the a priori knowledge can be formulated in terms of constraints which force the possible solutions to lie in closed convex sets (a set is *closed* if it contains all its limit points; it is *convex* if, for any f_1, f_2 belonging to the set, the new element obtained by the linear combination $\mu f_1 + (1-\mu) f_2$, $0 \leqslant \mu \leqslant 1$, also belongs to the set).

It can be proved (Youla and Webb, 1982; Youla, 1987) that an appropriate F can be constructed such that it effectively converges to a solution conforming to all given closed convex constraints, provided that the characterizations of the solution are consistent; that is, that they all can be met simultaneously. This solution, however, is not unique. It depends on the initial choice of f_0 and on the way in which F is constructed. This new restoration method, called POCS, has already provided many useful results in a variety of fields, including electron tomography.

Convex Sets and Their Projections The notion of a "projection" onto a closed convex set is a consequence of the unique nearest-point property of closed convex sets in a Hilbert space setting. Given an arbitrary element $f \in \mathscr{H}$ ($\mathscr{H}$ being here the Hilbert space of square-integrable functions $f(x, y, z)$ over $\Omega \subset R^3$), the projection of f onto the closed convex set C is that unique element $f_0 \in C$ that satisfies

$$\underset{f_C \in C}{\text{minimum}} \; \|f - f_C\| = \|f - f_0\| \tag{59}$$

In words, f_0 is that element in C that is nearest to f in norm. This rule, which assigns to every $f \in \mathscr{H}$ its nearest neighbor in C, defines the (in general) nonlinear projection operator $P_c : \mathscr{H} \rightarrow C$ without ambiguity.

The link between general constrained restoration methods and POCS rests on the realization that compositions of projection operators onto closed convex sets with a nonempty intersection constitute and exceptionally well behaved subclass of nonexpansive transformations known as *reasonable wanderers* (Youla and Webb, 1982). This link has permitted the exploitation of many useful results of Browder (1965) and Opial (1967) on the fixed points of nonexpansive maps.

In the following, we will treat in depth the method of projections onto closed convex sets, both because it is an interesting example of constrained restoration methods and because it has recently led to many important new developments and results in different areas.

At this point a semantic remark about the use of the term *projection* needs to be made. Transmission electron microscopy images are said to be projection images of the specimen because their intensities are associated with the result of strip integrals (or raysums) through the object density distribution (see Eq. 10). However, the term *projection* as used in this section only refers to the concept of mathematical projections onto sets defined by Eq. (59) as, for example, in the phrase "projection of a point onto a plane."

According to the discussion started before, there exists a class of image restoration problems in which many of the known properties of the (unknown) solution f can be formulated as constraints that restrict f to lie on well-defined closed convex sets C_j, $j=1,\ldots,k$, and the unknown solution structure $f \in \mathscr{H}$ must belong to $C_O \equiv \bigcap C_j$. C_O will certainly not be empty if all the specified properties of the object can be met simultaneously. All we need now is a practical way to reach C_O.

Evidently, C_O is also a closed nonempty convex set containing the unknown true structure f, and we shall denote the respective projection operators projecting onto C_O and C_j by P_O and P_j, $j=1,\ldots,k$. Clearly, f is a fixed point of P_O ($P_O f = f$), and of all the P_j. More generally, f is a fixed point of P_O and of every

$$T_j = 1 + \lambda_j (P_j - 1) \tag{60}$$

for any choice of relaxation constants $\lambda_1, \lambda_2, \ldots, \lambda_k$, $0 < \lambda_j < 2$. Hence, under the same conditions, f is a fixed point of P_O and the composition operator

$$T = T_k T_{k-1} \cdots T_1 \tag{61}$$

Conversely, and this is the most important aspect, *any* fixed point of T lies in C_O. Furthermore, the sequence $\{T^n f_0\}$ converges to a point of C_O as $n \to \infty$, irrespective of the initialization $f_0 \in \mathscr{H}$. We have therefore a practical iterative method to reach C_O, that is, to obtain solutions that simultaneously conform with all the given constraints. However, from the basic formulation of the restoration problem, it is clear that *any* element of C_O is a possible solution and is as desirable as any other element of C_O. Hence, the result is not generally unique. Using the formalism of measurement/null space, we can understand this situation as a lack of information that results in $M^\perp$ not being empty.

Therefore, for any given composition of projectors, the final point in C_o that will be reached after iterating depends on the initial estimate of f_0. Note that the projection of an element f in C is that element of C that is nearest to f in norm. If the composition of projectors were also a projector, we would then reach that element of C_o that is the closest to f_0 in norm. In general, that composition will not be a projector, but each of the elements of the composition certainly is, therefore, the point on C_o that is reached is somehow "close" to the initial estimate f_0 (Trussell and Civanlar, 1983). Consequently, f_0 should agree as much as possible with the already known properties of the true solution. In electron tomography, f_0 has been usually taken as the projection of f_0 onto M (that is, a 3D reconstruction considering only the measured data) (Carazo and Carrascosa, 1987a, b).

There are many properties of the desired solution that can be formulated as constraints which define closed convex sets (Youla and Webb, 1982; Sezan and Stark, 1982). Usually, they are presented by specifying the closed convex set and the corresponding projector operator. Among those that have been practically used in electron tomography (Carazo and Carrascosa, 1987a, b) we list

1. The set of all functions in $\mathscr{H}$ that vanish outside a prescribed region $\delta \in \Omega$. Its associated projection is realized by

$$P_1 f = \begin{cases} f & \text{if} \quad (x, y, z) \in \delta \\ 0 & \text{otherwise} \end{cases} \tag{62}$$

2. The set of all functions in $\mathscr{H}$ whose Fourier transform F over a closed region τ in XYZ Fourier space is described by G (known a priori). The associated projector is

$$P_2 f = \begin{cases} G(X, Y, Z) & \text{if} \quad (X, Y, Z) \in \tau \\ F(X, Y, Z) & \text{otherwise} \end{cases} \tag{63}$$

In many practical occasions τ is taken as the measurement region (see Section 3) and G as the experimental data obtained within τ. A further refinement would be to directly take into account the effect of the electron microscope instrumental degradations (modeled by the CTF introduced in Section 3.2) by considering G to be the experimental values divided by the CTF (alternatively, it could be easier to multiply F by the PSF than to divide G by the PSF).

This operator can be regarded as an explicit reinforcement of G, i.e., of the experimental data. The combined action of P_2 and P_1 is what has been called the Gerchberg-Papoulis restoration method (Gerchberg, 1974; Papoulis, 1975).

3. The set of all f's in $\mathscr{H}$ whose amplitude must lie in a prescribed interval $[a, b]$, $a \geqslant 0$, $b \geqslant 0$, $a < b$, over a closed region μ in $\mathscr{H}$. The projection is realized by

$$P_3 f = \begin{cases} a, & f(x, y, z) < a, \quad (x, y, z) \in \mu \\ f(x, y, z), & a \leqslant f(x, y, z) \leqslant b, \quad (x, y, z) \in \mu \\ b, & f(x, y, z) > b, \quad (x, y, z) \in \mu \\ f(x, y, z), & \text{otherwise} \end{cases} \tag{64}$$

Usually, transition regions are applied at the spatial limits of the object, so that smooth borders are created.

4. The set of all real-valued nonnegative functions in $\mathscr{H}$ that satisfy the energy constraint

$$\iiint |f(x, y, z)|^2 \, dx \, dy \, dz \leqslant E \tag{65}$$

where E is the upper energy bound. Letting f_1 be the real part of f ($f = f_1 + if_2$), f_1^+ the rectified portion of f_1 (i.e., $f_1^+ = f_1$ if $f_1 \geqslant 0$, otherwise $f_1^+ = -f_1$), and E_1^+ the energy of f_1^+, the associated projector is

$$P_4 f = \begin{cases} 0 & f_1 \leqslant 0 \\ f_1^+ & E_1^+ \leqslant E \\ \sqrt{\dfrac{E}{E_1^+}} \times f_1^+, & E_1^+ > E \end{cases} \tag{66}$$

5. The set of all real functions in $\mathscr{H}$ whose Fourier transform over the data region (see Section 3) are within a given distance σ from the noisy data. This constraint is related explicitly to the noise level of the input signal through σ (for the implementation of its associated projector see Sezan and Stark, 1983; Carazo and Carrascosa, 1987a).

Note that this last set is formed by 3D structures derived from images that *may* be different from the ones actually measured. Recalling what was said at the end of Section 4 about modifying the problem in order to get stable solutions, it is clear that this is precisely what the projector associated to C_5 does. It is, therefore, a regularizing constraint. The need of such a regularization stems from the fact that the intersection set C_O is supposed to be nonempty. However, with noisy data it could easily happen that $C_{o'} = C_1 \cap C_2 \cap C_3 \cap C_4$ might indeed be empty, since we are enforcing known properties of the true solution (not properties of the solution + the noise) on data that are certainly contaminated by noise, hence the need for allowing a certain discrepancy between the measured data and the a priori knowledge. The action of POCS in restoring missing information is of course dependent on which sets are used in the restoration process. Some work in electron tomography has already been done with the sets introduced so far.

Electron Tomography Applications. The first study of this type was carried out by Agard and Stroud (1982), who effectively used properties defined by sets C_1, C_2, and C_3 and the composite operator $P_3P_2P_1$ in the restoration of the 3D structure of bacteriorhodopsin from the incomplete data set obtained by Henderson and Unwin (1975). The goal of this work was to reveal the polypeptide regions connecting the seven α-helices present in the 3D structure of bacteriorhodopsin. Although the work of Agard and Stroud can be regarded as a special application of POCS, it was not originally based on the general framework of convex projections, but on an independent reasoning about the validity and usefulness of these three constraints. The lack of a more general theoretical framework was perhaps the reason why this truly pioneering work did not receive broader attention.

The first formal introduction of POCS into electron tomography was due to Carazo and Carrascosa (1987a), who performed the recovery of a computer-generated 3D object in a variety of missing data situations and signal-to-noise ratios. Their results were promising and led to a further study (Carazo and Carrascosa, 1987b) in which they first recovered an object from computer-simulated projection images under conditions similar to experimental ones, and then approached a real-data problem, namely, the solution of the 3D structure of the bacteriophage ϕ29 head-to-tail connector. The goal of this latter study was to determine whether the inaccuracies introduced in the practical implementation of a 3D reconstruction method, modeled as a sort of "intrinsic practical noise," could be handled by their implementation of POCS. Their conclusion was that POCS could indeed handle this portion of the noise, although not all missing data could be recovered. The result was thus similar to the one reported by Barth *et al.* (1988) using the maximum entropy restoration method. In particular, Carazo and Carrascosa compared the mean density value in each plane of the reconstruction of the known 3D object, from which the projection images were calculated, and the final reconstructed/restored 3D volume (Fig. 10). Recall that in the initial unrestored reconstruction there was no mean density modulation along z since all frequency data along the reciprocal axis $(0, 0, Z)$ were missing. After POCS restoration the shape of the mean density modulation was found to be partially recovered.

Carazo and Carrascosa (1987b) application of POCS to real data was performed in a way conceptually similar to the one already discussed in connection with the maximum entropy work of Lawrence *et al.* (1989), that is, they compared the final results with and without POCS (using a direct Fourier inversion in the later case), and concluded that the 3D volume after restoration "seemed" to be more in agreement with their structural expectations. Indeed, they obtained a modulation of the mean density along z that was in general agreement with the expected shape and boundaries of the specimen. However, they could not perform some important validity tests, such as, for example, comparing the result obtained by restoring a $\pm 60°$ limited reconstruction with an unrestored $\pm 80°$ limited

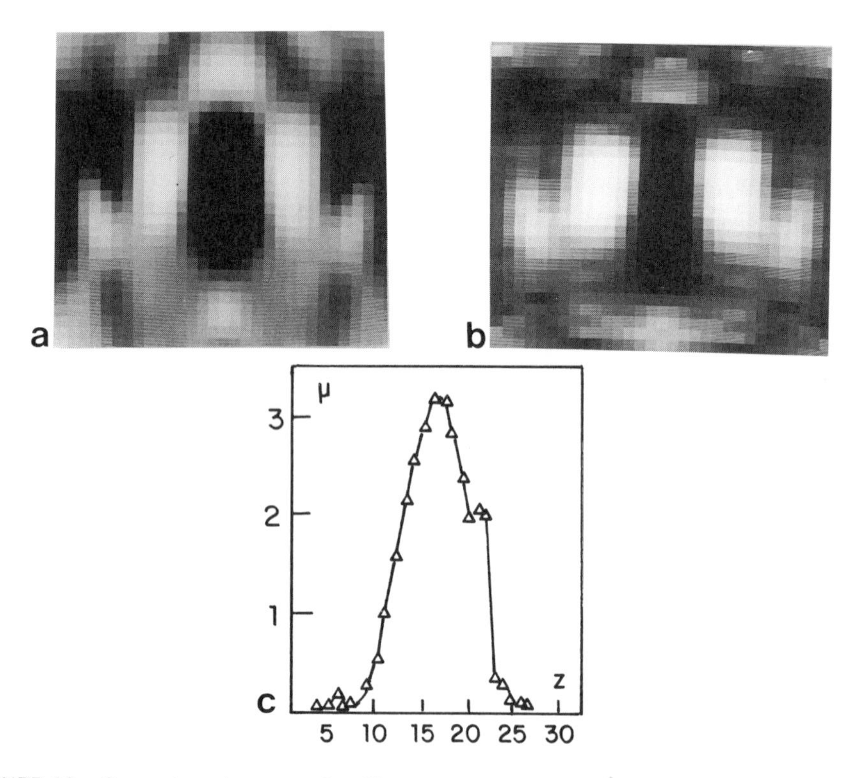

FIGURE 10. Comparison between the 3D structures of the bacteriophage $\phi 29$ head-to-tail connector obtained with and without using POCS. As in Fig. 9, the comparison is performed by showing central slices through the structure of the specimen. (a) Reconstruction obtained by the direct Fourier method from projections up to 60° tilt (FFT^{-1}). (b) Reconstruction plus restoration from the same projections as (a) but using the constraints onto closed convex sets C_1 to C_5 described in the main text (POCS). (c) Mean modulation along z of the density values obtained from the reconstructed and restored structure $[\mu(z)]$. (From Carazo and Carrascosa, 1987b, © Royal Microscopy Society, reproduced with permission.)

reconstruction, because they did not have high-tilt angle projections. Also, no instrumental degradations were modeled in this first study.

New Developments. One of the main advantages of POCS is that it is relatively easy to introduce new constraints in the recovery process. In fact, the formulation of useful characterizations of constraints as closed convex sets is a very active field of research.

Several recent types of constraints could lead to interesting developments in the near future. One type is based on the consideration that the measurement data g carry information not only about the true solution f but also about the noise n. It is therefore logical to introduce constraints on the noise, in order to solve the inversion problem. Another type of constraints seeks to overcome the effects of data manipulation (especially interpolations) during the recovery process. Finally, the introduction of adaptative space-variant regularization and artifact suppression constraints, coupled with the definition of the "partial Wiener set" and the proposal of reformulating POCS within a fuzzy sets context, all seem promising.

A method which uses the information about the noise statistics effectively is one that puts the appropriate constraints on the residual, defined by

$$r \equiv g - H\hat{f} \tag{67}$$

where $\hat{f}$ is an estimate of the original signal. It is reasonable to require the residual to have the same statistical properties of the noise since these conditions would indeed hold for the true solution. The statistical properties that have been so far characterized as closed convex constraints are (Trussel and Civanlar, 1984):

1. Variance of the residual: The set based on this constraint is

$$C_v = \{f \mid \|g - Hf\|^2 \leqslant \delta_v\} \tag{68}$$

   which is closed and convex. The set C_v defined by Eq. (68) requires only that the variance of the noise be less or equal to the estimated variance.
2. Mean of the residual: The noise is assumed to be zero mean. The set based on this constraint is

$$C_m = \left\{f \,\middle|\, \left|\sum_i (g_i - [Hf]_i)\right| \leqslant \delta_m\right\} \tag{69}$$

3. Outliers of the residual: It is desirable to exclude those individual measurements whose residuals deviate an unlikely amount from the mean. For the most common case of Gaussian noise, this set is defined as

$$C_o = \{f \mid |g_i - [Hf]_i| \leqslant \delta_o\} \tag{70}$$

   which is closed and convex.

Projectors onto these three sets are given by Trussell and Civanlar (1984). The power of this approach is to limit the solution space, allowing other constraints to be applied to the true solution, leading to more accurate results.

Historically, ART (algebraic reconstruction techniques) has been one of the first, and initially more popular, reconstruction methods (Gordon and Herman, 1971). It assumes that the images are true projection images of the object (i.e., with either negligible or correctable instrumental degradations). It is therefore important to observe that ART has long been realized as a POCS-type of algorithm (Herman, 1980). More recently, Oskoui-Fard and Stark (1988) have formally placed ART in the framework of POCS. In this way they note that every iteration of the additive weighted ART (each full cycle of corrections) is a series of L alternating projections onto L distinct convex sets. In terms of convex projections, we can describe the ART algorithm by

$$f_{k+1} = P_{\mathrm{ART}} f_k \tag{71}$$

where $P_{\mathrm{ART}} \equiv P_L \cdots P_2 P_1$, i.e., a composition of L projection operators that project onto the closed convex sets C_i, $i = 1,..., L$. Here each C_i is defined as the set of functions whose line integral in the ith direction renders the g_ith measurement (see Eq. (10). In the discrete case we have (Eq. 15 and 16) $g_i = l_i^{\mathrm{T}} f$. In this way the set defined by each measurement is

$$C_i = \{ f \mid l_i^{\mathrm{T}} f = g_i \} \tag{72}$$

which is a linear variety.

Noise can be explicitly taken into account in this derivation by allowing for some discrepancies between the measured data and the restored ones after each cycle of iteration. Formally, this can be done by reformulating the set defined by Eq. (72) in the following way:

$$C_i = \{ f \mid l_i^{\mathrm{T}} f - g_i \leqslant \varepsilon_i \} \tag{73}$$

A further refinement was introduced by Peng (1988), who reported that under certain conditions (which are met in electron tomography) the iterative ART method can be performed in just one step by applying a suitable convex projector; this reconstruction method has been called one-step projection reconstruction (OSPR). It is therefore possible to define a composition of projectors such that part of them actually perform the reconstruction (the finding of those components of the true solution f in the measurement space M) and the rest perform the restoration by applying constraints derived from the body of a priori knowledge. Within this scheme f_0 could either be a constant or, perhaps better, a trial structure where some of the pieces of a priori knowledge that cannot be expressed as convex constraints are introduced.

A further extension of POCS was developed by Civanlar and Trussell (1986), who proposed the use of fuzzy logic (Zadeh, 1965) into the formulation of the method. They pointed out that by using fuzzy sets it is possible to model partially defined information as well as exact knowledge. The intersection of all the fuzzy sets is termed the feasibility set. One of the key points of this fuzzy-set-based formulation is that it makes it possible to attach to every element $f \in \mathscr{H}$ a certain "membership function." This function can be interpreted as the strength of our belief that f has property A. This is particularly important in the study of the intersection set C_o of all possible solutions. It is clear that C_o will be another fuzzy set and, therefore, the value of the membership function of each element of C_o may be used as a criterion to choose a particular subset of solutions. In particular, the true solution must be a member of C_o characterized by a high membership value. The converse, however, is also true: any high-membership-valued element of C_o is a nonrejectable solution. Work on fuzzy set restoration has been so far limited to one-dimensional signals due to the numerical complexity of the approach. Prospects of its use in electron tomography are contingent on finding a sufficiently flexible and efficient implementation that allows the study of multidimensional signals.

Another interesting development has been provided by Goldburg and Marks (1985), who addressed the question of what happens if, in spite of all efforts to have a nonempty C_o, we still introduce inconsistent constraints. Conceptually, their answer to that question turned out to be very nice: they showed that if two inconsistent constraints are enforced, then it is still possible to reach a solution that is in agreement with one of the imposed constraints and it is as close as possible to the set defined by the other constraint. A unique generalization to the case of more than two constraints does not seem possible (Youla and Velasco, 1986), although some interesting geometric considerations can still be made.

Some of the most recent contributions to the body of useful projectors are due to Sezan and Tekalp (1990a) and Sezan and Trussell (1991). They introduced a new closed convex regularization constraint called the partial Wiener solution set. Projection onto this set forces the solution to be equal to the Wiener solution over a predetermined set of frequencies that lie outside of the neighborhoods of the zeros of the optical transfer function. They also developed an adaptative space-variant restoration algorithm incorporating both regularization and artifact suppression constraints.

Two final concluding remarks about POCS should be made. One refers to fundamental limits of POCS, the other to its practical implementation. By definition, POCS requires the a priori information to be coded as constraints onto closed convex sets and, although there are many properties that can be expressed this way, there are others that definitely cannot. In particular, POCS cannot be applied to the phase-retrieval problem because the set of functions with a prescribed value of the magnitude of the Fourier transform is not a convex set. The second remark concerns a practical limitation of POCS. So far, POCS has been formulated in a general function-theoretical way, without taking into account the discrete nature of any digitally based problem. However, the fact that a limited number of samples is the most that can be measured and treated by digital means, renders some of the basic theoretical foundations of POCS questionable. Fortunately, it can be proved that results obtained in the discrete case converge to those that would be obtained in the continuous under certain conditions (Schlebusch and Splettstösser, 1985; Zhou and Xia, 1989), at least for the multidimensional Gerchberg-Papoulis algorithm.

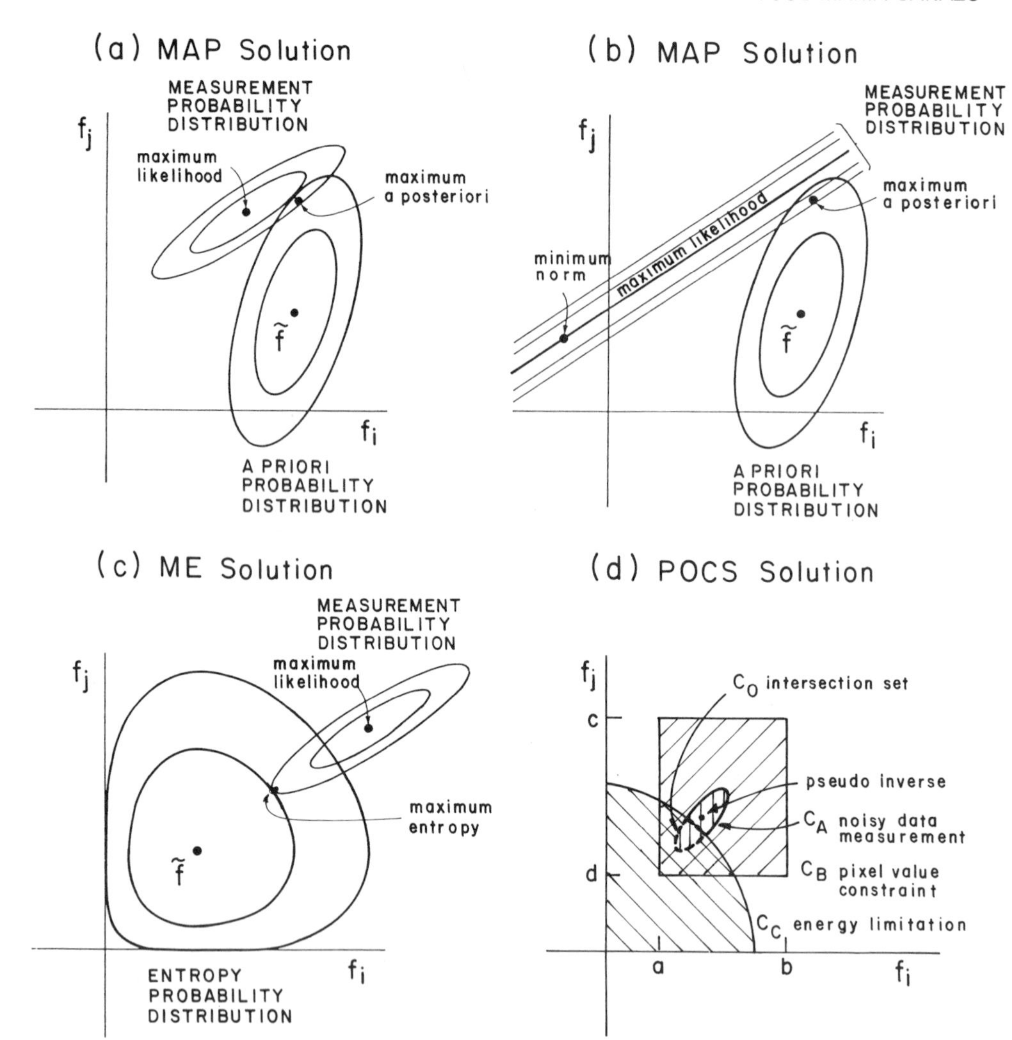

FIGURE 11. The space of possible solutions associated with different recovery methods as a function of two components of the image vector, f_i and f_j. (a) For the MAP method under the assumption that the probability distribution of f, $p(f)$ in Eq. (39), is Gaussian; contour plots for this probability distribution are therefore ellipsoids centered at the mean image $\tilde{f}$. The shape of contour plots associated with the maximum likelihood solution, corresponding to $p(f \mid g)$ in Eq. (39), is controlled by the noise covariance matrix R_{nn} and the measurement matrix H and may be therefore complex; here we are representing a simple case in which they are also ellipsoids. (b) As in (a), but the f_i-f_j plane now includes a component of the null space. There is no unique solution for f_i and f_j that satisfies the measurements. Every point on the solution line shown in (b) yields exactly the same measurement vector. The difference between the minimum norm solution and the maximum a posteriori solution is also graphically shown. (c) Conceptually as (a), but now the probability function $p(f)$ is the one associated with the maximum entropy solution of Trussell (1980) [modified in order to make the maximum of $p(f)$ coincide with the vector $\tilde{f}$ in (a) and (b)]. (d) The POCS solution considering some closed convex constraints; C_A is the set associated with the reinforcement of measured data within a certain noise variance limitation (also referred to as C_5 in the text), C_B is associated with a pixel value constraint (C_3 in the text), C_C is the energy limitation constraint (C_4 in the text). C_O is the intersection set containing all possible constraints. [(a), (b), and (c) from Hanson, 1987; © Academic Press, reproduced with permission.]

In conclusion, POCS is a rich method for signal reconstruction and restoration. It is also very easy to apply, once the sets and operators are defined. The solution, however, is not unique, reflecting the fact that the measurement data plus the a priori knowledge are not generally sufficient to uniquely determine the solution. Future lines of research will probably be centered on the characterization of more pieces of a priori knowledge as convex constraints (and obtaining their associated projectors), the study of the set of all possible solutions in order to decide which solutions are the "best" (perhaps with the use of fuzzy-set-based restoration), and further theoretical work on the connection between the continuous and the discrete case.

6. CONCLUSIONS

We have shown that, in general, it is impossible to uniquely determine a 3D structure from only a finite set of projection images. In fact, the space of all possible solutions can be expressed as the direct sum of the set of measurement-determined solutions (called *measurement space* and denoted by M) plus its orthogonal complement (the null space $M^{\perp}$), and that the whole point here is that $M^{\perp}$ is far from being empty. Restoration methods thus emerge as a way to obtain information about $M^{\perp}$ not from the measured images in themselves but from known general properties that the true solution is known to possess.

We have described several restoration methods, mainly those that seem to be most promising for electron tomography applications. However, restoration is a very active field, and other methods are emerging that have not been covered here. Among them we just want to mention Kalman-filtering-based methods (Woods, 1981), which may be of future use once the nontrivial extension to three dimensions is accomplished.

The restoration of 3D reconstructions is just starting and experimental results are scarce. It is therefore impossible now to describe the "standard" restoration technique in electron tomography. All the more, it is of great importance to have a general perspective of the field, as we have tried to provide in this work, that enables us to know the relations between the different methods and to assess their relative merits for solving a given problem. Figure 11 illustrates some of these relations for a simple two-point case.

ACKNOWLEDGMENTS

I wish to thank Drs. M. I. Sezan, Kodak Research Laboratories; H. J. Trussell, North Caroline State University; K. M. Hanson, Los Alamos National Laboratory; Martin Barth, European Molecular Biology Laboratory; and Joachim Frank, Michael Radermacher, and Paul Penczek, New York State Department of Health, for their invaluable help in preparing this work. Their careful reading of previous versions of this chapter has been very useful to me. I also wish to thank Mery Muyal, IBM Madrid Scientific Center, for her help in preparing Figs. 4 and 5. This work has been supported by Comisión Asesora para la Investigación Científica y Técnica through grant number PB87-0365. The institutional help of Fundación Ramón Areces to the Centro de Biología Molecular is also acknowledged.

REFERENCES

Agard, D. A. and Stroud, R. M. (1982). Linking regions between helices in bacteriorhodopsin revealed. *Biophys. J.* **37**:589–602.

Andrews, H. C. and Hunt, B. R. (1977). *Digital Image Restoration.* Prentice-Hall, Englewood Cliffs, NJ.

Barth, M., Bryan, R. K., and Hegerl, R. (1989). Approximation of missing-cone data in 3D electron microscopy. *Ultramicroscopy* **31**:365–378.

Barth, M., Bryan, R. K., Hegerl, R., and Baumeister, W. (1988). Estimation of missing data in three-dimensional electron microscopy. *Scanning Microsc. Suppl.* **2**:277–284.

Bricogne, G. (1988). A Bayesian statistical theory of the phase problem I. A multichannel maximum-entropy formalism for constructing generalized joint probability distributions of structure factors. *Acta Crystallogr. A* **44**:517–545.

Browder, F. E. (1965). Fixed-point theorems for noncompact mappings in Hilbert space. *Proc. Nat. Acad. Sci. USA* **53**:1272–1276.

Carazo, J. M. and Carrascosa, J. L. (1986). Towards obtaining more reliable 3D reconstructions: 3D restoration by the method of convex sets, in *XI Int. Congr. in Electron Microscopy,* Vol. 3, pp. 2383–2384.

Carazo, J. M. and Carrascosa, J. L. (1987a). Information recovery in missing angular data cases: an approach by the convex projections method in three dimensions. *J. Microsc.* **145**:23–43.

Carazo, J. M. and Carrascosa, J. L. (1987b). Restoration of direct Fourier three-dimensional reconstructions of crystalline specimens by the method of convex projections. *J. Microsc.* **145**:159–177.

Carazo, J. M., Wagenknecht, T., and Frank, J. (1989). Variations of the three-dimensional structure of the *Escherichia coli* ribosome in the range of overlap views. *Biophys. J.* **55**:465–477.

Carazo, J. M., Wagenknecht, T., Radermacher, M., Mandiyan, V., Boublik, M., and Frank, J. (1988). Three-dimensional structure of 50*S Escherichia coli* ribosomal subunits depleted of proteins L7/L12. *J. Mol. Biol.* **201**:393–404.

Chalcroft, J. P. and Davey, C. L. (1984). A simply constructed extreme-tilt holder for the Philips eucentric stage. *J. Microsc.* **134**:41–48.

Chiu, M. Y., Barrett, H. H., Simpson, R. G., Chou, C., Arendt, J. W., and Gindi, G. R. (1979). Three-dimensional radiographic imaging with a restricted view angle. *J. Opt. Soc. Am.* **69**(10):1323–1333.

Civanlar, M. R. and Trussell, H. J. (1986). Digital signal restoration using fuzzy sets. *IEEE Trans. Acoust. Speech and Signal Process.* **ASSP-34**(4):919–936.

Crowther, R. A., DeRosier, D. J., and Klug, A. (1970). The reconstruction of a three-dimensional structure from projections and its application to electron microscopy. *Proc. R. Soc. London A* **317**:319–340.

DeRosier, D. J. and Moore, P. B. (1970). Reconstruction of three-dimensional images from electron micrographs of structures with helical symmetry. *J. Mol. Biol.* **52**:355–369.

Erickson, H. P. and Klug, F. R. S. (1971). Measurement and compensation of defocusing and aberrations in Fourier processing of electron micrographs. *Phil. Trans. R. Soc. London B* **261**:105–113.

Frank, J. and Radermacher, M. (1986). Three-dimensional reconstruction of nonperiodic macromolecular assemblies from electron micrographs, in *Advanced Techniques in Biological Electron Microscopy* (J. Koehler, ed.), Springer-Verlag, Berlin.

Frieden, B. R. (1972). Restoring with maximum likelihood and maximum entropy. *J. Opt. Soc. Am.* **G2**:511–518.

Gerchberg, R. W. (1974). Super-resolution through error energy reduction. *Opt. Acta* **21**(9):709–720.

Glaeser, R. M., Tom, L., and Kim, S. H. (1990). Three-dimensional reconstruction from incomplete data: Interpretability of density maps at "atomic" resolution. *Ultramicroscopy* **27**:307–318.

Goldburg, M. and Marks II, R. J. (1985). Signal synthesis in the presence of an inconsistent set of constraints. *IEEE Trans. Circuits Syst.* **CAS-32**(7):647–663.

Gordon, R. and Herman, G. T. (1971). Reconstruction of pictures from their projections. *Comm. ACM* **14**:759–768.

Gulf, S. F. (1988). Developments in maximum entropy data analysis, in *8th MaxEnt Workshop*, Cambridge, UK.

Hanson, K. M. (1987). Bayesian and related methods in image reconstruction from incomplete data, in *Image Recovery: Theory and Application* (H. Stark, ed.), Academic Press, Orlando.

Hanson, K. M. and Myers, K. J. (1990). Comparison of the algebraic reconstruction technique with the maximum entropy reconstruction technique for a variety of detection tasks. *SPIE Proc.* **1231**.

Hanson, K. M. and Wecksung, G. W. (1983). Bayesian approach to limited-angle reconstruction in computed tomography. *J. Opt. Soc. Am.* **73**:1501–1509.

Henderson, R. and Unwin, P. N. T. (1975). Three-dimensional model of purple membrane obtained by electron microscopy. *Nature (London)* **257**:28–32.

Herman, G. T. (1980). *Image Reconstruction from Projections*. Academic Press, New York.

Howard, J. (1988). Tomography and reliable information. *J. Opt. Soc. Am. A* **5**(7):999–1014.

Johnson, R. W. and Shore, J. E. (1983). Comments on and correction to "Axiomatic derivation of the principle of maximum entropy and the principle of minimum cross-entropy. *IEEE Trans. Inf. Theory* **IT-29**(6):942–943.

Katz, M. B. (1978). Questions of uniqueness and resolution in reconstruction from projections, in *Lectures in Biomathematics*, Vol. 26, Springer-Verlag, Berlin.

Klug, A. and Crowther, R. A. (1972). Three-dimensional image reconstruction from the viewpoint of information theory. *Nature* **238**:435–440.

Lawrence, M. C., Jaffer, M. A., and Sewell, B. T. (1989). The application of the maximum entropy method to electron microscopy tomography. *Ultramicroscopy* **31**:285–302.

Lepault, J. and Pitt, T. (1984). Projected structure of unstained frozen-hydrated T-layer of *Bacillus brevis*. *EMBO J.* **3**(1):101–105.

McCaughey, D. G. and Andrews, H. C. (1977). Degrees of freedom for projection imaging. *IEEE Trans. Acoust., Speech and Signal Process.* **ASSP-25**:63–73.

Opial, Z. (1967). Weak convergence of the sequence of successive approximations for nonexpansive mappings. *Bull. Am. Math. Soc.* **73**:591–597.

Oskoui-Fard, P. and Stark, H. (1988). Tomographic image reconstruction using the theory of convex projections. *IEEE Trans. Med. Imaging* **7**(1):45–58.

Papoulis, A. (1975). A new algorithm in spectral analysis and band-limited extrapolations. *IEEE Trans. Circuits Syst.* **CAS-22**:735–742.

Pavlovic, G. and Tekalp, A. M. (1990). Restoration in the presence of multiplicative noise with application to scanned photographic images, in *Proc. Int. Conf. Acoust., Speech and Signal Processing.*

Penczek, P., Radermacher, M., and Frank, J. Three-dimensional reconstruction of single particles embedded in ice. *Ultramicroscopy* (in press).

Penczek, P., Srivastava, S., and Frank, J. (1990). The structure of the 70*S* *E. coli* ribosome in ice, in *Proc. XIIth Int. Congr. Electron Microscopy*, Vol. 1. pp. 506–507.

Peng, H. (1988). *Fan-Beam Reconstruction in Computer Tomography from Full and Partial Projection Data.* Ph.D. thesis, Rensselaer Polytechnique Institute, Troy, New York.

Polak, E. (1971). *Computational Methods in Optimization: A Unified Approach.* Academic Press, New York.

Radermacher, M. (1980). *Dreidimensionale Rekonstruktion bei kegelförmiger Kippung im Elektronenmikroskop.* Ph.D. thesis, Technische Universität München.

Radermacher, M. (1988). Three-dimensional reconstruction of single particles from random and nonrandom tilt series. *Elect. Micros. Tech.* **9**:359–394.

Radermacher, M. and Hoppe, W. (1980). Properties of 3-D reconstruction from projections by conical tilting compared to single axis tilting, in *Proc. Seventh European Congr. Electron Microscopy*, Vol. 1, pp. 132–133.

Rushforth, C. K. (1987). Signal restoration, functional analysis and Fredholm integral equations of the first kind, in *Image Recovery: Theory and Application* (H. Stark, ed.), Academic Press, Orlando.

Schafer, R. W., Mersereau, R. M., and Richards, M. A. (1981). Constrained iterative restoration algorithms. *Proc. IEEE* **69**(4):432–450.

Schelebusch, H.-J. and Splettstösser, W. (1985). On a conjecture of J. L. C. Sanz and T. S. Huang. *IEEE Trans. Acoust. Speech and Signal Process.* **ASSP-33**(6):1628–1630.

Sezan, M. I. and Stark, H. (1982). Image restoration by the method of convex projections. II: Applications and numerical results. *IEEE Trans. Med. Imaging* **MI-1**:95–101.

Sezan, M. I. and Stark, H. (1983). Image restoration by convex projections in the presence of noise. *Appl. Opt.* **22**:2781–2789.

Sezan, M. I. and Tekalp, A. M. (1990a). Adaptative image restoration with artifact suppression using the theory of convex projections. *IEEE Trans. Acoust. Speech and Signal Process.* **38**(1):181–185.

Sezan, M. I. and Tekalp, A. M. (1990b). A survey of recent developments in digital image restoration. *Opt. Eng.* **29**(5):393–404.

Sezan, M. I. and Trussell, H. J. (1991). Prototype image constraints for set-theoretic image restoration, in *IEEE Trans. Acoust. Speech Signal Process.* **39**(10):2275–2285.

Shore, J. E. and Johnson, R. W. (1980). Axiomatic derivation of the principle of maximum entropy and the principle of minimum cross-entropy. *IEEE Trans. Inf. Theory* **IT-26**(1):26–37.

Skilling, J. (1988). Classical MaxEnt data analysis, in *8th MaxEnt Workshop*, Cambridge, UK.

Skilling, J. and Bryan, R. K. (1984). Maximum entropy image reconstruction: general algorithm. *Mon. Not. R. Astronom. Soc.* **211**:111–214.

Smith, K. T., Solmon, D. C., and Wagner, S. L. (1977). Practical and mathematical aspects of the problem of reconstructing objects from radiographs. *Bull. Am. Math. Soc.* **83**(6):1227–1268.

Smith, P. R., Peters, T. H., and Bates, R. H. T. (1973). Image reconstruction from finite numbers of projections. *J. Phys. A* **6**:361–382.

Tam, K. C. and Perez-Mendez, V. (1981a). Limited-angle three-dimensional reconstruction using Fourier transform iterations and Radon transform iterations. *Opt. Eng.* **20**(4):586–589.

Tam, K. C. and Perez-Mendez, V. (1981b). Tomographical imaging with limited-angle input. *J. Opt. Soc. Am.* **71**:582–592.

Tikochinsky, Y., Tishby, N. Z., and Levine, R. D. (1984). Consistent inference of probabilities for reproducible experiments. *Phys. Rev. Lett.* **52**(16):1357–1360.

Trussell, H. J. (1980). The relationship between image restoration by the maximum a posteriori method and the maximum entropy method. *IEEE Trans. Acoust. Speech and Signal Process.* **ASSP-28**(1):114–117.

Trussell, H. P. (1984). A priori knowledge in algebraic reconstructions methods, in *Advances in Computer Vision and Image Processing*, Vol. 1, pp. 265–316, JAI Press.

Trussell, H. J. and Civanlar, M. R. (1983). The initial estimate in constrained iterative restoration, in *ICASSP'83*, pp. 643–646.

Trussel, H. P. and Civanlar, M. R. (1984). Feasible solution in signal restoration. *IEEE Trans. Acoust. Speech and Signal Process.* **ASSP-32**:201–212.

Welton, T. A. (1979). A computational critique of an algorithm for image enhancement in bright field electron microscopy, in *Advances in Electronics and Electron Physics* (L. Maston, ed.), Vol. 48, pp. 37–101. Academic Press, New York.

Whittaker, E. T. (1914). On the functions which are represented by expansion of the interpolation theory. *Proc. Roy. Soc. Edinburg* **A35**:181–194.

Woods, J. W. (1981). Two-dimensional Kalman filtering, in *Topics in Applied Physics* (T. S. Huang, ed.), Vol. 2. Springer-Verlag, Berlin.

Youla, D. C. (1987). Mathematical theory of image restoration by the method of convex projections, in *Image Recovery: Theory and Applications* (H. Stark, ed.), Academic Press, Orlando.

Youla, D. C. and Velasco, V. (1986). Extensions of a result on the synthesis of signals in the presence of inconsistent constraints. *IEEE Trans. Circuits Syst.* **CAS-33**(4):465–468.

Youla, D. C. and Webb, H. (1982). Image restoration by the method of convex projections. I: Theory. *IEEE Trans. Med. Imaging* **MI-1**:81–94.

Zadeh, L. A. (1965). Fuzzy sets. *Inf. Vontrol* **8**:338–353.

Zhou, X.-W. and Xia, X.-G. (1989). The extrapolation of high-dimensionality band-limited functions. *IEEE Trans. Acoust. Speech and Signal Process.* **ASSP-37**(10):1576–1580.

III

Methods

7

Tilting Stages for Biological Applications

James N. Turner and Ugo Valdrè

The specimen stage of the electron microscope is the most sophisticated mechanical subsystem in the instrument. It must position any portion of the specimen in the field of view, orient the specimen with respect to the optical axis, cool it to liquid nitrogen or liquid helium temperatures, and be stable to atomic dimensions for a minute or more. The stage must have two 3-mm translation motions, which at

James N. Turner • Wadsworth Center for Laboratories and Research, New York State Department of Health and School of Public Health Sciences, State University of New York at Albany, Albany, New York, 12201-0509 *Ugo Valdrè* • Centro di Microscopia Elettronica, Dipartimento di Fisica, Università Degli Studi di Bologna, 40126 Bologna, Italy

Electron Tomography: Three-Dimensional Imaging with the Transmission Electron Microscope, edited by Joachim Frank, Plenum Press, New York, 1992.

0.3 nm resolution corresponds to a stability of one part in 10^7. This degree of precision corresponds to a deviation of 40 m in the orbit of the moon. The stage must tilt the specimen $\pm 70°$ about at least one axis, but two orthogonal axes and $\pm 90°$ would be ideal. Stages for tomography have a precision of $\pm 0.03°$ in tilt angle, corresponding to one part in 2×10^3 (Turner *et al.*, 1988). These are especially demanding criteria since the device must operate in a high-strength magnetic field, which cannot be perturbed to one part in 10^6. The mechanism must be contained in a volume of about one to a few cubic centimeters in the middle of an electron optical column, and vibration problems can be severe (Turner and Ratkowski, 1982). Some vibration modeling has been done (Valle *et al.*, 1980), but is of limited value due to the extreme mechanical complexity. In spite of the stringent conditions, a number of successful stages have been constructed (Valdrè, 1979; Swann, 1979; Valdrè and Goringe, 1971; Turner *et al.*, 1989a).

Tilt stages and three-dimensional (3D) electron microscopy were featured by von Ardenne in 1940 (1940a, b, c) with the publication of tilt stage designs and stereo images. This chapter considers the design and operation of stages for the collection of 3D information.

1. NEED FOR SPECIMEN TILTING

1.1. Perspective and Overlapping Information

Electron microscopic specimens are by most measures "thin," ranging from individual molecules to whole cells or sections up to several micrometers thick. Compared to the lateral resolution, nearly all are "thick"; i.e., they possess 3D information. In a single image, structures separated along the direction of the optical axis (z direction) cannot be discerned due to image superimposition (Fig. 1). The only way to view the object's 3D structure is to alter the relative orientation of the electron beam and the specimen. This is done by mechanically tilting the

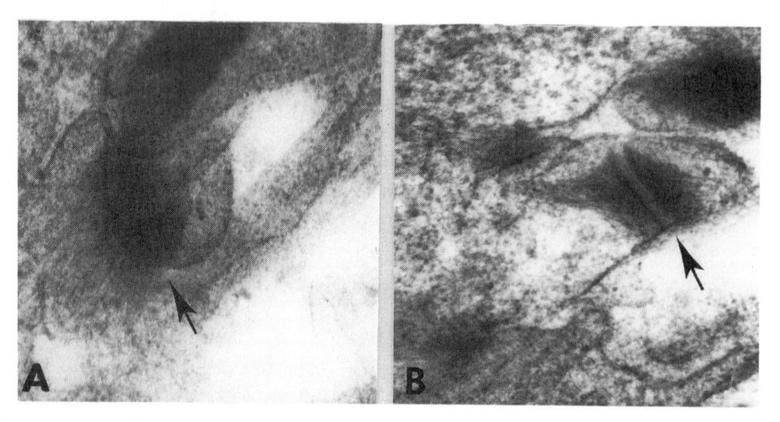

FIGURE 1. Image of a thick section (0.5 μm) recorded at 1.0 MV. (A) is untilted while (B) is tilted about two orthogonal axes (3.5° about one and 21.2° about the other) using a double-tilt stage. The arrow indicates the location of a desmosome, a cell junctional complex, seen in detail in (B). Bar = 0.25 μm.

specimen via a stage mechanism (Valdrè and Goringe, 1971; Valdrè, 1979; Swann, 1979; Turner *et al.*, 1989a, b).

The image in figure 1a of a thick section recorded in the high-voltage electron microscope (HVEM) shows a dense diffuse region. After tilting, the same region shows a desmosome (Fig. 1B) with characteristic dense parallel plates, tonofilament bundles, and a thin dense intercellular plate. Optimal alignment makes structures visible which were previously not observed, resulting in a totally different interpretation. Thus, there is a complex relationship between image resolution (distinct from instrumental resolution), specimen orientation, and image contrast.

1.2. Data Collection for 3D Reconstructions

A tilt series or group of images recorded at particular specimen-beam orientations (perspectives) provides the data for 3D reconstruction by tomography, and is recorded differently depending on the nature of the object. If a unique object is being reconstructed, the projections must be recorded from that single object (Crowther *et al.*, 1970; Frank *et al.*, 1987). However, if the sample occurs in many copies which are assumed to be identical, a few micrographs, or even a single micrograph at one tilt angle is all that is required (Radermacher, 1988).

A large number of projections recorded with high precision ($\pm 0.1°$; preferably $\pm 0.05°$) is required for tomography. The maximum tilt angle influences the resolution of the reconstruction by limiting the amount of Fourier space sampled (Crowther *et al.*, 1970; Klug and Crowther, 1972). Stage geometry and specimen mounting are also important. Two orientation motions (double-tilt, or tilt rotation) sample Fourier space more evenly than a single motion, but the mechanism may limit the tilt range (Valdrè, 1979; Turner *et al.*, 1990). The orientation of particular details relative to the tilt axes affects the collection of spatial frequencies that are critical to their reconstruction.

1.2.1. Single-Tilt Axis

Single-tilt is the simplest geometry and collection method. The specimen holder can be extremely thin, allowing an angular range encompassing a large portion of Fourier space ($\pm 70°$ or more) (Chalcroft and Davey, 1984; Turner *et al.*, 1988; Peachey and Heath, 1989). However, the data are incomplete, with a missing wedge of spatial frequencies (Fig. 2a). This situation can be improved by recording a second set of images after rotating the specimen by 90° and reinserting it in the microscope, or by using a stage with a second orientation motion. The former technique is cumbersome and imprecise.

1.2.2. Double Orientation Motions and Conical Data Collection

Stages with two orientation motions allow a wider range of projections to be recorded, corresponding to a larger portion of Fourier space. This is achieved by either a double tilt or by a single tilt and rotation mechanism (Valdrè and Goringe, 1971; Valdrè, 1979). These motions are equivalent if their mechanical limits are the same and if the axis of rotation corresponds to the specimen normal vector (Turner

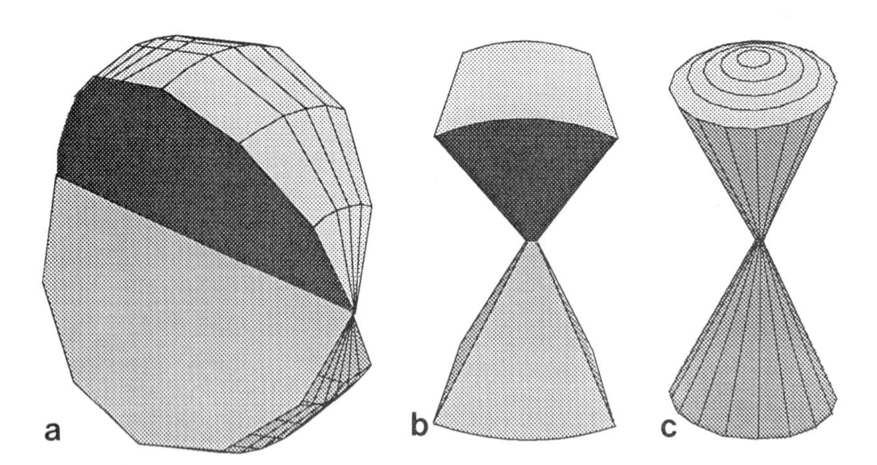

FIGURE 2. Missing regions of information in Fourier space for the (a) single-tilt (wedge), (b) double-tilt (pyramid), and (c) conical (cone) collection schemes.

et al., 1989b). If the rotation is about the optical axis, the orientation of the image relative to the photographic emulsion changes but the projection remains the same; i.e., no new information is obtained. The second motion must be built into the specimen holder and not into a larger more remote portion of the stage. The main reason for attempting this difficult task is to reduce the region of missing information in Fourier space. Unfortunately, the additional mechanism usually restricts the angle to which the specimen can be tilted. In spite of the difficulties, double orientation stages of the cartridge type have been built with tilt ranges as high as $\pm 60^\circ$ (Siemens' *Focus Information*, 1975).

The simplest double-motion collection procedure is to record a tilt series about both axes of a double-tilt stage. This reduces the missing region of information from a wedge (Fig. 2a) to a four-sided pyramid (Fig. 2b). By using a vector analysis of the double-tilt geometry, the specimen can be tilted about any arbitrary direction as the effective tilt axis, independent of the mechanical axes (Laufer and Milliken, 1973; Sykes, 1979; Arii and Hama, 1987; Chou, 1987; Turner *et al.*, 1989b). This approach can also optimize the observation of particular details, and a tilt series can be recorded about any azimuthal direction. For tomography, a set of projections corresponding to a conical data set can be recorded (Radermacher and Hoppe, 1980; Turner *et al.*, 1989b). That is, the specimen normal sweeps out a cone whose axis corresponds to the optical axis. Images can be recorded at any arbitrary increment about the surface of the cone. The missing portion of Fourier space is, thereby, reduced to the volume of this cone (Fig. 2c). The precision of the data is determined by the precision and stability of both tilt mechanisms. A disadvantage of this approach is that both mechanisms have to be driven in the forward and backward directions during the recording of a complete set of projections. This procedure entails backlash which can only be avoided by the use of complicated driving routines so that every position is approached from the same direction.

Conical data sets can also be recorded using a single-tilt and rotation stage by tilting to the maximum angle and rotating about the specimen normal. If the tilt angle is equal to the cone half-angle of the double-tilt geometry, the same projections will be recorded. The precision of the data is again influenced by the precision and stability of the two mechanisms, but the rotation mechanism can always be driven in the same direction, minimizing the effect of backlash and the need for changes in the tilt angle.

1.3. Measurement of Tilt Angle

While most biological specimens have a defined structure, few are periodic and even fewer are crystalline. Thus, it is normally impossible to determine the relative perspectives of the images of a tomographic data set from symmetry conditions observed in the image or the diffraction pattern. Thus, the overall reconstruction procedure is optimized when the specimen orientation angle, defining the perspective of each image, is accurately known from the tilting and rotating mechanisms. Although the influence of tilt angle inaccuracies on the reconstruction has not been experimentally determined, it is generally agreed that a precision of at least $\pm 0.1^\circ$ is required (Frank *et al.*, 1987).

Rotary shaft encoders can easily measure smaller angular increments and can be mounted on the stage control mechanism. For most stages, tilt angle measurements are indirect, that is, the encoder is not mounted on the specimen holder, but is mechanically coupled to it through another mechanism. The precision and motion of the coupling mechanism determines the overall measurement precision. The coupling mechanism in motor-driven stages includes the drive system, which has a large mechanical reduction and mounts the encoder on the motor shaft rather than on the specimen holder. If the coupling or stage motion is nonlinear, the relationship between the readout values and the specimen orientation angle is complicated. This is usually the situation when the tilt angle is controlled by a linear motion. Side-entry stage specimen rods provide both a linear relationship and a direct encoder mount. However, antivibration couplings in the rod may adversely effect angular precision due to insufficient torsional stiffness (Turner and Ratkowski, 1982) or imprecise universal joints.

The tilt and rotation mechanisms can be calibrated outside the microscope by reflecting a laser beam off a mirror attached to the specimen holder. If the position of the reflected beam is monitored a meter or two away from the stage, small angular increments produce relatively large displacements. A high-precision single-tilt stage has been calibrated to $\pm 0.06^\circ$ in this way (Turner *et al.*, 1988). A second tilt or rotation motion can also be monitored by this method.

Measurement of tilt angles is more difficult in top-entry stages due to the complex coupling mechanism. In addition, the relative motions of the control devices and the specimen holder are often nonlinear. The tilting and rotating mechanisms are also smaller and more delicate, resulting in a greater possibility of slippage in the mechanism.

The tilt angle can also be calculated from the images of a set of specimen points that can be followed as a function of tilt (Chalcroft and Davey, 1984). Colloidal gold particles deposited on both surfaces of the specimen are ideal, and can be used to accurately determine the tilt angle and to align the data set for reconstruction (Lawrence, 1983; Luther *et al.*, 1988; Lawrence, Chapter 8). This approach is completely general, and therefore attractive for biological objects, which are usually amorphous. However, the tilt angle of crystalline specimens can be calculated from symmetry conditions (Shaw and Hills, 1981), or from Kikuchi patterns (Waddington, 1974). These in-situ methods are important because they directly determine the angle of the imaged region of the specimen. A stage of extreme precision is useless if the specimen varies in an unknown manner during the recording of the tomographic data set. Sectioned material may alter its shape and angle relative to the beam as a function of irradiation (King, 1981). This effect must be monitored and corrected to ensure the accuracy of the reconstruction.

2. *KINEMATIC PRINCIPLES*

2.1. *Degrees of Freedom and Requirements*

Consider the general case of a specimen without any ordered structure at the working resolution. Five degrees of freedom (each related to a motion) are needed for the complete geometrical characterization of the specimen motion. Two translation movements position different sample regions under the electron beam along two axes (x, y) that define a plane normal, or nearly so, to the optical axis. A further linear motion along the objective axis (z axis) adjusts the specimen level, making it coincident with the object plane of the lens. The remaining two degrees of freedom position the specimen in any desired orientation relative to the beam. They correspond to two inclination axes, which usually do not coincide with the translation axes. The amount of inclination is specified by the azimuthal (θ) and longitudinal (ϕ) angular coordinates (Fig. 3).

The combined motions of the inclination degrees of freedom can be expressed as a single rotation or tilt (ω) about an "effective tilt axis" (T in Fig. 3), which is

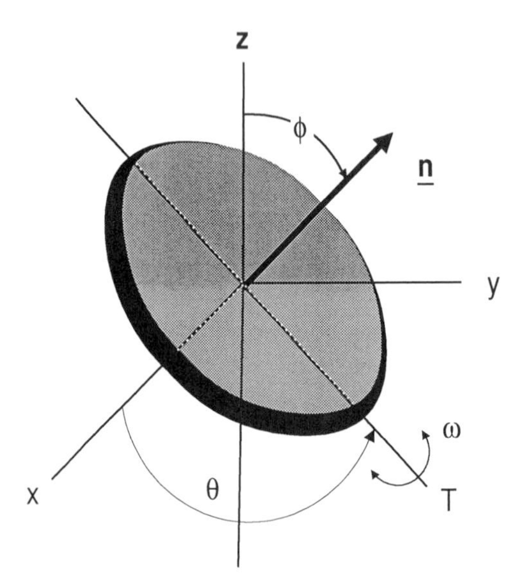

FIGURE 3. The five degrees of freedom required for defining the orientation of the specimen. x, y, and z are the translation axes, θ and ϕ are the azimuthal and longitudinal coordinates, respectively, θ is the angle between the x axis and an arbitrary direction within the specimen plane, **n** is the specimen normal unit vector, and ω is the angle of rotation about the effective tilt axis T.

independent of the stage's mechanical inclination axes. As a matter of definition ω denotes the tilt angle for all geometries, including single tilt and the situation where either ϕ or θ equal zero. Any specimen orientation can be obtained by performing appropriate rotations about the two axes. This approach is particularly useful for double-tilt mechanisms and can be used to position the specimen in a series of orientations corresponding to a conical data set (Section 1.2.2). The rotation motion of a tilt rotation mechanism can also be used to position structures relative to a tilt axes by direct observation. However, vector analysis can be effective in both stage types for viewing particular details after an initial reconstruction has been analyzed. The greater efficiency of positioning would minimize viewing time, contamination, and radiation damage.

An ideal tilt stage should fulfill the following requirements:

1. Highest tilt angle in all directions. The highest practical tilt angle reported is 70° (Turner *et al.*, 1989a), although a stage capable of 90° has been described (Chalcroft and Davey, 1984). As the angle of tilt increases the sample thickness penetrated by the beam increases; for $\omega = 60°$ the effective thickness is twice that of an untilted specimen and is 2.9 times for $\omega = 70°$. Special clamping devices must be adopted to avoid beam obstruction by the holder at high tilts.
2. High resolution. The stage should not reduce the resolution obtained by the instrument. However, due to increase in effective thickness, the image resolution is adversely affected at high tilt angles by multiple and inelastic scattering, unless energy filtration is performed.
3. High accuracy. The image of the specimen during tilting should move smoothly, continuously, and reproducibly up to at least 100,000 × magnification. There should be no backlash, inertial effects, or coupling to other motions (e.g., traverse, double-tilt, or rotation).
4. No specimen shift. The specimen should not translate nor change height as a function of tilt. The image of the observed area should only appear to shrink or expand in a direction normal to the effective tilt axis. Complete elimination of image shift facilitates the work of the microscopist not only during observation but also in subsequent image processing, because magnification, instrumental resolution, diffraction camera length, and rotation between image and diffraction pattern all remain constant. Of course, it is possible to account for these changes, but the necessary corrections would be tedious and time consuming, resulting in a buildup of contamination and radiation damage during long exposures.
5. Adjustment in z direction. Compensation for local variations of specimen height is desirable (see also Section 2.4).
6. Angular readout. All motions including those about any effective tilt axis should be linearly related to the readout.
7. Robustness. Little maintenance should be required.
8. Compatibility. There should be no restriction of the traverse movement compared to that of standard stages. The device should accept a wide range of specimen sizes and thicknesses, and should operate with an airlock, an anticontaminator, and a low-dose accessory.

Such an ideal stage does not exist. However, the next section will show to what extent practical stages meet these requirements.

There are three possible locations for specimen stages, the top plate of the objective lens (top stages), the objective pole piece gap (gap stages), and outside the column (appended stages). The first has the advantage of leaving more room in the region of the specimen holder for special treatments (e.g., for cooling and level adjustment). The second is helpful when a narrow but long space is required (e.g., for tensile deformation). In the appended stage, the mechanism is outside the column and the specimen is mounted at the end of a much smaller mechanism inside the gap. This last arrangement can be prone to vibration problems.

For top stages, the specimen holder is inserted into the stage from above the objective lens and for gap stages from its side. In the first case, the holders are called *cartridges*, and in the second they are referred to as *rods*, *specimen rods*, or *side-entry rods*. In some cases, the tip of the rod where the specimen seats is detachable and is "injected" into the stage.

2.2. Two Axes of Tilt

The most common tilting geometry is obtained with two perpendicular tilt axes built inside the traverse stage. The tilt axes are placed at 45° with respect to the traverse axes to allow incorporation of the actuating devices (tilt controllers). To access the specimen region for other purposes (e.g., decontaminator, x-ray detector), the tilt controllers may be incorporated into one unit (twin drive), which is placed in one objective lens port. This applies to both cartridges and rods.

For two axes a direct measurement of the tilt angles is not generally possible unless formulae relating the (angular) readout of the tilt controllers with the effective tilt angle are used (Turner *et al.*, 1989a). For crystalline specimens, direct tilt readout is usually not required since the specimen orientation can be accurately determined through diffraction patterns or Kikuchi lines (the latter applies to relatively thick and not heavily deformed specimens). However, most biological work requires direct knowledge of the tilt angles, and consequently, high-precision tilt stages have been developed (Turner *et al.*, 1988). Computer-controlled tilt stages have been constructed for monitoring and controlling specimen orientation (MacKenzie and Bergensten, 1989).

2.3. Tilt Rotation

The combination of tilting around one axis with rotation about an axis perpendicular to the specimen plane can be used to orient the specimen about any effective tilt axis. If the tilt axis is fixed with respect to the specimen holder (or to the traverse stage), specimen orientation motions result in changes in specimen position (x, y), rotation, and hight (Fig. 4). Image rotation is, of course, always present between tilted images except for stereo pairs once the stereo axis has been chosen. The different orientations of the images with respect to the operator at various tilts is sometimes confusing (Figs. 1 and 4), and for crystalline specimens, it is difficult to recognize the same area for work using reciprocal lattice vectors.

Rotation-free images can, however, be obtained for any specified specimen

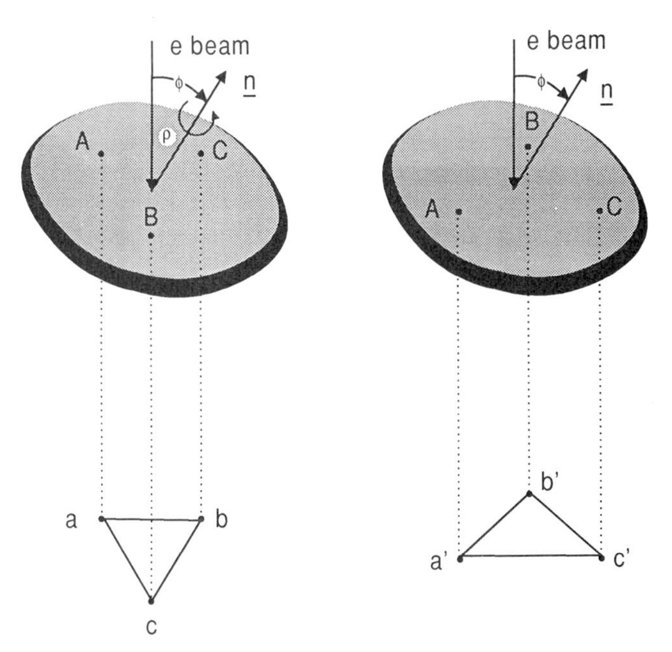

FIGURE 4. Change of orientation and relative distances of the images of specimen points *A*, *B*, and *C* upon performing tilt and rotation. Projected triangle *A*, *B*, *C* becomes *A'*, *B'*, *C'*. The electron (*e*) beam coincides with the *z* axis. ρ is the angle of rotation about the normal **n** to the specimen plane. ϕ is the angle between **n** and the optical axis.

orientation by rotation of the tilt axis (instead of the specimen) around the optical axis, followed by tilting about the newly oriented tilt axis. This requires decoupling the tilting device from the specimen holder, rotating the holder, and then reengaging it. This process is discontinuous and tedious and has been pursued only in a prototype (Lucas *et al.*, 1963).

The possibility of combining double tilting around two orthogonal axes with rotation has also been explored in the case where the rotation axis is perpendicular to the specimen plane (Valdrè, 1968); it is obtained by rotating the external support of a gimbal around the optical axis while ϕ (defined in Fig. 4) and ω are kept constant. There is no change of tilt during specimen rotation about any effective axis. This type of stage is particularly useful for aligning specimens with respect to a deflecting (electric or magnetic) field, a biprism for holography, a slit, or a row of thin sections cut from biological material with respect to the photographic plate.

2.4. Height Adjustment

A stage performing z motion is called a z stage or a lifting stage. Since tilting usually alters the specimen height (noneucentric stages, Section 2.5), it is necessary to reset the normal level. Refocusing after height changes affects magnification, rotation of the image with respect to the diffraction pattern, field of view, and instrumental resolution. The former two can be compensated in the photographic process. Irrespective of the general effect of tilt on specimen height, stages should be designed with provisions for centering the tilt axes (i.e., the tilt axes should intersect and their common point should lie in the specimen plane). Usually one relies on mechanical accuracy, which in practice is not entirely satisfactory. Furthermore, height adjustment is needed to account for specimens of irregular shape and for local level changes due to buckling. Other reasons for the adoption of a z stage are flexibility in focal length, by increasing the excitation of the objective lens, improving resolution, low-angle electron diffraction, and special conditions (e.g., cooling).

A z stage may be built either in the transverse stage or in the specimen holder. Large z movements, around a few centimeters, can be obtained only in top-entry stages. The central part of the traverse stage where the cartridge seats can be raised or lowered by external controllers via pinions and conical gears (Goringe and Valdrè, 1968; Valdrè, 1979; Sparrow *et al.*, 1984). In the case of side-entry stages, the z excursion is restricted to about 1 mm and is obtained by pivoting around the support of the rod.

One such cartridge stage combined double tilting with z motion operated from outside the microscope (Valdrè and Goringe, 1971); tilt by $\pm 25°$ and 7 mm of z traverse are provided by a gimbal arrangement whose tubular support is the means for changing the height. Some specimen holders incorporate provisions for the continuous adjustment of the specimen level to be performed with the specimen holder out of the microscope (Donovan *et al.*, 1983; Everett and Valdrè, 1985).

2.5. Axis-Centered and Eucentric Stages

The concept of eucentricity is often misunderstood. For any tilt stage, there are a number of important axes, two important planes and one important point. They are shown in Fig. 5 and discussed below.

- The objective lens axis, *ee*, is also referred to as the electron optical axis, the microscope (column) axis or the beam axis (in the case of untilted illumination).

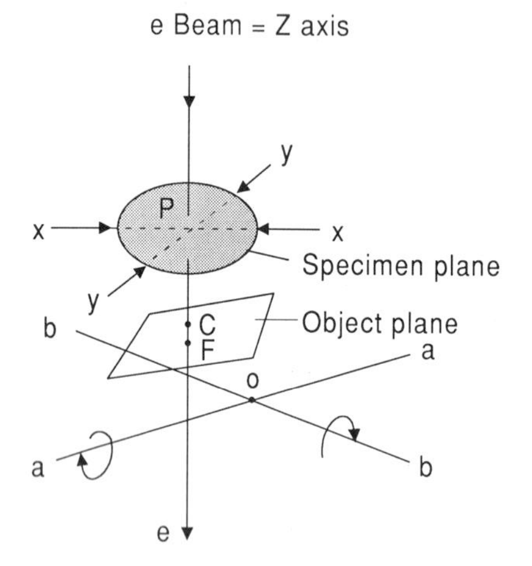

FIGURE 5. Essential geometric elements of a traverse and orientation stage. While *xx* and *yy* are the translation axes, *aa* and *bb*, the orthogonal tilt axes, intersect at *O*. *P* represents the region of specimen to be observed. For an ideal stage, *O* and *P* coincide at a point *C* on the optical axis defined by the mode of operation of the objective lens. *F* is the objective front focal point. The beam direction is parallel to the *z* axis.

- The two tilt axes, *aa* and *bb*, are (or should be) perpendicular to each other. The rotation axis of tilt rotation stages is considered a tilt axis. Ideally, the tilt axes intersect at a point O. This is not necessarily the case since even precision machining is not perfect, and an adjustment mechanism is suggested for work at high magnification. Ideally, they all intersect the microscope axis, but, again, in practice this is rarely the case. For side-entry rods, the tilt axis that is parallel (or nearly so) to the rod axis will be called the *principal tilt axis*.
- The two traverse axes, *xx* and *yy*, are not necessarily perpendicular to the microscope axis. The traverse stage translates the specimen. In practical cases, the traverse axes are either at 45° with respect to the tilt axes or nearly parallel to them. For side-entry rods the x axis is often the axis of the rod.
- The z or height adjustment axis is parallel to the microscope axis but not necessarily perpendicular to the (x, y) plane. A z motion is required to bring the specimen point of interest, P, into the objective lens object plane.
- The specimen plane is the beam exit surface of a flat specimen.
- The object plane is perpendicular to the optical axis, conjugate to the image plane, and close to the front focal plane of the objective.
- The point P imaged by the microscope is at the intersection of the beam axis and the specimen plane (point C of the object plane in normal operation), and is close to the front focal point F of the objective lens. Its image is formed on the optical axis at the center of the microscope fluorescent screen.

Eucentricity means that any region of the specimen can be observed and tilted without image shift or changes in focus. In principle, the following types of eucentricity may be achieved; each subsequent type includes the performance of the previous one and is more complex.

1. Point eucentricity. This applies to point O (Fig. 5), the intersection of the tilt axes, when it coincides with the image point P. Points O and P usually coincide with point C on the optical axis which defines the standard specimen level.
2. Line eucentricity. The specimen points lying on a line segment are eucentric. The segment is part of the tilt axis of stereo stages or of one tilt axis for double-tilt stages.
3. One-axis eucentricity, or axis-eucentric stages. Eucentricity is satisfied for all points of the specimen plane for one tilt axis. This has been achieved in side-entry stages.
4. Quasieucentricity. The tilt axes intersect, but the intersection point is not in the specimen plane.
5. Full eucentricity or eucentricity. All points in the specimen plane are eucentric irrespective of the tilt axis, e.g., for all effective tilt axes.

A tilt stage is eucentric if the traverse motions are within the tilting mechanisms. When this condition applies to only one tilt axis, the stage is axis-centered (number 3). Figure 6 shows schematically how this can be achieved in a side-entry stage. The axis of tube T defines the principal tilt axis. Inside T is a hollow shaft S whose inner hemispherical end seats on the slope of the tapered hole in T.

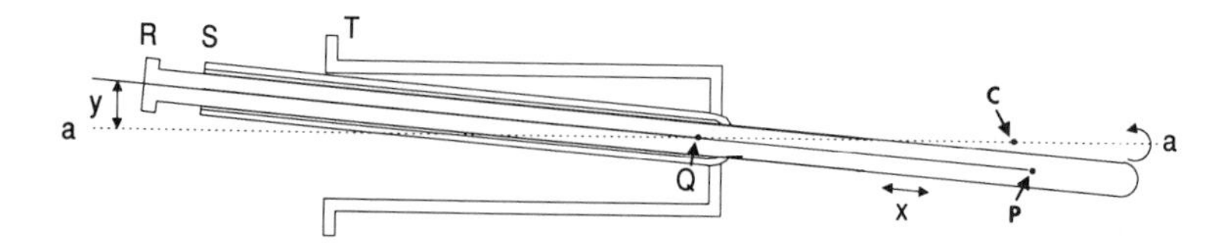

FIGURE 6. Schematic diagram of a practical side-entry axis-centered stage shown from above with the specimen traverse movements inside one tilt axis. The specimen rod R inserts into a tubular support S whose spherical internal end seats in a recess of the external tube T. By pivoting S about the point Q in orthogonal directions, y and z (perpendicular to the page and not shown) movements are obtained along arcs of circles. The rod R is advanced or withdrawn along its long axis to translate the specimen in the x direction. C is the intersection of the principal tilt axis with the optical axis. P is the specimen point to be imaged.

The shaft S is the coaxial support of the specimen rod R whose axis defines the x traverse, and S and R can pivot arount point Q. The free end of S can be moved perpendicularly to x along two orthogonal directions: y (in the plane of the figure) and z (perpendicular to the figure plane). Actually, y- and z traverse are arcs, but in practice are essentially straight lines. The principal tilt axis device incorporates x, y, z traverses, which makes any specimen point coincident with point C, the intersection of the optical axis and the principal tilt axis, *aa*. Tilting around *aa* will not produce image shift and loss of focus. This property is not shared, however, by the second axis, irrespective of whether it is a tilt or rotation axis. Axis-centered stages are available (Rakels *et al.*, 1968; Browning, 1974; Valle *et al.*, 1980). There are no axis-centered top-entry stages, because the small bore size of the top objective lens pole piece has discouraged their construction even for HVEM, where the bore is two to three times larger than for conventional instruments.

Only a few fully eucentric stages have been proposed (Leteurtre, 1972; Valdrè, 1967; Valdrè and Tsuno, 1986, 1988; Turner *et al.*, 1989a), but none of them are commercially available. A schematic drawing of a "mushroom" type eucentric stage is shown in Fig. 7. Tilting is obtained by sliding a concave platform, P, carrying the traverse stage, T, on a portion of a spherical surface, S, having its center at C. The specimen holder (H) of the cartridge type seats in the conical part of T. External means were contemplated to adjust the specimen height at the level of C. This design is feasible for an HVEM, but the maximum tilt angle in any direction of only

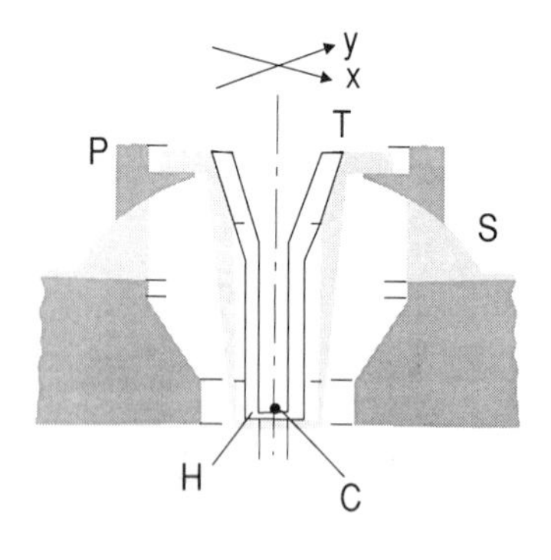

FIGURE 7. Schematic drawing of a top-entry cartridge goniometer with a traverse stage built inside. Ideally, the center, C, of the portion of the spherical surface S should lie in the specimen plane. T is the traverse stage, H is the specimen holder, and P is the tiltable platform.

15° (limited by geometrical factors) is too small. An approach to eucentricity is presented in Section 3.3.5.

2.6. Comparison

The various proposed solutions are compared on the basis of the requirements listed in Section 2.1 for an ideal inclination stage. Points 7 and 8 are usually met by all stages with the exception of the specimen size and traverse, which are in mutual conflict in top-entry holders. Either the traverse is reduced or, more typically, the specimen size is limited to a diameter of 2.3 mm by the bore diameter.

In side-entry stages, an axis of tilt is readily available, frequently with no restriction in the rotation of the rod; the useful tilt angle is, however, limited by design parameters such as material strength, specimen clamping, and the decontaminator blades. The second axis of tilt (or rotation) can be operated in a simple way (Section 3.2), and the mechanical angle of tilt may be $\pm 180°$. Reading of the principal tilt angle is direct, and a linear relationship is easily achieved for the second tilt angle. The effective tilt axis and orientation has to be calculated (Turner *et al.*, 1989a, b). The accuracy may be good, since the controllers operate either directly or via a small number of intermediate components. A precision of $\pm 0.06°$ has been obtained with such a design (Turner *et al.*, 1988). Designs based on friction for the second tilt may suffer from lack of reproducibility (by slipping contacts), or from backlash or rough motion. The specimen is susceptible to mechanical (mainly vibrations) and thermal instabilities due to the length of the rod, lack of symmetry with respect to the microscope axis, and the fact that the external tilt drives act directly on the specimen holder. Great attention must be paid to this problem for high-resolution stages. In fact, improvements have been obtained by constructing the rod from two detachable parts: only the tip carrying the specimen is loaded in the stage and is tilted by internal means. Then z motion is restricted to about 1 mm. As already mentioned, axis-centered stages have been satisfactorily developed for side-entry stages, although their cost is over 10 times that of line eucentric stages.

Top-entry cartridges do not suffer as much from vibration problems, have a symmetrical structure, and therefore, reach high resolution with less difficulty than the rod-entry counterparts. The ultimate resolution of the microscope is maintained (Bursill *et al.*, 1979; Donovan *et al.*, 1983). The maximum tilt angle is about 60° around both axes for double-tilt holders. In the case of cartridges designed for HVEM, full rotation around one axis is available. Because of the complexity of the tilt drive mechanism, direct angular readout is not possible, although linearity may be nearly satisfied. In addition, top-entry stages are more prone to play, backlash, and jerking motion. They are less accurate and less robust than the rod-entry stages. However, the specimen level can be changed easily by large amounts (several centimeters) (Section 2.4).

Tilt rotation holders are not popular because of the change in specimen orientation, although for some applications they are ideal, such as for stereo images (Section 2.3). They are more easily incorporated in rods than in cartridges, and direct reading is possible in the former case.

The best solution from the user's point of view is to make both top- and side-entry stages available and to select the appropriate specimen holder for each application. In some cases, it might be necessary to use both facilities, such as for a cold stage combined with an environmental cell (Tatlock *et al.*, 1980) (Sections 3.3.2 and 3.3.4) or for eucentric stages (Section 3.3.5) (Valdrè and Tsuno, 1986, 1988).

3. WORKING MECHANISMS AND MULTIPURPOSE HOLDERS

The practical application of the two basic principles for inclining a specimen (i.e., double tilt and tilt rotation) is performed in different ways, depending on the design used to provide the various mechanical axes and on the mechanisms used to produce and to transfer the movement (e.g., linear or rotation motions, positive control in forward and reverse tilt, etc.). The various solutions are a consequence of both the space available in the pole-piece region, and of the type of traverse stage chosen. Solutions suitable for an HVEM may not be appropriate for conventional (100 kV) microscopes because of space restrictions. Mechanisms suitable for cartridges may not be applicable to rods, and vice versa. Furthermore, the type of traverse stage and the available provisions for tilt control influence the design.

Below are a few examples selected for one or more of the following features: high tilt angle, simplicity of construction, accuracy, avoidance of backlash, resolution, and direct reading of the tilt angles.

3.1. Practical Top-Entry Designs

It is fair to say that practically all present-day cartridges utilize two mutually perpendicular axes of tilt. Tilt rotation, in addition to not being popular (Sections 2.3 and 2.6), is not simple to achieve in a cartridge. The two tilt axes can be realized with various kinematics, and can be classified into those that are "materialized," that is, the axes correspond to shafts or pins and are bound to their bearings, and those where the tilt axes are not physically identifiable.

Examples of the latter case are devices in which the tilting body carrying the specimen is a portion of a sphere that seats either on a spherical or in a conical bearing, or in a tubular support (Figs. 6 and 7) (Valdrè, 1962; Barnes and Warner, 1962; Valdrè *et al.*, 1970; Everett and Valdrè, 1985). These devices do not allow large effective tilt angles ($\leqslant 30°$), particularly in regions far from the center of the specimen holder. They require angular calibration, and the lack of shafts affects their accuracy. However, devices based on a tailed ball elastically retained in a spherical seat are very robust (Valdrè, 1962; Valdrè and Goringe, 1971) and have been efficiently applied to the study of crystalline specimens.

A typical example of physical tilt axes is provided by the gimbal arrangement. The gimbal, or universal suspension, is composed of a thin ring (R) (Fig. 8) whose axis of tilt is fixed and of a specimen platform (P) supported on pins by R. The axes of tilt are materialized either by minishafts, pins, or bearings made of hard stones (ruby or sapphire rods or balls). The shafts are spring loaded against V shaped grooves. Tilt may be obtained by means of two pairs of thin, braided wires anchored at opposite sides of both R and P. One wire of each pair is positively

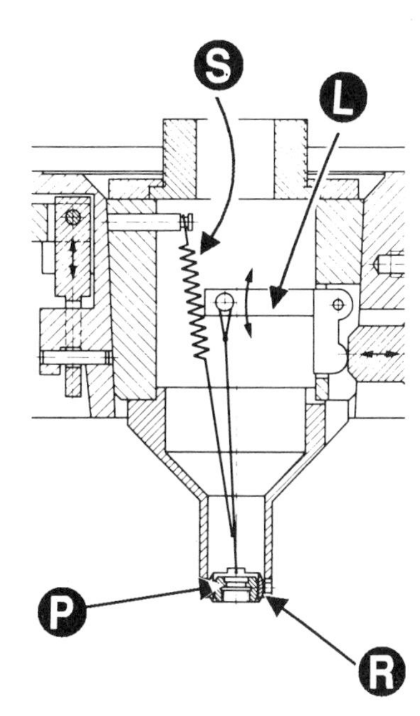

FIGURE 8. Example of a tilting cartridge of the gimbal type. *P* is the specimen platform, *R* the gimbal ring, *L* the actuating lever, and *S* the return spring. The second tilt controller is at 90° (not shown).

controlled by a lever (*L*) operated by micrometers outside the microscope. The other end of the second wire of each pair is fixed to the cartridge body via a coil spring, which provides the return motion and holds the shafts in position. The mechanism is simple, the movement is smooth, and the tilt angle can be as high as $\pm 60°$. For large pole-piece bores ($\geqslant 1$ cm), the tilt of the gimbal may be produced by the linear motion of elastically loaded strips engaged in recesses or grooves in the tiltable gimbal components. The connections between the strips and the gimbal are made via hard stones. These cartridges are delicate and easily damaged.

Another way to achieve tilting is to use translational movements in push-pull (Ward, 1965). Two pairs of strips or rods replace the wires and spring and independently drive the platform (*P*) that carries the specimen (Mills and Moodie, 1968). The linear motion of the strips is positive in both directions. The system is somewhat complex, and balancing the action of the pairs of strips is difficult. Although somewhat inaccurate, this device produces tilt angles of up to 40°, and in particular it provides a large clear solid angle to the specimen. The latter property is shared also by the hemisphere-on-tube kinematics (Everett and Valdrè, 1985) where a 2π free solid angle is available. This is an advantage for the collection of signals emitted from the specimen (Section 3.3.1).

A variant to the push-pull device is the replacement of the rods with sprung leaves coupled to the platform by means of ruby ball bearings (Donovan *et al.*, 1983). This arrangement improves the kinematics of motion between *P* and the tilt controllers.

3.2. Practical Side-Entry Designs

The protrusion of the specimen rod outside the microscope provides direct control of the principal tilt and a simple means of incorporating the second axis of tilt or rotation. In addition, the space available in the pole-piece gap allows the use of multispecimen rods (Section 3.3.1). In an HVEM, owing to the large size of the rods, the specimen can be manipulated through the rod tip, which terminates with a hollow tip supported by the *xy* stage (Swann, 1979). Controllers from outside the microscope enter the specimen region via the tip hole. A high-tilt ($\pm 70°$), high-precision ($\pm 0.06°$) single-tilt stage with *z* adjustment and 0.6 nm stability has been constructed (Turner *et al.*, 1988). It is important to note that the high tilt angle is obtained by mounting the specimen at the thin end of a cantilever. For single-tilt and thin supports, angles as high as $\pm 87°$ have been achieved (Horn *et al.*, 1991). This concept seems to be the only one that allows high tilt angles (Chalcroft and Davey, 1984), and it may be possible to incorporate a second high-tilt axis in the cantilever. The maximum tilt angle is limited by the specimen support and clamping mechanisms, if the increase in effective specimen thickness can be ignored.

In the case of tilt rotation, the specimen is mounted in a circular turret (in practice a pulley) (Fig. 9). Rotation is achieved by a spring-loaded tape, or braided cable, which is wound in the groove of the pulley. When the cable is pulled from outside the microscope, the specimen rotation is always about the specimen normal; the return movement is provided by a spring. The turret is trapped on top and bottom in a recess, and may be held against two pins to reduce friction and backlash (Fig. 9) (Turner, 1981). The system relies on the friction between cable and pulley being greater than that between the turret and its supports; it is, therefore, susceptible to slippage, which can alter the reading of the angle of rotation.

Positive drive in both directions of rotation is possible with a spur gear operated by a rack gear (Turner *et al.*, 1986). The latter is the specimen turret, and is held by two blocks, one of which is spring loaded. This arrangement minimizes friction and backlash. The rotation is operated via a plunger and a lever though the hollow tip of the rod (Swann, 1979). A stability of at least 0.6 nm has been demonstrated with rotational accuracy of $\pm 1.5°$.

The majority of double-tilt stages incline the specimen turret (*T*) (Fig. 10) by tilting around an axis (*bb*) that is perpendicular to the rod axis (*aa*) and lies in the specimen plane (Swann, 1972, 1979; Allinson and Kisch, 1972). The second tilt can be performed either by using the rod tip or the rod tail. Motion is usually coupled to the specimen turret by friction or a lever. In the first case, two rails advanced by an external controller rotate *T* about its axles so as to convert linear motion into specimen tilt. Two pulleylike grooved wings can be added to the turret and the rails spring loaded against them, providing better traction. The tilt angle is linearly

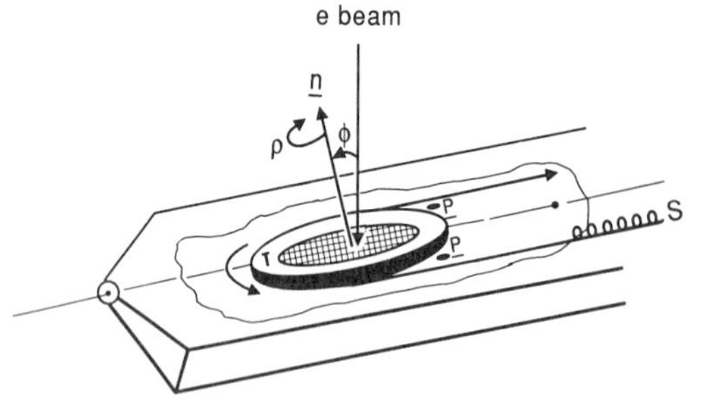

FIGURE 9. Schematic of a side-entry single-tilt and rotation stage. **n** is the specimen normal, ρ is the angle of rotation, ϕ is the longitudinal (tilt) angle. The rotating turret *T* is held by the retaining pins *P*. *S* is the return spring (Modified from Turner, 1981.)

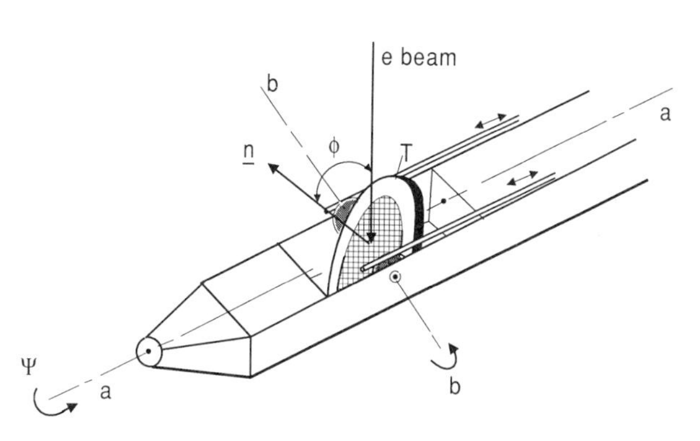

FIGURE 10. Schematic of a side-entry double-tilt specimen rod. *aa* is the principal tilt axis and corresponds to the long axis of the rod. *bb* is the second tilt axis. Ψ is the principal tilt angle, ϕ is the second tilt angle, and **n** is the specimen normal. The sliding rods, actuated by mechanisms incorporated in either end of the rod, tilt the specimen about the axis *bb* by rotating the wheels attached to the turret *T* via the axle along *bb* (Modified from Turner, 1981.)

related to the controller movement unless slipping occurs between the rails and shafts. A full 360° of rotation is available which can be useful in a scanning transmission instrument for observing specimen surfaces using secondary electrons (Craven and Valdrè, 1979). (In such cases, the top and bottom specimen surface structures could be correlated with the 3D reconstruction from tomography.) In the lever-coupled design, a small protrusion on the specimen turret perpendicular to the second tilt axis is coupled to a shaft pushed in and out by actuators outside the microscope. Alternatively, the protrusion engages a lever whose axis is normal to the principal tilt (rod) axis; the lever is tapered and rotated by a tapered shaft whose motion along the rod axis is controlled by a micrometer. The return motion in both designs is provided by springs.

Vibration coupling to long rods causes specimen instability and image degradation. Mounting the tip with a flexible coupling can improve stability by a factor of 3 to 10 while still providing the principal tilt (Turner and Ratkowski, 1982). However, a detachable tip is normally used for high-resolution imaging. In both cases, the conical end of the tip is forced into the seat by a clamping arrangement. The conical connection and clamp allow rotation of the tip for tilting. This concept was first used for cartridges where gravity seats the mechanism.

3.3. *Combination Tilting Holders*

Combination tilting holders are those that perform one or more functions in addition to single tilt. Double tilting, rotation, and double tilting and lifting stages are discussed in Sections 2.3 and 2.4. Other combination holders of interest to biologists incorporate multiple specimens, beam-current measurements, environmental control, cooling, or straining. The last is probably the most important. Also included in this category are tilting stages modified for the detection of radiation emitted by the specimen, namely x-rays, secondary electrons, and light (cathodoluminescence).

3.3.1. Multispecimen and Signal-Compatible Holders

Several specimens and a Faraday cup for beam-current measurements can be incorporated at multiple positions along the rod. For speed and convenience, a detent mechanism in the rod usually indexes the specimen or Faraday cup under the beam. This action can also be performed by a traverse stage if it has sufficient range of motion. An advantage for tomography may be the tilting of two or more similar specimens that are oriented differently with respect to the microscope. Angles not accessible in one specimen might be explored in the other. Rods carrying up to three specimens, or two specimens and a Faraday cup, have been constructed. The latter is important for measuring the dose when observing radiation sensitive materials or for correcting quantitative x-ray microanalysis data.

While several specimens may be mounted in a rod-entry stage, multiple specimen mounting is practically impossible in cartridges, although they are possible for HVEMs due to their size. However, a cartridge with a Faraday cage has been built that operates by deflecting the electron beam into a thick part of the holder with one pair or two pairs of minicoils (Capiluppi and Valdrè, 1987). The cartridge and traverse stage have electrical contacts that engage when the cartridge is inserted, and contact resistance is minimized by a loading mechanism. The current measurement accuracy is approximately 1%. Although presently incorporated only in a fixed cartridge, this capability could be implemented in tilting cartridges, particularly those based on the "tailed" sphere design (Section 3.1).

Signal-compatible holders used in conjunction with tomographic methods could provide additional information, e.g. chemical composition and surface topography. Modified holders have been optimized for the collection of x rays (Stols *et al.*, 1986), cathodoluminescence signals (Pennycook *et al.*, 1980), or secondary and backscattered electrons (Bleloch, 1989). These functions are best implemented in the scanning mode of TEMs, or when a small probe is used for spatially resolved elemental analysis. However, the variation of the take-off angle as a function of tilt makes x-ray collection difficult. A miniaturized scanning tunneling microscope (STM) head has been mounted in a side-entry rod to study the surfaces of bulk specimens in conjunction with reflection microscopy (Spence, 1988; Spence *et al.*, 1990). Similar devices could be combined with TEM studies of thin specimens providing they are stable under the STM tip. The combination of STM topography and tomographic 3D reconstructions could result in a more complete characterization of the specimen than either method alone.

3.3.2. Tilting and Environmental Cell

Environmental cells allow specimens to be studied in a controlled environment (gaseous, or liquid) while allowing the beam to traverse the specimen and its surrounding fluid without excessive scattering. This requires the volume around the specimen where the environment is produced and maintained to be small. In a gaseous environment, there is little degradation of image resolution and contrast. However, a liquid layer adds sufficient mass thickness to substantially reduce image contrast, although it does not necessarily affect diffraction patterns from crystals (Parsons *et al.*, 1974).

Both sealed and differentially pumped cells have been developed. In the former, the specimen is placed between two apertures sealed with a thin, electron transparent film with thickness ranging from 25 to 100 nm, depending on the required pressure and accelerating voltage. A fluid inlet is provided between the windows whose separation can be varied from 100 μm to about 1 mm. In differentially pumped cells, two pairs of coaxial apertures create three spaces in the objective lens volume. Each space is pumped separately, the inner space (specimen region) is connected to the fluid supply and is at the highest pressure (up to 1 atm). The intermediate chambers are separately pumped by an added vacuum system, while the third volume is part of the conventional vacuum system. The size and spacing of the apertures are very important (Parsons *et al.*, 1974). The design and operation of both types of cells has recently been reviewed (Turner *et al.*, 1989a).

Tilting mechanisms have been incorporated in environmental cells (Barnard *et al.*, 1986), and one design includes both tilting and cooling (Tatlock *et al.*, 1980).

Environmental cells can be used to study biochemical reactions, specimen preparation techniques, and the structure of biological specimens under more "natural" conditions than conventional procedures. Specimens imaged in controlled environments are more faithfully perserved and are expected to provide improved 3D reconstructions.

3.3.3. Tilting and Straining

This type of stage has been developed for work in the science of materials and physical metallurgy; however, it may be applied in the biological and polymer sciences for the study of polymer fibers, bones, muscle fibers, and cytoskeletal elements. Imaging and tomographic analysis of these specimens as a function of mechanically applied stress and strain would provide valuable insights to their function. For many of these specimens, the tilting and straining holder would have to be combined with an environmental cell. Most straining stages are designed for HVEMs because deformed structures of thick specimens closely resemble bulk deformed material, and because of the relatively large space available in the pole-piece region. Objective lenses used with in-situ specimen stages are designed for maximum space, limiting resolution to 0.5 nm. The application of these methods to biological materials is limited by specimen preparation techniques. Reviews of the design and operation of straining stages are available (Butler and Hale, 1981; Turner *et al.*, 1989a).

3.3.4. Tilting and Cooling

Several reasons suggest that combined cooling and tilting stages are important for tomography:

1. The study of specimens close to their native state requires cryofixation and observation at low temperatures.
2. Correlations among structure, composition, and function can be enhanced by 3D reconstruction, with compositional and chemical information being derived from the microanalysis of rapidly frozen specimens.
3. Reduction or suppression of carbon contamination.
4. Reduction of damage, particularly mass losses.

The first two reasons are related to new trends in the investigation of biological structures, and the symbiosis of tomography and cryomicroscopy is already taking place.

Since the same specimen region is exposed for a long time to obtain a tomographic data set, the possibility of contamination is much higher than in conventional observations. The specimen must remain uncontaminated (mechanically, thermally, and chemically), and structurally unchanged for long periods of irradiation. Although a cold specimen environment reduces carbon contamination, cooling the specimen itself also suppresses carbon contamination and may even produce carbon removal (Heide, 1962; Valdrè and Horne, 1975).

As for point 4, it is well known that there is a maximum electron dose that a specimen can sustain before its structure is irretrievably damaged, and that this dose is dependent on the structure of the specimen. It is clearly impossible to perform tomography on specimens such that their exposure, for recording an image at a given resolution, is equal to that causing structural changes. The size d of the smallest detectable object detail is related to the dose D (in C cm^{-2}) for complete damage of the structure under specified conditions (specimen nature, accelerating voltage, temperature, substrate, etc.) by (Glaeser, 1975; Reimer, 1989):

$$d\,(\text{nm}) > \frac{K \cdot 10^{14}}{C(fD/e)^{1/2}}$$

which is indicative of the resolution achievable by accounting for the spatial noise due to limited counting statistics. K is the Rose constant, usually taken equal to 5, e is the electron charge in coulombs, and f is the fraction of electrons impinging on an area of size d^2 which formed the image. Here, $C = (N - N_0)/N_0$ is the contrast obtained with N electrons incident on the area d^2, and N_0 is the number of electrons in the background. A structure giving rise to weak contrast is considered detectable at $C = 0.05$. Taking $K = 5$, $f = 1$, and $C = 0.1$:

$$d\,(\text{nm}) > \frac{0.2}{\sqrt{D}}$$

Since tomography requires a number of images (n), the size d of the finest details that can be recorded from a specimen before it is visibly damaged will be at least $\sqrt{n}$ times greater than the size of a detail observable if a single image was recorded:

$$d\,(\text{nm}) > 0.2\sqrt{\frac{n}{D}} \qquad \text{with } D \text{ in C cm}^{-2}$$

or

$$d\,(\text{nm}) > 5\sqrt{\frac{n}{D}} \qquad \text{with } D \text{ in electrons Å}^{-2}$$

In terms of the magnification M required to image such a detail, if we assume the resolving power of the naked eye to be 0.2 mm, $dM = 2 \cdot 10^5$. Hence,

$$M < 10^6 \sqrt{\frac{D}{n}} \qquad (D \text{ in C cm}^{-2})$$

or

$$M < 4 \cdot 10^4 \sqrt{\frac{D}{n}} \qquad (D \text{ in e} \cdot \text{Å}^{-2})$$

For instance, in the case of uranyl-stained catalase and adenosine, $D = 10^{-2}\,\text{C cm}^{-2}$ and by taking $n = 9$, we obtain $d > 2$ nm and $M < 30{,}000\times$.

Obviously one would like to improve the resolution limit by decreasing radiation damage. One effective method for reducing mass loss is to lower the specimen temperature to that near liquid nitrogen. Little gain is, however, obtained on the preservation of the crystalline structures, but in some cases an increase of lifetime by a factor of 4 to 20 is observed at liquid helium temperature (Chiu *et al.*, 1981; Aoki *et al.*, 1986).

Superconducting objective lenses provide the ideal cooling environment, but only a small number of instruments fitted with such lenses exist due to the operating inconveniences (Zemlin *et al.*, 1985). In addition, the space in the pole-piece region is smaller than in conventional lens microscopes, and no double-tilt facilitates are available. A different approach is offered by double-tilt cooling stages (helium or nitrogen) that have been developed for both conventional and high-voltage instruments. The range of second-tilt angles in liquid nitrogen stages can be as large as for room temperature devices (Swann, 1979).

The basic design of a cooling attachment depends on the type of specimen stage available (i.e., top or gap stage), on the space available in the objective pole piece, and on the temperature range of operation. The coolants used are liquid or gaseous helium, and liquid nitrogen, whose boiling points at atmospheric pressure are 4.21 K and 77.36 K, respectively. The cooling fluid may be either confined to a reservoir attached to the microscope or circulated through suitable tubing passing close to the specimen. A reservoir placed either inside or just outside the specimen chamber is convenient when a constant temperature must be maintained for long periods without the need of refilling; it is also less prone to vibration produced by the boiling liquid than dynamic systems. Temperature and mechanical stabilities are essential for achieving good resolution. However, reservoir stages have the disadvantages of requiring a long time (as much as several hours) to reach equilibrium, and of having a long heat path if mounted outside the microscope, which increases the minimum obtainable temperature.

Direct continuous coolant flow produces high cooling rates (e.g., from 300 to 10 K in $\sim$3 min) (Valdrè, 1964), but at the expense of temperature and mechanical stability. Coolant (He in particular) circulation causes vibrations mainly due to the transition from the liquid to the gaseous state, which takes place by nucleate boiling with a large volume change. To reduce this effect, phase separators may be used, or cooling may be performed with cold gas evaporated from the liquid in the main

Dewar by a small heater. The latter results in a somewhat higher minimum specimen temperature. A better method is to use superfluid helium (Fujiyoshi, 1990).

The parameters affecting the cooling efficiency are the latent heat of evaporation and specific heat of the liquid, and the thermal conductivity and viscosity of both liquid and gas. The specific heat of helium, particularly in the gaseous phase, is high compared to that of nitrogen. However, the latent heat of evaporation is low, 10 times smaller than for N_2. Thus, the evaporative cooling effect of helium, which can be stimulated by reducing the pressure above the liquid, is rather poor. These facts plus low viscosity make helium more suitable for gaseous continuous flow systems. However, for reasons of mechanical stability given above, He cold stages of the reservoir type are frequently used. Cooling by nitrogen is better done via a reservoir due to its high latent heat of evaporation (which provides the dominant heat transfer mechanism), high viscosity (at least three times that of He), and rather inefficient gaseous cooling due to its low specific heat.

Each stage design should ideally fulfill the following requirements:

- Specimen temperature should have the widest possible range. This depends on the coolant used; some He stages are designed with the possibility of pumping He so that the temperature in the reservoir can reach about 2 K. The use of liquid helium may not always be necessary, since liquid N_2 is usually sufficient to satisfy points 1–3. However, the high heat impedance added by a second tilt mechanism may increase the specimen temperature above that required to prevent sublimation or rearrangement of components.
- The chosen temperature should remain constant over the time period of the experiment. This is achieved with a low-power heater or by varying the flow of gaseous coolant.
- Specimen temperature should be measured as directly as possible, usually with a thermocouple or a resistor close to the sample and in intimate contact with the holder. Thermocouples of Au with 0.03 at.% of Fe versus chromel are used for liquid helium, and Cu versus constantan for liquid N_2. Small resistors (C, Ge) are also used; in this case it is important to account for the resistor's power dissipation. However, the actual temperature of the observed area can be measured only at specific temperatures, corresponding either to the condensation of gases under the electron beam, or to the temperature of phase transformations.
- The cooldown to equilibrium temperature should be fast, typically being from 3 to 60 min.
- Resolution should not be compromised. Factors affecting resolution are thermal drifts or instabilities (related either to heat transfer or to asymmetries in the stage), condensation of residual vapors on the specimen, and mechanical vibrations from circulating and boiling coolant. A resolution of 0.2 nm has been obtained with a reservoir-type stage (Fujiyoshi *et al.*, 1986; Fujiyoshi, 1990).
- Tilting facilities should have the highest possible angular range. A stage with $\pm 30^\circ$ along the rod and $\pm 60^\circ$ about the orthogonal axis has been developed (Swann, 1979).
- Specimen exchange via a cryotransfer device.

- Compatibility with normal use of the microscope. This is rarely achieved except in side-entry stages.
- Low coolant consumption for economy (particularly for He) and infrequent reservoir refilling. Consumption at the equilibrium temperature is 1.2 l of liquid He per hour for continuous-flow stages and 0.2–0.8 L h^{-1} for reservoir stages (Butler and Hale, 1981).
- Absence of vapor and carbon contamination. Good vacuum, clean surfaces, and a cold trap minimize these problems. The exhaust fluid is used to cool a thermal screen shielding the reservoir and the specimen stage from room-temperature radiation. This is particularly important for liquid helium operation.

The selection of the materials used is very important. The reservoir and holder should have high thermal conductivity and good machining properties, e.g., copper with tellurium. Surfaces should be smooth and gold plated to reflect infrared radiation. The number of components and mechanical contacts should be minimized, and all detachable parts should have large contact surfaces. Heat conductance between parts can be improved by a thin layer of vacuum grease, or by soldering with a low-melting-point alloy (e.g., In-Ga). Alternatively, a good contact is obtained by sandwiching a thin foil of soft metal such as indium between the parts.

Flexible couplings (e.g., bellows, braids, convolutions, or loops) avoid excessive mechanical stress to components without free ends as a result of temperature variations, particularly when large dimensions are involved. This also allows stage movement while effectively damping vibrations produced by the circulating coolant or by its boiling in the reservoir. Coolant transfer lines must be loosely coupled to the instrument.

Supports should be insulating materials such as nylon (which remains reasonably soft at liquid He temperatures), PTFE (Teflon), PVC, fused quartz, or sapphire to provide a thermal barrier between the stage and the microscope. The contact surfaces should be small, clean and rough.

Although the resolution was limited (3 nm), double-tilting liquid helium stages were used to perform diffraction contrast experiments on specimens undergoing phase transitions at low temperature (Valdrè, 1964). This device featured low specimen temperature (10 K), simple construction, easy operation and assembly, and fast cooling (3 min). The incorporation of cryotransfer devices took place later in conjunction with the application of cryotechniques for specimen preparation and microanalysis. A number of improved stages were subsequently built (Butler and Hale, 1981). The best performance is obtained with top-entry reservoir stages designed for intermediate-voltage electron microscopes (IVEMs) or HVEMs. Top-entry designs are thermally and mechanically more stable owing to their symmetry and are less prone to vibrations (Heide and Urban, 1972; Aoki *et al.*, 1986). The former used in an HVEM had 1 nm resolution at 8 K, a transfer device, and double tilting up to $\pm 30°$. A cross section of Fujiyoshi's (1990) very successful liquid helium stage is shown in Fig. 11. The stage consists of a liquid nitrogen reservoir, N (1400 cm^3), that cools the shield surrounding the specimen chamber, a liquid He reservoir, H (1200 cm^3), a pot, P (3 cm^3), connected to H via a capillary tube, T, having an empirically optimized impedance, and a specimen cartridge, C, seating in

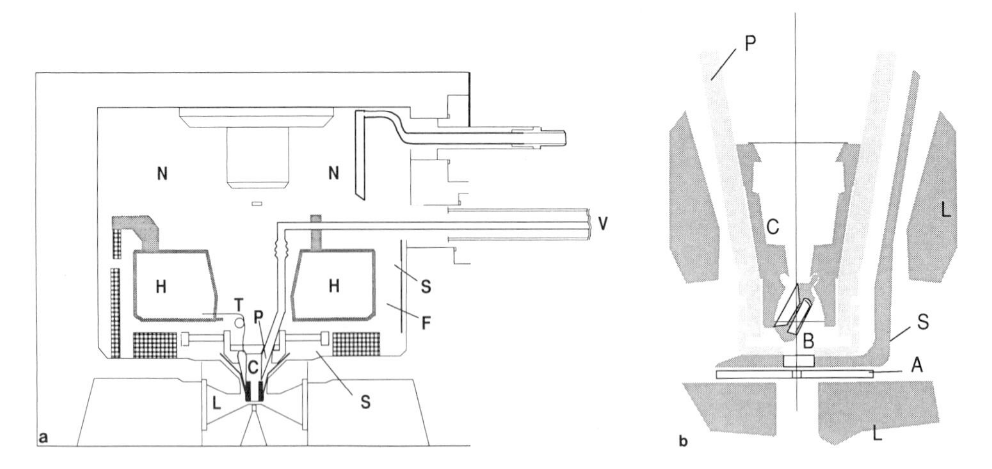

FIGURE 11. Cross section of a cold stage capable of working with superfluid helium. (a) General view but with air lock and liquid-nitrogen-cooled transfer device not shown. *N*, liquid nitrogen reservoir; *H*, liquid helium reservoir; *P*, pot; *T*, capillary tube; *C*, cartridge; *S*, thermal shield; *L*, objective lens; *V*, pumping line; *F*, copper braid. (b) Detail of pot, *P*, with conical seating for cartridge, *C*. *B*, spherically shaped specimen platforms; *S*, thermal shield; *A*, objective aperture; *L*, objective lens. (Courtesy of Y. Fujiyoshi, Protein Engineering Research Institute, Osaka, Japan.)

the pot. Helium in the pot is pumped to produce a superfluid cooling the pot to 1.5 K. The specimen temperature is 2.6 K under weak beam illumination. The superfluid improves mechanical stability since its evaporation causes no vibrations. To efficiently remove heat, the specimen is deposited on gold-coated microgrids (50-nm average hole diameter), which are held by silver grids (100-μm holes) that are glued to the specimen platform, *B* (Fig. 11*b*). The platform seats in a spherical recess in the cartridge and can be positioned at angles up to 60° before insertion in the microscope. In-situ specimen tilting is planned (Y. Fujiyoshi, personal communication). All contact surfaces have been lapped to increase the area of contact. This solution is better than the use of screws and more practical than low-temperature solders. The cool down time of the entire cryostat to 1.5 K is 7 h, and the equilibrium He consumption rate is 140 $cm^3 h^{-1}$. One He charge lasts more than 7 h. The specimen exchange time is 10 min. If the cartridge is precooled to 100 K, the pot temperature (as measured by a Ge resistance thermometer) increases from 1.5 to 30 K upon cartridge insertion, but returns to 1.5 K in 10 s. The pot temperature is constant to within 0.01 K at 1.15 K with the beam off. A resolution of 0.2 nm has been demonstrated at 4.2 K and 400 kV. At 8 K or less, the radiation damage of tRNA crystals was found to be reduced by a factor of 20, compared to that at room temperature (Aoki *et al.*, 1986).

Side-entry double-tilt stages sacrifice resolution for quick cooling, simplicity, and low cost. An early stage for an HVEM with $\pm 30°$ tilt about the rod axis and $\pm 60°$ about the transverse axis operated at -160 °C. The design was similar to that shown in Fig. 10 (Swann and Lloyd, 1974; Swann, 1979). A highly flexible copper braid connected the heat exchanger to the cradle, preventing vibrations and thermal contraction of coolant pipes from affecting the stability of the holder.

3.3.5. Eucentric Goniometer Stages: Integrated Design Approach

The most challenging goal for designers of electron microscope stages is to build a highly stable eucentric stage with high tilt angles (65° or greater). This is an ambitious task that can only be attained by designing the stage and objective lens in concert. It is not acceptable to limit stage function to achieve high resolution or to reduce resolution to enhance specimen manipulation (Valdrè and Tsuno, 1986, 1988). This design utilizes an upper pole piece with a wide bore (40 mm) and a relatively large gap (12 mm). The optics have been optimized by computer methods to produce high-resolution capability (0.3 nm at 200 kV) while maximizing the available space for the stage. Because every cubic millimeter is critical the stage is a combination of top- and side-entry mechanisms. This tentative design contemplates the insertion of a specimen holder via a side-entry air lock in the central body of a universal suspension by a clamp-and-release device. The normal *xy* translations are within the tilting body. The nine stage controllers act from the chamber above the upper pole piece, five being used to set the stage (i.e., to bring the point of intersection of the tilt axes to the optical axis and to adjust the specimen level to that point) and four for tilt and translation. The eucentric stage allows a 360° rotation around one tilt axis (useful tilt of $\pm 45°$) and $\pm 23°$ about the orthogonal tilt axis.

The stage is based on a miniaturized mechanical design and is, therefore, rather complex. The construction may be simplified by the adaption of electrical transducers and unorthodox mechanical approaches. However, this stage represents the first example of a collaborative approach between specialists in electron optics and stages in which the magnetic field of the objective lens has been shaped around a specified volume. It is also the first attempt to solve the longstanding problem of realizing a fully eucentric high-tilt goniometer.

4. FUTURE PERSPECTIVES AND CONCLUSIONS

The exploitation of electron tomography for the study of cells, cell components, and tissues requires the construction of specimen stages capable of tilting the sample to the highest possible angle in all directions with respect to the electron beam (Turner *et al.*, 1989b, 1990). The stage should be reproducible, accurate, easy to operate, work under constant electron optical parameters, and be quantitative (i.e., the specimen position and tilt should be measurable). All of these features should be incorporated in a eucentric goniometer. Due to the increased projected specimen thickness as a function of tilt, an HVEM is the instrument of choice for many specimens. Due to the larger volumes available in HVEMs, the constraints related to stage construction are less severe, making practical solutions for eucentricity feasible.

As tomography evolves toward higher resolution, other problems are encountered. One of these is the effect of specimen preparation (fixation, staining, embedding) on the preservation of structures. Cryopreparation offers the best way to avoid the artifacts introduced by conventional methods, which become noticeable at high resolution. Thus, the need for double-tilt cold stages (liquid nitrogen being acceptable in most cases) is expected to expand. The combination of tomography and analytical information from x-ray microanalysis, electron energy loss spectroscopy, backscattered electrons, or cathodoluminescence would be an important development and would also require cryotechniques to preserve chemical gradients, particularly those of diffusible ions. If the difficult development problems can be overcome, a powerful new method of analytical tomography will bring an additional dimension to the elucidation of structure and function. Since radiation damage will continue to be a serious problem, a future development is expected to be a double-tilt liquid (superfluid) helium stage with analytical capability.

Although high-tilt angle stages allow the recording of most specimen projections, some perspectives are missed. In addition, the specimen thickness along the beam varies as a function of tilt. Thus, the quality of the individual images of a tomographic data set vary due to multiple and inelastic scattering. Some specimen regions may be unsuitable for the collection of tomographic data, because some projections may be obscured by nearby structures overlapping the area of interest. A stage and specimen mounting geometry with cylindrical symmetry is being explored (Turner *et al.*, 1990). The prototype version uses a glass micropipette to hold the specimen. The pipette is held in a bobbin in an HVEM side-entry rod so that its long axis corresponds to the long axis of the rod (perpendicular to the beam). This arrangement has unlimited tilt, and the 1-MV beam can penetrate the thin glass wall forming an image of the specimen inside (Sachs and Song, 1987; Ruknudin *et al.*, 1991). This stage has been used to record a tomographic data set, which was used to compare reconstructions with missing wedges of different angular widths to the reconstruction formed using all perspectives (Barnard *et al.*, forthcoming; Turner *et al.*, 1991). The cylindrical symmetric approach holds promise for the future of tomography, but requires new specimen preparation procedures.

ACKNOWLEDGMENTS

This work was partially supported by the New York State Department of Health and United States Public Health Services grant RR01219 from the National Center of Research Resources supporting the Biological Microscopy and Image Reconstruction Resource as a National Biotechnology Resource. U. V. thanks Consiglio Nazionale delle Ricerche (CNR), Ministero Pubblica Istruzione (MPI), Rome, and CISM-INFM for financial support.

REFERENCES

Allinson, D. L. and Kisch, E. (1972). High voltage electron microscope specimen rod with demountable double tilting facilities. *J. Phys. E: Sci. Instrum.* **5**:205–207.

Aoki, Y., Kihara, H., Harada, Y., Fujiyoshi, Y., Uyeda, N., Yamagishi, H., Morikawa, K., and Mizusaki, T. (1986). Development of super fluid helium stage for HRTEM, in *Proc. XIth Int. Cong. Electron Microscopy*, (T. Imura, S. Maruse and T. Suzuki, eds.), Vol. 3, pp. 1827–1828. Japanese Society of Electron Microscopy, Tokyo.

Arii, T. and Hama, K. (1987). Method of extracting three-dimensional information from HVTEM stereo images of biological materials. *J. Electron Microsc.* **36**:177–195.

Barnard, D. P., Rexford, D., Tivol, W. F., and Turner, J. N. (1986). Side-entry differentially pumped environmental chamber for the AEI-EM7 HVEM, in *Proc. 44th Electron Microsc. Soc. Am.*, pp. 888–889.

Barnard, D. P., Turner, J. N., Frank, J., and McEwen, B. F. (forthcoming). An unlimited-tilt stage for the high-voltage electron microscope. *J. Micros.*

Barnes, D. C. and Warner, E. (1962). A kinematic specimen goniometer stage for the Siemens electron microscope. *Br. J. Appl. Phys.* **13**:264–265.

Bleloch, A. L. (1989). Secondary electron spectroscopy in a dedicated STEM. *Ultramicroscopy* **29**:147–152.

Browning, G. (1974). A new axis-centered stage, in *High Voltage Electron Microscopy: Proc. 3rd Int. Conf.* (P. R. Swann, C. J. Humphreys, and M. J. Goringe, eds.), pp. 121–123. Academic Press, New York.

Bursill, L. A., Spargo, A. E. C., Wentworth, D., and Wood, G. (1979). A goniometer for electron microscopy at 1.6 Å point to point resolution. *J. Appl. Crystallogr.* **12**:279–286.

Butler, E. P. and Hale, K. F. (1981). Dynamic experiments in the electron microscope, in *Practical Methods in Electron Microscopy* (A. M. Glauert, ed.), Vol. 9. North-Holland, Amsterdam.

Capilluppi, C., and Valdré, U. (1987). A specimen holder for brightness measurements in a TEM, in *ATTI XVI Congr. Microscopia Electtronica*, pp. 277–278. Societá Italiana di Microscopia Electtronica.

Chalcroft, J. P. and Davey, C. L. (1984). A simply constructed extreme-tilt holder for the Philips eucentric goniometer stage. *J. Microsc.* **134**:41–48.

Chiu, W., Knapek, E., Jeng, T. W., and Dietrich, I. (1981). Electron radiation damage of a thin protein crystal at 4 K. **6**:291–295.

Chou, C. T. (1987). Computer software for specimen orientation adjustment using double-tilt or rotation holders. *J. Electron Microsc. Tech.* **7**:263–268.

Craven, A. J. and Valdré, U. (1979). Visibility of diffraction patterns and bend contours in thick composite amorphous-crystalline specimens observed in STEM and CTEM. *J. Microsc.* **115**:211–223.

Crowther, R. A., DeRosier, D. J., and Klug, A. (1970). The reconstruction of a three-dimensional structure from projections and its application to electron microscopy. *Proc. R. Soc. London A* **317**:319–340.

Donovan, P., Everett, P., Self, P. G., Stobbs, W. M., and Valdré, U. (1983). High resolution, top entry goniometers for use in the JEOL transmission electron microscopes. *J. Phys. E: Sci. Instrum.* **16**:1242–1246.

Everett, P. G. and Valdré, U. (1985). A double tilting cartridge for transmission electron microscopes with maximum solid angle of exit at the specimen. *J. Microsc.* **139**:35–40.

Frank, J., McEwen, B. F., Radermacher, M., Turner, J. N., and Rieder, C. L. (1987). Three-dimensional tomographic reconstruction in high voltage electron microscopy. *J. Electron Microsc. Tech.* **6**:193–205.

Fujiyoshi, Y. (1990). High-resolution cryo-electron microscopy of biological macromolecules, in *Proc. XIIth Int. Congr. Electron Microscopy* (L. D. Peachey and D. B. Williams, eds.), Vol. 1, pp. 126–127. San Francisco Press, San Francisco.

Fujiyoshi, Y., Uyeda, N., Yamagishi, H., Morikawa, K., Mizusaki, T., Aoki, Y., Kihara, H., and Harada, Y. (1986). Biological macromolecules observed with high resolution cryo-electron microscope, in *Proc. XIth Int. Congr. Electron Microscopy* (T. Imura, S. Maruse, and T. Suzuki, eds.), Vol. 3, pp. 1829–1832. Japanese Society of Electron Microscopy, Tokyo.

Glaeser, R. M. (1975). Radiation damage and biological electron microscopy, in *Physical Aspects of Electron Microscopy and Microbeam Analysis* (B. M. Siegel and D. R. Beaman, eds.), pp. 205–208. Wiley, New York.

Goringe, M. J. and Valdrè, U. (1968). An investigation of superconducting materials in a high voltage electron microscope, in *Proc. 4th European Reg. Conf. on Electron Microscopy* (D. S. Bocciarelli, ed.), Vol. I, pp. 41–42. Tipografia Poliglotta Vaticana, Rome.

Heide, H. G. (1962). The prevention of contamination without beam damage to the specimen, in *Proc. 5th Int. Congr. Electron Microscopy* (S. S. Breese, ed.), Vol. I, p. A-4. Academic Press, New York.

Heide, H. G. and Urban, K. (1972). A novel specimen stage permitting high-resolution electron microscopy at low temperatures. *J. Phys. E: Sci. Instrum.* **5**:803–807.

Horn, E., Ashton, F., Haselgrove, J. C., and Peachy, L. D. (1991). Tilt unlimited: A 360 degree tilt specimen holder for the JEOL 4000-EX 400kV TEM with modified objective lens, in *Proc. 49th Electron Microsc. Soc. Am.* (G. W. Bailey and E. L. Hall, eds.), pp. 996–997. San Francisco Press, San Francisco.

King, M. V. (1981). Theory of stereopsis, in *Methods in Cell Biology*, Vol. 23, *Three-Dimensional Ultrastructure in Biology* (J. N. Turner, ed.), Academic Press, New York.

Klug, A. and Crowther, R. A. (1972). Three-dimensional image reconstruction from the viewpoint of information theory. *Nature* **238**:435–440.

Laufer, E. E. and Milliken, K. S. (1973). Note on stereophotography with the electron microscope. *J. Phys. E: Sci. Instrum.* **6**:966–968.

Lawrence, M. C. (1983). Alignment of images for three-dimensional reconstruction of non-periodic objects. *Proc. Electron Microsc. Soc. S. Afr.* **13**:19–20.

Leteurtre, J. (1972). Reported by B. Gentry in *Methodes et Techniques Nouvelles D'Observation en Métallurgie Physique* (B. Jouffrey, ed.), p. 29. Societé Francaise de Microscopie Electronique, Paris.

Lucas, G., Phillips, R., and Teare, P. W. (1963). A precision goniometer stage for the electron microscope. *J. Sci. Instrum.* **40**:23–25.

Luther, P. K., Lawrence, M. C., and Crowther, R. A. (1988). A method for monitoring the collapse of plastic sections as a function of electron dose. *Ultramicroscopy* **24**:7–18.

MacKenzie, J. M. and Bergensten, R. W. (1989). A five -axis computer-controlled stage for the transmission electron microscope, in *Proc. 47th Ann. Mtg. Elect. Microsc. Soc. Am.* (G. W. Bailey, ed.), pp. 50–51. San Francisco Press, San Francisco.

Mills, J. C. and Moodie, A. F. (1968). Multipurpose high resolution stage for the electron microscope. *Rev. Sci. Instrum.* **39**:962–969.

Parsons, D. F., Matricardi, V. R., Moretz, R. C., and Turner, J. N. (1974). Electron microscopy and diffraction of wet unstained and unfixed biological objects, in *Advanced Biology and Medical Physics* (J. H. Lawrence, J. W. Gofman, and T. L. Hayes, eds.), Vol. 15, pp. 161–262. Academic Press, New York.

Peachey, L. D. and Heath, J. P. (1989). Reconstruction from stereo and multiple tilt electron microscope images of thick sections of embedded biological specimens using computer graphic methods. *J. Microsc.* **153**:193–203.

Pennycook, S. L., Brown, L. M., and Craven, A. J. (1980). Observation of cathodoluminescence at single dislocations by STEM. *Philos. Mag. A* **41**:589–600.

Radermacher, M. (1988). Three-dimensional reconstruction of single particles from random and non-random tilt series. *J. Electron Microsc. Tech.* **9**:359–394.

Radermacher, M. and Hoppe, W. (1980). Properties of 3-D reconstruction from projections by conical tilting compared with single axis tilting, in *Proc. 7th European Congr. Electron Microscopy* (P. Brederoo and G. Boom, eds.), Vol. I, pp. 132–133.

Rakels, C. J., Tiemeijer, J. C., and Witteveen, K. W. (1968). The Philips electron microscope EM300. *Philips Tech. Rev.* **29**:370–386.

Reimer, L. (1989). *Transmission Electron Microscopy*, p. 440. Springer, Berlin.

Ruknudin, A., Song, M. J., and Sachs, F. (1991). The ultrastructure of patch-clamped membranes: A study using high-voltage electron microscopy. *J. Cell Biol.* **122**:125–134.

Sachs, F. and Song, M. J. (1987). High-voltage electron microscopy of patch-clamped membranes, in *Proc. 45th Electron Microsc. Soc. Am.* (G. W. Bailey, ed.), pp. 582–583. San Francisco Press, San Francisco.

Shaw, P. J. and Hills, G. J. (1981). Tilted specimen in the electron microscope: A simple holder and the calculation of tilt angles for crystalline specimens. *Micron* **12**:279–282.

Siemens' *Focus Information* (1975). Double tilt and lift unit for Elmiskop 101 and 102. Publication C73000-B3176-C2-1, pp. 1–18.

Sparrow, T. G., Tang, T. T., and Valdré, U. (1984). Objective lens and specimen stage for a versatile high voltage CTEM/STEM. *J. Microsc. Spectrosc. Electron* **9**:279–290.

Spence, J. C. H. (1988). A scanning tunneling microscope in a side-entry holder for reflection electron microscopy in the Philips EM400. *Ultramicroscopy* **25**:165–170.

Spence, J. C. H., Lo, W., and Kuwabara, M. (1990). Observation of the graphite surface by reflection electron microscopy during STM operation. *Ultramicroscopy* **33**:69–82.

Stols, A. L. H., Smits, H. T. J., and Stadhonders, A. M. (1986). Modification of a TEM-goniometer specimen holder to enable beam current measurements in a (S)TEM for use in quantitative x-ray microanalysis. *J. Electron Microsc. Tech.* **3**:379–384.

Swann, P. R. (1972). High voltage microscopy studies of environmental reactions, in *Electron Microscopy and Structure of Materials* (G. Thomas, R. M. Fulrath, and R. M. Fisher, eds.), p. 878. University of California Press, Berkeley.

Swann, P. R. (1979). Side entry specimen stages. *Krist. Tech.* **14**:1235–1243.

Swann, P. R. and Lloyd, A. E. (1974). A high angle, double tilting cold stage for the AEI EM7, in *Proc. Electron Microsc. Soc. Am.* (C. J. Arcenaux, ed.), pp. 450–451. LA.

Sykes, L. J. (1979). Computer-aided tilting in the electron microscope, in *Proc. Electron Microsc. Soc. Am.* (G. W. Bailey, ed.), pp. 602–603. Claitor's, Baton Rouge, LA.

Tatlock, G. J., Spain, J., Raynard, G., Sinnock, A. C., and Venables, J. A. (1980). A liquid helium cooled environmental cell for the JEOL 200A TEM, in *Electron Microscopy and Analysis*, pp. 39–42. *Int. Phys. Conf. Series N. 52*, Institute of Physics, London.

Turner, J. N. (1981). Stages and stereo pair recording, in *Methods in Cell Biology* Vol. 23, *Three-Dimensional Ultrastructure in Biology* (J. N. Turner, ed.), pp. 33–51. Academic Press, New York.

Turner, J. N., See, C. W., and Matuszek, G. (1986). A simple specimen rotation tip for CTEM and HVEM. *J. Electron Microsc. Tech.* **3**:367–368.

Turner, J. N., Barnard, D. P., Matuszek, G., and See, C. W. (1988). High-precision tilt stage for the high-voltage electron microscope. *Ultramicroscopy* **26**:337–344.

Turner, J. N., Barnard, D. P., McCauley, P., and Tivol, W. F. (1990). Specimen orientation and environment, in *Electron Crystallography* (J. Freyer, ed.). Kluwer, Boston.

Turner, J. N. and Ratkowski, A. J. (1982). An improved double-tilt stage for the AEI EM7 high-voltage electron microscope. *J. Microsc.* **127**:155–159.

Turner, J. N., Rieder, C. L., Collins, D. N., and Chang, B. B. (1989b). Optimum specimen positioning in the electron microscope using a double-tilt stage. *J. Electron Microsc. Tech.* **11**:33–40.

Turner, J. N., Valdrè, U., and Fukami, A. (1989a). Control of specimen orientation and environment. *J. Electron Microsc. Tech.* **11**:258–271.

Turner, J. N., Barnard, D. P., McCauley, P., and Dorset, D. L. (1991). Diffraction and imaging from all perspectives: Unlimited specimen tilting in the high-voltage electron microscope. *Proc. 49th Electron Microsc. Soc. Am.*:994–995.

Valdrè, U. (1962). A simple goniometer stage for the Siemens Elmiskop. I. *J. Sci. Instrum.* **39**:278–280.

Valdrè, U. (1964). A double-tilting liquid-helium cooled object stage for the Siemens electron microscope, in *Proc. 3rd European Reg. Conf. on Electron Microscopy*, Vol. A, pp. 61–62. Academy of Sciences, Prague.

Valdrè, U. (1967). Unpublished report at 126–129, in *Electron Microscopy in Materials Science* (E. Ruedi and U. Valdrè, eds.), Part I, pp. 113–114. Commission of the European Communities, Luxemburg 1975.

Valdrè, U. (1968). Combined cartridges and versatile specimen stage for electron microscopy. *Nuovo Cimento Ser. B* **53**:157–173.

Valdrè, U. (1979). Electron microscope stage design and applications. *J. Microsc.* **117**:55–75.

Valdrè, U. and Goringe, M. S. (1971). Special electron microscope specimen stages, in *Electron Microscopy in Material Science* (U. Valdrè, ed.). Academic Press, New York.

Valdrè, U. and Horne, R. W. (1975). A combined freeze chamber and low temperature stage for an electron microscope. *J. Microsc.* **103**:305–317.

Valdrè, U., Robinson, E. A., Pashley, D. W., Stowell, and Law, T. J. (1970). An ultra-high vacuum electron microscope specimen chamber for vapor deposition studies. *J. Phys. E: Sci. Instrum.* **3**:501–506.

Valdrè, U. and Tsuno, K. (1986). A possible design for a high tilt, fully eucentric, goniometer stage. *J. Electron Microsc.* (Suppl.) **35**:925–926.

Valdrè, U. and Tsuno, K. (1988). A contribution to the unsolved problem of a high-tilt eucentric goniometer stage. *Acta Crystallogr. A* **44**:775–780.

Valle, R., Gentry, B., and Marraud, A. (1980). A new side entry eucentric goniometer stage for HVEM, in *Electron Microscopy 1980*, Vol. 4., *Proc. 6th Int. Conf. HVEM* (P. B. Brederoo and J. van Landuyt, eds.), pp. 34–37.

von Ardenne, M. (1940a). Stereo-Übermikroskopie mit dem Universal-Elektronenmikroskop. *Naturwissenschaften* **16**:248–252.

von Ardenne, M. (1940b). Uber ein Universal-Elektronenmikroskop fur Hellfeld-, Dunkelfeld- und Stereobild-Betrieb. *Z. Phys.* **15**:339–368.

von Ardenne, M. (1940c). *Elektronen-Ubermikroskopie Physik, Technik.* Ergebnisse Verlag von Julius Springer, Berlin.

Waddington, C. P. (1974). Calibration of a Siemens Elmiskop 1b double tilting goniometer stage. *J. Phys. E: Sci. Instrum.* **7**:842–846.

Ward, P. R. (1965). A goniometer specimen holder and anticontamination stage for the Siemens Elmiskop I. *J. Sci. Instrum.* **42**:767–769.

Zemlin, F., Reuber, E., Beckmann, B., Zeitler, E., and Dorset, D. L. (1985). Molecular resolution electron micrographs of monolamellar paraffin crystals. *Science* **229**:461–462.

8

Least-Squares Method of Alignment Using Markers

Michael C. Lawrence

1. INTRODUCTION

An intermediate problem arises in the tomographic reconstruction of an object from a series of electron microscope exposures, namely that of determining the relationship between the axes of the individual digitized images, the tilt axis of the microscope, and a hypothetical internal coordinate system of the specimen. Accurate resampling of the tilt series images onto a common coordinate system is an essential prerequisite for any tomographic reconstruction algorithm. The lack of an immediately available common system of coordinates results from the microscope goniometer not being truly eucentric at high resolution. Thus, the goniometer control system is unable to provide the user with sufficient control at a subpixel level as the specimen is tilted in the microscope, and the precise relationship

---

Michael C. Lawrence • Electron Microscope Unit, University of Cape Town, Rondebosch 7700, South Africa *Present address* CSIRO Division of Biomolecular Engineering, Parkville, Victoria 3052, Australia.

Electron Tomography: Three-Dimensional Imaging with the Transmission Electron Microscope, edited by Joachim Frank, Plenum Press, New York, 1992.

between the microscope coordinate system and the specimen coordinate system is lost. In the case of photographic image recording, further inaccuracies can arise from the positioning of the plates in their holders or on the microdensitometer.

Methods described in the literature for determining a common coordinate system fall into two classes, those based on fiducial markers and those based on correlation functions. This chapter discusses the theory and practice of the method of fiducial markers. In this method, marker beads (typically colloidal gold particles) are distributed on the specimen and/or its support film during specimen preparation. The relationship between the specimen and digital raster coordinate systems is then determined from a least-squares analysis of the measured positions of the fiducial markers in the digital images. The images are transformed to a common coordinate system, and the pixel densities are then resampled by interpolation before the tomographic reconstruction can proceed. A comparison of the fiducial marker method with other alignment techniques will be given in Section 5.

2. *HISTORY OF THE FIDUCIAL MARKER METHOD*

Hart (1968) is credited with having first used colloidal gold particles for electron microscope image alignment in the method of "polytropic montage"—digital overlay of images taken at various tilts. Hart's algebra for the alignment of the images based on the fiducial marker positions is essentially correct, although it is not presented in a form readily applicable to tomography. (The idea of polytropic montage, however, appears crude in the light of modern three-dimensional reconstruction methods).

A description of the fiducial marker alignment algorithm, developed explicitly for tomographic reconstruction, is given in the next section. The formulation follows Luther *et al.* (1988) and Lawrence (1983), although similar algorithms have been developed independently by others (Olins *et al.*, 1984; Berriman *et al.*, 1984).

3. *MATHEMATICAL DESCRIPTION*

3.1. *Coordinate Systems*

Microscope, specimen, and image coordinate systems are defined as follows (see Fig. 1). The microscope z axis (z_m) is taken to be parallel to the microscope optical axis, with the primary goniometer tilting axis taken as the microscope y axis (y_m). At 0° primary tilt the specimen and microscope axes are defined to be coincident in orientation. The choice of location for the specimen coordinate system origin is dealt with later. The operation of tilting in the microscope (about any axis or combination of axes) corresponds to a rotation of the specimen axes relative to the microscope axes, and any translation of the specimen in the microscope likewise translates the specimen axes relative to the microscope axes. The raster (x^i, y^i) axes of each digitized tilt image lie in a plane perpendicular to z_m; the rotation and translation of the raster axes in this plane relative to the microscope system are

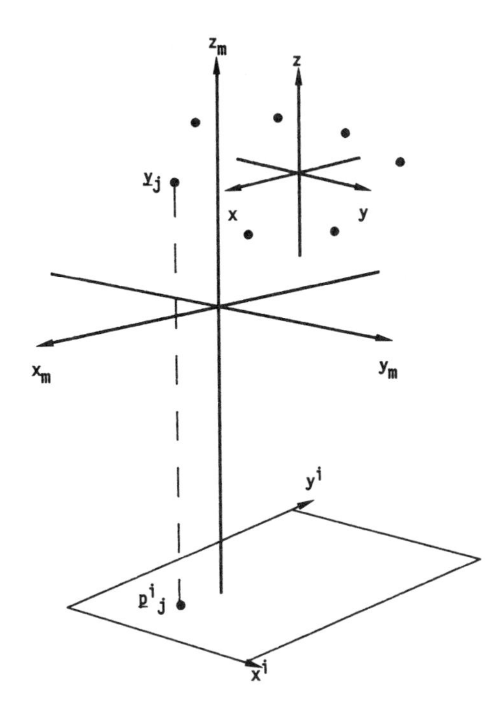

FIGURE 1. Definition of coordinate systems. (x_m, y_m, z_m): microscope axes, (x, y, z): specimen axes, (x^i, y^i): images axes, $i=1,\ldots, v$. The colloidal gold marker particles have coordinates $\mathbf{y}_j$ $(j=1,\ldots, s)$ in the specimen frame and have coordinates $\mathbf{p}^i_j$ in the image frames.

initially unknown. The problem is thus to determine the relationship between the specimen coordinate system and the digital raster coordinate systems of the individual micrographs.

3.2. The Imaging Operation

Assume that s distinct colloidal gold particles in the specimen are selected for use as fiducial markers and that there are v micrographs in the tilt series. The relationship between the colloidal gold particle coordinates in the specimen frame and their coordinates in the v image frames is described by the following system of equations:

$$\mathbf{p}^i_j = S^i P A^i M^i \mathbf{y}_j + \mathbf{d}^i, \qquad i=1,\ldots, v, \qquad j=1,\ldots, s \tag{1}$$

where

1. $\mathbf{y}_j$ are 3-vectors describing the coordinates of the colloidal gold particles in the specimen frame.
2. M^i are 3×3 diagonal matrices. These matrices are included to represent either dimensional changes in the specimen during imaging or small changes in microscope magnification in successive images.
3. A^i are 3×3 rotation matrices describing the tilting operation. In the case of single-axis tilting these matrices describe a rotation through an angle β_i about y_m.
4. P is a 2×3 matrix accomplishing projection down z_m,

$$P=\begin{pmatrix}1 & 0 & 0\\ 0 & 1 & 0\end{pmatrix}$$

5. S^i are 2×2 rotation matrices describing the rotation (through an angle α_i) of the digital image $x^i y^i$ axes with respect to the microscope $x_m y_m$ axes.
6. $\mathbf{d}^i$ are 2-vectors describing the combined effect of specimen translation and in-plane translation of the image frame origins with respect to the microscope $x_m y_m$ axes. These translations are generally inseparable.
7. $\mathbf{p}_j^i$ are 2-vectors describing the coordinates of the projected fiducial markers in the digital image frames.

If M^i, α^i, A^i, and $\mathbf{d}^i$ can be determined for each micrograph, then the digital images can be resampled via interpolation to lie on a common coordinate system.

3.3. Least-Squares Determination of Parameters

Let $\mathbf{p}_j^{i'}$ denote the measured values of $\mathbf{p}_j^i$ (methods of measurement will be discussed in Section 4). The least-squares approach seeks to minimize

$$f = \sum_{i,j} (\mathbf{p}_j^{i'} - \mathbf{p}_j^i)(\mathbf{p}_j^{i'} - \mathbf{p}_j^i) \tag{2}$$

as a function of the parameters M^i, α^i, A^i, $\mathbf{d}^i$, and $\mathbf{y}_j$. This function, although non-linear, is readily differentiable, and the minimization can proceed via an algorithm such as the method of conjugate gradients (Fletcher and Powell, 1963).

To obtain unique values for the parameters from the minimization, it is essential to define a reference projection r, usually that obtained from the untilted specimen, or the first exposure in the series. For this projection the matrices M^r and A^r are assumed to be identity matrices. The translation invariance of (1) also requires defining the specimen coordinate system origin. A convenient choice is to place it at the geometric centre of the colloidal gold markers, such that

$$\sum_j \mathbf{y}_j = \mathbf{O} \tag{3}$$

With this choice of origin, least-squares estimates of $\mathbf{d}^i$ can be obtained directly from Eqs. (1) and (3) as

$$\mathbf{d}^i = \sum_j \frac{\mathbf{p}_j^{i'}}{s} \tag{4}$$

Initial estimates for the remaining parameters are required before the minimization can proceed. The estimates $M^i = I$, $\alpha^i = 0$ and $A^i =$ (goniometer orientation matrix) for all $i \neq r$ are reasonable. The colloidal gold particles have an approximately planar distribution on the specimen grid, leading to the choice $(\mathbf{y}_j)_z = 0$ for the initial values of the z component of the $\mathbf{y}_j$. Initial estimates of $(\mathbf{y}_j)_{x,y}$ can then be obtained from Eq. (1) as

$$(\mathbf{y}_j)_{x,y} = [S^r(A^r M^r)_{22}]^{-1} (\mathbf{p}_j^{r'} - \mathbf{d}^r) \tag{5}$$

where the subscript 22 denotes the upper left 2×2 matrix of $(A^r M^r)$, $\mathbf{p}_j^{r'}$ has been substituted for $\mathbf{p}_j^{i}$, and $(\mathbf{y}_r)_z$ has been set to zero. These initial estimates for $\mathbf{y}_j$ are consistent with the constraint (3).

The constraint (3) can be rewritten as $\mathbf{y}_1 = -\sum_{j \neq 1} \mathbf{y}_j$, thus making $\mathbf{y}_1$ a function of the remaining fiducial specimen coordinates. This equation is added to the system (1), and the constraint is enforced during the calculation of f.

Digital resampling of the individual micrographs onto a common coordinate system can proceed once optimal values have been obtained for the alignment parameters. A convenient choice of common coordinate system is to place the new raster axes parallel to x_m and y_m, respectively, and to place the projection of the specimen system origin (defined by Eq. 3) at the center of the new raster. During the optimization of f, values are also obtained for $\mathbf{y}_j$, the three-dimensional coordinates of the marker particles in the specimen. These coordinates are of no further use in the image alignment or in the tomographic reconstruction, but may yield useful information about specimen thickness and shrinkage during beam exposure (Berriman *et al.*, 1984; Luther *et al.*, 1988 and Chapter 3, this volume).

3.4. Number of Fiducial Markers Required

The number of unknowns in the system of Eqs. (1) depends on

1. The number of tilting axes
2. Whether the tilt angles are treated as parameters to be refined
3. Whether the matrix M is restrained to represent isotropic scale change $(m_x^i = m_y^i = m_z^i)$

In the case of single-axis tilting, isotropic scale change and refined tilt angles, the number of unknowns is $5sv + 3s - 5$, compared to $2sv$ equations. If $s \geqslant 3$, then the number of equations will exceed the number of unknowns and the minimization of f can proceed. The author recommends using about 15 markers to ensure accuracy. After minimization the residuals $e_j^i = |\mathbf{p}_j^{i\prime} - \mathbf{p}_j^i|$ should be inspected; e_j^i should be less than one pixel for all i and j if the fiducial coordinates have been determined accurately and there has been no distortion of the surrounding support film.

4. MEASUREMENT OF THE FIDUCIAL POSITIONS

The success of the projection alignment is dependent on the fiducial coordinates being measured with sufficient accuracy. These measurements can be done manually on a hard copy of the digital image, interactively using a graphics cursor superimposed on a raster video display of the image, or entirely automatically using a peak-searching algorithm. The accuracy of fiducial coordinates determined manually or via an interactive graphics device can be improved by subsequent digital interpolation in the image itself. Care should be taken to select a set of markers that is spread evenly about the region of intersect and can be discerned in all projections. Alignment accuracy can also be improved by selecting markers lying near the periphery of the field of view, though this should be avoided if gross deformation of the support film is suspected.

The measurement technique adopted will depend on the available computer hardware and on the relative ease with which each marker image can be discerned from the background of surrounding pixels. Both thresholding and peak-searching algorithms can fail to detect fiducials if the colloidal gold particles are embedded in large amounts of negative stain.

5. COMPARISON WITH CORRELATION FUNCTION METHODS

Alternative approaches to the alignment problem are based on image cross-correlation functions (see Frank, 1980; Hoppe and Hegerl, 1980; Guckenberger, 1982). Traditionally, the alignment of a pair of images using correlation functions proceeds in two stages. The rotational misalignment is determined and corrected first; thereafter, the translational misalignment is determined and corrected. Rotational alignment is the absence of fiducial markers is problematic and involves determining the projected direction of the microscope tilt axis in every micrograph. Frank (1980) discusses three strategies for achieving the rotational alignment of a pair of images taken at the same tilt: iterative search, triangulation, and angular cross-correlation of autocorrelation functions (ACF). None of these methods is readily applicable to the rotational alignment of tilt series images, and in published tomographic reconstructions, rotational alignment has been achieved by examining markers, either colloidal gold particles or features intrinsic to the specimen.

Translational alignment of rotationally aligned images is relatively straightforward. Guckenberger (1982) has shown that the position (with respect to the origin) of the center of mass of the peak in the cross-correlation function of two rotationally aligned images in a tilt series is equal to the translational displacement between the objects. A change of scale normal to the projected tilt axis must be applied to each micrograph in the tilt series prior to cross-correlation to compensate for the varying tilt. Hence, it is imperative that the micrographs are rotationally aligned first.

In principle, correlational alignment has a number of advantages:

1. The method incorporates all the pixel image data rather than a few limited measurements of markers. It should be more accurate since it is based on more data.
2. The correlation can be performed using only the area of interest, whereas fiducial markers are usually distributed on the surrounding support film. Alignment based on fiducials can be susceptible to support film distortion during imaging.
3. The manual identification of fiducial and measurement of their coordinates are avoided.
4. The specimen preparation is not complicated by having to place colloidal gold particles on the support film.

There are, however, a number of serious disadvantages:

1. The cross-correlation functions are computed pairwise through the image stack. This can lead to the accumulation of error, particularly at high tilt (Frank *et al.*, 1987) or when the correlation function peaks are broad or

noisy. In contrast, the fiducial algorithm presented in Section 3 is a global procedure, and it distributes the errors in a proper fashion among all projections.

2. As mentioned above, the correlation method does not provide a straightforward approach to determining the direction α^r of the projected tilt axis in the reference micrograph or to determining change of scale. The least-squares algorithm refines these automatically.
3. The regions correlated should ideally consist of the specimen surrounded by a region of uniformly dense film (Guckenberger, 1982). This is not always easily achieved in negatively stained preparations.

Frank *et al.* (1987) used a hybrid stategy for image alignment: fiducial markers are used for rotational alignment and cross correlation-functions for the translational alignment (see Chapter 9). More recently, Dengler (1989) proposed an alignment scheme based on alternating alignment and reconstruction at ever increasing resolution. The image shifts are determined by correlating each image with its corresponding calculated projection obtained from the reconstruction itself. This technique is theoretically sound but computationally intensive. A quantitative comparison of this technique with the fiducial method and traditional pairwise image correlation applied to the same data would be instructive.

6. CONCLUSION

Tomographic image alignment strategies should fulfil two requirements: they should be accurate at the subpixel level of alignment, and they should require little manual intervention. Coupled with interactive computer image display, the fiducial alignment procedure satisfies both criteria. The method has proved successful in a number of reconstructions (Olins *et al.*, 1983; Skoglund *et al.*, 1986; Lawrence *et al.*, 1989). As the method is global across the entire tilt series, it does not suffer from error accumulation at high tilt, and computationally it is straightforward and quick. An alternative formulation of the method has recently been given by Jing and Sachs (1991). Their algorithm does not require all the markers to be imaged on every micrograph and represents a useful generalization of the techniques described in this chapter.

There are two drawbacks to the method of fiducial markers. The additional step of placing the colloidal gold on the specimen grid can unduly complicate the specimen preparation; whether this is so will depend on the precise nature of the specimen and its support film. Further, if the colloidal gold particles are close to or in contact with the specimen, there will be parts of the tomographic reconstruction volume that are severely affected by artefacts from the attempted reconstruction of the extremely optically dense gold particles. Considerable effort may thus be required to prepare or select a specimen with a suitably dispersed environment of fiducial gold particles.

Computer programs ALIGN for performing least-squares image alignment using markers and TILT for radially weighted back-projection tomographic reconstruction are available from the author.

ACKNOWLEDGMENTS

I thank Dr. Tony Crowther for guiding my initial development of the fiducial alignment method.

REFERENCES

Berriman, J., Bryan, R. K., Freeman, R., and Leonard, K. R. (1984). Methods for specimen thickness determination in electron microscopy. *Ultramicroscopy* **13**:351–364.

Dengler, J. (1989). A multi-resolution approach to 3D reconstruction from an electron microscope tilt series solving the alignment problem without gold particles. *Ultramicroscopy* **30**:337–348.

Fletcher, R. and Powell, M. J. D. (1963). A rapidly convergent descent method for minimization. *Comput. J.* **6**:163–168.

Frank, J. (1980). The role of correlation techniques in computer image processing, in *Computer Processing of Electron Microscope Images* (P. W. Hawkes, ed.). Springer-Verlag, New York.

Frank, J., McEwen, B. F., Radermacher, M., Turner, J. N., and Rieder, C. L. (1987). Three-dimensional tomographic reconstruction in high voltage electron microscopy. *J. Electron Microsc. Tech.* **6**:193–205.

Guckenberger, R. (1982). Determination of a common origin in the micrographs of tilt series in three-dimensional electron microscopy. *Ultramicroscopy* **9**:167–174.

Hart, R. G. (1968). Electron microscopy of unstained biological material: The polytropic montage. *Science* **159**:1464–1467.

Hoppe, W. and Hegerl, R. (1980). Three-dimensional structure determination by electron microscopy (nonperiodic structures), in *Computer Processing of Electron Microscope Images* (P. W. Hawkes, ed.). Springer-Verlag, New York, pp. 127–185.

Jing, Z. and Sachs, F. (1991). Alignment of tomographic projections using an incomplete set of fiducial markers. *Ultramicroscopy* **35**:37–43.

Lawrence, M. C. (1983). Alignment of images for three-dimensional reconstruction of non-periodic objects. *Proc. Electron Microsc. Soc. S. Afr.* **13**:19–20.

Lawrence, M. C., Jaffer, M., and Sewell, B. T. (1989). The application of the maximum entropy method to electron microscope tomography. *Ultramicroscopy* **31**:285–301.

Luther, P. K., Lawrence, M. C., and Crowther, R. A. (1988). A method for monitoring the collapse of plastic sections as a function of electron dose. *Ultramicroscopy* **24**:7–18.

Olins, A. L., Olins, D. E., Levy, H. A., Durfee, R. C., Margle, S. M., Tinnel, E. P., Hingerty, B. E., Dover, S. D., and Fuchs, H. (1984). Modeling Balbiani ring gene transcription with electron microscopy tomography. *Eur. J. Cell Biol.* **35**:129–142.

Olins, D. E., Olins, A. L., Levy, H. A., Durfee, R. C., Margle, S. M., Tinnel, E. P., and Dover, S. D. (1983). Electron microscope tomography: Transcription in three dimensions. *Science* **220**:498–500.

Skoglund, U., Andersson, K., Strandberg, B., and Daneholt, B. (1986). Three-dimensional structure of a specific pre-messenger RNP particle established by electron microscope tomography. *Nature* **319**:560–564.

9

Alignment by Cross-Correlation

Joachim Frank and Bruce F. McEwen

1. EFFECT OF ALIGNMENT ERRORS

Highest precision in the alignment of projections is one of the prerequisites of tomography. The effects of random translational alignment errors on the reconstruction can be appreciated by considering the much simpler situation where a given number of images containing the same motif are averaged.

Let the positions of the motif within the images be subject to random translational errors Δx, Δy following a joint Gaussian distribution with standard deviation σ:

$$E(\Delta x, \Delta y) = C \exp\left[\frac{-(\Delta x^2 + \Delta y^2)}{\sigma^2}\right] \tag{1}$$

The analysis shows that the average of the images contains a blurred version of the motif, with the blurring described by convolution with a point-spread function

Joachim Frank • Wadsworth Center for Laboratories and Research, New York State Department of Health, Albany, New York 12201-0509; and Department of Biomedical Sciences, School of Public Health, State University of New York at Albany, Albany, New York 12222 *Bruce F. McEwen* • Wadsworth Center for Laboratories and Research, New York State Department of Health, Albany, New York 12201-0509

Electron Tomography: Three-Dimensional Imaging with the Transmission Electron Microscope, edited by Joachim Frank, Plenum Press, New York, 1992.

$b(x, y)$. Alternatively, the resolution-limiting effect can be described by a multiplication of the Fourier transform of the uncorrupted average with the function

$$\mathscr{F}\{b(x, y)\} = B(X, Y) = \exp[-(X^2 + Y^2)\sigma^2] \tag{2}$$

where X, Y are the spatial frequency components. This function is analogous to the temperature factor used in x-ray crystallography to describe the effect of thermal motion on the measured structure factor. To exemplify the action of this term, assume that the standard deviation of the translational error is equal to the resolution distance. The effect of this term is to reduce the transform of the average to $1/e$ at the bandlimit; high resolution is thus strongly attenuated even when the errors are on the order of magnitude of the resolution. In a similar way, random errors in the alignment of projections lead to an attenuation of high-frequency information of the reconstruction. The above result shows, by 3D inference, that, in order to make maximum use of the information contained in the projections, one must keep the alignment errors substantially smaller than the resolution distance.

2. THEORY OF CROSS-CORRELATION ALIGNMENT

In the previous chapter, triangulation from positions of gold markers has been presented as the alignment tool in routine use in electron tomography. Here we describe another tool for alignment, which may be used either separately or in conjunction with marker triangulation: the two-dimensional cross-correlation function (CCF). Since the CCF alignment is based on an overall match between features intrinsic to the structure (as it appears in two projections), the potential for obtaining high accuracy is evident (see Frank *et al.*, 1987).

The CCF of two images represented by discrete arrays i_1 and i_2 is defined as (see Frank, 1980; appendix to Frank and Radermacher, 1986)

$$\phi_{lm} = \sum_j \sum_k i_1(j, k)\, i_2(j + l, k + m) \tag{3}$$

where $i_1(j, k)$ and $i_2(j, k)$ denote the measurements of the image optical density of the respective images at point $r = (x_j, y_k)$ of a regular sampling grid.

If the two images contain a common motif in the same orientation but different positions r_1 and r_2, then the CCF shows a peak in a position described by the shift vector $\Delta r = (r_1 - r_2)$. This peak is actually the center of the motif's autocorrelation function, which is the function describing the CCF of the motif with itself.

This property is put to use for alignment in the following way: the CCF of two images showing the same motif is computed, and the resulting array is searched for a high peak. The vector pointing from the CCF origin to the center of this detection peak is the desired shift vector. In order to bring the two motifs in register, the second image now has to be shifted by a vector that is of equal magnitude, but opposite direction, to that of the shift vector.

In practice, the computation of the CCF does not follow the route of Eq. (3) but makes use of the convolution theorem:

$$\phi_{lm} = \mathscr{F}^{-1}\{\mathscr{F}\{i_1\}\,\mathscr{F}^{*-1}\{i_2\}\} \tag{4}$$

where $\mathscr{F}$ and $\mathscr{F}^{-1}$ stand for the discrete forward and reverse Fourier transformation, respectively, and * stands for the use of the complex conjugate of the expression.

3. *ALIGNMENT OF PROJECTIONS*

The CCF can be used for alignment of different projections of an object, even though in this case the two images to be compared do not contain an identical motif. It can be easily seen (see Frank and Radermacher, 1986) by applying the projection theorem and considering the Fourier transform of the CCF that two projections of the same object are similar, in the sense that their Euclidean distance is small, and give rise to correlation (Fig. 1). However, as the angle between the directions of projection increases, two effects are seen: (i) the size of the correlation peak *decreases* relative to the noise background, increasing the risk that the position found by peak search is subject to random error, and (ii) its width *increases*, limiting the precision of alignment. For this reason, the correlation should be computed between projections from nearly the same angular direction whenever possible.

Guckenberger (1982) has investigated the question of how to apply the cross-correlation alignment to projections of a 3D object. According to his findings, the proper common alignment of the set of projections comprising a single-axis tilt series, relative to the object from which they originate, is found when each projection is stretched, prior to the computation of the CCF, by a factor $1/\cos\theta$ in the direction perpendicular to the tilt axis.

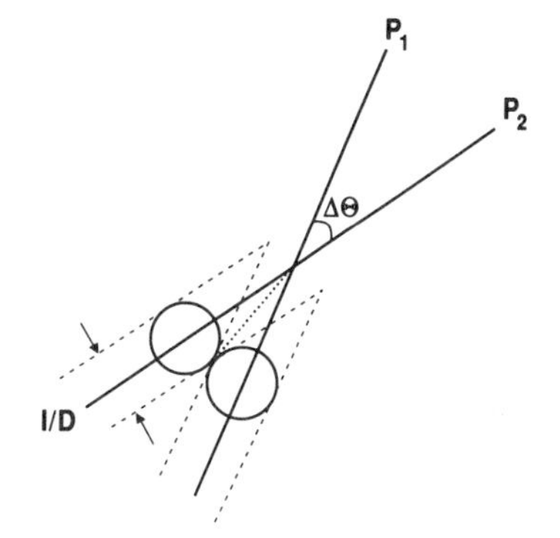

FIGURE 1. Correlation of two projections of an object, separated by angular distance $\Delta\theta$. In Fourier space, each projection is represented by a central section. The amount of information the two sections have in common, which gives rise to correlation, can be inferred from the geometry of this drawing, bearing in mind that Fourier components within a "circle of influence" with radius 1/2D are mutually correlated as a consequence of the finite size of the object.

4. IMPLEMENTATION

When implementing the CCF for alignment purposes, either one of the image files, or the CCF, should be both low- and high-pass-filtered in order to reduce noise sensitivity (Frank and Radermacher, 1986; Frank *et al.*, 1987) and to eliminate "rooftop" effects due to the overall shape of the specimen (Saxton and Frank, 1976). It is convenient to apply these filters to the Fourier transform of one of the image files because, as discussed, the CCF is computed from the Fourier transform. In addition, it is sometimes helpful to apply equatorial, and meridional, filtering to the Fourier transform before computing the CCF. These filters set the equatorial (or the meridional) line in the Fourier transform, along with a specified number of lines on either side of it, to zero. The effect of equatorial and meridional filtering is to dampen the parallel lines that sometimes arise in the CCF from scanning artifacts, e.g., due to imperfect balancing of diodes in the EIKONIX scanner. Examples of CCFs from the alignment of a tilt series can be seen in Fig. 2.

In order to obtain the desired shift vector, the exact position of the peak centroid of the CCF is determined by computing the peak's center of gravity within a specified radius of the maximum pixel. Since the size and shape of the CCF peak vary with the sample, and even with the viewing direction of the sample, as is evident from Fig. 2, the center of gravity radius should be adjusted for each data set. We have implemented an elliptical center-of-gravity computation, with two principal axes being specified, in order to give more flexibility for samples with non-spherical CCF peaks. In addition to the center-of-gravity axes, the peak search requires specification of the number of peaks to find and distances for nearest-neighbor and edge exclusions.

The component of the shift vector that is perpendicular to the tilt axis must be multiplied by the cosine of the tilt angle of the input file to compensate for the cosine stretching. The corrected shift vector is a record of the position of the input image relative to the reference image. The input image is aligned by shifting it in its frame by a vector of equal magnitude, but opposite direction, to the shift vector. Alternatively, the alignment can be accomplished by adjusting the data windows. Normally when an image is digitized, the area scanned is larger than the one desired for the 3D reconstruction. This is particularly true if fiduciary markers are used for the rotational alignment (see Chapter 8). A window of the total scan, containing the desired area, is isolated interactively and used as an input for the translational alignment scheme. If the coordinates of the window are recorded, or if the alignment procedure itself makes the first window, then the alignment can be achieved by reisolating the window at coordinates that are adjusted by the shift vector. This windowing method of alignment has the advantage that it eliminates wraparound, and the resulting loss of part of the image, that occur with simple "circular" shifting. Regardless of which method is used to align the input image, the shift vector and the size and coordinates of the final data windows should be recorded so that if the aligned files are lost they can be recreated without recomputing the CCFs and the peak searches.

The cross-correlation-based alignment is applied sequentially to neighboring images of a tilt series in order to minimize the difference in tilt angle between the input and the reference images. In such a scheme, the input file for one step of

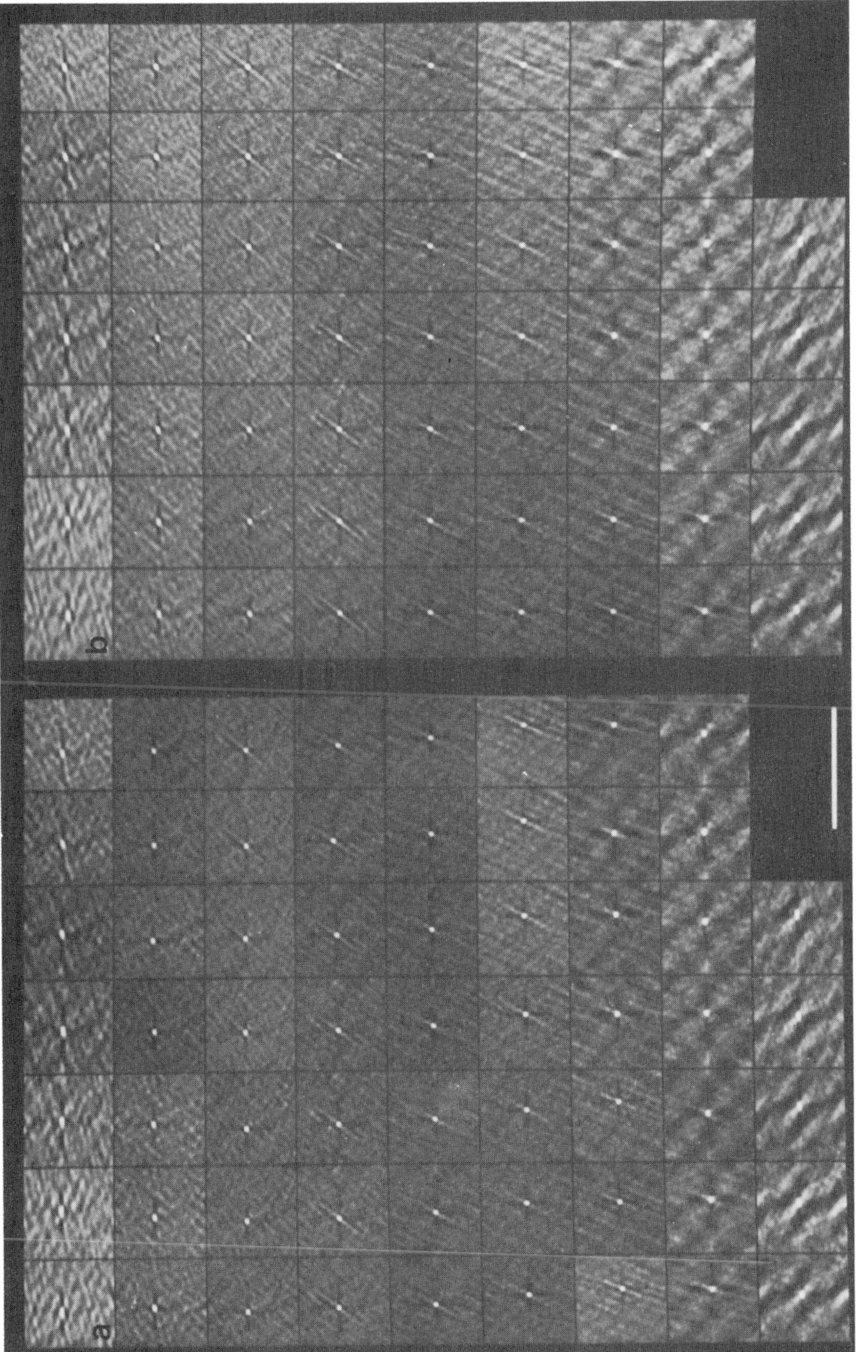

FIGURE 2. Cross-correlation functions from a tilt series alignment. The sample is an osmium-impregnated Golgi preparation from a bullfrog spinal ganglion. The tilt series range from $-62°$ to $60°$ in $2°$ intervals. The CCF from the $-62°$ tilt is shown in the upper left of each gallery, and the others are presented sequentially through the $+60°$ CCF at the lower right. The $\pm 2°$ CCFs are marked with an asterisk. There is no CCF for the $0°$ file because it is the reference to which the rest of the series is aligned (see text). (a) The set of CCFs from the initial alignment of the series. (b) The same set of CCFs but from the second pass of the tilt series through the alignment scheme. Bar $= 1.5\ \mu m$.

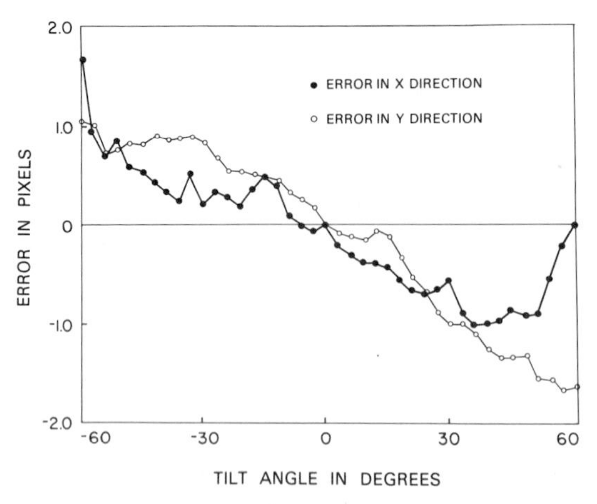

FIGURE 3. The accuracy of translational alignment by the cross-correlation-based scheme as tested on a model data set. The curves represent the error accumulation in the x and y directions as a function of tilt angle. (From Frank *et al.*, 1987.)

the alignment becomes the reference file for the next. We use the untilted image as the initial reference, and the alignment proceeds as two separate sequential branches over the negatively and positively tilted images. Thus, for a tilt series with a 2° angular interval, the 0° image serves as a reference for the 2° image and the aligned 2° image serves as the reference for the 4° image, etc. The 0° image also serves as a reference for the −2° file, which, when aligned, serves as a reference for the −4° image, etc. The schematic illustration of a computer implementation of this branched scheme is given in Fig. 3.

5. *A PROCEDURE FOR TRANSLATIONAL ALIGNMENT OF A TILT SERIES*

Enter the tilt series Here the names of the input (before alignment) and output (aligned) image files, the name of the untilted file, and the number of files in the tilt series are specified.

Enter input parameters Here the angular interval of the tilt series, the low- and high-pass filter radii, and the parameters for the peak search are specified.

DO LB2 Initiation of the outer of a nested pair of DO loops. The outer loop separates the tilt series into branches of negatively and positively tilted images. The loop repeats twice, once for each branch of the tilt series.

Set registers Registers are set so that file numbers and tilt angles are incremented in a forward or reverse direction in the inner loop, depending on which branch of the tilt series is being aligned. The untilted file is set as the first reference file and the number of images in the branch of the tilt series is computed.

DO LB1 Initiation of the inner DO loop. This loop iterates over the files in the current branch of the tilt series.

Cosine stretching Input and reference files are stretched along their x axis (orthogonal to the tilt axis) by a factor equal to the inverse of the cosine of their respective tilt angles.

Compute Fourier transform of the stretched input files Prepare for the following filtering step.

Filtering Apply high- and low-pass filters to the Fourier transform of the input image, using the specified values. Meridional and equatorial filtering (i.e., masking off coefficients along the X and Y axes in Fourier space, respectively) are also applied in order to eliminate possible scanning artifacts.

Cross-correlation Compute the CCF of the filtered, stretched input image relative to the stretched reference image.

Peak search Compute the x and y coordinates for the center of mass of the CCF peak and place these coordinates into registers.

Unstretch The x coordinate from the peak search must be multiplied by the cosine of the tilt angle (of the input file) in order to compensate for cosine stretching in the cross-correlation step. The corrected x coordinate and the y coordinate are the components of the desired shift vector.

Shift The input image is shifted in its frame by the negative of the shift vector (i.e., by the negative value of the x and y components of the shift vector).

Storage The shift vector is stored in a document file for possible future use.

Reduce CCF The CCF is reduced (by local averaging and sampling) to a smaller size for visual inspection and storage. The CCF is overwritten in successive passes of the DO loop.

Advance loop counters All counting registers are advanced so that (1) the aligned input file becomes the reference file for the next pass; (2) the next image in the tilt series becomes the input file; and (3) the proper tilt angle is assigned to the input and reference files.

LB1 Ending label for inner loop.

LB2 Ending label for outer loop.

EN Signal to end the procedure.

6. *EXPERIMENTAL RESULTS*

The CCFs from the alignment of a tilt series of an osmium-impregnated Golgi sample (see Chapter 13) are shown in the two galleries of Fig. 2. The gallery on the left is from a normal alignment of the files, and the gallery on the right is a second application of the scheme to the image files. The peaks for the right-hand gallery are all centered, while those of the left-hand gallery are not, indicating that our scheme found the files well aligned after the first pass. This demonstrates that the scheme is highly reproducible, and, in fact, there is no gain in accuracy by applying the scheme in multiple passes.

It is evident that the appearance of the CCF changes with tilt angle. There are two reasons for this. First, the signal-to-noise decreases with tilt angle, because of cosine stretching, resulting in the high-tilt images being noisier. For example, the 60° image is stretched by a factor of 2 along one axis and then a window of the original size is selected for computation of the CCF. This means that, for computing the CCF, the 60° image has only half the information of the 0° file. This, combined with the usual problems of focus and an increase in section thickness, explains why the high-tilt CCFs are noisier. The second difference arises because the CCF is influenced by the shape of the sample, and thereby the direction from which it is viewed. For this reason, the shape of the peak, and the manner in which the

CCF changes with tilt angle, is unique for each tilt series. An example of this behavior is seen in Fig. 2, where the CCFs from the high positively tilted images contain a set of ridges that are not seen in the negatively tilted branch of the scheme. Similar ridges do not necessarily appear in other CCFs from high-tilt images, even if they are from a similar Golgi preparation.

In general, the main position peak is strong enough that the peak search does not have trouble distinguishing it from ridges and other extraneous peaks. Occasionally the filtering and peak search parameters must be adjusted before the tilt series can be completely aligned, and, rarely, special attention must be given to align one or two of the high-tilt images.

We have tested the accuracy of the CCF-based alignment scheme on a model tilt series (Frank *et al.*, 1987). The projections for this series were computed from a 3D reconstruction, noise was added to them, and then they were shifted by random, but known, vectors. This data set was then aligned, using the scheme outlined in Section 5, and the cumulative error between the known and calculated shift vectors was plotted against tilt angle (Fig. 3). This test gives an overly pessimistic view of the accuracy of the method because the 3D reconstruction from which the projections were computed was itself computed from a tilt series with an angular range of only $-54°$ to $+50°$. Projections computed from outside this range will contain very little information and hence will have noisy CCFs, which could explain why the accuracy trails off at the higher tilt angles. Despite this limitation, the test shows that the cumulative shift error is less than one pixel for most of the tilt series. It would be instructive to repeat this test using a data set computed from a 3D reconstruction that was computed from a tilt series with a full 180° angular range (see Chapter 13).

A simple visual test for the alignment of a tilt series can be created by showing the aligned files in rapid succession on a graphics terminal. This produces a "movie" effect, and, if the series is well aligned, the sample will rotate smoothly about the tilt axis without jumps or in-plane rotations. Tests demonstrate that this visual inspection can detect errors in rotational alignment of 0.50° and errors in translational alignment of 2 pixels (unpublished results). We have used this visual inspection to evaluate the alignment of several tilt series from different types of samples (see Chapter 13), all of which were rotationally aligned by the method of Lawrence (see Chapter 8) and translationally aligned by the methods described here. In general, the alignments are quite good, except for cases when there were not enough clear fiduciary markers for the rotational alignment.

7. SUMMARY

The cross-correlation function is a tool that can be effectively used for the translational alignment of images. When it is used to align a tilt series, the CCF is applied sequentially over the negatively and the positively tilted branches of the series. This sequential application minimizes the angular difference between the projections compared in each step. While it has been suggested that the CCF is not as accurate an alignment tool as the use of fiduciary markers (Skoglund and Daneholt, 1986), our experience (Frank *et al.*, 1987) suggests that it is quite effective, especially in combination with fiduciary markers to determine the rotational alignment. Tests indicate that most of the files of a tilt series can be aligned to within one pixel. In addition to its accuracy, the CCF-based alignment scheme is well grounded in theory, it is relatively easy to implement, and it uses only a moderate amount of computing resource. These characteristics make the CCF-based alignment scheme an important component of tomographic 3D reconstruction methodology.

ACKNOWLEDGMENTS

This work was supported by National Institutes of Health grants GM40165 and RR01219.

REFERENCES

Frank, J. (1980). The role of correlation techniques in computer image processing, in *Computer Processing of Electron Microscope Images* (P. W. Hawkes, ed.), pp. 187–122. Springer-Verlag, Berlin.

Frank, J., McEwen, B. F., Radermacher, M., Turner, J. N., and Rieder, C. L. (1987). Three-dimensional tomographic reconstruction in high voltage electron microscopy. *J. Electron Microsc. Tech.* **6**:193–205.

Frank, J. and Radermacher, M. (1986). Three-dimensional reconstruction of nonperiodic macromolecular assemblies, in *Advanced Techniques in Biological Electron Microscopy* III (J. K. Koehler, ed.), pp. 1–72. Springer-Verlag, Berlin.

Guckenberger, M. (1982). Determination of a common origin in the micrographs of tilt series in three-dimensional electron microscopy. *Ultramicroscopy* **9**:167–174.

Saxton, W. O. and Frank, J. (1976). Motif detection in quantum noise-limited electron micrographs by cross-correlation. *Ultramicroscopy* **2**:219–227.

Skoglund, U. and Daneholt, B. (1986). Electron microscope tomography. *Trends in Biochemical Sciences* **11**:499–503.

10

Computer Visualization of Volume Data in Electron Tomography

ArDean Leith

1. INTRODUCTION

Visualization of the results of electron microscopic (EM) tomography is crucial to extracting the most meaningful biological information from a reconstruction. Choosing the most appropriate technique allows the user to draw better conclusions about structures. For some types of EM tomography such as reconstructions from tilted thick sections, the display of results is the most time-consuming step of the reconstruction. The choice of visualization methods will also determine the format

ArDean Leith • Wadsworth Center for Laboratories and Research, New York State Department of Health, Albany, New York 12201-0509.

Electron Tomography: Three-Dimensional Imaging with the Transmission Electron Microscope, edited by Joachim Frank, Plenum Press, New York, 1992.

in which the reconstruction can be viewed and published. For these reasons methods for the display and visualization of reconstructions are of great interest to investigators who use EM tomography as a tool for their research.

Visualization hardware and techniques have advanced rapidly during the last five years (Barillot *et al.*, 1988). Systems have changed from graphic display terminals driven by minicomputers in the 1980s to the workstations with/without imaging accelerators in use today. A study by Goldwasser (1986) listed speed, storage, and display goals for an ideal biomedical visualization workstation. The workstation should be able to analyze a full-size biomedical reconstruction using different modes of volume and surface rendering. It should be able to slice and rotate the reconstruction, and display the results in color at full-scale resolution. Only four years later, hardware and software are available at less than $40,000, which achieve most of these goals. Indeed the focus of the field has turned to work on editors for measuring and interacting with the volume data, improved interfaces, and making real-time renderings for medical applications.

Electron microscopic tomography produces a three-dimensional (3D) data set. In most EM tomographic procedures the data set has the same units in all three dimensions. Each cubic volume element (voxel) from this data set contains a density value. This value is derived from the electron transparency of the corresponding material in the original sample. Electron microscopic tomography can be done using positively, negatively, or unstained specimens (Frank, 1989). Thus the transparency of a voxel may correspond to the relative presence or absence of actual material in the structure being imaged. A simple contrast reversal can always be applied to the data to invert the transparency, if desired.

The earliest EM tomographic studies often used a montage of the individual slices to represent the reconstruction. However, it is usually difficult to visualize the 3D structure of an object from its component slices. Some type of 3D rendering of the volume is necessary in order to appreciate the 3D connectivity of object features (Vigers *et al.*, 1986; Namba *et al.*, 1985; Luther, 1989).

2. RENDERING

Rendering is the process whereby the data is converted to a visible format. Rendering usually attempts to provide depth and texture clues using hidden surface removal, shading, and texturing. These clues help to show the structure of the object. Currently there are two general methods of rendering in common use for tomographic reconstructions: surface and volume rendering. In addition, there are several other methods available which are of lesser utility.

Surface rendering reduces the volume data to a set of surfaces which describes the boundaries between the components present in the volume. These surfaces are then illuminated and shaded. Volume rendering retains the volume data and sums it from back to front with a weighted projection technique so that no binary surface decisions are necessary. Information from all of the voxels is potentially incorporated in the rendering.

Recently, there has been considerable controversy over which of these two methods provides better results, particularly in medical applications (Frenkel, 1989;

Udupa and Herman, 1989; Ney *et al.*, 1990; Tiede *et al.*, 1990; Udupa and Hung, 1990). Some of the advantages of these methods in EM applications will be discussed. There is considerable variation in methodology within each of the two methods. The descriptions of volume rendering methods (Schlusselberg *et al.*, 1986; Levoy, 1988; Drebin *et al.*, 1988; McMillan *et al.*, 1989; Upson and Keeler, 1988) differ in the actual algorithms applied, the use of color, the use of surface extraction before projection, and the type of perspective used. Indeed some volume rendering methods appear very similar in results to certain surface rendering methods.

3. SURFACE RENDERING

3.1. Surface Definition

The initial step in most surface renderings is to define which voxels are at the desired surface(s) of the object. This step essentially requires 2D or 3D segmentation of the volume. In EM tomographic reconstructions the densities are usually scaled to arbitrary values, and there are no a priori levels for a surface. There are three general approaches to defining the surface.

The first approach is to choose a density threshold level for the surface. The choice of density threshold is usually somewhat arbitrary and is done by trial and error. A particular threshold is chosen for the surface and a rendering is calculated at that threshold level. The rendering is examined, and the surface density threshold is altered to display more or less of the structure(s) of interest. By altering the density threshold, one can "peel away" the less dense areas. In some case one has a priori knowledge that portions of the structure (such as RNA) have a higher mass density or opacity, and this technique can be used to localize the dense structures within the reconstruction.

The second approach for defining surfaces is more rigorously analytical. Instead of setting the surface at a particular density "threshold," the surface is set at a particular level of the density "gradient." The surface is placed wherever the density is changing most rapidly. To do this, the original density volume is converted to a gradient magnitude volume by finding the change of density at each voxel. The gradient magnitude can be calculated slice by slice or in 3D. Three-dimensional gradients are preferred since more of the voxels context is used. A histogram of the values in the gradient volume provides a guide for choosing the gradient threshold where the surface is set. Viewing slices of the gradient volume may also be helpful in determining the threshold. The gradient magnitude volume is used for surface extraction once an appropriate gradient threshold has been determined. This method is especially appropriate for reconstructions from negatively stained objects since there is a strong gradient between the natural boundary of the object and the surrounding stain (Verschoor *et al.*, 1984). However, this method may not be appropriate for objects which lack strong density contrast. Also, thin layers or shells of material in an object sometimes develop artifactual holes in the reconstruction.

The third approach for defining surfaces utilizes contour- or surface-following algorithms (Liu, 1977; Cappelletti and Rosenfeld, 1989). A contour-following algorithm is less desirable since it works slice by slice and fails to take advantage of the 3D connectivity present in the volume. Some surface-following algorithms are interactive, in that they require the user to identify a seed voxel on the desired surface. If there are many unconnected surfaces, interactive selection of seed voxels is difficult and algorithmic selection then becomes necessary. The algorithm attempts to move away from the seed voxel following surface connectivity and places the surface voxels into a binary surface volume. Methods for determining connectivity use information from the density values, density gradient values, texture filters (Lenz *et al.*, 1986), as well as knowledge-based methodologies (Suh *et al.*, 1990; Garcia *et al.*, 1990, Dhawan *et al.*, 1990). If the surface-following algorithm is sufficiently clever, it may be able to overcome the problem of "holes" developing in materials which form thin layers or shells in the volume. Surface-following algorithms are useful for noisy density volumes or volumes made from unevenly stained material. However, there is no general method of determining connectivity which works for all reconstructions. Surface-following algorithms are usually integrated with the surface modeling step described below. Although surface-following algorithms have been applied to medical tomography, apparently they have not yet been used in EM tomography.

3.2. Surface Modeling

Once suitable levels have been established for the surfaces, the rendering algorithm must actually derive a model for the surface. Most approaches to surface rendering derive a 3D surface model, which can then be viewed from any desired direction and with arbitrary selection of shading parameters. By deriving this surface model, the original data can usually be compressed to a fraction of the size of the original volume data. A few algorithms for surface rendering (Radermacher and Frank, 1984) use an alternative method in which the viewing angle is set before deriving a 2D surface model. This model is only valid for the chosen viewing angle. This method is rapid if a single viewpoint is needed, but is generally slower when many successive viewpoints are used.

There are several methods commonly used for deriving a 3D surface model. The cuberille method (Herman and Liu, 1979) steps through the volume, placing square tiles around the boundary of any voxel that is on the desired surface. Algorithms for determining cuberille surface tiles are given by Artzy *et al.* (1981). Heffernan and Robb (1985), and Gordon and Udupa (1989). The resulting surface model is thus made up of tiles that are parallel to one of the three mutually perpendicular planes which were present in the original volume. Since polygons from adjacent cubes can often be combined to form a single polygon, the cuberille method allows a very large compression in the data base for the model. When this model is rendered using simple shading methods (described below), the reconstructed object appears to be built from little cubes. The final shaded cuberille rendering can be passed through a low-pass filter to blur the cube surfaces (Herman and Liu, 1979). This blurring causes a loss of resolution and still does not remove all of the "blockiness." There are several shading methods which can be used to decrease the blockiness of the final rendering. However, reconstructions from cuberille surface models seldom appear as visually satisfying as those made from other methods.

The marching-cubes method (Lowrensen and Cline, 1987; Cline *et al.*, 1988) creates a smoother surface model than the cuberille method. In the standard cuberille method a binary decision is made whether to include a whole voxel as part of the surface. In the marching-cubes method a surface tile can subdivide a voxel, depending upon the plane of intersection of the desired surface density with the voxel. To calculate this intersection, the eight voxel densities at the vertices of a cube coming from two consecutive slices are considered. The algorithm determines whether each of the eight densities is above, on, or below the desired surface. The algorithm then creates triangular tiles by interpolating the portions of the cube which lie on the surface. One disadvantage of this method is that it produces a large number of tiles. The number of tiles depends upon the surface area and can be as large as several hundred thousand. This large number often results in slow rendering speeds. However, adjacent tiles can often be averaged and combined, with some loss of resolution, to decrease the number of tiles and speed up rendering.

The contouring method (Cook *et al.*, 1983; Pizer *et al.*, 1986) usually provides very smooth reconstructions. However, contouring may require manual interaction with the data. An isovalue contouring algorithm is used to draw smooth contours at the desired surface density level inside a density or density gradient volume. This contouring is usually done slice by slice through the volume. The algorithms (Winslow *et al.*, 1987) are similar to those used for automatically drawing topographical maps from altitude data. When the data have variations in the staining density or when it is difficult to set a specific surface density, the automatic contouring method may fail. In these cases contours can be manually traced around the profiles of interest with a contour-tracing program (Marko *et al.*, 1988; Johnson and Capowski, 1983; Young *et al.*, 1987). Simple montages composed of slice by slice isovalue contour maps have often been used to display EM tomographic reconstructions (Oettl *et al.*, 1983; Lepault and Leonard, 1985).

A tiling algorithm (Fuchs *et al.*, 1977; Christiansen and Sederberg, 1978; Shantz, 1981; Winslow *et al.*, 1987) can then be used to connect nearby points on successive contours to form a surface of triangular tiles. Unfortunately, biological structures often have complex free-form surfaces with several contours on each slice. This may cause connectivity problems for tiling algorithms. There may be several alternative tilings available which provide vastly different surfaces. There is no general algorithm available for tiling successive complex branching contours which sometimes enclose holes. When connectivity problems arise, the user may have to intervene and guide the tiling process. This intervention is often difficult, even if the user knows the probable shape of the biological structure.

The orientation chosen for slicing the volumes before contouring can help overcome problems with contour connectivity. The slicing orientation also influences the shape of the final surface. The contouring method usually produces a smooth surface with fewer tiles than the marching-cubes or cuberille algorithms; thus, subsequent rendering procedures are faster.

A surface-rendered reconstruction is not limited to a single surface level. Most surface rendering programs allow one or more options for visualizing internal surfaces. A common method is to make the outer surface partially transparent (Jiez *et al.*, 1986). Transparency can be simulated by dithering the pixels on the display between showing the outer surface and showing a inner hidden surface. This dithering allows the viewer to see the internal structure. However, this method seldom works well for more than two surface levels. Another approach is to section the volume or cut windows out of parts of the outer surface. A third option is to display a video loop series of renderings at different surface levels and allow the mind to integrate the display into a 3D model.

3.3. Surface Rendering Implementations

Once the surface has been described in a mathematical form the reconstruction can be rendered. The rendering consists of rotating the model to the desired angle, determining which surfaces are hidden, and calculating the shading for the model surface. In most modern rendering implementations these steps are usually integrated to speed up the process. There are various rendering implementations available which are usually limited in the type of hardware they require. Graphics hardware often contains specializations, such as a Z buffer for hidden surface removal, which are useful for speeding up the rendering. A Z buffer is a stack of memory planes which are used to implement the "painters algorithm" for hidden surface removal. When nearby objects are written into the nearer planes of the buffer, they hide objects which are most distant in the buffer.

Surface rendering can be done by a general-purpose surface rendering implementation such as MOVIE.BYU (Christiansen and Stephenson, 1984) or by an implementation which is specialized for the data and problems that usually occur in biomedical tomography. General-purpose rendering implementations usually operate on the surface model itself. They were originally designed for rendering polygonal or patch graphics data and emphasize photorealistic rendering with true perspective. Patch graphics uses a mesh or patches of splines to represent the surface. General-purpose renderers often do not contain routines for obtaining a surface model from a tomographic density volume. The rendering implementations may often restrict the number of tiles, the shape of tiles, the order of listing of tile vertices, etc. Thus a rendering implementations should be evaluated not only from the rendering capability but also from its ability to accept the surface model chosen for the reconstruction. MOVIE.BYU was an early general-purpose rendering implementation that contained a contour tiling routine. Many newer general-purpose implementations have the ability to accept the MOVIE.BYU data format and convert it to their own preferred data formats.

Most general-purpose graphics rendering implementations have the capability to display a tiled surface model (whether from cuberille or contoured data) as a wire-frame drawing, with or without hidden-line removal (Figs. 1 and 2). Wire-frame drawings are useful as they are quick to compute and one can rotate the reconstruction around to find optimal viewpoints. Modern graphics hardware is often fast enough to interactively rotate such models in real time.

3.3.1. Surface Shading Algorithms

The shading step in the rendering is important in determining the degree of smoothness and realism of appearance. Various algorithms which differ in surface smoothness are available for shading the reconstruction (Greenberg, 1989). The

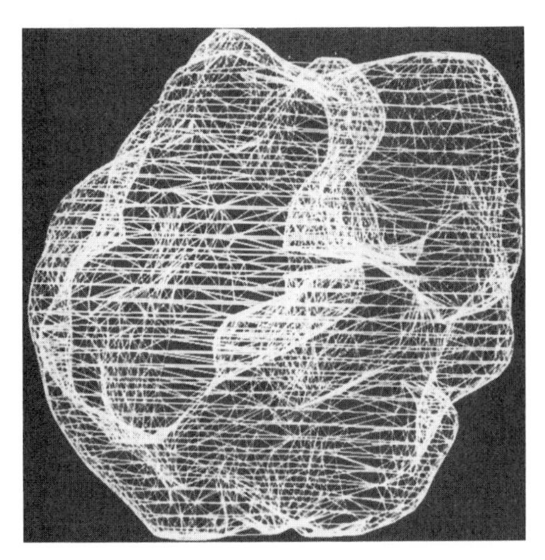

FIGURE 1. Polygonal surface rendering of 70S *E. Coli* ribosome displayed as a wireframe. Made from isovalue contours which were connected with a tiling routine. (Ribosome data courtesy of P. Penczek, S. Srivastava, and J. Frank, Wadsworth Labs, Albany, NY.)

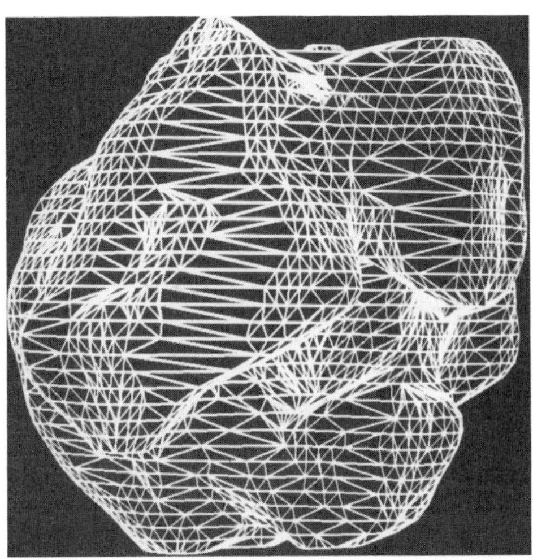

FIGURE 2. Polygonal surface rendering of ribosome of Fig. 1 displayed with hidden-line removal.

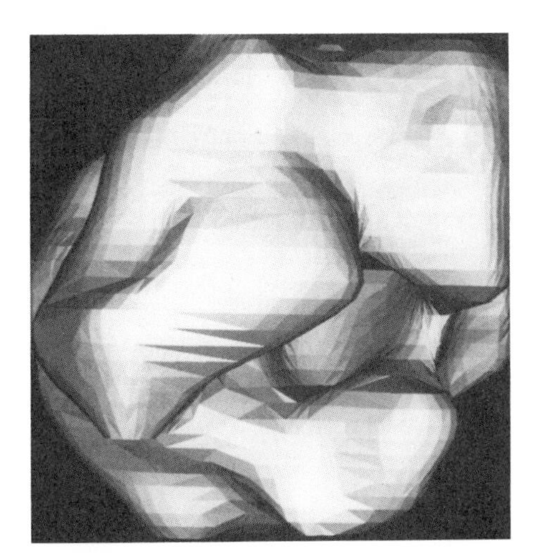

FIGURE 3. Polygonal surface rendering of ribosome of Fig. 1 displayed with uniformly shaded surfaces. Single light placed at eye of observer.

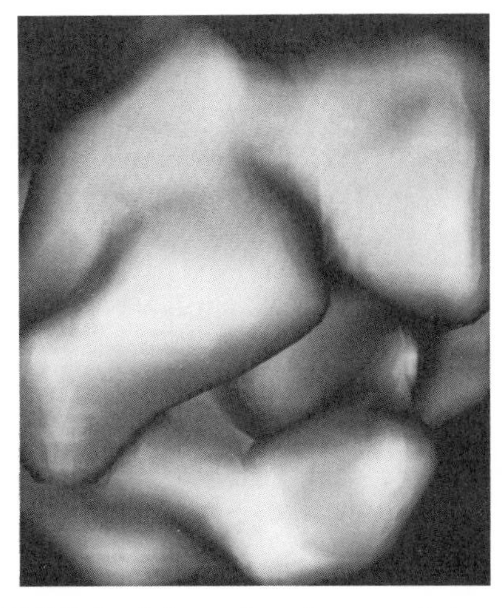

FIGURE 4. Polygonal surface rendering of ribosome of Fig. 1 displayed with smoothly shaded surfaces. Single light placed at eye of observer.

following three algorithms are used in both general-purpose and biomedical surface rendering implementations.

In the simplest shading algorithm, a single shade or intensity is assigned to each surface tile (Fig. 3). The intensity of illumination for each tile is calculated with a cosine rule by considering the angle between the incoming light and the normal to the plane of the tile. When this algorithm is used for a rendering, each tile of the rendering appears as a separate facet and can usually be distinguished from its neighbors. This is especially distracting for cuberille-modeled surfaces.

Gouraud (1971) shading yields a smoother appearance. The intensity is calculated at each vertex of the tile, and then the intensity is interpolated between the vertices. This shading is smooth, but banding artifacts may appear and some surface features may be lost unless tiles are subdivided.

Phong (1975) shading overcomes some of these disadvantages and provides a more realistic rendering (Fig. 4), although it is slower. In Phong shading the normal vectors, not the intensities, are interpolated between vertices. Some rendering implementations (Pizer *et al.*, 1986) choose between Gouraud and Phong shading, depending upon the angle between the tile and adjacent tiles. If tiles are small and numerous, Gouraud shading appears almost as realistic as Phong shading.

3.3.2. General-Purpose Surface Rendering Implementations

More complex and "realistic" rendering implementations have additional features to approximate the effect of light on real objects (Fig. 5). These implementations provide reconstructions which differ in surface smoothness, lighting options, color options, transparency options, texture options, perspective, etc. In some implementations the object can be illuminated with multiple lights of various

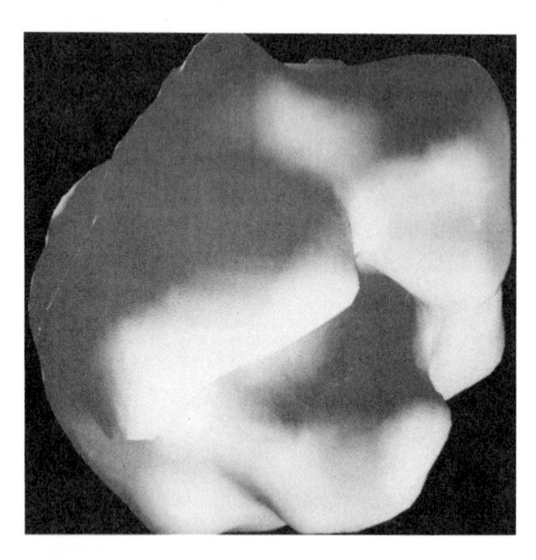

FIGURE 5. Polygonal surface rendering of ribosome of Fig. 1 displayed with smoothly shaded surfaces. A strong light placed at lower right and another light placed behind the ribosome.

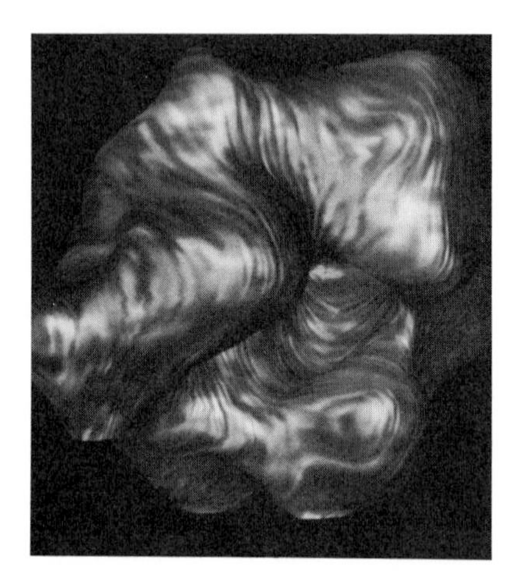

FIGURE 6. Polygonal surface rendering of ribosome of Fig. 1 displayed with smoothly shaded surfaces and a wood-grained texture map. Single light placed at eye of observer.

colors, intensities, and beam shapes. The illumination may allow shadowing and self-shadowing of the object as well as hazy or foggy backgrounds. As mentioned earlier, most implementations allow the user to section or cut away surfaces in the model and view previously hidden surfaces. The object may be rendered with a partially transparent surface. A texture map may be draped over the surface of the object to add high-frequency detail to the surface (Fig. 6).

While these features may be important for achieving photorealism in automobile advertisements, they are probably of little value for increasing the understanding of biological structure from an EM tomographic reconstruction. Contemporary general-purpose rendering implementations provide such a multitude of choices for variation of appearance of a completed reconstruction that there is seldom time or incentive to take full advantage of these features. Nevertheless, there has been a rapid increase in the sophistication of renderings of tomographic reconstructions in recent years. Simple depth-shaded displays have been replaced by smooth shaded surface renderings with shadows, etc. There appears to be a developing consensus (Levoy *et al.*, 1990) to display medical tomographic reconstructions using a reflective metallic color or texture. This coloration allows easy detection of small features. This convention may also be appropriate for EM tomographic reconstructions. However, many areas of choice still remain for a personal variation of renderings.

3.3.3. Specialized Biomedical Surface Rendering Implementations

The cuberille method of surface modeling is often favored since it is relatively fast, easily implemented and does not encounter contour connectivity problems. However, the reconstruction often appears blocky when rendered with general-purpose rendering implementations. To overcome this blockiness, several special-purpose rendering implementations have been developed for biomedical data. Many of these implementations do not actually derive a cuberille surface model. They instead use a binary data volume which is formed by the surface definition step of the rendering. In practice, the binary surface volume may be stored in various compressed formats rather than an actual volume. Instead of modeling the cuberille voxel surfaces as reflective planes, other properties of the surface voxels can be used to find the reflectance.

One simple approach uses depth shading of the voxels (van Heel, 1983; Vannier *et al.*, 1983; Borland *et al.*, 1989). Each surface voxel is given an intensity declining according to its distance from some arbitrary viewing plane (Fig. 7). If the scale for the image is large, a single value of intensity can be used for all exposed faces of a single voxel. Simple depth shading is quick, but gives low resolution (Chen *et al.*, 1985).

Another approach used by Radermacher and Frank (1984) calculates the reflectance of light by Gouraud shading off planes defined by the depth of the voxel and the depth of two of its neighbors. This surface reflection intensity is then added to a depth-shading intensity to give the final rendering (Fig. 8). This method is rapid, but creates artifacts in areas of high slopes. Improvements on this method with fewer artifacts are given by Gordon and Reynolds (1985) and Saxton (1985).

Hohne and Bernstein (1986) have proposed using the density gradient around the voxel to determine a reflective plane instead of the voxel surface. This algorithm

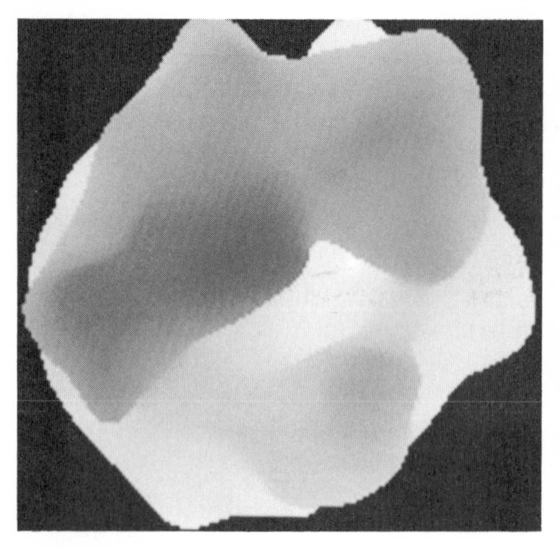

FIGURE 7. Depth-shaded surface rendering of ribosome of Fig. 1.

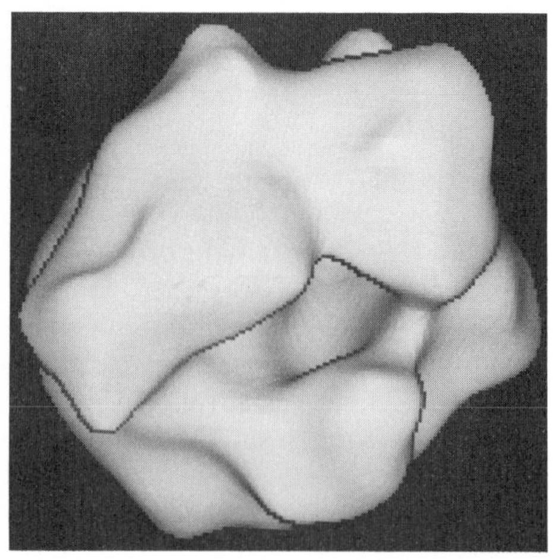

FIGURE 8. Depth- and surface-shaded rendering of ribosome of Fig. 1.

requires that the surface model carries along added density information. Data compression is less than with some other approaches, but the surface appears quite smooth.

In most of these implementations the hidden-surface removal and shading are calculated using orthogonal perspective so that these renderings do not give true 3D appearance. Furthermore, these variations seldom allow for transparency of the surface rendering.

4. VOLUME RENDERING

Volume rendering operates in volume/image space so that until the final projection step the whole volume of data is usually present. There are several steps involved in the rendering process. As discussed, there are differences in the meaning of the term *volume rendering*, and the following description corresponds to the implementation described by Drebin *et al.* (1988) and Ney *et al.* (1990b).

The first step in volume rendering is classification of the input density data. Classification is essentially the same as 3D segmentation of the data, but the segmentation is fuzzy rather than binary. Classification converts a density volume into a component percentage volume. There are various classification strategies which can be followed, depending upon the data characteristics and whether the rendering is to be in color. Classification can be determined by density values, by the gradient of density values, by masking using texture filters, interactively by region growing, or by any other method which assigns components to voxels based upon the original density values. In x-ray medical tomography, air, fat, soft tissue, and bone are probabilistically assigned to different data densities values. If the renderer operates in color, these components can be assigned different colors. Each voxel of the volume is assigned a color, depending upon the mapping between the density and the components with that density. Additionally each voxel is assigned an opacity. The opacity for each pure component is arbitrarily selected based upon what components are to be emphasized in the rendering. The opacity for a voxel is assigned by summing the percentage of each component in the voxel times the opacity of the pure component.

For EM tomographic data the classification mapping between the data densities and the components is not obvious a priori. There are obvious differences in mapping between reconstructions from negatively and positively stained specimens. In the absence of any conventions or further guidelines for the assignment of components to data densities, a conservative approach seems advisable. According to this approach, the densities are mapped directly to the component percentage, and color is not used in the rendering. Alternatively, one can arbitrarily map higher densities to a color which contrasts with that chosen for the lower densities and assign the higher densities a high opacity. This classification mapping will give a transparent coating on a dense interior backbone.

The final result from a volume rendering is usually a shaded surface projection. However, most renderers can omit the surface extraction step to produce unshaded projections (Figs. 9 and 10). Ney *et al.* (1990a) recommend the use of unshaded projections in some medical applicatons. This option should be investigated for rendering data with high contrast among opacity ranges for classified components and for filamentous data having large areas of background between structural elements.

If shaded surface projections are desired, then a surface extraction step is necessary. This step uses a 3D gradient of the refractive index volume. This refractive index volume is often the same as the opacity volume discussed above, but it can vary. Refractive index values can be arbitrarily established for the different components in the volume. Components with large differences in refractive indices have distinct surfaces at their interfaces. The refractive index for each voxel is the

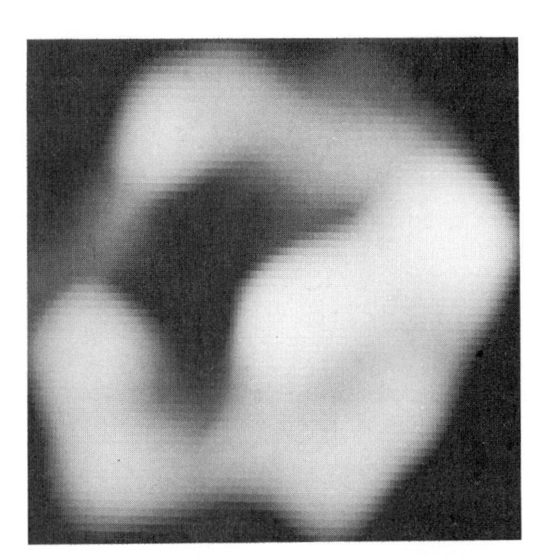

FIGURE 9. Unshaded volume rendering of ribosome of Fig. 1.

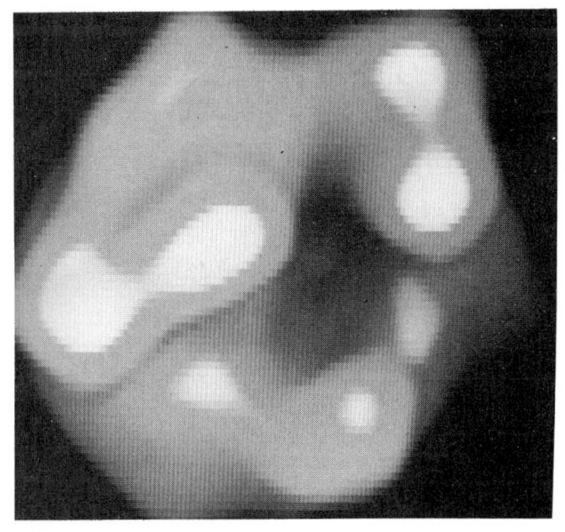

FIGURE 10. Unshaded volume rendering of ribosome of Fig. 1. High data densities set to a lighter color.

sum of the refractive indices of the pure components times the amount of the component in the voxel. The direction of the refractive index gradient in contained in a surface normal volume. The magnitude of the density gradient is contained in the surface intensity volume. These volumes are used by the lighting model to assign reflectivities to the surface which is present in each voxel. In volume rendering there is a surface value available for each voxel of the volume, not just at interfaces.

The lighting model used in volume rendering follows the path of a ray of light along a column of voxels through the pixel volume. All of the rays are parallel, so the eye of the observer is located at infinity. Light is emitted by a voxel corresponding to its opacity and color. For surface shading, an external light is aimed at the object from the desired direction. This light is not attenuated as it passes into the volume, but reflects off the surfaces it encounters. The light reflects with an intensity which depends upon the angle of incidence between the ray and the surface present in each voxel. A shaded surface volume is thus calculated from the density gradient volumes and the color/opacity volumes.

The shaded surface volume (or color/opacity volume if shaded surfaces are not desired) is then rotated to the desired viewing angle. In the simplest geometry the rotation occurs within each of the original slices. The data is then projected through the plane of the slices. Thus, if the original data consisted of XY slices the reconstruction would be viewed from the edge of the Z stack.

The final step is projection through the volume using the merge-over projection operator (Porter and Duff, 1984). Projection takes place from back to front to obtain a sum of color or intensity weighted by the opacity. A whole column of voxels is used to provide an individual pixel in the resulting 2D image. Any voxel within the column can contribute to the final pixel intensity, depending upon its relative opacities and surface reflections (Figs. 11 and 12).

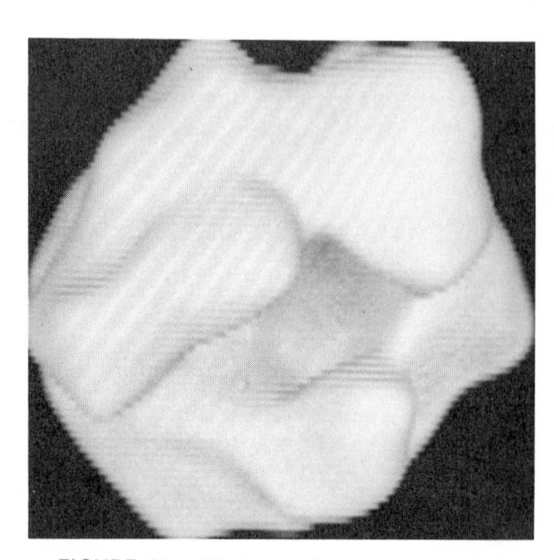

FIGURE 11. Shaded surface volume rendering of ribosome of Fig. 1. Single almost opaque surface. Light is placed at eye of observer.

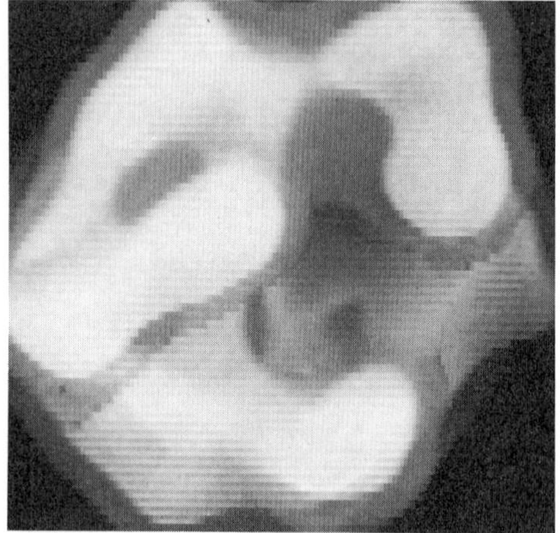

FIGURE 12. Shaded surface volume rendering of ribosome of Fig. 1. Has an partially transparent outer surface over an almost opaque inner surface set at a higher data density level. Light is placed at eye of observer.

In some volume rendering implementations (Drebin *et al.*, 1988) the rendering goes slice by slice through the original volume. The final 2D image is built up scan line by scan line from each successive slice. Other implementations operate iteratively plane by plane from the back of the volume to the front by replacing each plane with a newly calculated projection as the summation works forward through the volume. If this rendering can be viewed plane by plane as it is built up, the user can often gain an added appreciation of the hidden structural details.

With the exception of early work by Schlusselberg *et al.* (1986) most volume renderers use orthogonal perspective when calculating visibility and shading. Schlusselberg used the slower, but more realistic, true 3D perspective.

A disadvantage of the volume rendering implementation of Drebin is that small data sets appear blocky. Upson and Keeler (1988) have proposed a method to reduce the blockiness by allowing the data values to vary continuously within a voxel.

4.1. Volume Rendering Implementations

Commercial volume rendering hardware and software implementations are available from Vicom, Fremont, CA; Vital Images, Inc., Fairfield, IA; Sun Microsystems, Mountainview, CA; and Stardent Corporation, Newton, MA. A software package that is capable of running on several different Unix workstations is described in Robb and Barillot (1989).

5. OTHER RENDERING METHODS

Isovalue contour stacks have been used for many years to display reconstructions (Fig. 13). The contours can be filled to give hidden-line–surface-removed contour stacks (Fig. 14). These methods are quick and often effective at presenting structural information (Leith *et al.*, 1989).

An effective method for displaying reconstructions of filamentous objects, where the cross section of the objects is small relative to the area of the background, is simple projection through the volume. Both additive (Fig. 15) and maximum density projections (Wallis *et al.*, 1989) have been used. Maximum density projection does not sum the density along the column of voxels but uses the maximum density found anywhere in the column (Fig. 16). These techniques are very useful in displaying reconstructions of nerve networks. Summed projections have also been used for EM tomographic reconstructions of a ribosome subunit (Lenz *et al.*, 1986), but the rendering is inferior to other, newer methods.

The most photorealistic renderings are done by true ray-tracing implementations (Whitted, 1980; Fujimoto *et al.*, 1986). These implementations follow the paths of rays of light within the structure and model the interaction of the rays with each of the surfaces that the rays strike. New rays are spawned at reflective surfaces. True ray tracing follows all the rays which interact with the structure and end at the location of the observer's eye. Ray tracing should not be confused with the ray casting used in volume rendering where only rays passing along a single column of voxels are considered. True ray tracing is particularly effective at rendering highly

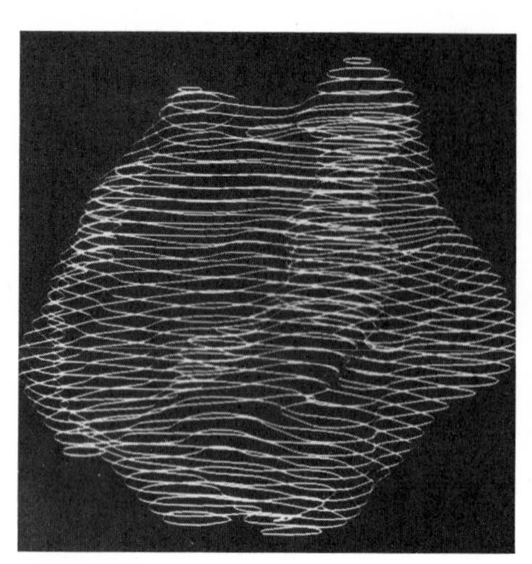

FIGURE 13. Isovalue contour stack rendering of ribosome of Fig. 1. The viewpoint is about 12° higher than other renderings so that contours are not viewed edge-on.

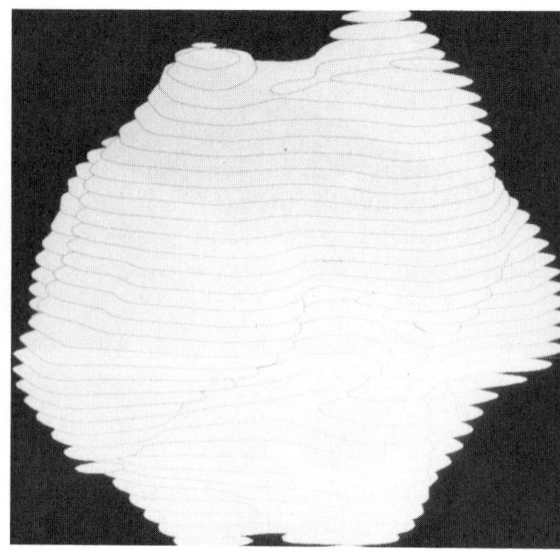

FIGURE 14. Isovalue-filled contour stack rendering of ribosome of Fig. 1.

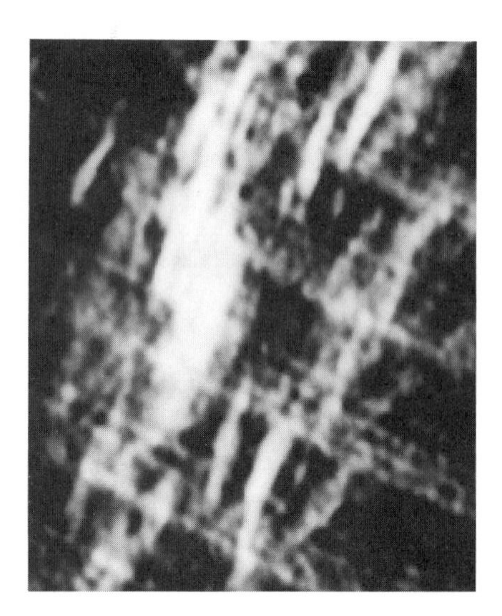

FIGURE 15. Simple additive projection rendering of developing bone reconstruction. (Data courtesy of W. Landis, Childrens's Hospital, Boston, MA, and B. McEwen, Wadsworth Labs, Albany, NY.)

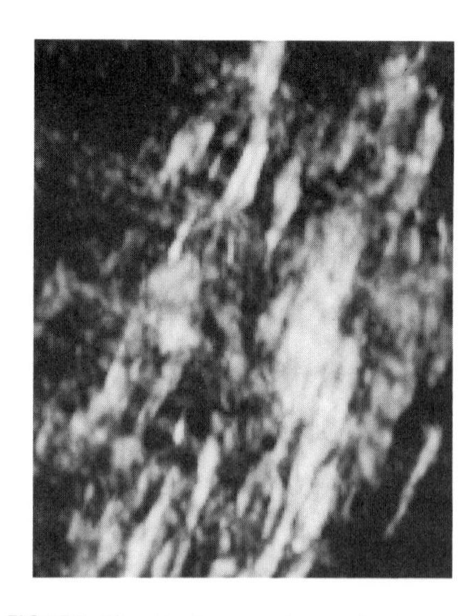

FIGURE 16. Maximum value projection rendering of developing bone reconstruction of Fig. 15.

reflective and transparent surfaces and gives realistic shadows and reflections. Such renderings are usually very time consuming, and it seems unlikely that they will add anything to the interpretation of biological information from EM tomographic reconstructions.

Another approach to defining the surface is to model the surface as a cloud of points (Hersh, 1990). A variation on this method is commonly used to represent molecular structure and provide reconstructions which resemble the common space-filling plastic models. This method is usually computationally expensive and has not been used for EM tomographic applications, but seems to provide an opportunity for further development.

6. VIDEO LOOPS

Video loops consist of a series of views from a rotating reconstruction (or a series at different surface levels). A rotating video loop provides motion parallax which is very useful for visualizing 3D structure. At present, a fast graphics workstation such as a Silicon Graphics Iris, (Mountainview, CA), or an ICAR Workstation, from ISG Technologies (Rexdale, Ontario, Canada), may be able to provide real-time rotation for a fairly simple surface reconstruction. However, more complex surface renderings and all volume renderings require precomputation of the individual rotation views. These precomputed views can be video recorded to provide a real-time rotation loop. Alternatively, imaging systems with a large video memory such as a Pixar Image Computer (Vicom Corp., Fremont) can be loaded with a series of precomputed views which can then be displayed in real time.

If video recording is necessary a video camera can be aimed at the monitor. Video cameras which can directly accept the video signals from high-resolution workstation monitors are not yet available. The options are to use a genlock board in the workstation to convert the video output to a video camera acceptable input signal, or to use a scan converter on the video output from the workstation. In any case a graphics workstation monitor has far better resolution than currently commercial video systems, so there is a considerable loss of detail in the recorded video loop.

7. MODELS

Physical models provide an informative method of displaying 3D EM tomographic data. Physical modeling techniques were highly developed during the 1950s and 1960s and are commonly used by crystallographers to present their findings. For years, results from 3D reconstructions in electron microscopy were similarly displayed (DeRosier and Klug, 1968; Skoglund *et al.*, 1986; Harauz and Ottensmeyer, 1984). Recently these methods have been neglected in the rush to computerized video displays.

However, models offer several advantages over video displays. Model construction is usually "low tech," and the equipment is relatively inexpensive. Models can often be transported to sites lacking compatible display methods and can be discussed in small group settings. They are useful for studying binding and docking of reconstructed objects. They also provide easily assimilated 3D understanding. Many people who have poor shape/form conceptualizing ability never obtain similar levels of understanding from a photograph or video display. With good lighting a model can often be photographed to provide a 2D representation of the reconstruction which is as informative as any video display.

Models do have some distinct disadvantages. They usually require a large amount of time and labor for their construction and they are often fragile. They are not appropriate for demonstration to large groups. They are static presentations at a defined surface level. If the modeler wants to ask "what would the structure be like if I alter the threshold level for the surface by 10%?" then a totally new model must be constructed. They are usually not appropriate for fibrous structures or structures with low solid content such as the cilium (McEwen *et al.*, 1986).

In many cases the best strategy is to use computerized video modeling in the early stages of studying a reconstruction. The researcher can then make a model of the object for added insight or display.

Traditional modeling techniques are quite similar in practice. However, the same methods can be used with many media to produce very different models. The initial step is to choose threshold levels for the surface of the model. This step is the same as the step necessary for surface rendering discussed above, and it gives similar disadvantages. If the model is made of a transparent medium, it may be possible to show multiple surface levels in the same model.

The volume of density data is contoured slice by slice using a contouring program that produces isovalue contours. These contours are transferred onto the modeling medium. The model is usually constructed by stacking up many layers of the media. A common approach is to direct the output of the contouring program into an *XY* plotter. The contours can be plotted on paper, transparent media, or layers of opaque media such as Styrofoam. The contoured area can then be cut out, stacked together with proper separation, and permanently attached in register.

With transparent media such as plastic sheets, it may not be necessary to cut out the contours. If a smooth surface is desired the model can be sanded or machined, depending upon the medium. The model can be used as a mold to make additional copies of the model in different media. The end result is a very effective representation of the reconstruction at the chosen isovalue contour level. However, as stated before, the process is usually very labor intensive.

Recently here have been several attempts to mechanize the model-building process. While these attempts are aimed toward the production of industrial prototypes or medical prosthesis models, they are also useful for our purposes.

The output of isovalue contouring programs or even the 3D density volumes can be entered into a numerically controlled milling machine that directly carves out a model from a block of suitable medium. Medizin Elektronik Gmbh (Kiel, Germany) currently is marketing such a system for medical applications. The complexity of the model that can be formed by such a device depends upon the number of degrees of freedom of the machine's milling arm. Many reconstructions of biological structures such as cilia and ribosomes at high threshold levels have complex invaginations and channels which may be impossible to machine from a single block. In such cases the model has to be built up from multiple layers. In the extremes of complexity this method may offer only minor improvement in speed over the methods described previously.

A recent innovation in computerized model building called *stereolithography* turns the model building process inside out (Wood, 1990). The model is built up by adding material layer by layer instead of by removing extraneous material from the layers. In stereolithography a computer-controlled laser shines on the surface of a vat of photopolymerizable plastic. The laser scans across the surface of the plastic, depositing a solidified layer of plastic on a flat support. When a full layer of the model has been formed, the support sinks deeper into the vat and the old layer of plastic provides a support for the attachment of the next layer of material. The model has excellent resolution and accurately reproduces complex convoluted shapes with internal cavities. The plastic can be somewhat translucent, which allows a degree of study of internal detail. At present the main disadvantage of this method appears to be the high cost of the system. Stereolithographic systems are available from 3D Systems (Valencia, CA).

8. STEREO

The use of stereo techniques to display a rendering usually results in substantial improvements in realism and ease of understanding structural detail. Most stereo display techniques require the formation of two renderings of the object from about 5° to 10° horizontal difference in viewpoint (Ferwerda, 1982). These two renderings are then presented to different eyes of the viewer. Various methods (Hodges and McAllister, 1989) are available for presentation of images using video and vibrating mirrors. Probably the most convenient method is to use the equipment from StereoGraphics Inc. A small modification is made to a graphics monitor to allow it to display the two renderings and synchronize with a pair of goggles containing liquid crystal shutters.

For hard-copy presentation of stereo pairs in publication the usual method is to print the two renderings side by side (Love, 1989). The pair of images is then viewed with a stereo viewer or by crossing one's eyes. This method is limited in the size of presentation by the separation between the human eyes. The left-right selection of the images will depend upon which of the two viewing methods is to be used.

Stereo presentation has two major disadvantages. The first is the requirement for specialized viewing technology. The second is that a sizable fraction of the population is unable to see stereo with any given method of presentation.

9. DISCUSSION

The accompanying figures demonstrate many of the points raised. Fig. 14 shows that a rapidly generated filled contour stack provides useful structural information, but is limited to views more than 10° off the equatorial axis of the stack. The renderings from tiled contours (Figs. 2–6) all require tedious interaction with the tiling software (MOVIE.BYU) in order to establish the geometric model.

Once the model was available, polygonal wire-frame rendering with hidden line removal provides useful rapid structural information (Fig. 2). The polygonal surface renderings from tiled contours (Figs. 4, 5, 6) are visually attractive. Simple depth shading (Fig. 7) gives very poor renderings, but adding surface shading (Fig. 8) gives a very realistic rendering. The unshaded volume rendering with little transparency (Fig. 9) has a lot of surface structural information in it, but does not appear visually satisfying. The shaded surface volume rendering is very good at showing surface details (Fig. 11). The partially transparent shaded surface rendering with an inner opaque shell (Fig. 12) gives significant information on two levels of surface with only a small loss of information on the outer surface. All of the renderings are more useful if shown as a video loop of the revolving object. Stereo views are always more informative, providing the user can perceive stereo.

For reconstructions of objects such as fibrous networks with small structural detail on large areas of background (Fig. 17), simple rendering methods are often very effective. For such data, maximum value projection renderings (Fig. 16) and shaded surface volume renderings (Fig. 18) are usually superior to simple additive projections (Fig. 15).

Methods of showing the uncertainty present in a reconstruction need further development. A current method is to use simple slice-by-slice contour maps with multiple contour levels at different degrees of uncertainty. Another second possibility is shaded surface volume rendering with partially transparent surfaces used to indicate uncertainty. Another possibility would be to drape a "bump map" or "fur map" over a surface rendering to add high-frequency relief to the surface. However, the viewer would have to be instructed as to the meaning of this detail.

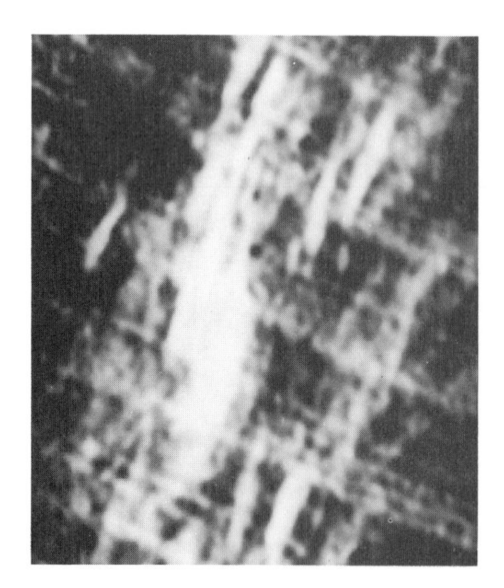

FIGURE 17. Projected volume rendering of developing bone reconstruction of Fig. 15.

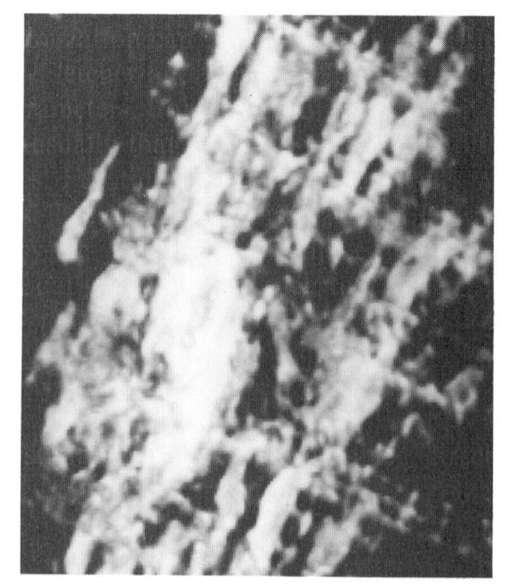

FIGURE 18. Shaded surface volume rendering of developing bone reconstruction of Fig. 15. Light is placed at eye of observer.

10. FUTURE DEVELOPMENTS

Over the last decade computing power per dollar has been doubling every two years. Visualization techniques have been a prime benefactor of this progress. For the next decade a similar rate of progress seems likely. Thus real-time rendering of both surface and volume models will be possible. Interactive real-time classification and surface emphasis in volume renderings will be available.

In the near future there will be improvements in tools which use the reconstructions. At present there is active work on editors for reconstructions (Rusinek and Mourino, 1989; Kaufman *et al.*, 1990). These editors will allow the user to slice the volume, move around pieces within the volume, and otherwise interact with the volume. Other work that is in progress focuses on methods of using volume data for measurement (Becker and Barett, 1990). Stracher *et al.* (1989) have proposed methods for visualizing distances of separation of components within a volume.

Finally, there will be continuing improvements in interfaces for the rendering process. Levoy and Whitaker (1990) are working on using a head-mounted pointer to look at areas within a volume. Kaufman *et al.* (1990) present an editor which is directed by a "data glove."

Sometime in the next century we can look forward to a three-dimensional data base of magnitudes adequate to represent structural levels within an organism (Leith *et al.*, 1989). This data base would be constructed by integrating EM tomographic data with other sources of biomedical 3D data. The user will then be able to interact with the data base by moving around in 3D within the model, at the same time zooming up or down in scale from the submicroscopic to the macroscopic level.

REFERENCES

Artzy, E., Frieder, G., and Herman, G. T. (1981). The theory, design, implementation, and evaluation of a three dimensional surface detection algorithm. *Comput. Gr. Image Process.* **15**:1–24.

Barillot, C., Gibaud, B., Lis, O., Min, L. L., Bouliou, A., Certen, L. G., Collorec, R., Coatrieux, J. L. (1988). Computer graphics in medicine: A survey. *CRC Crit. Rev. Biomed. Eng.* **15**:269–307.

Becker, S. C. and Barrett, W. A. (1990). Interactive morphometrics from three-dimensional surface images, in *Proc. First Conf. on Visualization in Biomedical Computing*, pp. 418–425. IEEE Computer Society Press, Los Alamitos, CA.

Borland, L., Harauz, G., Gahr, G., and van Heel, M. (1989). Three-dimensional reconstruction of a human metaphase chromosome. *Eur. J. Cell Biol.* **48**(Suppl. 25):149–152.

Cappelletti, J. D. and Rosenfeld, A. (1989). Three-dimensional boundary following. *Comput. Vision, Gr. Image Process.* **48**:80–92.

Chen, L., Herman, G. T., Reynolds, R. A. and Udupa, J. K. (1985). Surface shading in the cuberille environment. *IEEE Comput. Gr. Appl.* **5**:33–43.

Christiansen, H. and Sederberg, T. W. (1978). Combination of complex contour line definitions into polygonal element mosaics. *Comput. Gr.* **12**:187–192.

Christiansen, H. and Stephenson, M. (1984). Overview of the MOVIE.BYU software system, in *Proc. 5th Int. Conf. Vehicle Structural Mech.*, pp. 117–185. S. A. E., Warrendale, PA.

Cline, H. E., Lorensen, W. E., Ludke, S., Crawford, C. R., and Teeter, B. C. (1988). Two algorithms for the three-dimensional reconstruction of tomographs. *Med. Phys.* **15**:320–327.

Cook, L. T., Dwyer, S. J., Batnitzky, S., and Lee, K. R. (1983). A three-dimensional display system for diagnostic imaging applications. *IEEE Comput. Gr. Appl.* **3**:13–19.

DeRosier, D. J. and Klug, A. (1968). Reconstruction of three-dimensional structures from electron micrographs. *Nature* **217**:130–134.

Dhawan, A. P., Misra, S., Thomas, S. R. (1990). Knowledge-based analysis and recognition of 3D image of human chest-cavity, in *Proc. First Conf. on Visualization in Biomedical Computing*, pp. 162–169. *IEEE* Computer Society Press, Los Alamitos, CA.

Drebin, R. A., Carpenter, L., and Hanrahan, P. (1988). Volume rendering. *Comput. Gr.* **22**(4):65–74.

Ferwerda, J. G. (1982). *The World of 3D, a Practical Guide to Stereo Photography*. Nertherlands Society for Stereo Photography, Haven, Netherlands.

Frank, J. (1989). Three dimensional imaging techniques in electron microscopy. *BioTechniques* **7**:164–173.

Frenkel, K. A. (1989). Volume rendering. *Comm. ACM* **32**(4):426–435.

Fuchs, H., Keder, Z. M., and Uselton, S. P. (1977). Optimal surface reconstruction from planar contours. *Comm. ACM* **20**:693–702.

Fujimoto, A., Tanaka, T., and Iwata, K. (1986). ARTS: Accelerated ray tracing system. *IEEE Comput. Gr. Appl.* **6**:16–26.

Garcia, E. V., Herbst, M. D., Cooke, C. D., Ezquerra, N. F., Evans, B. L., Folds, R. D., and De Puey, E. G. (1990). In *Proc. First Conf. on Visualization in Biomedical Computing*, pp. 157–161. IEEE Computer Society Press, Los Alamitos, CA.

Goldwasser, S. M. (1986). Rapid techniques for the display and manipulation of 3-D biomedical data, in *Proc. NCGA Computer Graphics '86*, Anaheim, CA.

Gordon, D. and Reynolds, R. A. (1985). Image space shading of 3-dimensional objects. *Comput. Vision, Gr. Image Process.* **29**:361–376.

Gordon, D. and Udupa, J. K. (1989). Fast surface tracking in three dimensional binary images. *Comput. Vision, Gr. Image Process.* **45**:196–214.

Gouraud, H. (1971). Continuous shading of curved surfaces. *IEEE Trans. Comput.* **C-20**:623–628.

Greenberg, D. P. (1989). Light reflection models for computer graphics. *Science* **244**:166–173.

Harauz, G. and Ottensmeyer, F. P. (1984). Nucleosome reconstruction via phosphorus mapping. *Science* **226**:936–940.

Heffernan, P. B. and Robb, R. A. (1985). A new method for shaded surface display of biological and medical images. *IEEE Trans. Med. Images* **MI-4**:26–38.

Herman, G. T. and Liu, H. K. (1979). Three-dimensional display of human organs from computed tomograms. *Comput. Gr. Image Process.* **9**:1–21.

Hersh, J. S. (1990). A survey of modeling representations and their applications to biomedical visualization and simulation, in *Proc. First Conf. on Visualization in Biomedical Computing*, pp. 432–441. IEEE Computer Society Press, Los Alamitos, CA.

Hodges, L. F. and McAllister, D. F. (1989). Computing stereographic views, in *ACM Siggraph '89 Course Notes*, pp. 4.1–4.31. ACM.

Hohne, K. H. and Bernstein, R. (1986). Shading 3D-images from CT using gray-level gradients. *IEEE Trans. Med. Imaging* **MI-5**:45–47.

Jimenez, A., Santisteban, A., Carazo, J. M., and Carrascosa, J. L. (1986). Computer graphic display method for visualizing three-dimensional biological structures. *Science* **232**:1113–1115.

Johnson, E. M. and Capowski, J. J. (1983). A system for the three-dimensional reconstruction of biological structures. *Comput. Biomed. Res.* **16**:79–87.

Kaufman, A., Yagel, R., and Bakalash, R. (1990). Direct interaction with a 3D Volumetric environment, Computer Graphics **24**:33–34.

Leith, A., Marko, M., and Parsons, D. (1989). Computer graphics for cellular reconstruction. *IEEE Comput. Gr. Appl.* **9**:16–23.

Lenz, R., Gudmundsson, B., Lindskog, B., Danielsson, P. E. (1986). Display of density values. *IEEE Comput. Gr. Appl.* **6**:20–29.

Lepault, J. and Leonard, K. (1985). Three-dimensional structure of unstained, frozen-hydrated extended tails of bacteriophage T4. *J. Mol. Biol.* **182**:431–441.

Levoy, M. (1988). Display of surfaces from volume data. *IEEE Comput. Gr. Appl.* **8**:29–37.

Levoy, M., Fuchs, H., Pizer, S. M., Rosenman, J., Chaney, E. L., Sherouse, G. W., Interrante, V., and Keil, J. (1990). Volume rendering in radiation treatment planning, in *Proc. First Conf. on Visualization in Biomedical Computing*. IEEE Computer Society Press, Los Alamitos, CA.

Levoy, M. and Whitaker, R. (1990). Gaze directed volume rendering. *Comput. Gr.* **24**:217–223.

Liu, H. K. (1977). Two- and three-dimensional boundary detection. *Comput. Gr. Image Process.* **6**:123–134.

Lorensen, W. E. and Cline, H. E. (1987). Marching cubes: A high resolution 3D surface construction algorithm. *Comput. Gr.* **21**(4):163–169.

Love, S. (1989). Three-dimensional hardcopy, in *ACM Siggraph '89 Course Notes*, pp. 5.1–5.32. ACM.

Luther, P. (1989). Three-dimensional reconstruction of the *Z*-line in fish muscle. *Eur. J. Cell Biol.* **48**(Suppl. 25):154–156.

Marko, M., Leith, A., and Parsons, D. (1988). Three-dimensional reconstruction of cell from serial sections and whole-cell mounts using multilevel contouring of stereo micrographs. *J. Electron Microsc. Tech.* **9**:395–411.

McEwen, B. F. G., Radermacher, M., Grassucci, R. A., Turner, J. N., and Frank, J. (1986). Tomograhic three-dimensional reconstruction of cilia ultrastructure from thick sections. *Proc. Nat. Acad. Sci. USA* **83**:9040–9044.

McMillan, D., Johnson, R., and Mosher, C. (1989). Volume rendering on the TAAC-1. *SunTech J.* **2**:52–58.

Namba, K. and Caspar, D. L. D. (1985). Computer graphics representation of levels of organization in tobacco mosaic virus structure. *Science* **227**:773–776.

Ney, D., Fishman, E. K., and Magid, D. (1990a). Three-dimensional imaging of computed tomography: Techniques and applications, in *Proc. First Conf. on Visualization in Biomedical Computing*. IEEE Computer Society Press, Los Alamitos, CA.

Ney, D., Fishman, E. K., Magid, D., and Drebin, R. A. (1990b). Volumetric rendering of computed tomography data: Principles and techniques. *IEEE Comput. Gr. Appl.* **10**:24–32.

Oettl, H., Hegerl, R., and Hoppe, W. (1983). Three-dimensional reconstruction and averaging of 50*S* ribosome subunits of *Escherichia coli* from electron micrographs. *J. Mol. Biol.* **161**:431–450.

Phong, B. T. (1975). Illumination for computer generated pictures. *Comm. ACM* **18**:311–317.

Pizer, S. M., Fuchs, H., Mosher, C., Lifshitz, L., Abram, G. D., Ramanathan, S., Whitney, B., Rosenman, J. G., Staab, E. V., Chaney, E. L., and Sherouse, G. (1986). 3-D shaded graphics in radiotherapy and diagnostic imaging, in *Proc. NCGA Computer Graphics '86*. Anaheim, CA.

Porter, T., and Duff, T. (1984). Compositing digital images. *Comput. Gr.* **18**(3):253–259.

Radermacher, M. and Frank, J. (1984). Representation of three-dimensionally reconstructed objects in electron microscopy by surfaces of equal density. *J. Microsc.* **136**:77–85.

Robb, R. A. and Barillot, C. (1989). Interactive display and analysis of 3-D medical images. *IEEE Trans. Med. Imaging* **8**:217–226.

Rusinek, H. and Mourino, M. (1989). Interactive graphic editor for analysis and enhancement of medical images. *Comput. Biomed. Res.* **16**:79–87.

Saxton, W. O. (1985). Computer generation of shaded images of solids and surfaces. *Ultramicroscopy* **16**:387–394.

Schlusselberg, D. S., Smith, W. R., and Woodward, D. J. (1986). Three-dimensional display of medical image volumes, in *Proc. NCGA Computer Graphics '86*. Anaheim, CA.

Shantz, M. (1981). Surface definition for branching, contour defined objects. *Comput. Gr.* **15**:242–259.

Skogland, U., Anderson, K., Stranberg, and Daneholt, B. (1986). Three-dimensional structure of a specific pre-messenger RNP particle established by electron microscope tomography. *Nature* **319**:560–564.

Stracher, M. A., Goiten, M., and Rowell, D. (1989). Evaluation of volumetric differences through 3-dimensional display of distance of closest approach. *Radiat. Oncol. Biol. Phys.* **17**:1095–1098.

Suh, D. Y., Mersereau, R. M., Eisner, R. L., Pettigrew, R. I. (1990). Automatic boundary detection on cardiac magnetic resonance image sequences for four dimensional visualization of left ventricle, in *Proc. First Conf. on Visualization in Biomedical Computing*, pp. 149–156. IEEE Computer Society Press, Los Alamitos, CA.

Tiede, U., Hoehne, K. H., Bomans, M., Pommert, A., Reimer, M., and Weibecke, G. (1990). Investigation of medical 3D-rendering algorithms. *IEEE Comput. Gr. Appl.* **10**:41–53.

Udupa, J. and Herman, G. (1989). Volume rendering versus surface rendering. *Comm. ACM* **32**:1364–1367.

Udupa, J. K. and Hung, H. (1990). Surface versus volume rendering: A comparative assessment, in *Proc. First Conf. on Visualization in Biomedical Computing*. IEEE Computer Society Press, Los Alamitos, CA.

Upson, C. and Keeler, M. (1988). V-Buffer: Visible volume rendering. *Computer Gr.* **22**(4):59–64.

van Heel, M. (1983). Stereographic representation of three-dimensional density distributions. *Ultramicroscopy* **11**:307–314.

Vannier, M. W., Marsh, J. L., and Warren, J. O. (1983). Three dimensional computer graphics for craniofacial surgical planning and evaluation. *Comput. Gr.* **17**(3):263–273.

Verschoor, A., Frank, J., Radermacher, M., Wagenknecht, T., and Boublik, M. (1984). Three-dimensional reconstruction of the 30*S* ribosomal subunit from randomly oriented particles. *J. Mol. Biol.* **178**:677–698.

Vigers, G. P. A., Crowther, R. A., and Pearse, B. M. F. (1986). Location of the 100 kd–50 kd accesssory proteins in clathrin coats. *EMBO J.* **5**:2079–2085.

Wallis, J. W., Miller, T. R., Lerner, C. A., and Kleerup, E. C. (1989). Three-dimensional display in nuclear medicine. *IEEE Trans. Med. Imaging* **8**(4):297–303.

Whitted, T. (1980). An improved illumination model for shaded display. *Comm. ACM* **23**:343–349.

Winslow, J. L., Bjerknes, M., and Cheng, H. (1987). Three-dimensional reconstruction of biological objects using a graphics engine. *Comput. Biomed. Res.* **20**:583–602.

Wood, L. (1990). Let your computer do the molding. *Inf. Week*, April 16:38–40.

Young, J. J., Roger, S. M., Groves, P. M., and Kinnamon, J. C. (1987). Three-dimensional reconstruction from serial micrographs using the IBM PC. *J. Electron Microsc. Tech.* **6**:207–217.

11

Analysis of Three-Dimensional Image Data: Display and Feature Tracking

Zvi Kam, Hans Chen, John W. Sedat, and David A. Agard

1. INTRODUCTION

Microscopes have been used for almost two centuries to produce a wealth of biological information based on direct observation and photography. Recent advancements in digital image processing and display have created a revolution in biological and medical imaging. Digital methods have lead to significant improvements in the ease of image acquisition and data quality for both light and electron microscopy. It is now possible to routinely obtain quantitative image data and not simply pictures. The advent of quantitative microscopy is a major step forward in the application, and utilization of structural information for biological problems at the level of cells and cell components. Perhaps most significant has been the possibility of examining complex, noncrystalline objects such as supra-

Zvi Kam • Department of Chemical Immunology, Weizmann Institute of Science, Rehovot 76100, Israel *Hans Chen, John W. Sedat, and David A. Agard* • Department of Biochemistry and Biophysics and The Howard Hughes Medical Institute, University of California at San Francisco, San Francisco, California 94143-0448

Electron Tomography: Three-Dimensional Imaging with the Transmission Electron Microscope, edited by Joachim Frank, Plenum Press, New York, 1992.

molecular assemblies in three dimensions. Combined with powerful new probes to examine specific molecular components, three-dimensional imaging can provide new insights into the spatial localization patterns of specific molecules and their redistribution, for example during the cell cycle or development.

Electron microscope tomography methods aim to fill the very wide gap between atomic structures defined by x-ray crystallography and the more global patterns of organization observed in the light microscope. The goals of tomographic projects range from reconstructing the three-dimensional surfaces of small particles, such as ribosomal subunits, to examining much larger structures, such as cilia, chromosomes, or even entire neurons. Unlike more conventional structural methods that implicitly depend on symmetry or averaging over a very large number of examples, microscopic imaging methods typically examine individual structures. As a consequence, there can be considerable variation in the details for each specimen examined. This variability results from intrinsic variations in the structures themselves, and variability resulting from inadequate preservation, heterogeneity in staining, and electron beam damage. The smaller, particulate structures such as ribosomal subunits are undoubtedly considerably more homogeneous than much larger structures such as chromosomes. Such structural variability extends to the organellar level where the details of Golgi or endoplasmic reticulum structure are dynamic and hence not precisely reproducible.

Much of the challenge in structural cell biology is to be able to extract biologically relevant structural information from the vast amount of data present in a three-dimensional reconstruction. For many cell biological problems (using either EM tomography or three-dimensional light microscopy), what is desired is the elucidation of general patterns of three-dimensional organization. Depending upon the problem, accomplishing this goal may require analysis at many different levels. In some cases, being able to trace the paths of structures in three dimensions is the best way to extract the relevant information. For other problems, different forms of semiquantitative analysis such as surface area or volumetric measurements, pairwise distance measurements, etc., may be the most important. While for still other problems a qualitative examination may be all that is required; thus, simply providing stereoscopic images may be sufficient.

In this chapter, we will emphasize approaches that provide the capability for a quantitative analysis of three-dimensional volumetric data in addition to the more conventional display aspects. A direct method of quantitatively analyzing the complex three-dimensional images typical of biological structures is to first identify the structural component of interest using continuous intensity images and to then construct graphical objects or models depicting features of interest. The kind of modeling required will vary in its geometrical (topological) nature, depending on the biological problem being investigated. The structure of chromosomes and the tracking of cytoskeletal structures can be well represented by 1D (linear or branched) tracks in 3D images. The topology of membranes (nuclear envelope, cells plasma membrane) and the definition of inside/outside relations of a 3D body require the modeling of 2D surfaces, whereas tracking 3D structures in time requires analysis of 1D or 2D elements within the context of a 4D space. Once a simplified representation of the volumetric data is developed, it can be used directly to guide any desired form of quantitative analysis. Discerning geometrical or topological patterns within complex structures is greatly simplified by having first abstracted the relevant information in the form of a model. In addition, once defined, models can be used to trace intensity profiles along the model or to directly unfold the image intensity in the region surrounding the model and to display it stretched out. Another very important use of models is to facilitate detailed correlations between features from one specimen to another.

Despite the immense progress in interactive computer power and graphics workstations, and the associated development of artificial intelligence and expert systems for machine vision and image understanding, our experience with structural studies of chromatin suggests that the complexity of biological images and the variability of the structural features studied represents a new class of problems that have only begun to be investigated. It has been our experience, that in general simple three-dimensional display approaches such as the three-dimensional iso-surface cage contours routinely used by the x-ray crystallographers (FRODO; Jones, 1982) or single-contour surface rendering approaches are unable to convey the richness and subtlties inherent in biological three-dimensional reconstruction data. As discussed below and elsewhere in this volume, the use of projection methods that maintain the gray-level density representation provide a powerful and versatile alternative to contouring approaches. Unfortunately, it has been our experience that many biological problems, especially those resulting from EM tomographic reconstructions of large, nonparticulate objects, are too complex to be readily visualized in their entirety. Thus, we think that it is necessary to develop an environment for the interactive display and analysis of complex volumetric image data that allows a variety of visualization methods to be used in concert on interactively chosen subregions of the object being displayed in arbitrary orientations. Furthermore, it is important to design this visualization environment from the outset with the capability for interactive or semiautomated three-dimensional model building. In this chapter we discuss our efforts to develop such a display environment and modeling capability. In addition, we will cover the work of others and ourselves on the challenging problem of automated or semiautomated modeling.

2. THE IMAGE DISPLAY AND PROCESSING ENVIRONMENT

In our laboratory, we are interested in understanding higher-order chromosome structure and the spatial and temporal dynamics of three-dimensional chromosome organization throughout the mitotic cell cycle. Toward this end, we have been developing data collection and reconstruction methods for both IVEM tomography as well as optical sectioning light microscopy. Due to the difference in resolution, nature of the optics, and sources of contrast, electron and light microscopes (EM and LM) are different, but complementary in their structural information contents. In spite of the differences, there are many commonalties in the analysis of both data types. As a necessary part of our overall structural effort, we have also been developing a general purpose display environment and structural analytic methods designed for the examination of complex cellular structures derived from both light and electron microscopic reconstructions. The key features are (i) an image display environment able to interactively manipulate and display multiple sets of three-dimensional data. Using a multiple-window approach, various combinations of section data and projection data (either raw or image processed) can be viewed simultaneously, and the windowing system keeps track of the geometrical relationship between images; (ii) interactive 3D model building using cursor-controlled graphics overlayed on the original or preprocessed data. Modeling can be performed in multiple windows which greatly facilitates tracing paths in complex objects; and (iii) semiautomatic and automatic modeling based on algorithms which utilize the characteristic features of the structures under study. In this and following sections we will describe this image display environment and modeling methods in more detail.

Before proceeding to discuss the display system, we give a brief overview of important features of the data collection and handling. Regardless of modality (electron microscope tomography or optical section microscopy), high resolution, quantitative three-dimensional microscopy requires the best possible data. We have taken great care in the design of our microscopic imaging systems which employ cooled, scientific-grade charge-coupled device (CCD) cameras having a wide dynamic range (12 bits), exceptional linearity, and geometric stability (Hiraoka *et al.*, 1987). The CCD systems on either the light microscope or our Philips EM430 IVEM are built around compact, MicroVaxIII workstations that use a high-performance array processor (Mercury) and Parallax display system. All aspects of microscope operation are computer controlled. The IVEM system employs a 20-μm-thick YAG single-crystal scintillator (Gatan) directly coupled to a 1024×1024 pixel CCD by fiber optics. The array processor permits on-the-fly gain and offset corrections for the individual pixels. Rapid image processing or Fourier transform calculations are useful for on-line critical evaluation of microscope focus or aberrations. For the EM system, we have a computer-controlled ultrahigh-tilt stage ($\pm 80°$), and complete computer control of x, y, z motor stages, image and beam deflection, and lens excitation parameters. For the light microscope we have employed computer-controlled barrier and excitation filters and double- or triple-wavelength dichroic mirrors to provide rapid wavelength switching free from image translation artifacts. Such problems are inevitable if dichroic mirrors have to be switched. Multicolor data acquisition has been employed in fixed samples and recently also in live developing embryos, which were co-injected with two molecules labeled with fluorescein and rhodamine. This enables the spatial relationships and interactions of several molecular components to be examined both as a static and as dynamic processes changing through time.

In addition to attempting to optimize the quality of the three-dimensional data collected, we have spent considerable effort on developing reconstruction algorithms for both the optical sectioning data (Agard *et al.*, 1989) and for EM tomography (Abbey and Agard, forthcoming). In general, these methods all share a common theme—they seek a solution which is everywhere positive, appropriately spatially bounded and that when projected (in the case of the EM) or convolved with the point-spread function (for the light microscope) will match the observed data. The use of such straightforward a priori information as constraints imposed during the iterations can have a profound effect on the stability of otherwise ill-posed reconstruction algorithms, both ameliorating the effects of missing data as well as conditioning against noise.

Convenient manipulation and trouble-free handling of a large number of three-dimensional data sets requires that the full volumetric data be kept within a single file, with relevant information maintained in a file header. For all of our studies we utilize images files written in the MRC format (for description see Chen *et al.*, 1990). The file begins with a header that contains parameters specifying the dimensions, geometrical scales, origins, orientation, mode of storage (bytes, integer*2, real, complex), data type (sections, projections, tilt series in mono or stereo, multi-wavelength data, and time lapse 3D series), range of intensities, and text strings describing the file history. The header record, which can be extended for individual picture attributes, is followed by a direct access data structure which is easily used to store 2D images, stacks of serial sections of 3D data, time series of 3D data, and multiple color pictures, etc. A complete set of subroutines is available to facilitate file setup and manipulation. Importantly, the header contains sufficient information so that files generated from different geometric manipulations of the same data (e.g., translations, rotations, projections, subsamplings, magnification changes, etc.) can be related to one another so that models can be displayed in the correct perspective view.

Our three-dimensional imaging and modeling package (PRISM, Chen *et al.*, 1990) employs a high-resolution (1280 × 1024, 60 Hz, noninterlaced) video board

TABLE 1. Functions Available within the PRISM DISPLAY Program

Display functions:
- Load a series of images into a window.
- Look-up table manipulations: linear, clipped and power LUTs.
- Zoom, pan, move window position and size.
- Manual paging of images associated with one or all windows.
- Movie mode for one or more windows (like stereo pair movie).
- Stereographics display (top-bottom stereo encoding to be used with the Stereographics polarizing screen on glasses)
- Photography (computer-controlled exposure with film recorder) and low-resolution video recording for videotape and videodisk.

Image processing functions employing the MERCURY array processor:
- Convolutions, Fourier transforms, and filtering.
- Local contrast enhancement.
- Interactive 3D projections and sections through a cube of data.

Modeling and presentation utilities:
- Point: interactive, mouse-selected set of 3D points overlayed on image stacks.
  - Intensity, distance and angle calculations between points picked by the cursor. Vertical and horizontal intensity profiles through the cursor.
- Label: overlay text, scale bars, arrows, and circles; in different colors.
- Montage: interactive pasting of images from all windows into one page.
- Superposition: Allows RGB color components to be translated independently in color images.
- Spawn: generate a subprocess.
- Model and model Show: an extensive menu for interactive and automatic 3D line model building: Add, delete, branch, modify, cut and paste. Models are composed of linear "objects" that can each be labeled, colored, turned off or displayed on top of the images, and interactively rotated in 3D.

Interactive operations within PRISM can be recorded in a journal file for automatic repetitive execution or for fast recall of complex multiwindow environments. Displayed screens can be saved as image files for future reference, reproduction, and photography.

equipped with 12 Mbytes of memory (Parallax Graphics, Inc. Santa Clara, CA), and implements a very flexible multiple windowing architecture. Based on a combination of software and fast video processing hardware, each window can be arbitrarily sized and positioned, and the images can be independently zoomed. Because there is so much memory available on the display board, it is possible to preload a large number of images. This permits dynamic display under interactive user control (or digital movies under computer control) of preprocessed sets of images like sections through a 3D cube at any angle, tilted projections, and stereo pairs. Based on the file header information, windows presenting different processing manipulations of the same data can be geometrically related. Several windows, each associated with series of images can be activated so that sections and projections from different directions could be viewed and shown in movie mode.

In conjunction with an array processor (Mercury Computer Systems Inc., Lowel MA), PRISM implements on-line image processing (Fourier transforms, Fourier filtering, convolutions, local contrast enhancement, scaling) and projection and sectioning of gray-scale 3D data. Images can be presented in a single window or as stereo pairs. Interactive presentation of three-dimensional gray-level data is a most powerful tool for perception and understanding of these images.

In addition to the ability to present and compare data from different sets of images in simultaneously displayed windows, PRISM contains a true color option

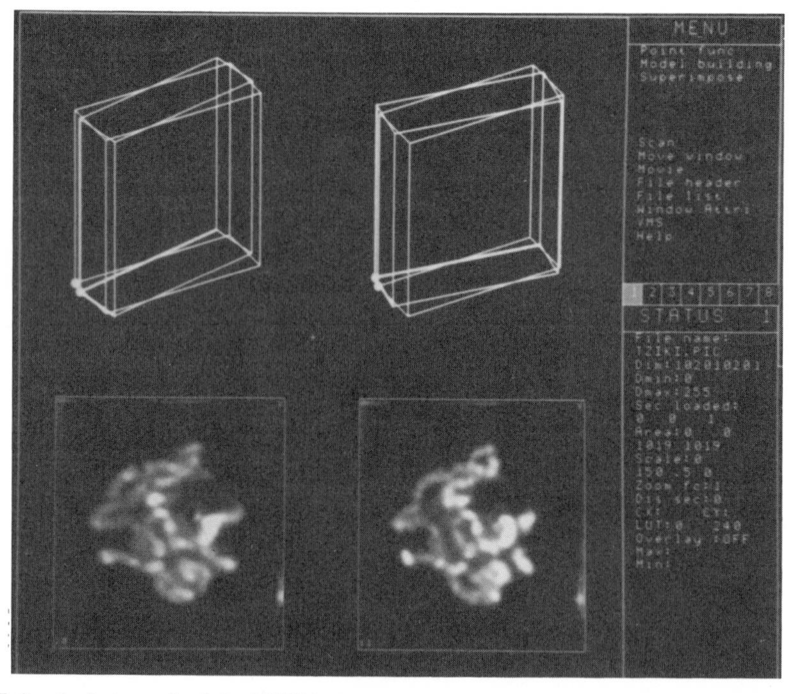

FIGURE 1. A photograph of the PRISM screen while displaying stereo pairs using the on-line 3D rotation capability. The actual rotated data is shown as a stereo pair in the two lower windows, while the two top windows show the orientation of the displayed data with respect to the original data. To the right of the display windows the menu window with mouse-activated commands is shown.

FIGURE 2. Example of another method for the interactive viewing of three-dimensional image data. In this approach, three orthogonal sections from the 3D data are presented in perspective on the faces of a cube. The sections can be moved in or out interactively, thereby slicing the cube in a varying manner, and displayed instantaneously to create an excellent perception of the volumetric data.

that allows two or three images to be superimposed, independently shifted to align them, and to independently tune the relative intensity and contrast of each color component. This is extremely useful for display and analysis of samples in which several molecules (like DNA, and antibodies to proteins) are all labeled simultaneously using different dyes. A listing of PRISM function options is given in Table 1. Examples of the interactive 3D rotation and 3D cube slicing functions are shown in Figs. 1 and 2.

3. PROCESSING DATA FOR DISPLAY

The initial aim in image processing is to present the data to the scientist with optimal contrast, and a minimum of interfering background and noise. A wide variety of filters have traditionally been used for processing two-dimensional images.

An important feature of our software is the ability to process a sequence of two-dimensional images as a true three-dimensional cube of data taking into account differences in scale and resolution along the axes. A list of typical processing routines is shown in Table 2.

A fundamental difficulty is developing means to adequately convey the information present in a complex three-dimensional reconstruction. The conversion of volumetric data into images on a two-dimensional raster screen is called volumetric rendering. Typically, such rendering techniques fall into two categories: those that extract a surface and then render it using shaded polygons, and those methods that perform various types of projection operations on the voxel data and maintain the gray-level information. Since this topic is covered extensively in Chapter 10, only a cursory overview will be given here.

For relatively simple objects, we have found the most generally informative approach for presenting the data with minimal processing is to use a series of rotated projections (Chen *et al.*, 1990). In this approach, a series of two-dimensional projections of the three-dimensional data are formed at different angles of rotation about a single axis. These projections are rapidly and sequentially displayed to form a digital movie. The apparent back-and-forth rocking of the data gives a realistic perception of the three-dimensional density. This volumetric perception can be further enhanced by using stereo pairs. For more complex images, small subregions need to be chosen and to preserve interpretability. The projections themselves can

TABLE 2. Useful Image Processing Methods for 3D Data

Linear Fourier filters (smoothing, enhancement) by high, low, and bandpass Gaussian profile 3D filtering in Fourier space.

3D Convolution filters for smoothing, and for enhancement by subtraction of the smoothed data from the original picture. The advantage of these simple local filters is the existence of fast algorithms (the complexity of the algorithm is proportional to the total number of voxels, and is independent of window dimensions. A fast 3D median filter was written by extending the 2D algorithm employing sweeping update of the histogram and the corresponding median (Huang *et al.*, 1979).

3D gradients, and edge-enhanced pictures (Chen *et al.*, 1990):
The modulus of the three-dimensional gradient of an image is computed and multiplied by the original image. The new image is a linear combination of the original and the weighted gradient image. This provides significant edge enhancement without building up noise.

3D interpolation of sections from 3D data cube and rotation of a cube of data by arbitrary angles.

3D projections of a slice from 3D data cube at any angle, and fast rotations about the X axis or Y axis for movies of tilt series and stereo pairs.

Entropy maximization filters.

3D alignment and matching of two sets of images, including translations, rotations, and magnification as free-fitting parameters.

Specialized image processing algorithms for optical and electron microscopic 3D reconstructions include:

Deconvolutions of the microscope CTF to eliminate out of focus contributions. The nearest-neighbor approximation to the deconvolution problem is implemented in near real time during data collection. More sophisticated full 3D algorithms and deconvolution of projections are batch processed (Agard *et al.*, 1989; Hiraoka *et al.*, 1990).

Tomographic reconstructions from EM tilt series. (Back projections, iterative, constrained methods) (Abbey and Agard, forthcoming).

be computed along an arbitrary view angle. The use of a pure projection operator (sum along the view line) or even selection of the maximum value tends to give the resultant images a transparent, ethereal quality. The perception of solidity can be added by preferentially weighting the images in front of the stack or, more accurately, by considering that each pixel is opaque in proportion to its intensity. Based on the properties of fluorescent objects, Van der Voort and his colleagues (1988) developed an approach that combines aspects of ray tracing with volumetric rendering to provide dramatic images having strong shadows suspended above a uniform background. Because of the shadowing, this technique is probably less useful for very complex images, yet it should find application for some problems in EM tomography.

Edges within images can be effectively enhanced using an approach developed at PIXAR (PIXAR Corp., San Rafael; Driebin *et al.*, 1988). The starting volumetric data is processed by multiplying the image pixel by pixel with the modulus of the three-dimensional gradient of the image. As this greatly enhances surface features, it is advisable to also incorporate opacity into the projection operator. Another approach is, instead of just taking the modulus of the gradient, to take the dot product of the gradient with a directional light source. This procedure produces a pleasant and fast surface rendering that importantly does not rely on the selection of a single contour threshold.

Improved background selection can be done utilizing a procedure called a 3D cookie cutter. This method uses an edge-detection scheme based on the PIXAR approach. By choosing a threshold, a surface boundary is defined and subsequently smoothed. The boundary is then used to zero-out background from the region of interest and thus present the object with better contrast and definition.

Three-dimensional surface rendering algorithms are useful for presenting the shape of solid bodies. Surface polygonization can be carried out by choosing a contour threshold to define the surface and then asking where that surface crosses through the 12 possible edges of each voxel in the image. These intersection points are determined by interpolation and then converted into vertices of a surface triangle (Chen *et al.*, 1990). The triangular skeleton can be presented as line graphics, or the shaded surfaces can be displayed on a raster display. Shading incorporates background light, as well as diffuse and reflective illumination resulting from arbitrarily positioned light sources. Alternatively, ray tracing algorithms can also be used to enhance surface reflectivity and to more easily show overlapping objects by the use of transparent surfaces (Paddy *et al.*, 1990).

4. *IMAGE MODELING, ANALYSIS, AND QUANTITATION*

4.1. *Interactive 3D Model Building Using PRISM*

Interactive modeling is probably the most effective mode of extracting general structural data from 3D image data. As discussed in the introduction, interactive modeling provides a means to trace the three-dimensional path of one or more objects that can be visualized within a reconstruction. The general approach used is to be able to step through a stack of images representing a volumetric data set and to use a cursor to sequentially mark locations along the path of the object being traced.

Within PRISM, a model is defined as a collection of objects where each object is a line drawing representation of the path of a structure in three-dimensional space. Each object is defined as a set of connected nodes, each of which which represent a single branch in the structure. A tree structure is used to store the connectivity of all these nodes, and a position array is used to store the coordinates of every node in real space. No limitations aside from total available memory space are imposed on the depth of the tree or on the number of entries at any level. For versatility and clarity, each object has its own set of display attributes which PRISM uses to determine its visibility and color, and any of six predefined marks can be attached to a node. Thus when it is displayed, the attached mark will be shown at the same location and in the same color as the object to which it belongs. While building the model, PRISM maintains a data pointer within the tree structure that determines the node at which modeling commands take place. There are four basic commands for updating a model: ADD, DELETE, BRANCH, and MODIFY. ADD will add a node right after the active node, and make the new node active. DELETE will remove the active node from the structure, and make the previous node active. BRANCH will create a new branch substructure at the active node and set the first node in the new branch as the active node. MODIFY will not affect the tree structure; it simply alters the coordinates of the current active node, allowing the user to reposition the selected point. In addition to these basic building commands, PRISM also provides a set of commands for moving pointers within the tree structure so that any part of model can be retrieved easily.

Building a model is accomplished by determining the locations of all the nodes. When displaying section images, a data point on the screen will correspond to a unique location in the data sampling space, as each image represents a slice in the sample. For such data, model building is accomplished by picking data points displayed on the screen. Stepping forward or backward in sections (which occurs rapidly because they have been preloaded) provides the third coordinate. Being able to rapidly step from one section to another is crucial for ease of use. Preprocessing the data to improve the clarity and contrast of structures being traced can be quite important for successful modeling. In our modeling of optical section data which suffers from nonisotropic resolution due to a missing cone of data, in much the same way as can EM tomographic data, we have found that use of the PIXAR image*gradient edge-enhancement approach greatly minimizes modeling problems.

In optimal cases, it may only be necessary to step through a stack of images in one, fixed orientation. However, for more complex images, especially where non-isotropic resolution may contribute to overlap in some directions, it is desirable to be able to model-build into the same data set that has been rotated in different ways. Thus, what appears ambiguous from one direction may be clear from another viewing direction. Although one could first model-build into data oriented in one direction and then correct the model by using another data orientation, it is greatly preferable to model-build while simultaneously viewing the data from several different viewing angles. If the views are related by 10–15° tilt about the vertical axis, they will comprise a set of stereo pairs. Alternatively, orthogonal views can be used to allow precise modeling in the perpendicular direction. Especially in the case of

EM tomographic data, it can be important to also show the model superimposed upon the original projection data (or calculated projection data from a subregion). For this reason, PRISM has been designed to allow simultaneous modeling into several windows that are geometrically related. Thus, in the model building function, PRISM will choose what section to display in every window so that they all contain the same point in three-dimensional space as defined by the cursor and the section number in the active window as well as the projected view on top of 3D projections. This approach to window management and model building allows the user to simultaneously construct models from multiple viewing angles.

Models are easy to analyze for structural statistical properties, such as length, correlation, or anticorrelation of particular loci, approximate symmetry, and comparison of shapes. For quantitative analysis of fluorescence intensity in images, we employ routines that facilitate interactive and automatic definition of volumes and surfaces within which data should be evaluated. For example, for fast computations of volume and area averages, closed contour lines generated as a series of ordered contour points, can be converted to a run-length list (that is, first and last pixels of the area defined along each line), and, vice versa, the sequential list of border points enclosing an area required, for example, for contour plots, can be computed from the membership list of area patches. The area enclosed by a polygon is calculated by the equivalent of Green's theorem for contour integrals (M. Koshy, private communication):

$$\mathrm{AREA} = \mathrm{SUM}\{[x(i) + x(i+1)]/2^*[y(i+1) - y(i)]\}, \qquad i = 1, 2, \ldots, N-1$$

where $x(i)$, $y(i)$ are the coordinates of vertex i, and there are N vertices. Patching, sorting, and counting of blobs can be done in one pass through the pixels in an image using the Union Find algorithm (Duda and Hart, 1973). This readily allows calculation of shape properties like the centers of mass, eccentricities, and the direction of the principal axes of the second moment ellipsoid (Castleman, 1979).

4.2. Computerized Recognition and Tracking of Objects

Where repetitive modeling of a particular kind of data is required, it is highly desirable to use an automatic or semiautomatic modeling scheme. This is a challenging task, that probably must be optimized for each type of data being analyzed. In general, the more a priori information that can be used to specify the behavior of the model, the more robust will be the entire procedure. Again, preprocessing can dramatically simplify the design and application of the necessary expert algorithms. Since many of these algorithms are not linear, their results can depend on data preprocessing for fast convergence to a visually acceptable estimate of the model. Proper treatment of the image to flatten or minimize background variation and enhance object discrimination will dramatically simplify the task of feature recognition. A trivial example of this is the ability to use threshold methods to define objects, or boundaries. The task of finding a suitable contour threshold can be greatly simplified by preprocessing to edge enhance and then smooth the data to maximize edge continuity. Having the ability to interactively process the data while testing various threshold choices is obviously quite desirable, since it permits an optimal staring point to be selected for more sophisticated but time-consuming algorithms.

Automatic structural analysis can require complex logistics. It is necessary to reduce the dimensionality and complexity of the algorithms or to restrict them to the relevant area of interest. This is particularly important in three-dimensional image processing given the vast size of the analyzed data sets. Image segmentation can be based on gray levels, gradients, or generalized "energy" functions, which are associated with each voxel and are evaluated in a volumetric neighborhood of arbitrary size. Neighborhood operators provide a method for segmentation based on local properties or averaging to provide smoothing. Analyses at various resolutions—multigrid methods—(Brandt, 1986) not only save computational time by locating the region of interest at low resolution and then confining high-resolution processing to that area, but they also provide a potentially very powerful method for combining local and global interpretation of images.

4.2.1. Ridge Methods for Line Tracking: Tracking of Chromosomes in 3D Light Microscopic Data

A central unsolved problem in modern biology is understanding the complex structure of chromosomes. Related to this problem are questions pertaining to the general architecture of the cell nucleus. For example: How are chromosomes organized and folded within the tight confinement of the spherical nucleus? How does chromosome packing change as a function of the cell cycle or development? Is there a relationship between transcription and the spatial location of a gene? The problem of chromosome structure needs to be approached at two levels. At high resolution, which is best addressed by EM tomographic methods, the question is to determine how DNA and protein become organized into successively higher order chromosomal structures. Also of central importance is understanding how the higher-order structural organization changes during condensation, decondensation, transcription, and replication. At the more global level, which is best examined by 3D light microscopy, are questions of spatial organization of chromosomes within the nucleus and their dynamic behavior during the cell cycle. Structural features that are of importance are the localization of genes along the chromosomes (mapping complex chromosome paths to straight linear structures for comparison with genetic maps and to determine architectural features such as compaction) and the correlation of spatial position with function. Such an analysis would, for example, include examining the association of specific genetic loci with the nuclear membrane, spatial correlation of the two homologous genes, active genes and puffs in some cases, or inactive genetic regions like heterochromatin.

In our light microscopic work, 3D images of nuclei are collected from intact *Drosophila* embryos that were fixed and fluorescently stained using DNA specific dyes, labeled hybridization probes, or antibodies to DNA associated proteins (Hiraoka *et al.*, 1987; Agard *et al.*, 1988; Rykowski *et al.*, 1988; Paddy *et al.*, 1990). Alternatively, many sets of 3D data can be collected as a function of time from living embryos microinjected with fluorescently labeled molecules such as tubulin, topoisomerase II, or histones, which then become incorporated into spindle or chromosomal structures (Minden *et al.*, 1989; Hiraoka *et al.*, 1989). The microscopic images recorded at a series of focal planes are processed to eliminate out of focus contributions, and projected tilt series and orthogonal section series are computed to facilitate visualization (Agard *et al.*, 1989). The processed 3D series are then displayed in separate windows in PRISM. Models are constructed by interactively positioning the cursor along the chromosome paths as judged by the user based on the multifaceted presentation of the 3D density.

Semiautomatic modeling allows processing of a large number of nuclei for statistical analysis. This is especially important when trying to derive what are the biologically significant features of a modeled organization. The semiautomatic algorithms that we have employed use local contrast filtering and three-dimensional gradients to define a sausage-like corridor. Within this corridor, the central ridge of

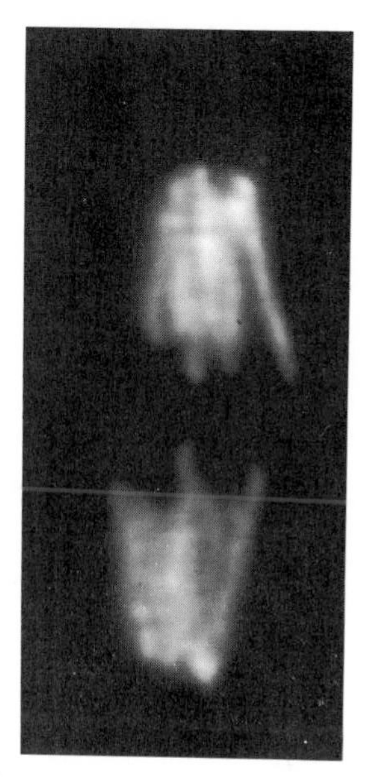
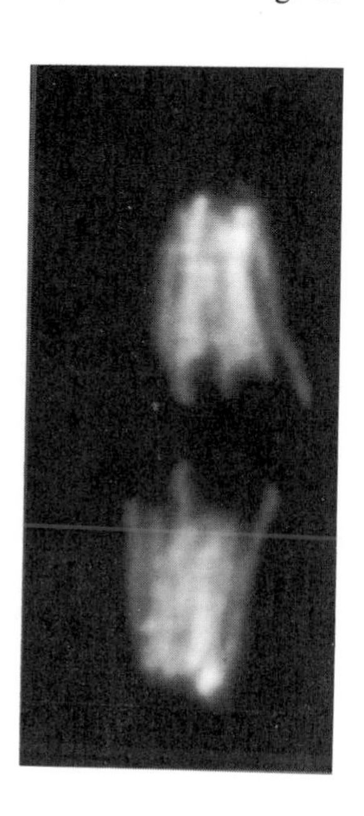

FIGURE 3. An example of the result of semiautomatic modeling of chromosomes. A stereo pair of an anaphase chromosome figure is shown at top, and the models are shown below. The user initiates the automatic tracking by seeding the model. Models created by random seeding at image voxels having an intensity above a set threshold demonstrate that most of the tracks accurately follow the chromosomal paths. The tracks that did not follow the chromosomes were a result of confusion caused by a combination of overlaping of two chromosomal arms and the lower resolution along the light microscope optical axis.

maximum density is determined. In order to avoid the accumulation of background and noise from regions outside the chromosome track, the data are preprocessed using a three-dimensional cookie cutter. Using spline routines to account for persistence of direction, a suggestion for the track of the chromosome is then made. Crosses between chromosome paths and chromosomal bands which introduce a somewhat fragmented staining pattern can be bridged by interactive editing, based on the known chromosome structure, or by "bridging" the gaps based on persistence of directions from both sides. An example of tracking seeded at a number of randomly chosen points along the chromosomes and automatically drawn from that point to in two directions is shown in Fig. 3.

An automatic fiber-tracking algorithm was developed by Borland *et al.* (1988) to identify chromatin fibers in EM tomographic reconstructions from expanded mitotic chromosomes. Given an appropriate starting point, their approach is to use the density gradient computed within a small region to determine the appropriate direction of the fiber center. Once the center is determined, they then attempt to travel along the central ridge of maximum density. With their algorithm, closely juxtaposed fibers whose densities overlap will be treated as a single entity. They use a threshold as a way of separating chromatin fibers from background. Although it seems that much manual intervention may have been required, using this approach they have been able to successfully track long segments of loosely packed chromatin fibers within metaphase chromosomes.

Houtsmuller *et al.* (1990) have described a semiautomated chromosome-tracing approach based on the idea of a "homing 3D cursor." When positioned near a chromosome, the cursor automatically attempts to home-in on the center of the chromosome. This is accomplished by convolving the voxel data with a small sphere centered at the current cursor position. The cursor is then stepped to the location of maximal overlap, and the process repeated until convergence is reached. Furthermore, the user can specify enough parameters (direction, distance, etc.) to allow the cursor to proceed automatically along a region of the chromosome and then return to manual mode. Perhaps the most serious drawback of this approach is the convolution operator used to center the cursor in the density. Since the sphere needs to be somewhat larger than the diameter of the object being traced for this method to work, other objects must be spaced further away than the diameter of the sphere. This will obviously pose problems when examining complex structures having numerous regions in close proximity. However, it should be possible to utilize preprocessing methods such as the cookie cutter to predefine spatial boundaries within which the homing cursor must operate.

4.2.2. Matched Filter Methods for Identifying Objects in Complex Images

Tracking Microtubules in Nerve Growth Cones Microtubules are the most dynamic component of the cell cytoskeleton. In order to study their role in cell locomotion and nerve growth, tubulin can be fluorescently labeled and micro-injected into cells. Its subsequent incorporation into microtubules allows them to be directly visualized by fluorescent microscopy (Kellogg *et al.*, 1988). Because of their long axonal processes and well-differentiated cell bodies, neurons are excellent cells for examining the behavior of microtubules. Since living nerve cells are very sensitive to the fluorescence excitation light level, images need to be recorded at very low light levels and thus tend to be quite noisy. Obtaining quantitative information on the rate of growth, directionality, etc., requires that the pictures be "distilled" into linear graphs presenting the microtubules. This can be readily achieved by using a matched filter that recognizes rods. The algorithm developed by Soferman (1989) tests for rods in all orientations by calculating the correlation of a box about each image point with rods in a discrete set of orientations. The result can be enhanced by subtracting the correlation obtained with two parallel rods displaced sideways by the characteristic line thickness to be traced. The original pixel is then replaced by the best correlated rod, scaled by the original image intensity. Although slow and time consuming in real space, this process is efficiently performed in Fourier space. The results of this approach shown in Fig. 4 demonstrate a better pattern of continuity along the track than obtainable using edge-enhancement algorithms which tend to produce highly fragmented paths along the microtubules due to noise.

Locating Beads in Electron Micrographs Another example of using matched filters to extract objects from complex images involves finding small beads in electron micrographs. A necessary step in computing tomographic reconstructions from tilt series of a thick sections is to bring the different images to a common coordinate system, so that relative displacements and rotations introduced by not precisely

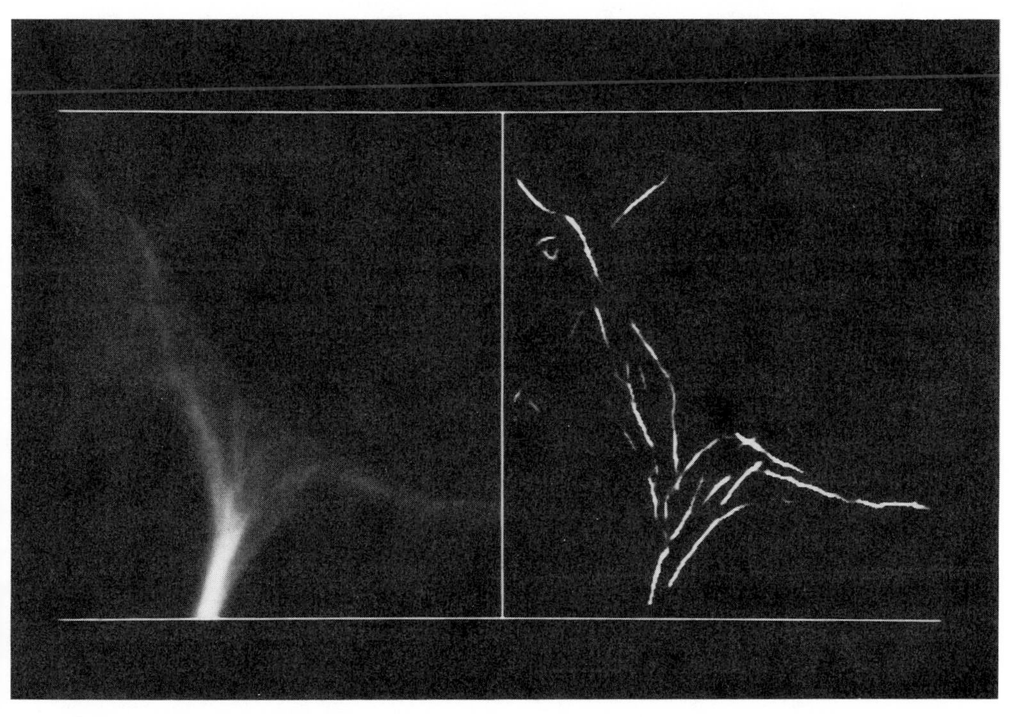

FIGURE 4. An example of the use of a matched filter (based on oriented rods, Soferman, 1989) for tracking microtubules dynamics in the tips of growth cones. (Taken from E. Tanaka, unpublished data.)

eucentric tilting can be compensated. The most reliable and internally consistent algorithm for tilt series alignment is to record the positions of small colloidal gold beads which are dispersed on the surface of the sample, and to use the projected coordinates to refine values for (x, y, z) bead position, tilt angle, tilt axis, translation, and magnification (see appendix in Luther *et al.*, 1988). Locating the small beads within complex images can be challenging due to their relative low contrast (especially at high tilt angles). The matched-filter approach followed by pattern recognition and classification methods provides a very reliable method for accurate and automatic identification. The matched filter is performed as a Wiener-type inverse filter:

$$\text{Filtered image} = \mathrm{FT}^{-1}\left(\frac{\mathrm{FT(image)}}{\mathrm{FT(bead)}+k}\right)$$

where FT = Fourier transform, FT^{-1} = inverse Fourier transform, and k is a user-chosen constant to determine the balance between sharpness and noise. The peak positions, indicating sites of best matching to the bead template, are then selected based on local contrast and finally screened on the basis of size and shape. Finally, a procedure has been developed which automatically matches the beads from all of the sections in the tilt series, producing a list of beads that is consistent throughout the data set and can be used for alignment. Images of the original digital micrograph and the processed images are shown in Fig. 5.

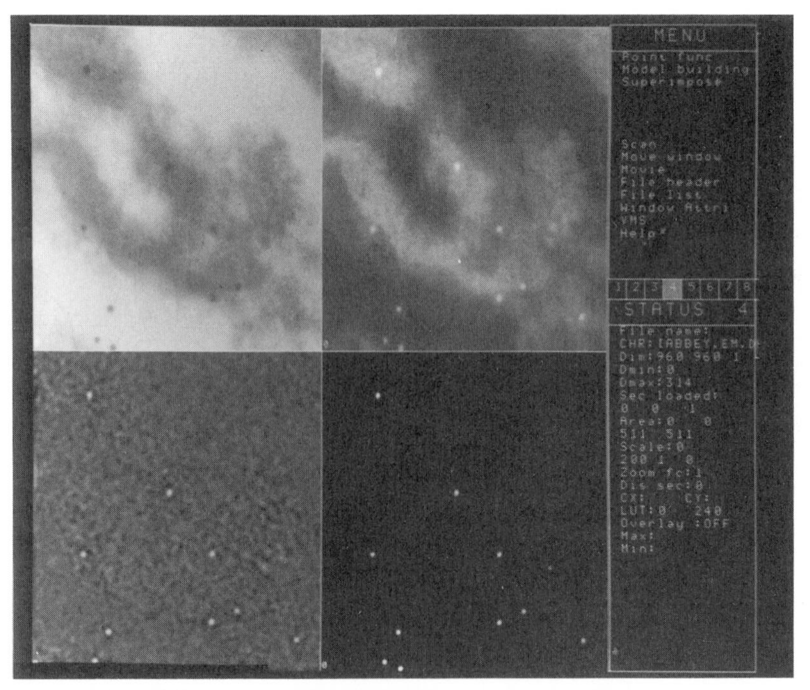

FIGURE 5. The identification of gold beads in electron micrographs using a matched Wiener filter. (a) shows the raw EM data taken from the CCD, (b) is the mass normalized image to convert the data to electron absorbance, (c) is the Wiener matched filter, and (d) is the filtered result after thresholding, demonstrating the high contrast achieved. This approach can accurately detect beads even when there is very little contrast difference with respect to the rest of the image.

4.2.3. Tracking Nuclei in Four Dimensions: A 3D Time Series

Tracking cell lineages in embryonic development is of interest in order to study the mechanisms of segmentation of new tissue, development of organs, and the underlying processes of cell differentiation. One of the unique features of the light microscope is its ability to acquire images from living biological specimens. To determine patterns of cell lineage in *Drosophila melanogaster*, we followed a large field of cells in a developing embryo that had been injected with fluorescently labeled histones (Minden *et al.*, 1989; Hiraoka *et al.*, 1989). The histones become incorporated into chromosomes during subsequent rounds of nuclear division, and act as a vital nuclear stain.

During embryonic development there are dramatic changes in the positions of nuclei. For the *Drosophila* embryo at the relevant stages, the cells move in all three dimensions. Because the embryo is quite thick, and the movements are large, three-dimensional imaging is required to follow the large number of nuclei that migrate in and out of any given focal plane. Thus, collecting data during embryonic development necessitates the acquisition of a time series of three-dimensional image stacks, which can be called a four-dimensional stack.

The problems of tracking objects in time series have been intensively studied in relation to fields like robotics, artificial intelligence, and automatic navigation. The special property of the time axis with respect to the other three spatial coordinates is that if the data are collected sufficiently rapidly, the changes in the images are bounded and can be used to identify the motion of defined objects from one time frame to the next. For example, to trace a single bright point in a dark field requires only searching for the object within a certain radius of its position in the previous frame, the radius being defined by the object's velocity and the time increment between the frames.

This approach, unfortunately, cannot be easily exploited when examining images of more diffuse objects. An example would be the definition of the velocity field of clouds, as recorded by a time series of satellite images, where evolving continuous patterns and textures need to be followed rather than a set of isolated points. Correlation methods can be applied to track defined regions which are moving and changing slowly with time. By comparing time sequences at reduced resolutions, the computational load can be reduced, and, more significantly, high contrast can be achieved for specific moving patterns or features within an evolving scene (Baker and Bolles, 1989). Still, the higher the resolution the details of interest, the closer in time the images have to be taken, so that correlations between consecutive time points do not decay for that resolution.

Unfortunately, for imaging developmental events in *Drosophila* embryos, a variety of instrumental limitations place a limit on how fast the three-dimensional data stacks can be acquired. For this reason, the nuclei traced were found to move distances greater than their own diameters between consecutive frames during daughter chromosome separation in anaphase. The conventional approach using correlations would thus not be able to track the lineage of cells through several mitotic divisions. An alternative approach is to first identify the objects to be traced in each time point, based on their known geometric properties, then apply point tracking.

A three dimensional patching routine was written, based on a 2D algorithm developed by Soferman (1989) that can be described as the emergence of mountain tops in a landscape during the decline of the water after a flood. All the voxels in the image are first sorted by height (intensity). Then, starting at the highest, and going down the list, they are in turn "patched" by assigning to each voxel the number of a touching patch or a new patch number for isolated voxels (new tops of hills). As patches grow, the newly added voxel may connect two existing patches. At that point, a decision is made to merge small patches (resulting from noise) into large ones or to leave them as separate patches if both are large enough. Patches can be restricted to some maximum size, although in order to prevent new patches from appearing it is necessary to assign the neighboring voxels to them as the

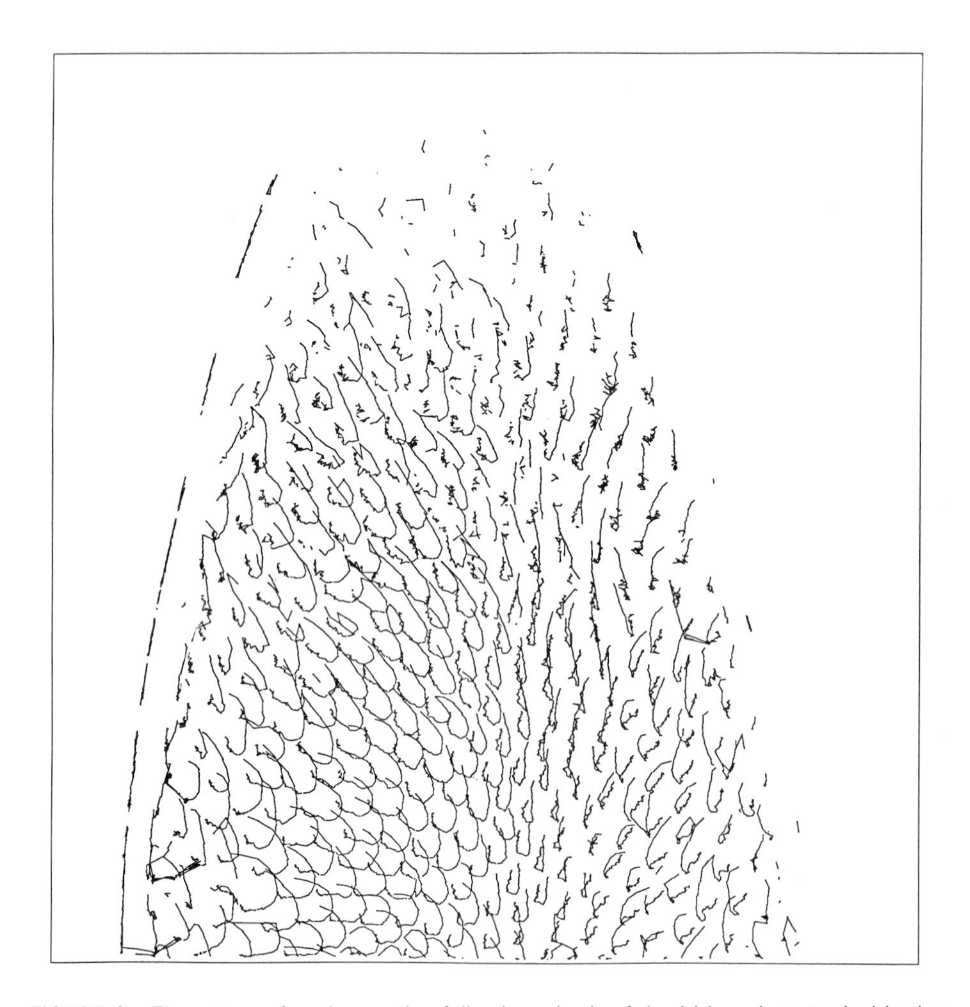

FIGURE 6. The pattern of nuclear motion following mitosis of the 14th cycle as tracked in time from images of *Drosophila melanogaster* embryos which were injected with fluorescently labeled histones (Minden *et al.*, 1989; Hiraoka *et al.*, 1989). Surprisingly, the patterns of motion were found to be coordinated over quite large areas. Mirror symmetry about the dorsal-ventral line can also be seen.

"water" level continues to descend. This approach allows all nuclei to be identified as local peaks with a typical size, irrespective of the background level or their peak height. The algorithm can be stopped at any fraction of the total number of voxels, corresponding to the fractional volume of the structure of interest (size of a nucleus). The original 2D lake algorithm (Soferman, 1989) exhausted all pixels and registered the lines of patch meetings which are the ridges in the intensity landscape (valleys for descending water). The network of lines, their intersects, and the regions generated in one sweep have interesting topological properties for the purpose of segmentation of pictures.

In our case, the isolated patches represent nuclei and define their center of mass in each time slice. Once each nucleus has been identified, tracking the paths in time was easy to achieve by point proximity criteria. To visualize mitotic events, the analysis must also allow for binary branched paths. The derived nuclear tracks can be used to assign lineages or patterns of parenthood, which is very important for understanding the biology. Interestingly, the tracked motion reveals patterns of motion (see Fig. 6) and changing orientations of cell division which reflect differentiated domains in the developing embryo (symmetry about dorsal-ventral line).

5. SUMMARY

Three-dimensional image reconstructions have the potential for dramatically extending our understanding of cell structure and behavior. Of central importance is the ability to extract relevant information from the complex voxel data derived from light and electron microscopic reconstructions. In general, for biological data this is an extremely difficult problem that is just beginning to be approached.

We have discussed the importance of utilizing a powerful and versatile interactive image display and processing system for extracting relevant information from the voxel data. In addition, the hardware and software that we have been developing and using for these purposes have been described. In many cases, the relevant biological information can be in the form of a line drawing model that represents the path of the structure in three-dimensional space. The key to dealing with such models and voxel data simultaneously are an image file structure that maintains geometrical data for multidimensional images and a display and modeling system that can utilize this information to allow orientation and processing-independent superimposition of model and data. Multiple windows enable simultaneous modeling in several orientations. A variety of approaches for data preprocessing as well as automatic and semiautomatic methods for model generation were also surveyed.

ACKNOWLEDGMENTS

This work was supported in part by NIH GM25101 and NIH GM32803 to JWS and NIH grant GM31627 to DAA, and by the Howard Hughes Medical Institute. D.A.A. was also supported by an NSF presidential young investigator grant.

REFERENCES

Agard, D. A., Hiraoka, Y., and Sedat, J. W. (1988). Three-dimensional light microscopy of diploid Drosophila chromosomes. *Cell Motility Cytoskeleton* **10**:18–27.

Agard, D. A., Hiraoka, Y., Shaw, P., and Sedat, J. W. (1989). Fluorescence microscopy in three dimensions. *Meth. Cell Biol.* **30**:353–377.

Baker, H. H. and Bolles, R. C. (1989). Generalized epipolar-plane image analysis on the spatiotemporal surface. *Int. J. Comput. Vision* **3**(1).

Borland, L., Harauz, G., Bahr, G., and van Heel, M. (1988). Packing of the 30nm chromatin fiber in the human metaphase chromosome. *Chromosoma(Berlin)* **97**:159–163.

Brandt, A. (1986). Algebraic multigrid theory: the symmetric case. *Appl. Math. Comput.* **19**:23–56.

Castleman, K. R. (1979). *Digital Image Processing.* Prentice-Hall, Englewood Cliffs, NJ.

Chen, H., Sedat, J. W., and Agard, D. A. (1990). Manipulations, display, and analysis of three-dimensional biological images, in *Handbook of Biological Confocal Microscopy* (J. B. Pawley, ed.), Chap. 13, pp. 141–150. Plenum, New York.

Duda, R. O. and Hart, P. E. (1973). *Pattern Classification and Scene Analysis.* Wiley-Interscience, New York.

Dreibin, R. A., Carpenter, L., and Hanranhan, P. (1988). Volume rendering. *Comput. Gr.* **22**:64–75.

Hiraoka, Y., Minden, J. S., Swedlow, J. R., Sedat, J. W., and Agard, D. A. (1989). Focal points for chromosome condensation and decondensation revealed by three-dimensional in-vivo time-lapse microscopy. *Nature* **342**:293–296.

Hiraoka, Y., Sedat, J. W., and Agard, D. A. (1987). The use of charged-coupled device for quantitative optical microscopy of biological structures. *Science* **238**:36–41.

Hiraoka, Y., Sedat, J. W., and Agard, D. A. (1990). Determination of the three-dimensional imaging properties of an optical microscope system: partial confocal behavior in epi-fluorescence microscopy. *Biophys. J.* **57**:325–333.

Houtsmuller, A. B., Oud, H. T. M., van der Voort, W. M., Baarslag, M. W., Krol, J. J., Mosterd, B., Mans, A., Brakendorf, G. J., and Nanninga, N. (1990). Image processing techniques for 3D chromosome analysis. *J. Microsc.* **158**:235–248.

Huang, T. S., Yang, G. J., and Tang, G. Y. (1979). A fact two-dimensional median filtering algorithm. *IEEE Trans. Acoust. Speech, Signal Process.* **ASSP-27**: In *Digital Image Processing and Analysis* (R. Chellappa and A. A. Sawchuk, eds.), Vol. 1. IEEE Computer Society Press, 1985.

Jones, T. A. (1982). In *Computational Crystallography* (D. Sayre, ed.), pp. 303–317. Oxford University Press, Oxford, England.

Luther, P. K., Lawrence, M. C., and Crowther, R. A. (1988). A method for monitoring the collapse of plastic sections as a function of electron dose. *Ultramicroscopy* **24**:7–18.

Kellogg, D., Mitchison, T. J., and Alberts, B. (1988). Behavior of microtubules in living *Drosophila* embryos. *Development* **103**:675–686.

Minden, J. S., Agard, D. A., Sedat, J. W., and Alberts, B. M. (1989). Direct cell lineage analysis in *Drosophila* melanogaster by time-lapse, three-dimensional optical microscopy of living embryos. *J. Cell Biol.* **109**:505–516.

Paddy, M. R., Belmont, A. S., Saumweber, H., Agard, D. A., and Sedat, J. W. (1990). Nuclear envelope lamins form a discontinuous network in interphase nuclei which interact with only a fraction of the chromatin in the nuclear periphery. *Cell* **62**:89–106 (1990).

Rykowksi, M. C., Parmalee, S. J., Agard, D. A., and Sedat, J. W. (1988). Precise determination of the molecular limits of a polytene chromosome band: regulatory sequences for the notch gene are in the interband. *Cell* **54**:461–472.

Soferman, Z. (1989). *Computerized Optical Microscopy*, Doctoral Thesis, Weizmann Institute, Rehovot Israel.

van der Voort, H. T. M., Brakenhoff, G. J., and Baarslag, M. W. (1988). Three-dimensional visualization methods for confocal microscopy. *J. Microsc.* **153**:123–132.

12

Holographic Display of 3D Data

Kaveh Bazargan

1. INTRODUCTION TO 3D PERCEPTION

Holography is a powerful optical technique for recording and displaying 3D information. In this chapter we discuss the application of holography in the display of tomographic data. It helps to have a good understanding of 3D perception in order to determine whether holography can be used beneficially in the display of scientific and medical data. For this reason, this chapter begins with an overview of the mechanisms which help us perceive the world in three dimensions. Later, we shall refer to these mechanisms to evaluate the various holographic techniques available.

The brain and eye work together to decipher the huge amount of optical information received by the eye. The information contributing to 3D perception, or depth perception, can be divided into many "depth cues." These depth cues can be

Kaveh Bazargan • Focal Image Limited, London W11 3QR, England

Electron Tomography: Three-Dimensional Imaging with the Transmission Electron Microscope, edited by Joachim Frank, Plenum Press, New York, 1992.

further divided into physiological, and psychological ones, although some cues fall into a gray area in between. As a rule, physiological depth cues are those derived from the physical motion of (or other physical changes in) the eyes, and psychological depth cues are those derived by the brain after processing the retinal images of the eye(s).

1.1. Physiological Depth Cues

Physiological depth cues are generally considered to be the most important ones for depth perception. The strongest such cues are listed below.

Accommodation Accommodation is the reshaping of the lens in the eye by the ciliary muscles in order to bring the image of a scene into sharp focus on the retina. The state of the lens at any given time is a cue for depth perception. In the normal eye, when the muscles are in a relaxed state, objects at infinity are in focus. To focus closer objects, the lens is made more convex by the ciliary muscles, thus shortening its focal length (Fig. 1). This physiological transformation is a cue to the brain about the distance of an object. On its own, accommodation is a weak cue, and is more effective when combined with other cues. It is a short-range cue, as changes in the focal length of the lens are minimal when objects are more than a few feet away.

Convergence When the brain concentrates on viewing a scene, the eyes are arranged such that the point of interest in the scene is focused onto the most sensitive spot in both eyes, i.e., the fovea. For objects at infinity, the axes of the two eyes are parallel. For objects at closer distances, the eyes converge such that there is an angle between the two axes (Fig. 2). This angle is another physiological depth cue called *convergence*. As in the case of accommodation, this cue is more effective at closer distances, because the variation of angle with distance diminishes at long distances. Accommodation and convergence are both weak depth cues on their own, but are significant when working together. (In fact, there is some interaction

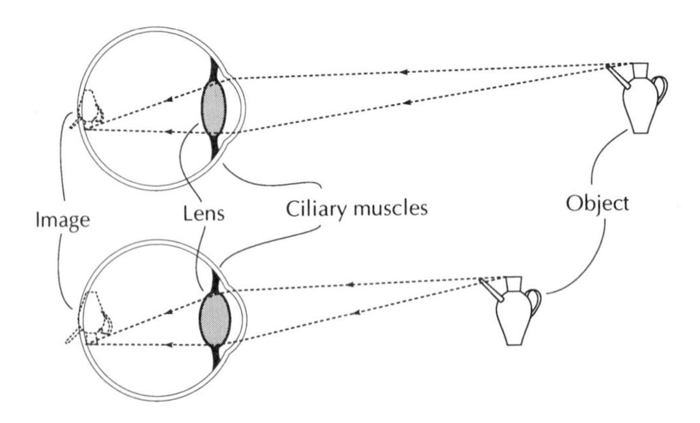

FIGURE 1. Accommodation of the eye. For nearer objects the lens is made more convex.

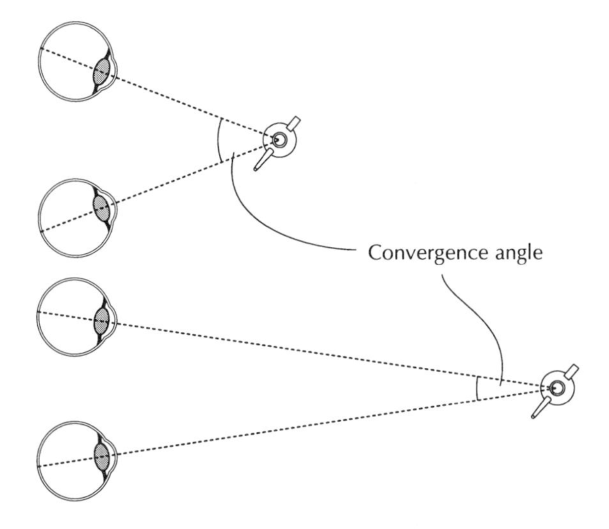

FIGURE 2. Convergence.

between the two, so that, for example, a variation in the focal length of the lens in one eye causes an involuntary convergence of the eyes, and vice versa.)

Binocular Disparity As mentioned in the case of convergence, concentration of the brain on a point in a scene causes the angle between the eyes to be such that the point of interest is focused onto the fovea in each eye. Points that lie approximately the same distance away from the eye, and are near the point of interest, are focused onto corresponding positions on the retina in both eyes. However, points that are nearer to or further from the eye than the point of interest are not focused onto corresponding positions in each eye, and are perceived as blurred double images. The angular separation of the two images gives rise to a cue called binocular disparity. This cue has a longer range than the first two, and is considered to be the most important for general depth perception.

Motion Parallax When an observer moves while viewing a scene, objects in the scene seem to move relative to one another. Closer objects seem to move faster than distant ones, thus constituting a depth cue. This effect is called motion parallax. It is present even when the observer is stationary—as the eyes are moved from side to side, the position of the lens of the eye changes, thus the scene is viewed from slightly differing positions.

1.2. Psychological Cues

While physiological depth cues are derived directly from the eyes, psychological ones are "higher-level" cues that are the result of the processing of retinal image by the brain. There are a multitude of different cues picked up by the brain in this way, and many of them are difficult to categorize. Listed below are some of the most important.

Hidden Surfaces A simple but powerful depth cue is the overlapping of objects in a scene. When the fields of view of several objects overlap, the nearer objects totally or partially hide the further ones. The simulation of this effect in computer graphics is called *hidden-surface removal.* This is a strong depth cue.

Image Size In everyday experience we learn the physical sizes of common objects. This information, combined with the size of the retinal image of the object gives a direct clue to the distance of the object from the eye. Of course this cue only applies to familiar objects.

Linear Perspective As objects move further from the eye, the size of the image of the object decreases. This effect is called linear perspective. It is most apparent in a photographs of familiar scenes with, for example, regular arrays of buildings. Farther buildings of similar size get smaller and smaller in the photograph.

Aerial Perspective A more subtle effect is the gradual reduction of contrast in distant objects due to atmospheric scattering (e.g., mist or dust).

Lighting We generally see objects by their reflection of light from sources in one or more positions. The gradual change of surface color, and the appearance of shadows on objects are strong depth cues.

2. INTRODUCTION TO HOLOGRAPHY

Having reviewed depth perception, we now take a look at different types of holograms and how they are made. For a deeper insight into various techniques in holography the reader is referred to textbooks on the subject, such as Hariharan (1984) and Syms (1989).

2.1. Laser Transmission Holography

A transmission hologram is one in which the light used to reconstruct the recorded image is on the opposite side of the hologram as the observer, and therefore passes through the plate in the reconstruction process. Transmission holograms are generally considered to be the easiest to record. The simplest hologram, both theoretically and practically, is probably the laser transmission hologram (Leith and Upatnieks, 1964), so called because a laser or other coherent light source is needed to view the recorded image.

Figure 3 shows the basic arrangement for recording a laser transmission hologram. The coherent beam from a laser source is split into two using a beam splitter. (The simplest beam splitter is a piece of high-quality thick float glass.) The beams are directed around the recording setup using a series of small mirrors. One of the beams, usually the more powerful one, is used to illuminate the object. As the laser beam is small in diameter, typically 3 mm, it has to be expanded before falling onto the object. This is done using a lens system.

The second beam emanating from the beam splitter is used as the reference

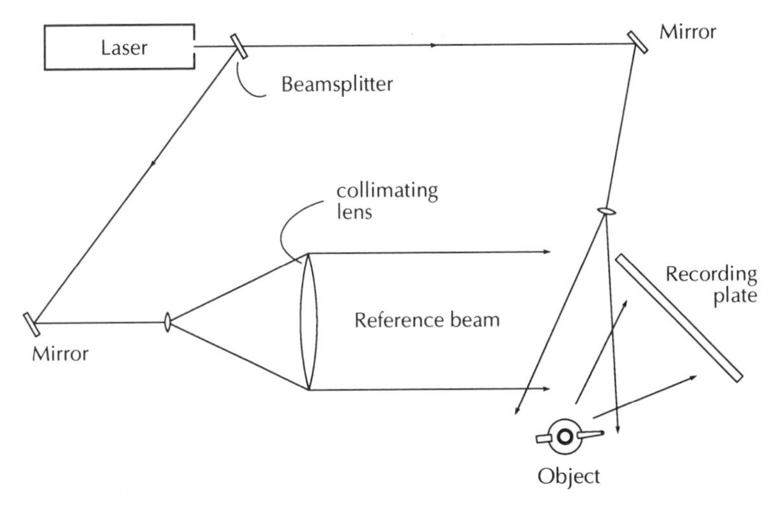

FIGURE 3. Recording arrangement for a laser transmission hologram.

beam. This time it is directed toward the recording plate. As before, it is expanded using a lens system. In this case, the beam is also collimated—made parallel—by introducing a large lens. (This is not essential to recording, but usually simplifies proceedings, especially if two-step procedures described later are employed.) The recording plate (or film) consists of a photosensitive material coated onto a base such as glass or polyester. The most common photosensitive material is the photographic (silver halide) emulsion which has the advantage of relatively high sensitivity. What distinguishes it from normal photographic material used in photography is the very high resolution it needs to have. Other suitable recording media include dichromated gelatin (DCG; Lin, 1969; Shankoff, 1968) and photopolymer.

According to the above description there are two beams falling onto the recording plate: One is the reference beam, and the other is the light scattered from the object, or the object beam. It is the optical interference of these two beams that is recorded on the photosensitive emulsion. The recorded interference patterns take the form of microscopic, meandering lines and contain, in an encoded form, all the optical information relating to the object. We can now see why the reference beam is so named. It is, in effect, the yardstick against which each portion of the complex scattered beam from the object is measured.

One way of looking at the recording process is to consider the light emanating from the object as consisting of a large number of rays of light, each with a different direction and intensity. They are scattered quasirandomly, and with no apparent order. The reference beam, on the other hand, consists of a set of rays, all with approximately the same strength, and all diverging from the same point. In the recording process, we use the reference beam to record the direction and intensity of each and every ray in the object beam.

Having recorded the interference fringes in the photosensitive material, the material usually has to be processed so that the fringes can be used in the reconstruction stage. If the material is of the silver halide variety, then the processing is similar to that in conventional photography, i.e., developing and fixing. In fact the material is usually bleached to improve the brightness. The processing of silver halide materials for holography is a vast field of theoretical and empirical research, the goal being the production of the brightest and cleanest holograms.

After processing, the complex interference fringes are transformed into regions of different optical density or different refractive index.

In order to reconstruct the recorded image, the processed plate is replaced in its original position, the object beam is eliminated (e.g., by blocking its path), and the plate is illuminated by a replica of the reference beam (now called the reconstruction beam) (Fig. 4). The optical phenomenon that now occurs is called diffraction and is, in a way, the opposite of interference: The reconstruction beam falls on the numerous fringes recorded on the plate and is diffracted. In other words, instead of continuing its path through the plate, some of the light is deflected into other directions. In particular, a major part of the diffracted light leaves the plate in the same direction as the original object beam. In fact, the rays making up this part of the diffracted light are exact replicas of the original rays that were scattered by the object during the recording stage. Following our ray model discussed above the hologram is, in effect, reconstructing all the rays incident on it from the original object beam, both in direction and in intensity. It follows, then, that if an observer looks in the direction of the original object, he or she will not be able to distinguish the reconstructed image from the original object, as the rays are identical in both cases. Clearly, all depth cues are present in the image.

Laser transmission holograms were the first type of display holograms to be made, and still produce the most realistic images which amaze even the most hardened of holographers. The main disadvantage is that a laser or another coherent optical source is required for the reconstruction of the image.

2.2. *White-Light Transmission Holography*

A major goal in the field of holography has been, and still is, the production of high-quality images that can be viewed using noncoherent light sources (such as

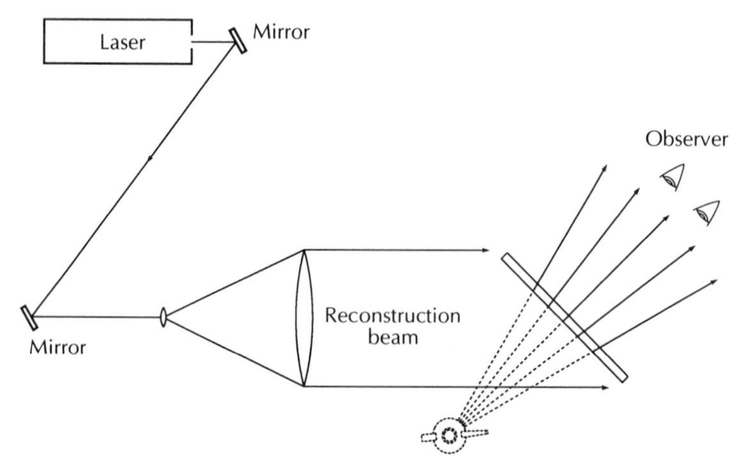

FIGURE 4. Reconstruction of a laser transmission hologram.

white light) for image reconstruction. Perhaps the most popular method is to use the reflection hologram recording technique (see below). Several methods have been devised to allow white light viewing of transmission holograms. Let us briefly look at these, starting with an explanation of why simple transmission holograms cannot be viewed with white light.

2.2.1. Why Can't We Use White Light?

Let's assume that we have produced a laser transmission hologram according to the above procedure. Why can we not use an ordinary light source such as a white spot lamp to view the image? Figure 5 illustrates the problem. White light is composed of a continuum of wavelengths from the blue end to the red end of the spectrum. An undistorted image is only obtained when the wavelength of the reconstructing beam is equal to that of the recording beam. The angle through which a beam is diffracted is dependent on its wavelength, shorter wavelengths being deviated less than longer ones. Consequently, each wavelength produces its own image at a different position. This effect is called *chromatic dispersion.* The result is a spectral blur of images, generally unacceptable for viewing.

2.2.2. Image-Plane Holography

As can be appreciated by examining Fig. 5, the magnitude of spectral blurring is proportional to the distance of the image from the hologram. It can therefore be minimized by reducing this distance. In fact, the spectral blurring would be totally eliminated if the image could appear in the plane of the hologram, resulting in a sharp, achromatic (black and white) image. However, the simple recording geometry of Fig. 3 does not allow this. If the object is placed too close to the recording plate, it impedes the path of the reference beam. The answer is to use a more complex, two-step technique for recording which allows the image to be positioned not only near, but even cutting through the hologram plate (Rotz and Friesem, 1966).

The first stage in the two-step process involves making a "master" hologram by a process identical to that for the laser transmission hologram. The second stage is shown in Fig. 6. The processed hologram is illuminated by a reconstruction beam

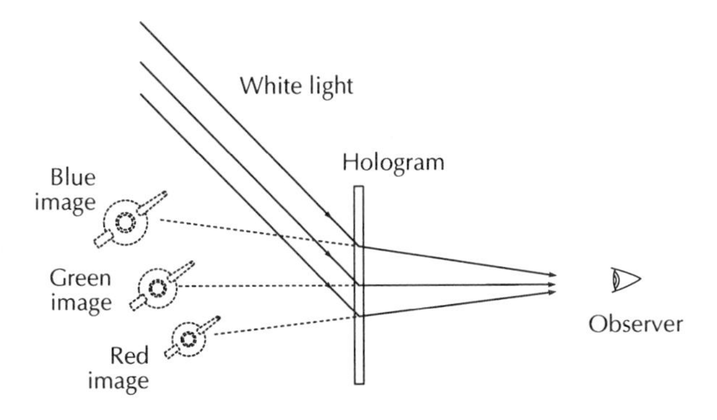

FIGURE 5. Chromatic dispersion in a hologram.

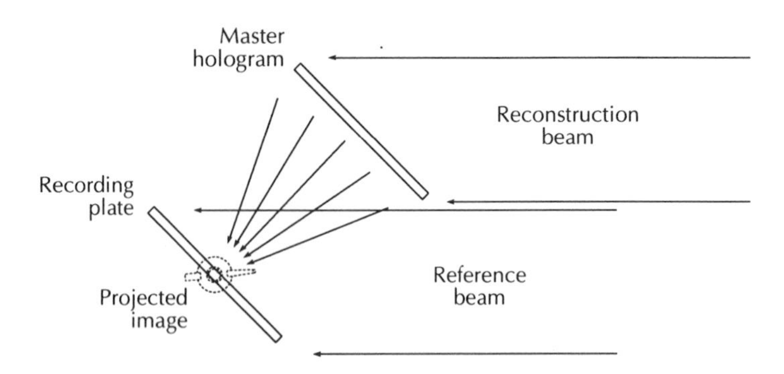

FIGURE 6. Recording an image-plane hologram.

that is "conjugate" to the original reference beam. In other words, all the original rays in the reference beam are reversed in direction. If the original reference beam had been diverging, then the conjugate beam is converging. As we have used a collimated beam in the first step, the conjugate beam is also collimated. The image from the hologram is now projected into space in front of the hologram. This image suffers from the curious effect of "pseudoscopy" or depth reversal. In other words, the shapes appear "inside out," with convex shapes becoming concave. However, this is only an intermediate image and will be corrected in the second stage.

A second hologram can now be recorded by placing a recording plate near, or even within, the projected image (something clearly impossible with a real object) and adding a reference beam as before. We can reconstruct an orthoscopic (as opposed to pseudoscopic) image from this hologram by using the conjugate of the second reference beam to reconstruct the image (Fig. 7). The image appears to protrude from the holographic plate and to project partially. This is called an image-plane hologram. As the image points are very close to the hologram plane, the image can be viewed with minimal chromatic dispersion. Points in the image plane will be absolutely sharp, and points in front and behind the plate will be blurred proportionally to their distance from the plate. [By an optical technique the plane of zero dispersion can be shifted (Bazargan and Forshaw, 1980).]

The disadvantage of the image-plane technique is that the depth of the image is severely limited, typically to some 20–30 mm.

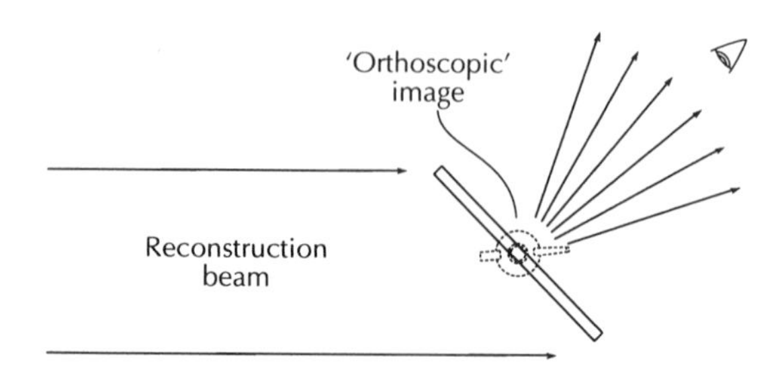

FIGURE 7. Viewing an image-plane hologram.

2.2.3. "Rainbow" Holography

The "rainbow" or "Benton" hologram (Benton, 1969) is an extension of the image-plane technique that dramatically increases the usable image depth in white light by sacrificing vertical parallax. The recording procedure is similar to that of the image-plane hologram, except that in the second stage the height of the master hologram is limited to a few millimeters. When the final image is viewed, it appears in a range of spectral colors, depending on the position of the viewer. As the eyes are moved in a vertical direction the color of the image changes from the blue end of the spectrum to the red. The perspective of the image, however, does not change. In other words, one cannot look over or under foreground objects to see background ones. The unorthodox recording geometry also means that there are distortions and aberrations present unless the image is viewed from a specific position in space.

The changing colors can be distracting to the viewer, but fortunately there are methods to achromatize the image (Benton, 1978), and the limitation is not a serious one. The limited parallax, however, is a more fundamental limitation and renders the method unsuitable for many scientific applications. The inherent distortions and aberrations must also be taken into account. Notwithstanding these problems the rainbow hologram has proved to extremely popular, especially in display and artistic applications.

2.2.4. Dispersion-Compensated Holography

A simple but powerful white-light display method is the dispersion-compensated technique. The idea is simply to cancel out the chromatic dispersion at the hologram by predispersing the light in the opposite sense before it reaches the hologram. The best way of achieving this predispersion is to use a diffraction grating with the same fringe spacing as the average fringe spacing of the hologram (Burckhardt, 1966; De Bitetto, 1966; Paques, 1966). Such a diffraction grating can be made by recording the interference pattern between two collimated beams. As in the case of the rainbow hologram, the method extends the effective depth of the image-plane hologram. Figure 8 shows the reconstruction of the image in a dispersion-compensated system. The collimated light from a white light source falls onto a diffraction grating. Some of the light is diffracted and is chromatically dispersed.

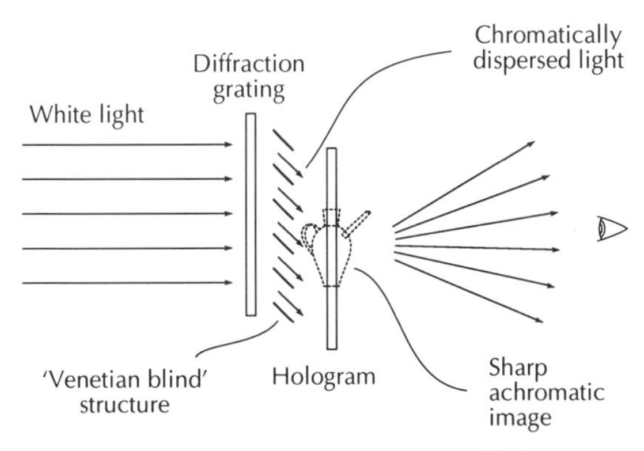

FIGURE 8. A dispersion-compensated hologram.

(The undiffracted light is blocked by a miniature venetian blind structure in order that it does not reach the observer.) The diffracted light is used as the reconstruction beam for the hologram which produces an equal and opposite dispersion. The resultant image is sharp and achromatic.

The distortions and aberrations in such a dispersion-compensated image are very small, and the image is comfortable to view. Moreover, the image retains full parallax and is therefore better suited than the rainbow hologram for scientific applications. The obvious disadvantage is the need for the diffraction grating and the venetian blind structure. One solution is to incorporate these and the light source into a desktop viewer (Bazargan, 1985). This has been tested and found to work well, with the viewer being mass producible.

2.3. Reflection Holography

A different approach to producing white-light-viewable holograms is to use the "reflection" recording technique (Denisyuk, 1962). The essential difference between this technique and the transmission one is that in reflection holography the object and the reference beams fall onto opposite sides of the recording plate. Figure 9 shows the basic arrangement. The recording medium in this case must be a "volume" material; i.e., the fringes must be recorded in the depth of the material rather than on the surface. Figure 10 shows a schematic representation of fringes recorded in a volume material. The fringes can now be regarded as 3D surfaces in the volume of the emulsion rather than 2D lines on its surface.

The image from a reflection hologram is reconstructed as before by illuminating it with a replica of the original reference beam. This time the viewer observes the image from the same side of the hologram as the reconstruction beam. The basic principles of interference and diffraction apply as before, with one qualification: the fringes have, in this case, recorded not only the complete optical information about the object, but also the wavelength used in recording. This means that, in general, the hologram will only respond to a reconstruction wavelength near the recording

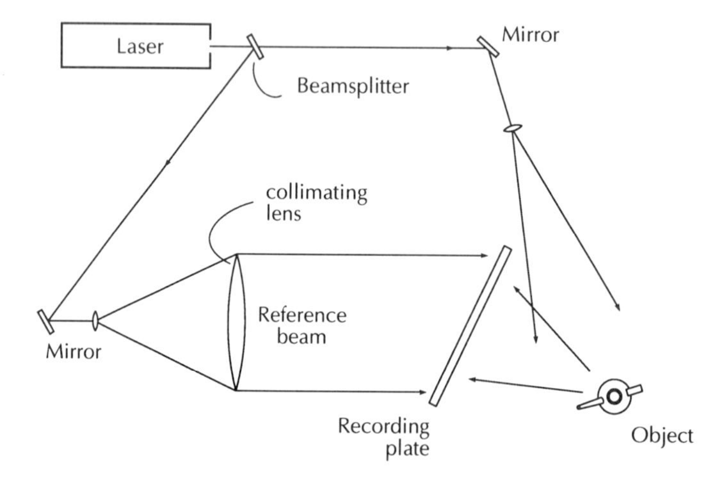

FIGURE 9. Recording a reflection hologram.

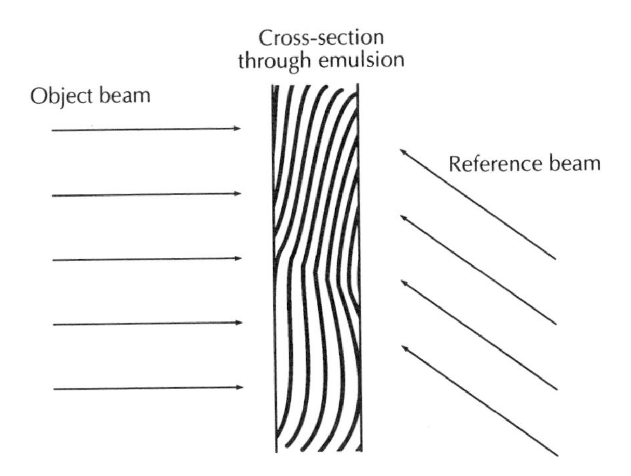

FIGURE 10. Schematic representation of fringes in a reflection hologram.

one. If white light is used in the reconstruction, only a narrow band of wavelengths near the recording one will take part in the reconstruction. This means that the chromatic dispersion will be small and that an image of considerable depth can be viewed in white light.

Reflection holography can, of course, be combined with image-plane holography to produce sharp, low-dispersion holograms. Usually a transmission master hologram is used, as described above.

One disadvantage of the reflection hologram is that if the observer gets too close to the hologram to examine the image the reconstruction beam is impeded and the image disappears.

2.4. *Multicolor Holography*

The above techniques use a single wavelength to record the hologram. The image produced is therefore in a single color. Even when multiple colors are present, as in the rainbow hologram, the colors seen bear no relation to the colors of the object. In order to record the color of the object, two or more laser wavelengths are required. They are a large number of techniques available to produce color holograms (Bazargan, 1983, 1986; Hariharan, 1983), and a detailed description of these methods is beyond the scope of this chapter.

3. *METHODS OF PRODUCING HOLOGRAPHIC DISPLAY FROM 3D DATA*

In all the previous discussions we have assumed that a real object is available at the recording stage. With the tremendous computing power now available, computers are used to store, manipulate, and display a wide range of 3D data. Displaying the data presents a fundamental problem as the display medium—e.g., the computer screen, printer output, etc.—is invariably two dimensional. Ingenious solutions have been found for displaying such data, especially in the field of 3D modeling and rendering, but the fact remains that an observer cannot see around foreground objects or artifacts simply by moving his or her head. Holography has the potential for displaying data in true 3D, and there are three broad approaches to the problem, namely *fringe writing*, *holographic stereography*, and *volumetric multiplexing*. These are discussed in detail below.

3.1. Fringe Writing

Let us take the most fundamental approach to creating a hologram artificially. For simplicity we can consider a laser transmission hologram. The information in such a hologram is recorded in the minute fringes in the recording material. The physics of the formation of such fringes is well known and, in principle, can be computed for every point on a hologram. The fringes usually vary gradually between totally light and totally dark. For the purposes of this discussion, however, we can assume them to be binary; i.e., each point on the hologram is either transparent (light) or opaque (dark). The basic rule is that the state of the interference pattern for each point on the hologram depends on the sum of the phases of all the rays incident on it. The reference beam can be considered as contributing one ray to each point, but the object contributes a tremendous number. In normal holographic recording the fringes are effectively calculated by a hugely parallel optical computer. When fringe writing, these calculations have to be made on conventional electronic computers. With the high density of points on a typical hologram, the number of calculations turns out to be prohibitively large. The fringe density in a typical hologram is about 1 million points/mm^2. For each of these points the contributions of all points of interest in the object must be calculated (Brown and Lohman, 1966). Even with the most powerful parallel processors the time taken to calculate the fringe structure for one hologram is impractically long.

There are ways of reducing the information content of the hologram (Barnard, 1988), but these invariably result in lower-quality holograms or limited fields of view.

Even if the computing problems for fringe writing were overcome, only half the problem is solved. It is necessary to write the fringes onto a suitable material. The only process currently available for writing such small data on a two-dimensional surface is photolithography, normally used for producing integrated circuits. Unfortunately, the hardware is designed specifically for producing shapes with rectilinear geometry, and is not suited to the undulating shapes of typical holographic fringes.

Fringe writing is a useful method for producing holographic optical elements used, for example, in optical testing. For the moment, however, it must be considered impractical for producing realistic three-dimensional images. If the two technological problems are resolved, then fringe writing has the potential advantage of producing totally realistic holographic scenes, with all depth cues present.

3.2. Holographic Stereograms

We now examine in detail one of the two practical methods for producing holographic images from 3D data. The most popular is the holographic stereogram

(Benton, 1982; De Bitetto, 1969; McCrickerd and George, 1968). There are many variations to this technique, but the general procedure is as follows: (1) The computer is used to produce a large number of perspective views of the 3D data. (2) With a laser beam, each image is projected onto a diffusing screen and a narrow-strip hologram is made of each view. (3) The strip holograms are treated as a master hologram, and all the views are simultaneously transferred onto a secondary recording plate. (4) The secondary plate is illuminated, and the composite 3D image is viewed.

Figure 11 shows the details of one geometry for recording the master hologram. Different views of an object are computed, and recorded on a film strip. Typically, the number of frames is 200, and the first and last views are angularly separated by some 90°. In a film recorder is not available as a computer peripheral, then the computer monitor can be photographed with a pin-registered camera. The film strip is placed into an optical system which projects each frame onto a diffusing screen, using a laser source. The diffusing screen is then treated as the object, and a hologram is recorded onto a narrow strip of the master plate. The position of the strip corresponds to the perspective view recorded in the frame. A moving slit controls the position of the strip. It starts at one extreme of the plate, and moves to the other.

After the plate is processed as usual, the image is transferred onto a second plate by a technique similar to the image-plane method. Figure 12 shows the arrangement for this stage of the recording. All perspective views are simultaneously recorded on the transfer plate, using a reference beam as before. The image can now be viewed by illuminating the transfer hologram with the conjugate of the reference beam used in the recording process (Fig. 13). In the reconstruction stage, all 2D views recorded are reconstructed simultaneously, but each view is visible only if viewed through the corresponding narrow slit which is now projected onto the viewer space. When the observer places the eyes within these projected slits, each eye sees a different perspective of the image. The depth cue in operation is clearly that of binocular disparity. This is a strong cue, and the sensation of depth is

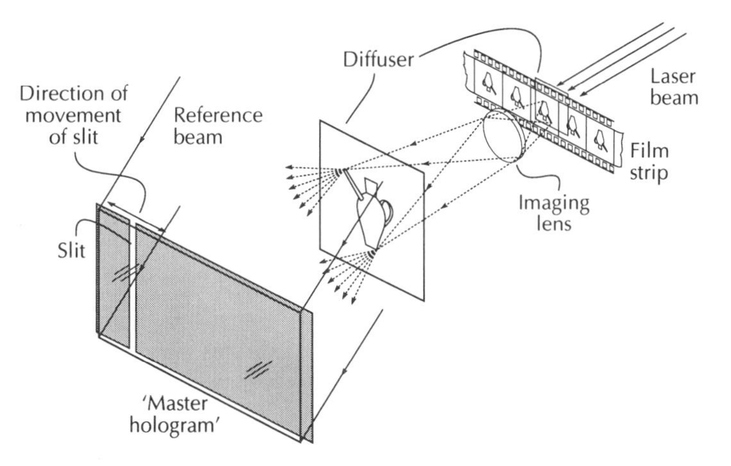

FIGURE 11. Recording the master for a holographic stereogram.

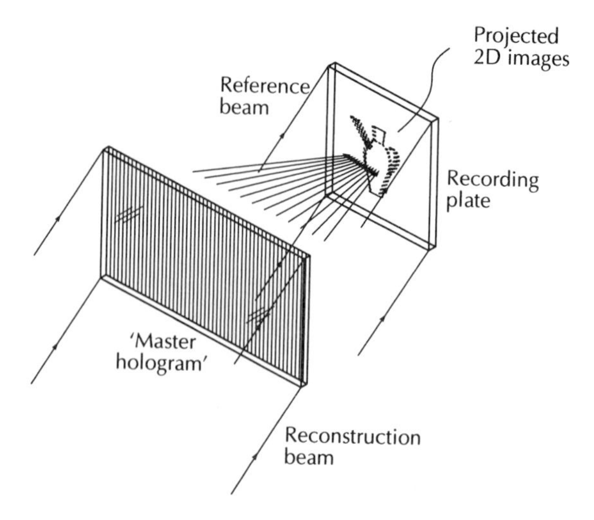

FIGURE 12. Recording a white-light-viewable holographic stereogram.

present. The convergence cue is also present as the observer concentrates on different parts of the image. As the observer moves laterally, each eye looks through a different slit, and a different view is therefore selected. This means that motion parallax, another strong depth cue, is also present. If the recording procedures have been carefully followed, then the transition from one view to the next will be smooth, and the depth sensation is further enhanced.

3.2.1. Advantages of Holographic Stereograms

Each image in the sequence is a 2D view of a 3D scene, calculated by a computer. Consequently, all the powerful techniques currently available in 3D modelling and rendering can be brought into play. This includes hidden surface

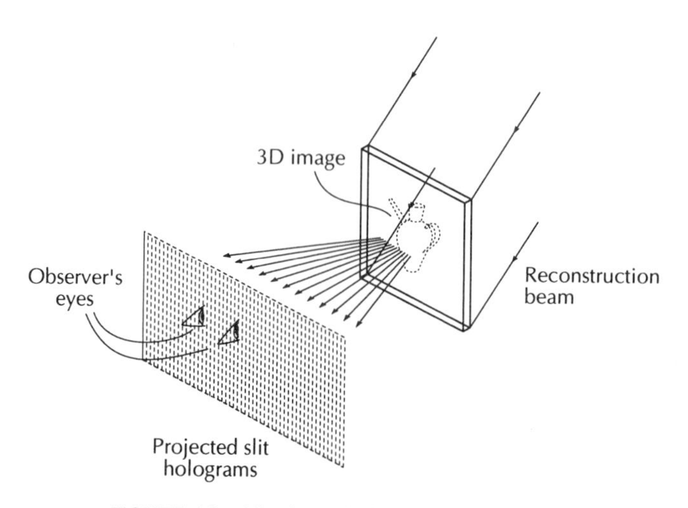

FIGURE 13. Viewing a holographic stereogram.

removal, texture mapping, Phong shading, etc. The method is ideally suited to computer-aided design (CAD) applications which use most of these techniques. The idea of 3D hard copy is an attractive proposition for the visualization of CAD-type images.

A holographic stereogram of a wire-frame image, i.e., one with no hidden-line or hidden-surface removal, can be perceived to have a perfectly understandable 3D form, even though the individual frames may have ambiguities about depth—the 3D depth cues present in the stereogram aid understanding of the data.

3.2.2. Drawbacks of Holographic Stereograms

It is important to realize that although holographic stereography is a powerful technique, it lacks the accommodation cue which is important for images near the observer. Also, the basic ingredient of a holographic stereogram is a set of perspective views. There are many types of data which cannot unambiguously be produced in this format. Tomographic data is a good example. In order to produce perspective views of such data, we have a choice of simply stacking the data together and adding the contributions from all slices, or we can attempt to extract 3D shapes from the data, and perform rendering operations on a computer. In the former case, the amount of data could be such that the overall picture for each view becomes almost a single shade of gray. In the second case, it may be very difficult to extract the shapes unambiguously. Indeed, many forms of data are "cloudy" in nature, and any attempt at shape extraction may destroy the data. In the case of medical tomography, the added dangers of manipulating the data are obvious.

3.2.3. Lack of Vertical Parallax

A major disadvantage of holographic stereography is the lack of vertical parallax in the image. With the recording system described, the perspective in each strip hologram is fixed. In other words, the observer cannot look over or under foreground objects. This can be a serious limitation for the display of scientific data.

It is possible to extend the technique to produce full parallax, but the number of frames required becomes tremendous, and the technique becomes impractical. For example if 200 views are used for a single parallax stereogram, then 40,000 will be required for the corresponding full parallax recording.

Because the normal holographic stereogram is single parallax, it is ideally suited to white-light display in a rainbow format.

3.2.4. Using a Spatial Light Modulator (SLM)

In the above procedure the perspective images are first recorded onto a photographic material which has to be processed and then placed in the master recording rig. Apart from the lengthy process, the film transport for recording and displaying the images must be pin registered, because the smallest misregistration will be visible in the final hologram. Furthermore, it is impractical to automate the above procedure because of the many mechanical steps involved. A great improvement would be to have a system in which the images are projected directly from the computer. There are different ways of achieving this, but in general what is needed is a spatial light modulator (SLM). This is a 2D screen that changes in transparency in response to an external signal (usually electrical). An example of an SLM in common use is the liquid crystal screen display on some portable computers. Most such SLMs do not have the required resolution and contrast ratio for use in recording stereograms, but, fortunately, SLM research and development field is currently very active, and some of the high-end products can produce very high quality images.

3.2.5. Multicolor Images

Most computer systems designed for serious 3D modeling and rendering work in several colors. It is obviously desirable to use these colors in the stereogram. Most techniques of color holography can readily be extended to stereograms. Again, there are many different techniques available (Bazargan, 1983; Hariharan, 1983; Walker and Benton, 1989), and the detailed explanation is beyond the scope of this chapter.

3.3. Volumetric Multiplexing

We now come to the technique which is perhaps the most suitable for tomographic data. Like holographic stereography, this technique relies on combining a number of 2D data sets to produce a composite 3D holographic image. The difference is that in the case of volumetric multiplexing the 2D data sets are not views of the whole object, but slices through it (Hart and Dalton, 1990; Johnson *et al.*, 1982; Keane, 1983). These slices may be tomographic data. Let us first look at the problem of image understanding in 3D tomographic data:

3.3.1. Image Understanding in 3D Tomographic Data

When a set of tomographic slices are placed side by side, the brain finds it difficult to deduce shapes or 3D patterns except for the very simplest of data sets. The obvious solution is to reorder the slices into the original 3D shape. This could be done, for example, by forming a transparency from each slice and physically mounting them in correct relation to one another. Apart from the inconvenience of building a cumbersome structure, the opaque areas on the front slices tend to obscure the back slices, and only a few slices can be accommodated usefully. This is because such a physical structure is a "subtractive" system. In other words the dark areas absorb ambient light, and the light areas transmit. In a volumetrically multiplexed hologram, the dark areas are "voids," and the light areas are analogues to sources of light, so there is no "hiding" of the back slices by the front ones, and all the information is visible simultaneously.

The tomographic data are in no way altered, and all the original data are recorded in the hologram. This is especially important in the case of medical data. Other 3D display techniques, including most computer rendering techniques and stereography, have to make a priori assumptions about the data in order to create different perspective views. Usually this involves looking for 3D surfaces in the data and removing hidden surfaces. The problem is that much tomographic data cannot be reduced to a set of surfaces, and any attempt to treat the data as such may distort it.

3.3.2. Recording a Volumetrically Multiplexed Hologram

Figure 14 shows the basic arrangement for recording a volumetrically multiplexed hologram. The projection system is identical to the case of holographic stereograms, with the different "frames" recorded on a film strip. The moving part in this arrangement is not a narrow slit, but the projection assembly. As each slice of the image is projected onto the screen, the screen and the projection system are moved towards or away from the hologram to a position corresponding to that slice. (If a collimating lens, such as a large Fresnel lens, is placed just before the projection screen, then it is possible to move only the screen and keep the projecting assembly fixed.) All the component holograms are recorded onto the same area of the plate, so each area of the plate has many different fringe patterns superimposed.

When the composite hologram is illuminated by a reconstruction beam, all component holographic images are simultaneously reconstructed, each slice of data in its correct relative position in space. The effect is that the brain reconstructs the original 3D object using the depth cues present. It is clear that this method is ideally suited to the display of tomographic data, as such data can be used almost without modification.

As in previous cases, it is convenient to produce a white-light-viewable version of the composite hologram. This can be done by following the procedure for recording image-plane holograms as described above. Figure 15 shows the setup for transferring the image.

3.3.3. Using Photographic Transparencies

The simplest way to project the 2D images onto the screen is to photograph the images consecutively, directly from the monitor. The film used should be a

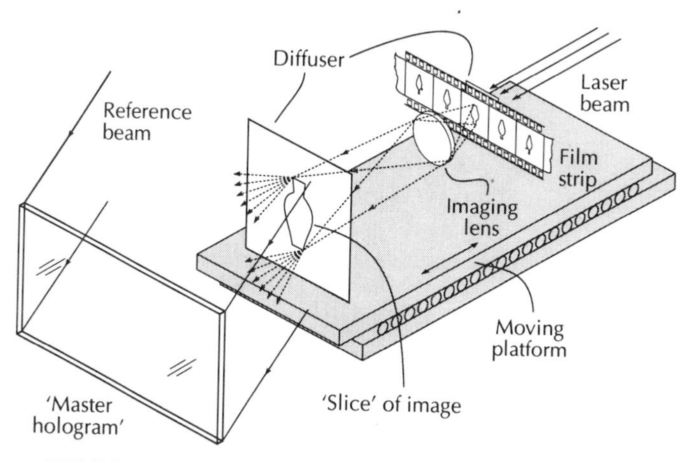

FIGURE 14. Recording a volumetrically multiplexed hologram.

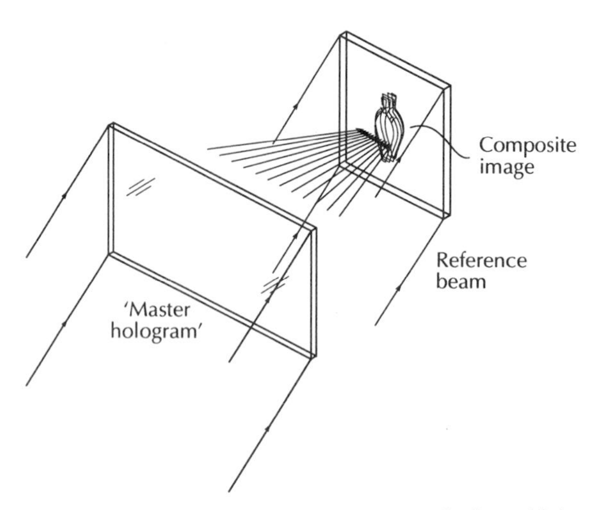

FIGURE 15. Recording a white-light-viewable volumetrically multiplexed hologram.

positive transparency, either color or monochrome. One problem that must be addressed carefully is that of registration. It is essential that the 2D images be recorded in correct registration to one another. The best way to achieve this is to ensure that in all the steps involved in the recording and projection of the images the frames are registered. The monitor displaying the images must be fixed with relation to the camera. Furthermore, the camera must have a pin-registration mechanism so that each frame is in register with the sprockets or with some other physical mark on the film. (There are commercially available 35 mm cameras with pin-registered backs.)

When recording the composite hologram, two approaches can be taken to projecting the images onto the diffusing screen: the transparencies can either be mounted in conventional slide mounts or they can be left in the form of a film strip. In either case, it is essential that the frames are projected in registration to one another. If the slides are mounted, then a slide projector can be adapted so that a laser is used to project the images rather than the normal projection lamp. Only a few commercially available projectors can be used for this purpose, as most cannot project with repeatable positional accuracy. It is also an advantage to choose a projector that can be interfaced to a computer so that the whole procedure can be automated.

The contrast of the holographic image is invariably reduced because of unwanted noise in the recording and reconstruction stages, so it is important that the quality of the original transparencies be high. It is usually best to increase the contrast in the transparencies in order to offset the noise in the final hologram. This can be achieved by adjusting the monitor or using a high-contrast film-processor combination.

3.3.4. Using a Spatial Light Modulator

As in the case of holographic stereography, the photographic stage can be eliminated by using a suitable SLM to produce the 2D sections directly from the computer. At present, transmitting LCD panels are the best types of SLMs available for this purpose. The most important parameter is the contrast ratio of the LCD panel (the ratio of the transmittance of the transmitting to the opaque pixels.)

3.3.5. Other Practical Considerations

The importance of positional stability in any form of optical holography cannot be overemphasized. It is essential that specialized vibration isolation units be used. Homemade tables can also be used for such vibration isolation. The details of such devices is beyond the scope of this book, and the reader is referred to any standard practical book on holography.

For practical reasons, it is best to control the whole process of recording by interfacing the laser (or shutter), film projection unit, and the moving platform to a microcomputer.

3.3.6. Depth Cues Available

Each recording in a volumetrically multiplexed hologram is a conventional laser transmission one, with the whole of the recording plate exposed. All the depth cues are therefore present, including accommodation which is absent in holographic stereography. When the composite image is viewed, the brain can therefore perceive the 3D data and extract shape information even with complex, irregular data. The limitation of single parallax is not present, and the observer can freely look around and over foreground objects.

3.3.7. Limitations of Volumetric Multiplexing

Let us look at some of the inherent drawbacks in using this technique. An important consideration is the absence of hidden-surface removal. In other words, nearer slices cannot obscure back ones. As the slices are recorded sequentially and independently, there is no mechanism for effecting hidden-surface removal. Although this can be a distinct advantage for medical data, it is a serious drawback for CAD applications.

The number of exposures that can be made on a single recording plate is not limitless. In general, as the number increases, the brightness of the composite image (diffraction efficiency) decreases, and the unwanted background intensity (noise) increases. The maximum number that can be combined depends on the characteristics of the recording medium, the maximum noise tolerable, and, to some extent, on the type of data. It is certainly possible to obtain useful images with 100 exposures (Hart and Dalton, 1990). One simple method of increasing the effective number of exposures is to distribute them among several holograms and then to use these holograms to record the final composite one (Johnson *et al.*, 1985). For example, if 500 data slices have to be recorded, exposures 1 to 50 could be made on the first plate, 51 to 100 on the second, and so on. The images from the 10 multiply exposed holograms could then be transferred to the final hologram. The clarity of the final image would be better than if 500 exposures were made onto the sample plate. This process is, of course, rather tedious and is difficult to automate.

Another problem is that equal exposure times given to all slices does not result in equal brightness. The earlier exposures tend to come out brighter (Bazargan *et al.*, 1988; Johnson *et al.*, 1985; Kostuk, 1989; Kostuk *et al.*, 1986). This can be corrected by gradually increasing the exposure through the series of recordings.

4. CONCLUDING REMARKS

The technique of volumetric multiplexing is ideally suited to the reconstruction of 3D images from a set of tomographic 2D images. The technique does not interfere with the original data, and the 3D image produced contains the full set of depth cues. This means that a large amount of information is available for the brain to understand the data. The hologram produced is a kind of 3D hard copy that can be viewed at leisure, without the need for sophisticated equipment. It can be kept conveniently with the patient's notes and need not be handled with any special care.

Current improvements in laser and LCD technology is making the technique more readily available and more reliable.

REFERENCES

Barnard, E. (1988). Optimal error diffusion for computer-generated holograms. *J. Opt. Soc. Am. A* **5**:1803–1817.

Bazargan, K. (1983). Review of colour holography, in *Proc. SPIE 391* (S. A. Benton, ed.), pp. 11–18. Los Angeles, CA.

Bazargan, K. (1985). A practical, portable system for white-light display of transmission holograms using dispersion compensation, in *Proc. SPIE 523* (L. Huff, ed.), pp. 24–25. Los Angeles, CA.

Bazargan, K. (1986). A new method of colour holography, in *Proc. SPIE 673* (J. Ke and R. J. Pryputniewicz, eds.), pp. 68–70. Beijing, China.

Bazargan, K., Chen, X. Y., Hart, S., Mendes, G., and Xu, S. (1988). Beam ratio in multiple-exposure volume holograms. *J. Phys. D: Appl. Phys.* **21**:S160–S163.

Bazargan, K. and Forshaw, M. R. B. (1980). An image-plane hologram with non-image-plane motion parallax. *Opt. Comm.* **32**:45–47.

Benton, S. A. (1969). Hologram reconstructions with extended incoherent sources. *J. Opt. Soc. Am.* **59**:1545–1546.

Benton, S. A. (1978). Achromatic images from white-light transmission holograms. *J. Opt. Soc. Am.* **68**:1441.

Benton, S. A. (1982). Survey of holographic stereograms, in *Proc. SPIE 367* (J. J. Pearson, ed.), pp. 15–19.

Brown, B. R. and Lohman, A. W. (1966). Complex spatial filtering with binary masks. *Appl. Opt.* **5**:967–969.

Burckhardt, C. B. (1966). Display of holograms in white light. *Bell. Syst. Tech. J.* **45**:1841–1844.

De Bitetto, D. J. (1966). White-light viewing of surface holograms by simple dispersion compensation. *Appl. Phys. Lett.* **9**:417–418.

De Bitetto, D. J. (1969). Holographic panoramic stereograms synthesized from white light recordings. *Appl. Opt.* **8**:1740–1741.

Denisyuk, Y. N. (1962). Photographic reconstruction of the optical properties of an object in its own scattered radiation field. *Sov. Phys. Dokl.* **7**:543–545.

Hariharan, P. (1983). Colour holography, in *Progress in Optics 20* (E. Wolf, ed.) North-Holland, Amsterdam.

Hariharan, P. (1984). *Optical Holography*. Cambridge University Press, Cambridge.
Hart, S. J. and Dalton, M. N. (1990). Display holography for medical tomography, in *Proc. SPIE 1212* (S. A. Benton, ed.), pp. 116–135. Los Angeles, CA.
Johnson, K. M., Armstrong, M., Hesselink, L., and Goodman, J. W. (1985). Multiple multiple-exposure hologram. *Appl. Opt.* **24**:4467–4472.
Johnson, K. M., Hesselink, L. and Goodman, J. W. (1982). Multiple exposure holographic display of CT medical data, in *Proc. SPIE 367*, pp. 149–154.
Keane, B. E. (1983). Holographic three-dimensional hard copy for medical computer graphics, in *Proc. SPIE 361* (E. Herron, ed.), pp. 164–168.
Kostuk, R. K. (1989). Comparison of models for multiplexed holograms. *Appl. Opt.* **28**:771–777.
Kostuk, R. K., Goodman, J. W., and Hesselink, L. (1986). Volume reflection holograms with multiple gratings: an experimental and theoretical evaluation. *Appl. Opt.* **25**:4362–4369.
Leith, E. N. and Upatnieks, J. (1964). Wavefront reconstruction with diffuse illumination and three-dimensional objects. *J. Opt. Soc. Am.* **54**:1295–1301.
Lin, L. H. (1969). Hologram formation in hardened dichromated gelatin films. *Appl. Opt.* **8**:963–966.
McCrickerd, J. T. and George, N. (1968). Holographic stereogram from sequential component photographs. *Appl. Phys. Lett.* **12**:10–12.
Paques, H. (1966). Achromatization of holograms. *Proc. IEEE* **54**:1195–1196.
Rotz, F. B. and Friesem, A. A. (1966). Holograms with non-pseudeoscopic real images. *Appl. Phys. Lett.* **8**:146–148.
Shankoff, T. A. (1968). Phase holograms in dichromated gelatin. *Appl. Opt.* **7**:2101–2105.
Syms, R. R. A. (1989). *Practical Volume Holography*. Oxford University Press, Oxford.
Walker, J. L. and Benton, S. A. (1989). In-situ swelling for holographic color control, in *Proc. SPIE 1051* (S. A. Benton, ed.), pp. 192–199. Los Angeles, CA.

IV

Applications

13

Three-Dimensional Reconstructions of Organelles and Cellular Processes

Bruce F. McEwen

1. INTRODUCTION

In the field of structural biology, the size of a structure often dictates the approach that must be used to study it. This is true for the three-dimensional (3D) reconstruction studies described in this chapter, which include the cilium, the kinetochore, sites of vertebrate calcification, the dendrite, the Golgi apparatus, patch-clamped membranes, and puff ball spores. All of these structures, as well as most of those discussed in Chapter 14, have diameters or thicknesses ranging from about 0.1 to 5.0 microns. The methodology used to compute these 3D reconstructions is quite different from the one used for the smaller and less complex structures described in Chapter 15. The size difference between the two groups of structures is illustrated in Fig. 1, where the 3D reconstructions of a cilium and a 50S ribosomal subunit are shown side by side on the same scale. Note that the cilium is in fact the smallest object considered in the present chapter. This size difference

Bruce F. McEwen • Wadsworth Center for Laboratories and Research, New York State Department of Health, Albany, New York 12201-0509

Electron Tomography: Three-Dimensional Imaging with the Transmission Electron Microscope, edited by Joachim Frank, Plenum Press, New York, 1992.

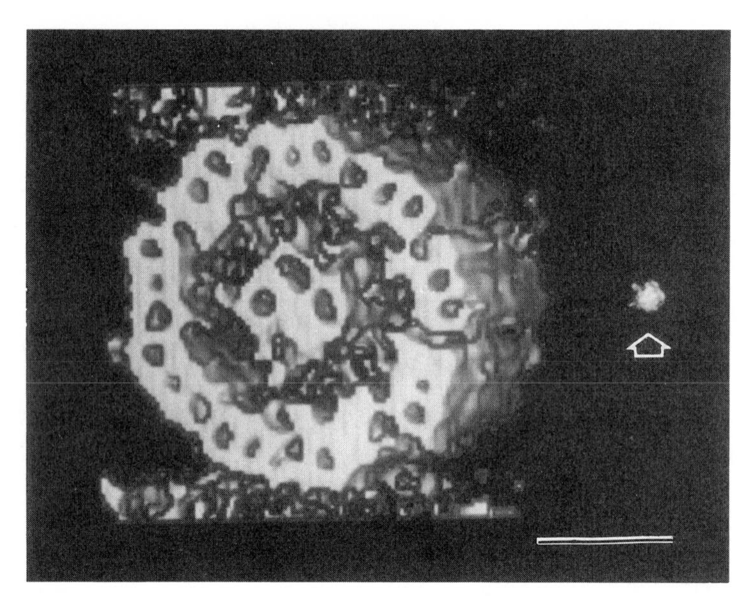

FIGURE 1. Cilium and 50S ribosomal subunit. Shaded surface representations of the 3D reconstructions of newt lung cilium and *E. coli* 50S ribosomal subunit (arrow). The representations are shown at the same scale in order to emphasize the difference in size between objects that have been investigated by single-axis tomography and those that have been investigated by single-particle reconstruction methods (McEwen, unpublished). Bar = 100 nm.

has two important consequences for the 3D reconstruction problem: (1) Generally, the individual specimens of the larger objects are not identical while those of smaller objects often are; and (2) the larger objects have a greater structural complexity at the limiting resolution level.

Initially, the high-resolution 3D reconstructions determined by electron microscopy were limited to ordered arrays of identical units, such as helical fibers (Klug *et al.*, 1958; Klug and Berger, 1964) and 2D crystals (Unwin and Henderson, 1975). In effect, these studies use a modification of the methodology developed for structural determination by x-ray crystallography. Frank and co-workers extended this approach to objects that do not form ordered arrays by using correlation functions and statistical methods (see Chapter 15). In both cases, the implicit assumption is that each individual unit structure is identical to all the others so that the signal-to-noise ratio can be enhanced by averaging over many repeating units of the structure. This condition also allows one to collect the different views required for a 3D reconstruction from different individual particles, or different arrays of particles, and hence minimize the radiation dose.

This approach clearly will not work if individual particles of the sample are not identical (to within the resolution limit of the study). The cilium is a well-ordered structure with no apparent variation in the motif between individuals so that combining different unit structures is possible, at least in principle. The structure is so large, however, that each individual tends to have a unique set of bends, twists, and staining variations, and many often show different amounts of fragmentation (especially if the membrane has been removed). Furthermore, the larger size and complexity makes aligning different cilia a difficult if not impossible task. With the rest of the 3D reconstructions described in this chapter, there is little doubt that each individual unit structure is unique.

The major consequence of not having identical unit structures is that each 3D reconstruction must be calculated from several views of a single object. This means that the radiation dose absorbed by the structure is much higher for a tomographic reconstruction. This problem is compounded by the greater structural complexity of the larger objects, which requires that more views be collected in order to achieve the desired resolution (Crowther *et al.*, 1970; Frank *et al.*, 1987). Thus, while 13 views suffice to give a satisfactory 3D reconstruction of a 50S ribosomal subunit, even 61 views of the cilium are inadequate. Finally, since at each tilt angle there is only one copy of the structure contributing to the image, the electron beam must be kept at a rather high level because there is no signal-to-noise enhancement from averaging over many repeating units. Because of the large amount of radiation damage, and the lack of signal-to-noise enhancement from averaging, the feasible resolution a tomographic 3D reconstruction is generally limited to 5.0 nm or more, depending upon the size of the object. It also appears that it will be difficult to apply tomography to frozen hydrated or frozen sectioned material in the near future due to the radiation sensitivity of those preparations.

Another consideration with larger objects is that the tomographic 3D reconstruction data sets usually have to be collected on a high-voltage electron microscope. Most of the samples in the size range under consideration are prepared for electron microscopy by embedding and sectioning. Sections that are thin enough to be imaged well in a conventional electron microscope are generally too thin to contain a useful amount of depth information. In addition, the path length of the electron beam through the sectioned samples increases with tilt angle so that it is double at 60° and triple at 70° of what it is in the untilted view. In the present studies, the cilia, kinetochores, and calcification sites were viewed in sections 0.25–0.50 μm thick, while the dendrite and Golgi specimens were cut into 3.0–μm-thick sections. The patch-clamped membrane and puff ball spore preparations have diameters of 2–5 μm. Since the path length of an electron through a 0.25–μm-thick section tilted to 70° is 0.75 μm, all of these samples are well beyond the range of a conventional electron microscope. While intermediate-voltage electron microscopes (accelerating voltages of 200–400 kV) might be able to handle 0.25–μm sections, clearly the dendrite, Golgi, patch-clamp membrane, and puff ball spore studies can only be carried out with a high-voltage electron microscope (accelerating voltages of 1 MV or more).

Because of their increased structural complexity, more computing power and disk storage space are required to compute the 3D reconstructions of the larger structures. Increased structural complexity implies that there is more to look at in a given 3D volume, which in turn requires that the pictures be scanned with a finer element size to preserve the required resolution level. Hence, a larger number of pixels or voxels are required to represent these images and 3D volumes (see Frank *et al.*, 1987, for a discussion of this issue). Thus, while the 50S ribosomal subunit can be adequately represented by a 3D volume that is $64 \times 64 \times 64$, the structures considered in this chapter require a volume of $128 \times 128 \times 128$ or even $256 \times 256 \times 256$. When each voxel is stored as a 4-byte word, a cubic volume that is 64 voxels on a side requires about 1 Mbyte of disk space, a 128 cube about 8 Mbytes, and a 256 cube of 64 Mbytes. Of course, the number of calculations required for the reconstruction algorithm and other image manipulations also increases with an increase in the number of pixels, voxels, and input projections. It is easy to see why tomography has become a practical approach only in the latter half of the past decade as larger and more powerful computers became available to research biologists.

Increased structural complexity also presents a problem for the analysis and representation of the reconstructed 3D volume. The shaded surface representation used in Fig. 1 is satisfactory for the 50S ribosomal subunit because the surface contours of the structure are its most interesting part. With the cilium, however, there are many internal features which one wants to examine. Furthermore, because the substructure is so complex, a surface threshold value that is optimal for some features is not optimal for others; in fact, most investigators find that the shaded surface representation is inadequate for many of the views they wish to present. The situation has resulted in a search for alternative strategies for representing 3D volumes (see Chapters 10–12) and new developments in this area could greatly increase the effectiveness of 3D analysis in the near future.

Thus, principally as a result of the common size and complexity of the structures involved, the studies described here require higher accelerating voltages in the electron microscope, more computing power for processing, a common approach for computing 3D reconstructions and more elaborate strategies for representing the 3D volume. These biological preparations share little else in common, however, and the successful application of tomography to all of them illustrates the versatility of the method. Such a diversity of preparations inevitably means that each study also has its own unique set of challenges which must be considered on an individual basis, so that the application of the method is still somewhat of an art. The size range under consideration contains numerous important biological structures, substructures, and organelles. This fact, along with its great versatility, makes the tomographic approach an extremely important method for obtaining 3D reconstructions.

2. *THE CILIUM*

The 3D reconstruction of a newt lung cilium (McEwen *et al.*, 1986) presented a turning point in the application of tomographic methods to electron microscopy because (1) it was the first study to combine tomography with HVEM, and (2) it was the first study of a well-known biological object that could serve as a test of the technique. The value of the cilium as a test object for developing new methodologies in electron microscopy is exemplified by its use in the pioneering studies of Manton and Clarke (1952), Fawcett and Porter (1954), Afzelius (1959), Gibbons and Grimstone (1960), Grimstone and Klug (1964), Allen (1968), Tilney *et al.* (1973), Amos and Klug (1974), and Goodenough and Heuser (1985). The chief advantage of the cilium is that it is a well-ordered arrangement of readily identifiable components, which include outer doublet microtubules, central pair microtubules, dynein arms, radial spokes, central sheath material, and interdoublet linkers. The morphology of these components has been reasonably well characterized by a large number of previous studies (for reviews see Warner, 1974; Gibbons, 1981). In addition, there have been numerous studies of the cilium waveform (e.g., Brokaw, 1965, 1983), of the biochemical mechanisms responsible for this waveform (Satir, 1974), and of cilia's biochemical composition (e.g., Huang *et al.*, 1982; Linck and Langevin, 1982; Luck, 1984). Genes for many of the cilia proteins have been cloned (e.g., Williams *et al.*, 1986). Since the cilium components are easily recognized and well studied, the effectiveness of the tomographic approach can readily be ascertained by how well these components are represented in the 3D reconstruction. The tomographic 3D reconstruction of cilia was not, however, undertaken solely as a test study. Rather, since none of the previous structural studies have analyzed whole cilia with a true 3D technique, the tomographic reconstructions of cilia are expected to eventually make significant contributions toward elucidating the organelle's ultrastructure and the structure-functional relationships of its components. For example, the in situ configuration of many of the components is still controversial, and little is known about how different components interact inside the cilium during the beat cycle. The basal body and the transitional zone (between the basal body and the cilium proper) have not been well studied, precisely because this area lacks repeating features. For the same reason, the ciliary membrane has been virtually ignored. All of these features are amenable to an analysis from tomographic 3D reconstructions.

The initial reconstructions (McEwen *et al.*, 1986) demonstrated the utility of the approach. A shaded surface representation of two reconstructions from this work is shown in Fig. 2. In this representation, a selected surface of the 3D density distribution is defined by a threshold value. A viewing plane is chosen, and shading is applied to each element of the surface according to its inclination to a virtual light source (see Radermacher and Frank, 1984; McEwen *et al.*, 1986; and Chapter 10). It is clear that, on the level of gross morphology, the 3D reconstruction accurately

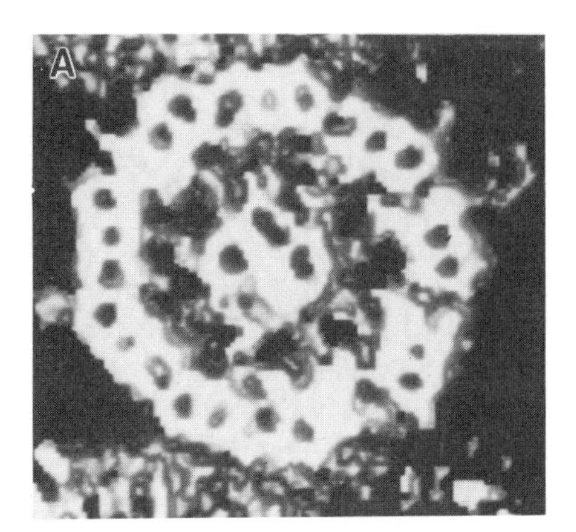

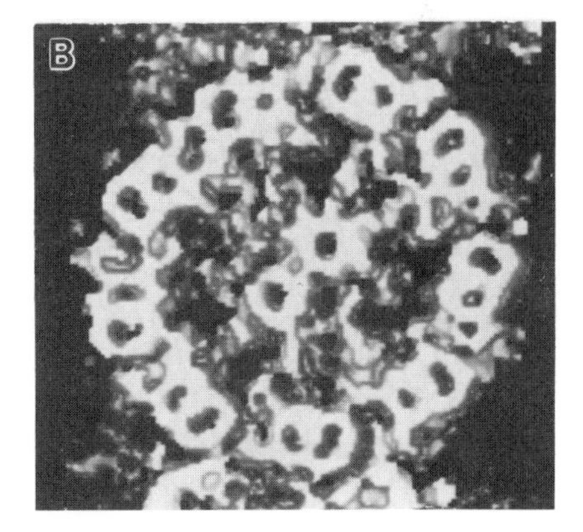

FIGURE 2. The effect of increasing the number of projections on the 3D reconstructions. Shaded surface representations of 3D reconstructions of two different newt lung cilia: (A) reconstruction from 53 projections with an angular range of 104° (2° angular interval); (B) reconstructed from 109 projections with an angular range of 108° (1° angular interval) (McEwen and Frank, unpublished). Bar = 100 nm.

portrays the cilium structure. The outer doublet and central pair microtubules are easily recognized, while the counterclockwise attachment of the dynein arms to the outer doublets, along with the slight skew of these doublets, establishes that the cilium in Fig. 2*a* is being viewed from the tip to the base (Warner, 1974; Gibbons, 1981). The cilium in Fig. 2*b*, on the other hand, is being viewed from the base to the tip. Dynein arms, radial spokes, and the central sheath are all clearly visible, even though the resolution of the reconstruction in Fig. 2*a* is only 12 nm. However, because these structural elements are only 5–7 nm in diameter, they appear blurred and their exact positions in the reconstruction are uncertain.

The periodicity of radial spokes and material connected to the central sheath were determined from the reconstruction (McEwen *et al.*, 1986). The radial spoke periodicity indicated the more common pattern of 3 spokes/96 nm rather than 2 spokes/96 nm as is found in Chlamydomonas (Warner, 1974; Gibbons, 1981). This fact had not been previously established for newt lung cilia. Still the observed periodicity of 3 spokes/90–96 nm agrees well with measurements from cilia of other species. This finding indicates that section shrinkage, at least in this preparation, was minimal (less than 10%). Another quantitative measurement the authors made was of a 5° rotation of the central pair of microtubules relative to the outer ring of doublets (Frank *et al.*, 1986). This rotation had been previously observed by Omoto and Kung (1980). These determinations illustrate how precise 3D measurements are possible once a 3D reconstruction is computed.

Despite the overall success of the cilium reconstructions, it is frustrating that the fine structure (i.e., dynein arms, central sheath material, and radial spokes), although visible, does not appear consistently or with a reproducible shape. In addition, we were unable to detect more than a vague hint of the 24-nm longitudinal repeat of the dynein arms, even though this spacing is similar to the one between radial spokes (McEwen *et al.*, 1986). The possible reasons for this include degradation of resolution in the longitudinal direction (i.e., along the z axis), due to the use of a tilt range of less than 180°, and the morphology of the dynein arm, which appears to be considerably spread out along the z axis direction (Johnson and Wall, 1983; Goodenough and Heuser, 1984; Avolio *et al.*, 1984). Despite these difficulties, the dynein periodicity would undoubtedly have been detected if the fine structure was sharper in the reconstruction and our representation of it.

A superior representation of the fine structure cannot be obtained by simply increasing the number of input views in order to increase the resolution of the reconstruction according to the formula of Crowther *et al.* (1970, see Chapter 5 in this volume). This is apparent from Fig. 2*b*, which shows a reconstruction computed from twice as many projections, and hence having nominally twice the resolution as the reconstruction in Fig. 2*a*. The fine structure, however, is no more distinct in Fig. 2*b* than it is in Fig. 2*a*. This means that the representation illustrated in Fig. 2*a* is the best that could be done with the methods used at that time, and that something other than the number of projections was limiting the detection of fine structure.

One factor that might limit the attainable resolution is alignment errors. The alignment of a data set can be accurately assessed by displaying the individual images sequentially on a graphics display screen to form a "movie" of the object being tilted up to the limits of the available angular range. This form of display is extremely sensitive to alignment errors because the human eye readily detects even small rotational or translational shifts in the data set when it is presented as a time sequence. With this test, small errors were detected in the rotational alignment of the data set used to compute the reconstruction represented in Fig. 2*a*. These errors were not present in data sets of other projects, which were aligned after we implemented the procedure of Lawrence (see Chapter 8) for rotational alignment. We were unable to realign the cilium data set using Lawrence's procedure because of a lack of suitable fiduciary markers. Presently we are collecting a new tilt series, with a greater angular tilt range, in order to evaluate the effect of the increased angular range, the refined alignment, and other improvements to the methodology which are discussed here.

Fine structure in a 3D volume can also be lost in the shaded surface rendering because there is a large reduction of data when all of densities in the volume are assigned to one of two values (corresponding to structure and background). This data reduction is less of a problem if the structure is evenly stained, a condition that often holds for small macromolecular assemblies, but rarely for structures as complicated as the cilium. As a result, threshold levels that are optimal for showing the connectivity of radial spokes and other fine features leave too much material on the outer doublets while threshold levels that are optimal for the microtubules leave the fine features barely detectable. This problem is illustrated in Fig. 3 by a series of shaded surface representations in which the threshold level is varied. Volume rendering (see Chapter 10), an alternative representation for 3D reconstructions, retains all densities above a specified threshold, and incorporates a certain amount of transparency into the representation (Fig. 4). This method appears to be superior to the shaded surface for preserving the dynein repeat in cilia, but other features, such as the radial spokes, are now less clear because the semitransparency results in features from different depths superimposing upon one another. Clearly, what is needed is a battery of rendering and representation techniques for 3D volumes so that an investigator can find the one that is most useful for each situation encountered. Some of the currently available methods are discussed in Chapters 10–12.

Degradation of resolution along the axis of the electron beam (z axis), which is caused by the limited angular range over which projections can be collected in the electron microscope, is another reason why the cilium fine structure is not well represented in the 3D reconstructions. The original cilium data set was collected over an angular range of only 104°, but subsequent modifications to the Albany HVEM tilt stage now allow the routine collection of data sets with a tilt range of over 120°. This increased tilt range should noticeably improve the z-axis resolution of the reconstruction (see Chapter 5). Some investigators, working with negatively stained macromolecular arrays, (i.e., Baumeister *et al.*, 1986) have achieved even greater tilt ranges and have shown this to result in a significant improvement in the z-axis resolution. This approach, however, is not attractive for thicker sections because of the increased path length of the electron beam with tilt angle ($2\times$ at 60°, $3\times$ at 70°, and $4\times$ at 75°). In addition, the amount of focus change across the surface of the section and overlapping from material neighboring regions become prohibitive at larger tilt angles. Finally, even with a greater tilt range, the resolution still varies with the viewing direction; this effect is disturbing to an investigator who

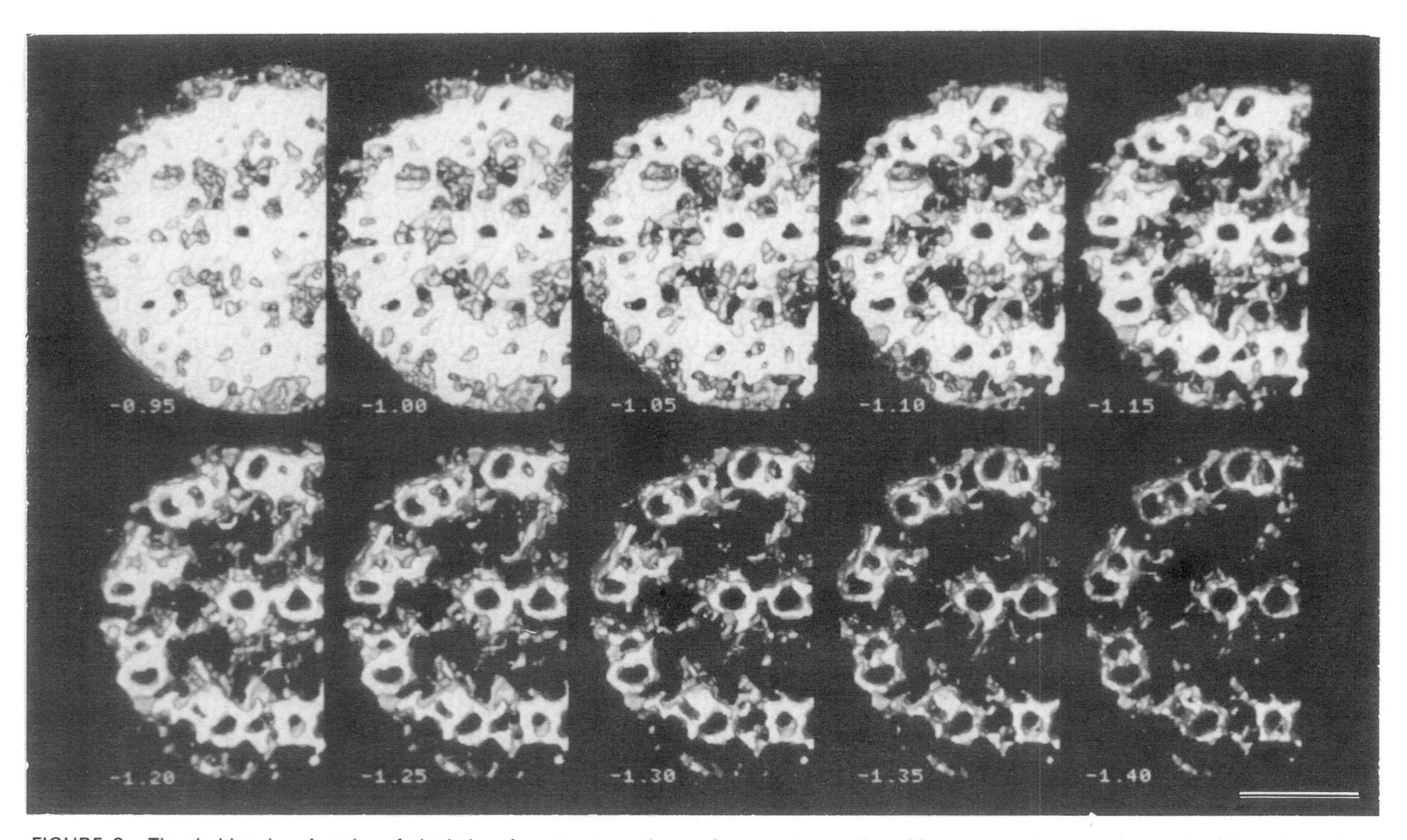

FIGURE 3. Threshold series. A series of shaded surface representations, of a newt lung cilium 3D reconstruction, were created with an increasing threshold for the boundary between the object and background. Numbers indicate a relative threshold scale (McEwen and Frank, unpublished). Bar = 100 nm.

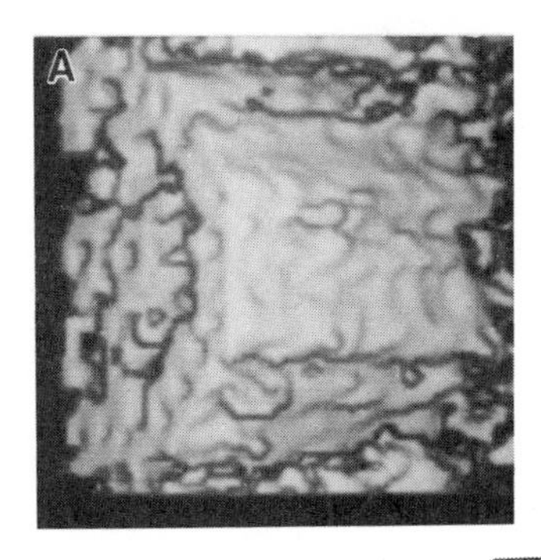

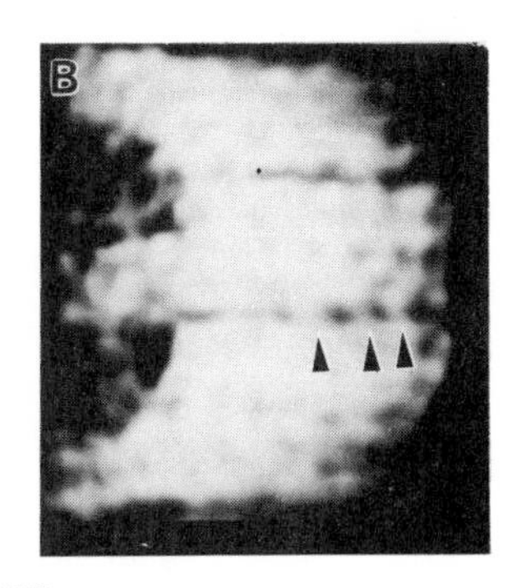

FIGURE 4. A comparison between shaded surface and volume rendering. The 3D reconstruction of a newt lung cilium was represented by (A) shaded surfaces; and (B) volume rendering. The shaded surface gives a better representation of the periphery of the microtubules, but the volume rendering reveals a periodic interdoublet connection (arrow heads) not seen in (A) (McEwen, Leith, and Frank, unpublished). Bar = 100 nm.

is trying to obtain meaningful biological information from the reconstruction. Ideally, the resolution in every direction of the reconstruction would be the same as is present in the original electron micrographs.

There are two promising approaches to overcoming the effect of the missing angular information: (1) mathematical restoration techniques (see Chapter 6), which use a prior knowledge to compute a most probable structure in which directional differences in resolution are minimized; and (2) mounting the sample in such a way that it is possible to collect a full angular tilt range. The latter can be achieved by placing the specimen in a glass micropipette as is described below for the patch-clamped membrane and the puff ball spore reconstructions. The cylindrical geometry of the micropipette allows the sample to be viewed from any direction. We have developed a stage capable of mounting a micropipette and rotating it through a full 180° tilt range. The two approaches are not mutually exclusive because the restoration techniques can be used to improve the resolution of the reconstruction in the "good" directions as well. Thus, the combination of a cylindrical mounting geometry and mathematical restoration techniques could come close to the ideal situation outlined in the last paragraph. In any case, these methods, along with a battery of different rendering techniques, should allow tomography to become a powerful tool for future investigations into the ultrastructure and structural-function relationships in cilia. In turn, cilia reconstructions should provide a good indication of how effective newly developed techniques are for helping to find new morphological information.

3. THE MAMMALIAN KINETOCHORE

The kinetochore is a specialized region of chromosomes that is responsible for attaching them to the mitotic spindle. Electron microscopy has established that, in most vertebrates, the kinetochore is a trilaminar disk, of variable (0.2–2.0 μm)

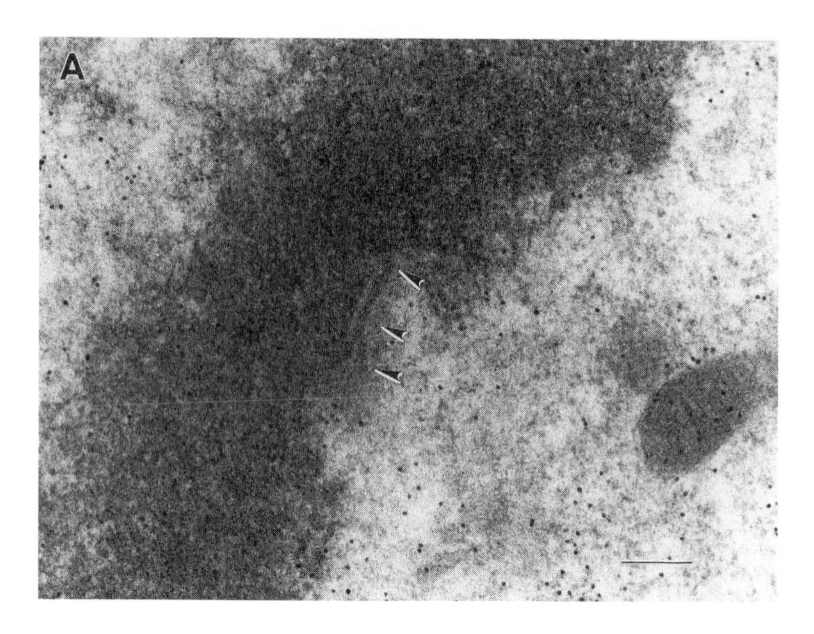

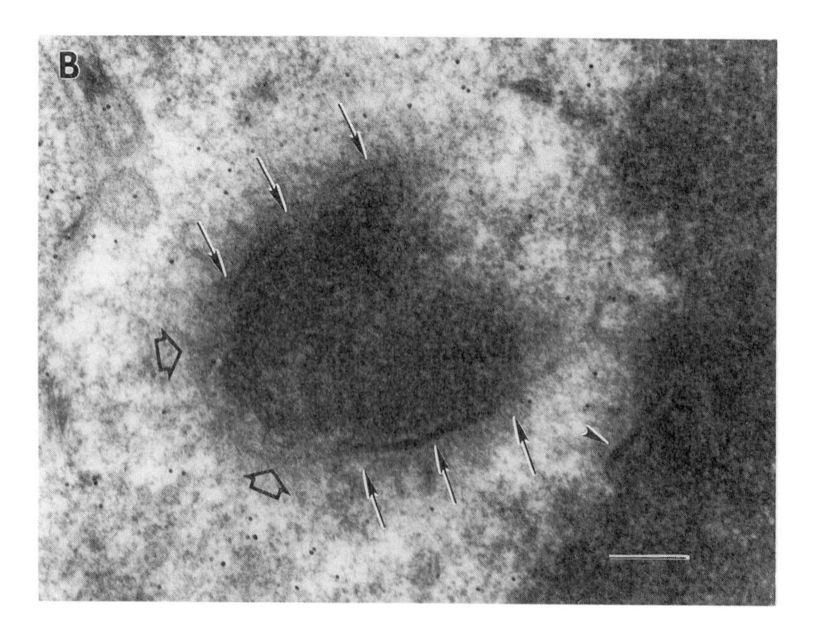

FIGURE 5. Electron micrographs of kinetochores on PtK cell chromosomes. Cells were grown, fixed, and embedded by standard methods. Images were recorded with the HVEM, at 1.0 MV, from 0.25-μm-thick sections stained with uranyl acetate and lead citrate. Colloidal gold particles (approximately 15 nm diameter) were added to both surfaces of the sections to serve as fiduciary markers for alignment procedures. (A) Longitudinal view of a chromosome with attached kinetochore (arrowheads). (B) Cross-sectional view of a chromosome with sister chromatids (arrows), which appear to have a thread of material connecting them (short, wide arrows). Another kinetochore is visible on a neighboring chromosome in longitudinal view (arrowhead) (McEwen and Rieder, unpublished). Bars = 250 nm.

diameter, which is closely opposed to the heterochromatin at the primary construction of the chromosome (reviewed by Ris and Witt, 1981; Rattner, 1986; Brinkley *et al.*, 1989; Rieder, 1990). Several recent studies have demonstrated that the force(s) for both prometaphase and anaphase chromosome movement are generated at the kinetochore (Gorbsky *et al.*, 1987; Koshland *et al.*, 1988; Mitchison, 1988; Nicklas, 1989; Rieder and Alexander, 1990). Models for chromosome movement to the pole include ATPase motors and microtubule depolymerization (reviewed by Mitchison, 1989). The work of Rieder and Alexander (1990) indicates that the opposing force (causing the equatorial movement known as congression) is also generated by the interaction of the kinetochore with microtubules.

This dual role as the site for both attachment and locomotion of chromosomes on the spindle establishes the kinetochore as one of the key structural components of mitosis. Despite its great importance to eukaryotic biology, little is known about the kinetochore's ultrastructure. The kinetochore first appears as a distinctive platelike structure after the onset of prometaphase, and it is considered to be fully differentiated by metaphase (the structure of the kinetochore during the different stages of the mitotic cycle, including interphase, is reviewed by Rieder, 1982). The mature kinetochore is often described as a trilaminar structure, with the outer, middle, and inner layers having thicknesses of 30–40 nm, 15–35 nm, and 20–40 nm, respectively (see Rieder, 1982, and Fig. 5). The outer plate is the only distinctive structure, however, since the middle layer appears to be an empty space between the other plates and the inner plate to be a specialized region of the chromosome. The latter stains more heavily than adjacent chromatin, possibly due to the presence of specific protein antigens, but the boundary between it and the rest of the chromosome is often obscured, and very little is known about its ultrastructure.

The outer plate is covered with a fibrous material known as the corona, which extends a variable distance (up to 250 nm) from the plate. The corona, which diminishes after the binding of microtubules, appears to be the site of initiation of microtubule polymerization. Different models for the ultrastructure of the outer

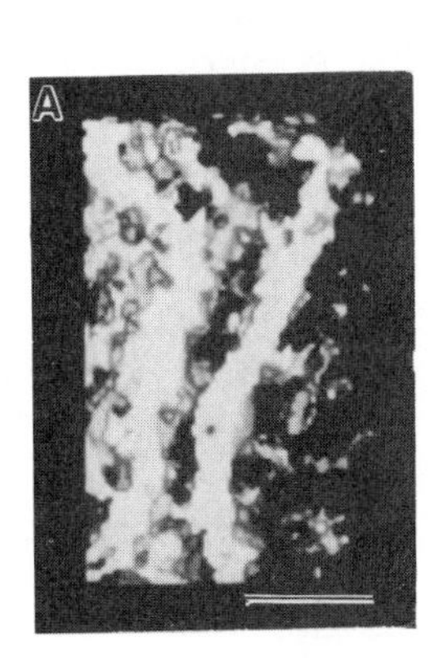

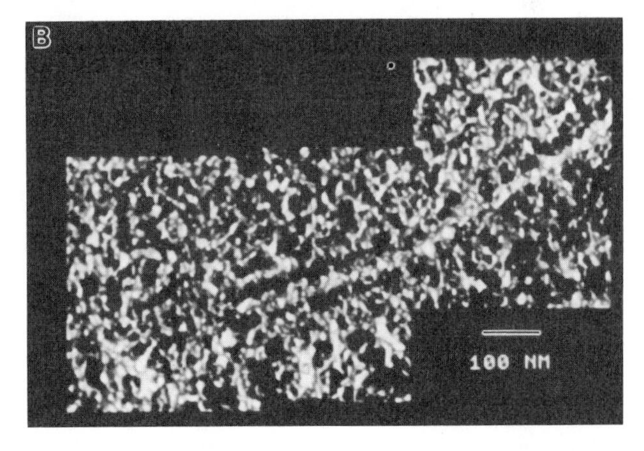

FIGURE 6. Kinetochore 3D reconstructions. Shaded surface representations of two of the kinetochores shown in Fig. 5: (A) the kinetochore indicated in Fig. 5A (from McEwen *et al.*, 1987); (B) the lower of the sister kinetochores indicated in 5B. (From McEwen and Frank, 1990.) Bars = 100 nm.

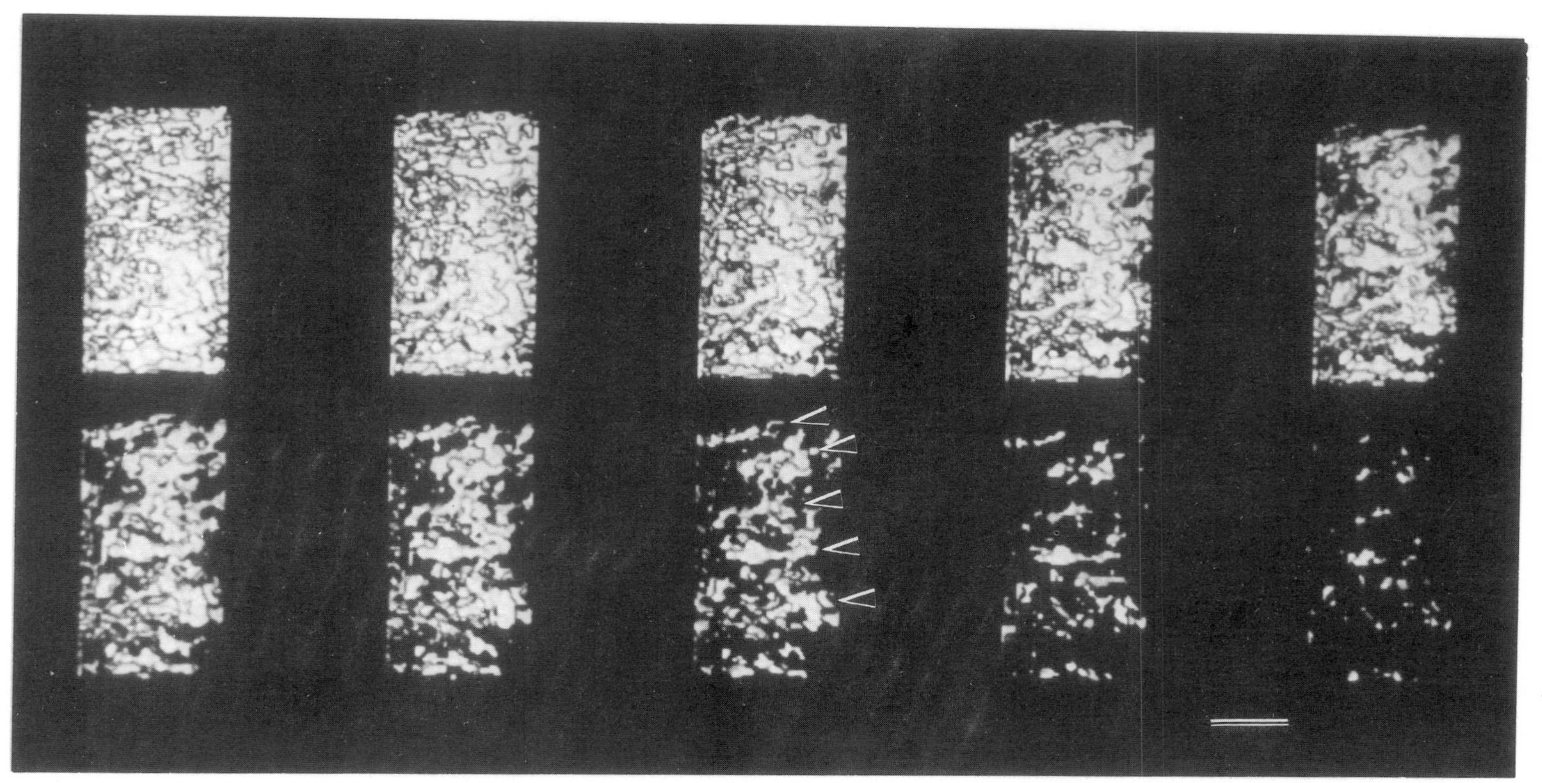

FIGURE 7. Threshold series of the en face view of a kinetochore reconstruction. The reconstruction represented in Fig. 6A was rotated 90° clockwise about the vertical axis, and the threshold was increased stepwise as in Fig. 3. Threads of higher-density material, possible corresponding to fibers, are indicated by arrowheads (McEwen and Frank, unpublished). Bar = 100 nm.

plate have been proposed by Roos (1977), Ris and Witt (1981), Rattner (1986), and others. The model of Roos is based upon observations of a preparation that was spread onto distilled water, but these conditions are known to alter the kinetochore structure (Ris and Witt, 1981). Ris and Witt studied the kinetochore in 0.25- and 0.50-μm-thick sections, which were imaged with by HVEM and viewed in stereo. While stereo viewing aided their analysis, there was not enough depth information to follow the structural elements far enough to verify their model. Rattner's model is compellingly simple and is based on the 30-nm fiber, instead of a 10-nm unit, but again the preparation technique, digestion with nucleases, could have severely altered the structure.

The first tomographic 3D reconstructions of kinetochores (McEwen *et al.*, 1987a, b) were from mammalian (PtK) cells treated with colcemid. Shaded surface representations of these reconstructions are shown in Fig. 6. While these initial efforts confirmed earlier reports (i.e., Brinkley and Stubblefield, 1970; Rieder, 1979; Ris and Witt, 1981) that the outer plate is directly connected to adjacent chromatin, they also found that these connections were usually made to the inner plate. This was somewhat surprising because earlier studies had indicated that the inner plate is not present in centromeres of colcemid-treated cells (reviewed by Rieder, 1982). One reason why this structure is easier to find in the 3D reconstructions is that even simple image-processing methods, such as the use of thresholds and shaded surface renderings, can help to increase the contrast between the inner plate and adjacent chromatin. More important, however, is the ability to extract views of the structure where the distinction between the inner plate and the adjacent chromosome is the greatest because the inner plate is difficult to detect in many regions of the 3D reconstruction. The detection of the inner plate in the colcemid-treated cells forces a reconsideration of the postulates of Roos (1973), who concluded that its appearance correlates with the acquisition of microtubules by the kinetochore, and of Pepper and Brinkley (1977), who suggested that its formation was blocked by the interaction of colcemid with kinetochore-associated tubulin.

The 5–10-nm-diameter fibers of the outer plate, which were described in several previous studies (e.g., Brinkley and Stubblefield, 1966, 1970; Comings and Okada, 1971; Ris and Witt, 1981), are not readily apparent in the shaded surface representations (e.g., Fig. 6). The outer plate appears as a continuous structure in these representations, but a direct examination of the 3D volume reveals threads of a higher density embedded in the outer plate. These structures, which could correspond to the previously reported fibers, can be viewed by a shaded surface rendering if a higher threshold is imposed (Fig. 7). Several studies show that both DNA (Pepper and Brinkley, 1980; Ris and Witt, 1981; Rattner and Bazett-Jones, 1989) and different protein components (Pepper and Brinkley, 1977; Brenner *et al.*, 1981; and Brinkley *et al.*, 1989) are present in the outer plate. Since the DNA is likely to be in a fiber form and stains more darkly than the protein components, which are expected to form a more diffuse structure, it is not surprising to find threads of higher density embedded in an amorphous plate of lighter staining material. The problem then becomes one of how to represent such a structure in a way that allows a meaningful 3D analysis. The shaded surface representation has the drawback that the fibrous nature of the outer plate cannot be shown in the

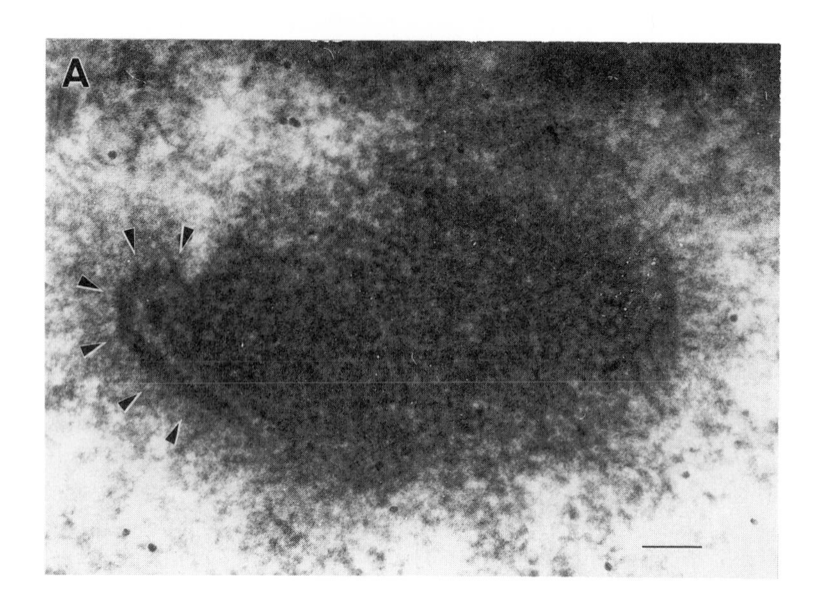

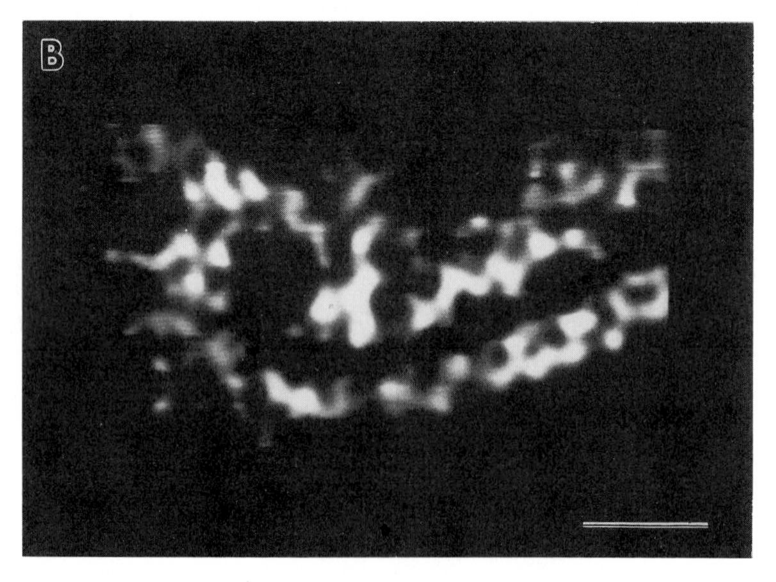

FIGURE 8. Connections between the inner and outer kinetochore plates. (A) An electron micrograph of the area shown in Fig. 5B but in a neighboring serial section and viewed at a high tilt angle (58°). The curious curved structure (arrowheads) was only visible at high tilt angles (>50°). (B) Representation of the 3D reconstruction of the area indicated in A. A tracing algorithm was used to isolate the inner and outer plates from the background and the rest of the chromosomal material. A subsequent analysis of the plates revealed that the looped feature was only present for a small depth in the reconstruction. The representation is a projection through the subvolume where the feature is present (Liu, McEwen and Frank, unpublished). Bars = 100 nm.

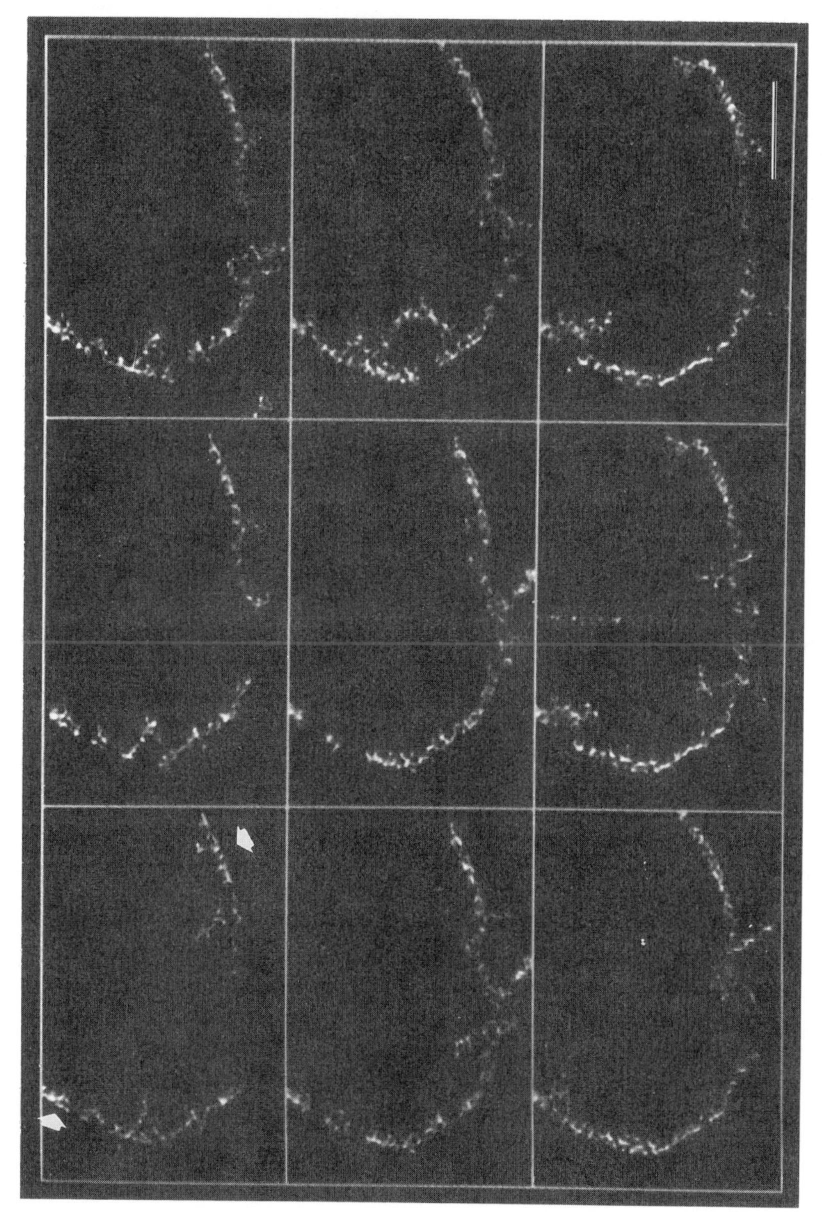

FIGURE 9. Connections between sister chromatids. The tracing algorithm used in Fig. 8B was also used to investigate possible connections between the sister kinetochores that are indicated in Fig. 5B. The algorithm was applied to 35-nm-thick slices of the 3D volume; a gallery of these tracings is presented here. The series demonstrate that at some levels there are continuous connections between the sister chromatids, which are indicated by arrows in the first slice (Liu, McEwen and Frank, unpublished). Bar = 250 nm.

context of the whole structure. The view in Fig. 7 has the additional disadvantage that it is along the direction of missing information, and hence it is the poorest view of the structure. The first problem should be solved by using a volume rendering with two density thresholds, and making the structure corresponding to the lower density value semitransparent. The suboptimal viewing direction can be corrected by reconstructing a kinetochore whose outer plate is orthogonal, rather than parallel, to the tilt axis. We are presently implementing both of these changes in our latest set of kinetochore reconstructions.

In subsequent 3D reconstructions, Liu *et al.* (1990) reported that the kinetochore was more complex than previously envisioned. They found instances where the outer plate at one end of the kinetochore would curve around to meet the inner plate (Fig. 8). These connections were only seen at one corner of the 3D volume, and they suggest that the outer plate is formed from the inner one. Liu *et al.* also found areas where the outer plate splits into two or more branches, with each branch having an approximate diameter of 10 nm. While these branches could represent material that grew on to the surface of the outer plate at a later time, it is also possible that they represent the unraveling of the fibrous substructure discussed above. Presently we are trying to discover how these fibers are packed in the outer plate. Finally, Liu *et al.* (1990) also reported a connection between the kinetochores of sister chromatids that had not previously been reported (Fig. 9). The function of these connecting fibers is not clear.

The tomographic 3D reconstructions of the kinetochore are only a first attempt, and many factors can be modified to refine them. For example, to obtain a complete view of all the connections between the outer plate and the chromosome (i.e., the inner plate), the whole kinetochore should be contained in a single section, which would probably have to be about 1 μm thick (unless an end face view can be recognized). Such a reconstruction would show the difference between the different kinds of connections observed in 0.25–μm sections. To study the fibrous fine structure of the outer plate, however, it would be more useful to use thinner sections (0.25–0.50 μm) that have been stained specifically for DNA. There are also selective stains for specific protein antigens, which would be useful as would removing the kinetochore from the rest of the chromosome (e.g., Brinkley *et al.*, 1989). Finally, the kinetochore needs to be studied under conditions where spindle fibers are present.

Thus, despite the preliminary nature of the initial tomographic 3D reconstructions, they have already made a substantial contribution to our knowledge of kinetochore structure, and they promise more for the future.

4. THE CALCIFICATION OF VERTEBRATE SKELETAL TISSUE

The skeletal system of the vertebrates is composed of hard tissue formed by the deposition of hydroxyapatite crystals (inorganic calcium phosphate molecules) into an extensive organic extracellular matrix of which type I collagen is the principal constituent (reviewed by Hodge and Petruska, 1963; Glimcher and Krane, 1968, for a general description of collagen structure and mineralization). Although bone and other vertebrate calcifying tissues have been well studied, the processes by which mineralization is initiated and regulated at the molecular level are incompletely understood.

One of the models for studying vertebrate calcification is the leg tendon of the domestic turkey, *Meleagris gallopavo* (Landis, 1986). While most vertebrate tendon tissue normally is unmineralized, certain tendons from a variety of species will calcify. The advantages of the tendon as a model system for investigating vertebrate calcification are that the tissue is highly regular in its collagen organization and architecture; the tendon mineralization proceeds in a comparatively ordered manner; and the tendon allows relatively direct and easy access to both spatial and temporal analysis (Landis, 1985). The latter is possible because mineralization progresses with time as streaks that develop in a distal-to-proximal direction parallel to the

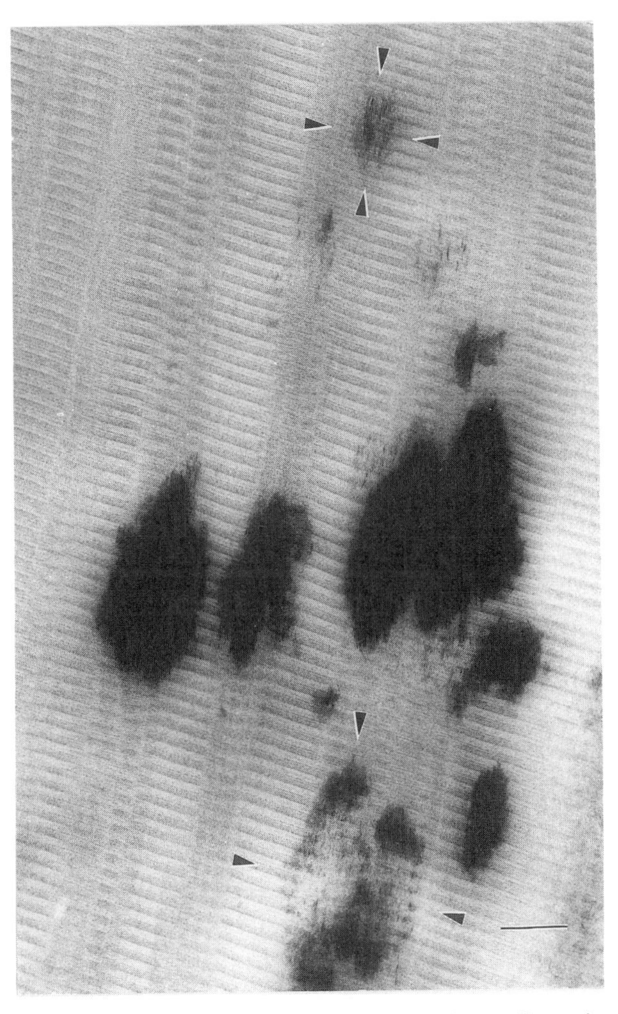

FIGURE 10. Calcification of turkey tendon. Tissue was prepared according to Landis (1986) and 0.50-μm sections were imaged in the HVEM as described for Fig. 5. Tomographic 3D reconstructions were computed from the two areas of the field, which are approximately delineated by the arrowheads (Song, McEwen, and Landis, unpublished). Bar = 250 nm.

tendon long axis. To follow the temporal sequence of events, one has only to examine the tendon long axis along the vicinity of the irregular, but rather abrupt, mineralization junction.

We have recently computed tomographic 3D reconstructions of sites along the turkey tendon that are in the early phases of mineralization (Landis *et al.*, 1990). At these sites, several small, isolated crystals are present, as seen in Fig. 10, which shows an overview of one representative region. Among a number of important findings we uncovered from the tomographic reconstructions, the most obvious is that individual crystals are shaped as thin plates rather than needles (Figs. 11, 12). These plates are oriented along the collagen fibril with one edge pointing along the fiber axis and the other edge pointing roughly in a radial direction. From selected slices in the 3D volume, such as shown in Fig. 13, it appears that individual plates can grow longer than one repeat distance along the collagen molecule and hence are not limited to one hole zone. Finally, closer examination of individual plates indicates that they are composed of many units seemingly fused together in a coplanar fashion (Fig. 12).

Many of these features were conjectural before the tomographic reconstructions were computed. The new and direct visual information gained by the reconstructions is now expected to be invaluable in helping to elucidate the initial events of

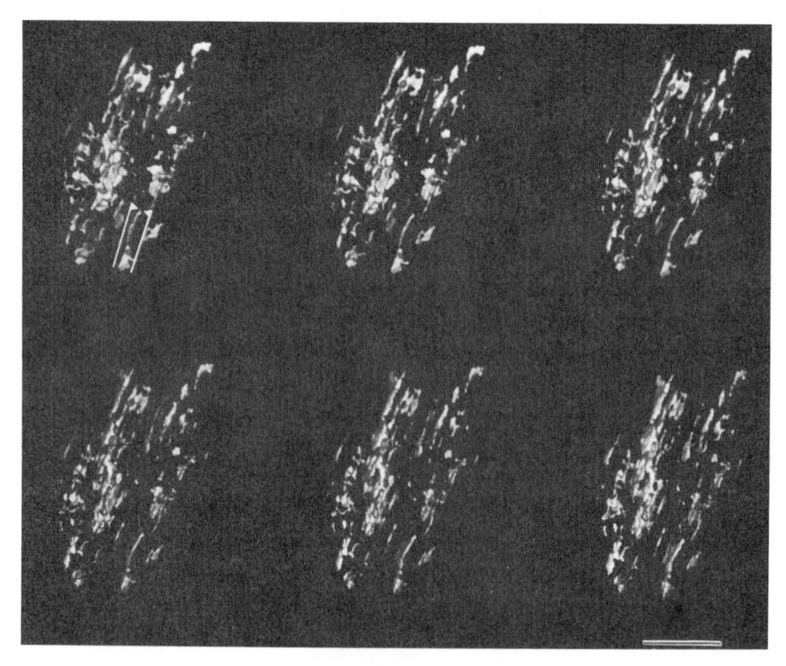

FIGURE 11. Stereo views of the 3D reconstruction of the calcium crystals. A gallery of shaded surface representations was created by rotating the viewing angle from −10° to 10° in 5° increments. Neighboring views in the gallery form stereo pairs, and the unrotated view is repeated at the end of the first line and the beginning of the second. The reconstruction was computed from the upper one of the areas delineated in Fig. 10. The subvolume indicated by the brackets in the first image is viewed separately from the rest of the 3D volume in Fig. 12 (McEwen, Song, and Landis, unpublished). Bar = 100 nm.

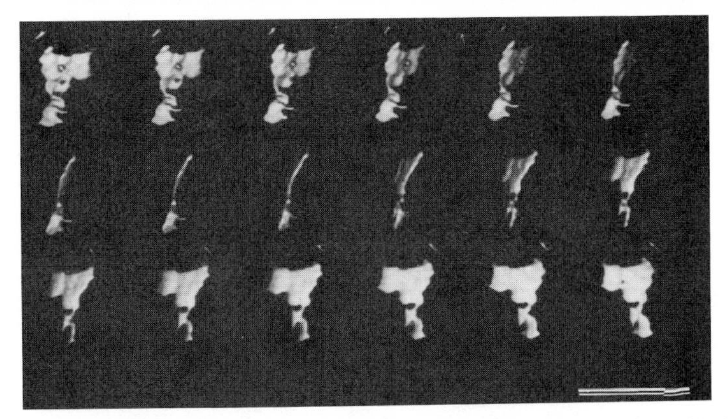

FIGURE 12. Rotation gallery of a small subvolume of the 3D reconstruction. Shaded surface representations of the subvolume indicated in Fig. 11 are shown over a 180° range of viewing angles in 10° increments. Neighboring views will form stereo pairs. It is apparent that individual crystals are flat, platelike structures and it appears that neighboring plates can fuse together (McEwen, Song, and Landis). Bar = 100 nm.

calcification. Thus, the turkey tendon project has been extremely productive in a short amount of time. It is worth considering why the method was so successful in this case. First, there was a clear idea of the goal of the project: a determination of the size, shape, and location of the hydroxyapatite crystals. Such a focused concept seems to produce quicker results than investigating a structure simply to see what is looks like in 3D, because the most time-consuming part of the project, which by far is examining the computed 3D volume, is much more efficient when one is looking for specific features. This does not mean that all projects should only seek to answer specific, preformulated questions, but rather that projects seeking unspecified information can be expected to be more time consuming and open ended.

A second reason for the rapid success of the turkey tendon project is the nature of the object itself and the information we were seeking. There seems to be little complex substructure in the crystals, while their general shape and location are interesting by themselves. As the cilium reconstructions demonstrate, even with a suboptimal angular tilt range the tomographic 3D reconstructions are able to reveal an accurate representation of the general shape of an object—it is the fine structure which gets lost. In addition, the plates are simple structures which, in the turkey tendon, are aligned in a nearly regular array. This organization tends to simplify the examination of the 3D volume when compared to the kinetochore, where one has to analyze many complex and irregular features. Furthermore, the relatively high electron scattering of the crystals helps to isolate them from their surroundings, a condition which is ideal for tomography (see below).

In this way, the turkey tendon project has readily demonstrated how powerful the 3D reconstruction can be and how the nature of the material examined, and the questions asked, can affect the success of the effort. In regard to the last points, however, it should be remembered that the methodology is continually being upgraded in several laboratories, specifically to extend its effectiveness over a wider range of morphological problems.

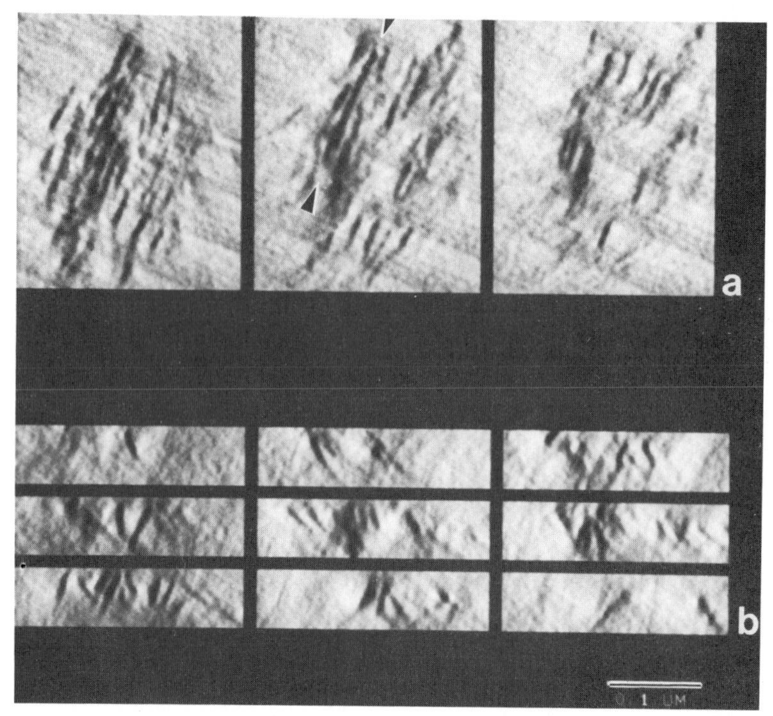

FIGURE 13. Individual slices of the calcium crystals. These were taken from the reconstruction presented in Fig. 11. (a) Slices taken from the volume in the same orientation as it was in the untilted view of the original data (upper region in Fig. 10). The crystal indicated by the arrowhead is as long as three periods of the collagen banding. (b) Slices viewed from a direction orthogonal to the viewing direction in (a) and parallel to the vertical axis in Fig. 10. That the crystals are platelike rather than needles is evident by their height in (b). Only about one-fourth the total height of the original reconstruction volume is shown in (b) because the crystals were at a single depth along the top surface of the volume (McEwen, Song, and Landis, unpublished). Bar = 100 nm.

5. *SELECTIVELY STAINED TISSUES: THE DENDRITE AND THE GOLGI APPARATUS*

Selective staining methods have been developed by several laboratories for the purpose of contrasting specific structures in biological systems (e.g., Peachey, 1982; Lindsey and Ellisman, 1985a, b; and Wilson, 1987). Often selective staining is used to acquire information about the 3D distribution of a complicated subsystem of a larger structural complex. Tomographic reconstruction methods have the potential to greatly enhance the effectiveness of this approach to 3D morphological analysis. Yet it is equally true that selective staining methods have the potential to greatly enhance the effectiveness of tomographic reconstruction methods. The reason for this is that tomography assumes that the micrographs contain projections of the object being reconstructed, which in turn assumes that the object is imaged without any interference from background density. Selective staining produces a contrast that is high enough to effectively isolate stained objects from their surroundings, as illustrated in Fig. 14.

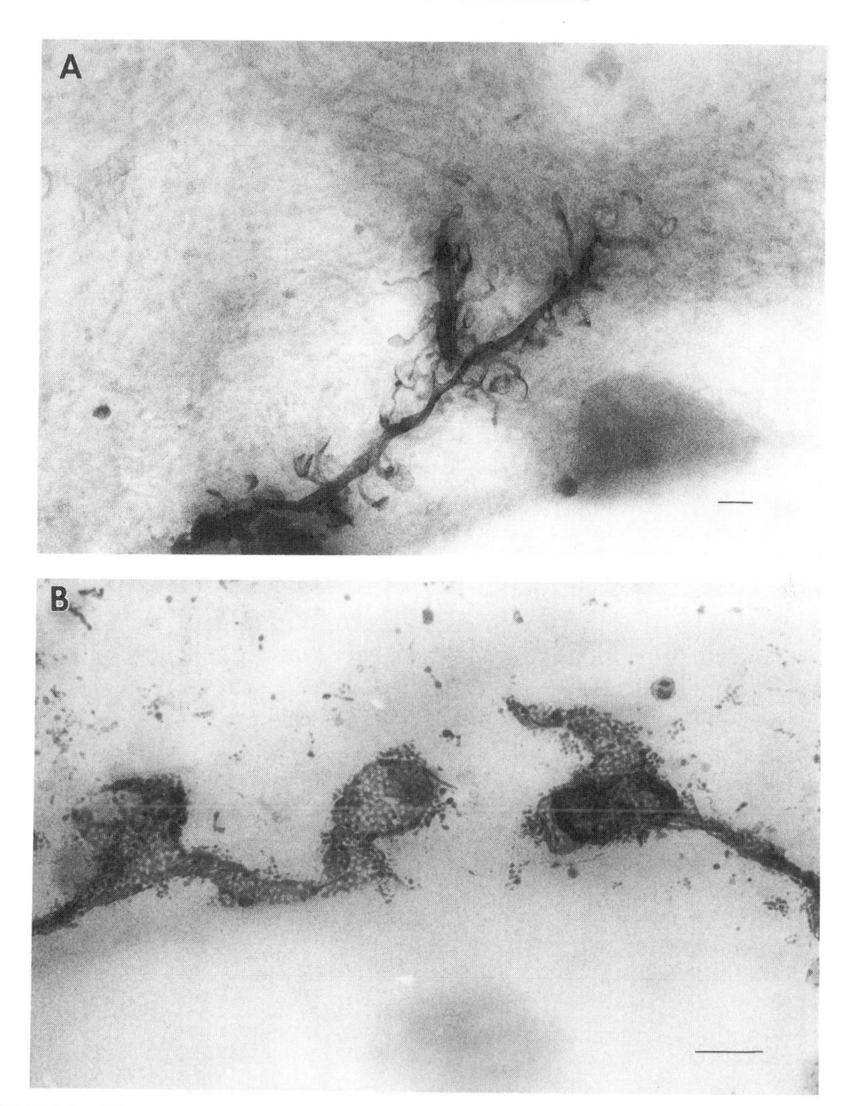

FIGURE 14. Micrographs of selectively stained preparations. (A) A spiny projection neuron from the rat neostriatum. Contrast is provided by the horseradish peroxidase reaction as described by Wilson (1987). The section is 3.0 μm thick and was imaged with the HVEM at 1.0 MV (Wilson and McEwen, unpublished). (B) Cis face of the Golgi apparatus from a bullfrog spinal ganglia. Selective staining was accomplished by osmium impregnation according to Lindsey and Ellisman (1985a, b). The section thickness was 3.0 μm and was imaged with the HVEM as in (A) (Ellisman and McEwen, unpublished). Bars = 1.0 μm.

Of course, in all successful tomographic reconstructions, the object is isolated from its surroundings by some means: the cilium protrudes out from the cell and hence is an isolated object along most of its length; the high electron opacity of the hydroxyapatite crystals and the localization of stain in collagen bands effectively make the initial crystallizations sites in turkey-tendon-isolated objects; and the micropipette, of the patch-clamped and puff ball spore reconstructions (see below), is mounted in open space and thus is an isolated object in the most literal sense. The use of selective stains, however, offers two advantages not found in these studies: (1) selective stains can contrast organelles, such as the Golgi apparatus and dendrites, which would otherwise not be isolated from the surrounding cellular components, and thereby broaden the applications of tomography; and (2) typically much thicker sections, up to 3.0 μm, can be viewed because the background tissue remains unstained. In addition, certain tomographic reconstructions, such as the kinetochore, could be enhanced or augmented by selective staining. For example, a selective stain for DNA would allow one to trace the fibrous DNA component in the kinetochore. Immunolabeling could also be used to selectively stain specific molecules.

Thus far, tomographic reconstructions have been reported for two selectively stained structures: (1) a dendrite of a spiny projection neuron from rat *neostriatum* (Mastronarde *et al.*, 1989) and (2) the cis face saccules of the Golgi apparatus from bullfrog spinal ganglia (Ellisman *et al.*, 1990). The former structure was selectively stained by microinjection of the neurons with horseradish peroxidase (Wilson, 1986, 1987), the latter by osmium impregnation (Lindsey and Ellisman, 1985a, b). Both structures were viewed as 3.0-μm-thick sections in the HVEM, and tilt series with a minimum angular range of 120° were recorded. Both reports were preliminary and more concrete conclusions from the studies are expected to follow. The dendrite study will attempt to determine the surface area of the spiny projections which is an important variable in functional theories (Wilson *et al.*, 1987, 1992). A computer graphical scheme developed by Mastronarde (Kinnamon, 1990; Wilson *et al.*, 1992) is being used in an attempt to find boundaries and minimize the effect of the distortion from the missing angular information upon this measurement. The initial representations of the Golgi 3D reconstruction (Fig. 15) indicated, among other

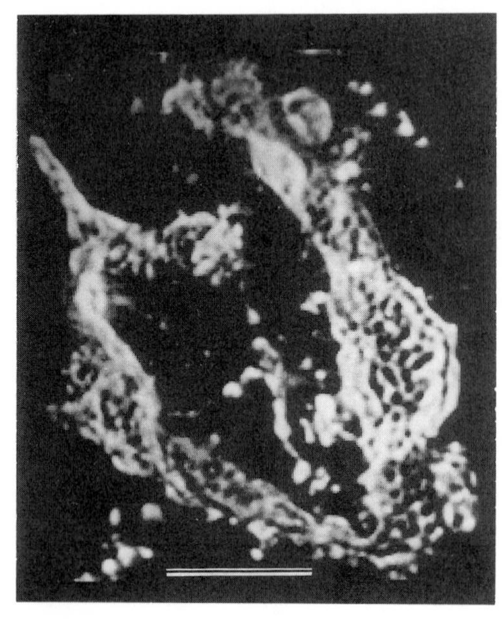

FIGURE 15. Shaded surface representation of a Golgi apparatus 3D reconstruction. The 3D reconstruction was computed from a tilt series (+60° in 2° intervals) that was taken from a preparation like that described in Fig. 14B. (From McEwen and Frank, 1990.) Bar = 1.0 μm.

things, the existence of a filamentous scaffold-like structure forming the boundary of the cis face saccule and mediating the connection between the cis face and the varicose tubule network. Other new structural information is expected to emerge as this study continues.

6. *RECONSTRUCTION FROM A CYLINDRICAL GEOMETRY: PATCH-CLAMPED MEMBRANES, PUFF BALL SPORES, AND THE COLLECTION OF AN UNLIMITED ANGULAR TILT RANGE*

Patch-clamp techniques have revolutionized the field of cellular electrophysiology (Sachs and Auerbach, 1984). In this method, a small patch of the cell membrane is sucked into a glass micropipette and electrical recordings are measured across the patch. The membrane patch is typically 0.5 to several micrometers in diameter and is attached to the glass by a seal which is electrically, diffusionally, and mechanically tight. Because of the tight electrical seal, the noise level is low enough to allow the activity of single ion channels to be recorded over a time scale extending from 10 μs to days.

Despite the importance of patch-clamped membranes, nothing was known about their ultrastructure until Sachs and Song successfully imaged the micropipettes using the Albany HVEM (Sachs and Song, 1987; Ruknudin *et al.*, 1989, 1991). These studies were followed by the report of a tomographic 3D reconstruction of the patch-clamped membranes (McEwen *et al.*, 1990). The membrane used for the reconstruction was from a *Xenopus* oocyte and was prepared for viewing in the HVEM according to the dry-mounting technique of Sachs and co-workers (Sachs

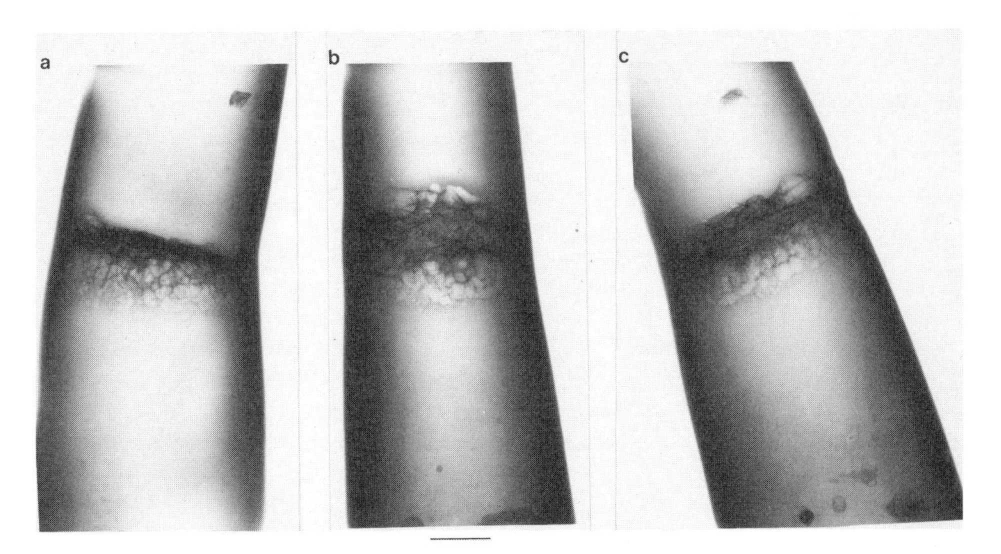

FIGURE 16. Three views of a patch-clamped membrane preparation. The sample was prepared for viewing in the HVEM according to the dry-mounting technique of (Ruknudin *et al.*, 1989) and the tilt series recorded at 1.0 MV. The micrographs were taken at tilt angles of (a) $-66°$, (b) 0° (untilted), and (c) $+66°$ (Song, McEwen, and Sachs, unpublished). Bar = 2.0 μm.

and Song, 1987; Ruknudin *et al.*, 1989, 1991). HVEM images of this preparation are shown in Fig. 16. A small subvolume, containing the pipette-spanning disk, was extracted from the full 3D reconstruction and rotated to give a cross-sectional view (Fig. 17). The fibrous nature of the disk is evident. The geometry of the micropipette lends itself to 3D analysis by cylindrical projections as is illustrated by the series of such projections in Fig. 18. The shorter projections on the top of the pyramid come from the inner radii of the micropipette (where the circumference is smaller), while the longer ones at the bottom of the pyramid come from the outer radii along the wall of the micropipette. The projections from the inner radii show the thin profile of the disk, while those at larger radii show the distribution of features along the wall of the micropipette. The blurred area in the middle (and at the ends) of each cylindrical projection is caused by the missing angular range, while the blank area in the longest projection is a result of the slightly elliptical cross section of the micropipette.

The patch-clamped membrane reconstruction was computed from a tilt series with a 2° tilt angle interval and an angular range from −64° to 66°. Normally, the

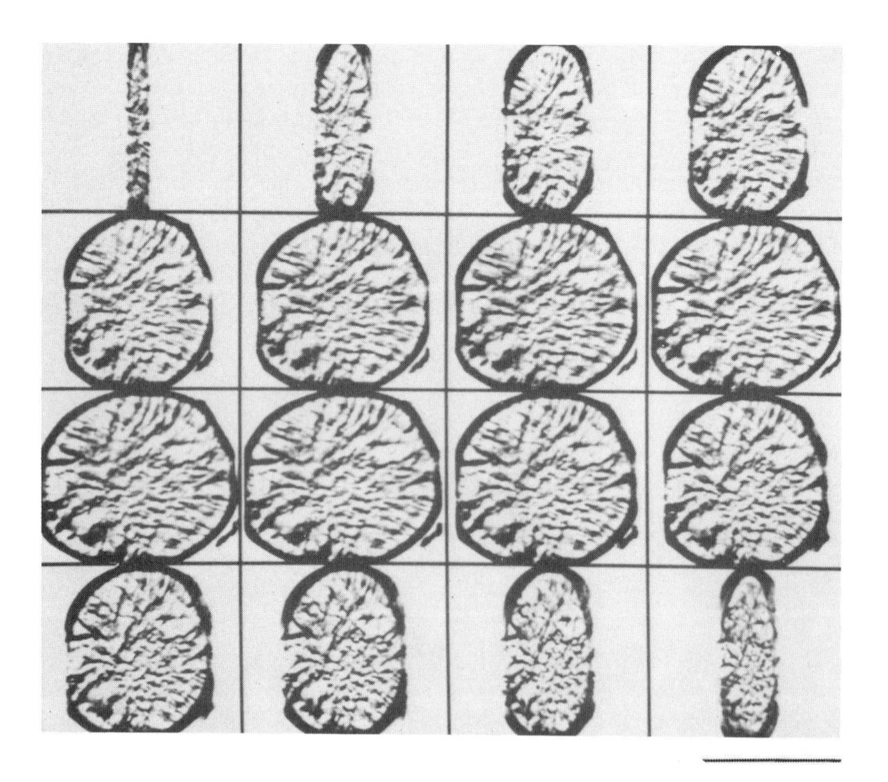

FIGURE 17. Rotation series of a subvolume of the patch-clamped membrane 3D reconstruction. The reconstruction was computed from the tilt series illustrated in Fig. 16 and rotated until the disklike membrane which spans the micropipette was coplanar with the computer screen. The subvolume was then isolated and rotated through an angular range o −80° to 70°, in 10° increments. The gallery presents a simple projection of the structure at each of these angles. The fibrous nature of the patch is evident. (From McEwen *et al.*, 1990.) Bar = 5.0 μm.

FIGURE 18. Cylindrical projections of the patch-clamped membrane 3D reconstruction. The projections were computed from a series of 0.26-μm-thick cylinders taken from the reconstruction at radii varying from 1.04 to 2.86 μm in 0.26-μm increments. Before the computation, the reconstruction was rotated until the micropipette cylindrical axis was coaxial with the vertical axis. The projections at the bottom of the gallery are a profile of the distribution of material along the wall of the micropipette, while those at the top are a good measure of the thickness of the disk. The blurred region at the center, and on the sides, of each projection is caused by the missing angular range in the input data set. (From McEwen *et al.*, 1990.) Bar = 5.0 μm.

images at the high tilt ends of the series are somewhat degraded by the increased thickness of the specimen, overlap from neighboring material, the rapid focus change across the field of the specimen, and the compression that occurs in some sectioned material. The striking feature of the patch-clamp tilt series is that these problems are absent, and the high-tilt data are of the same quality as the low or untilted views (Fig. 16). This, of course, is a result of the cylindrical geometry, which is much more favorable for collecting tilt data than the rectangular geometry of conventional sections or negatively stained grids.

For some time we had been considering ways to mount specimens which would accommodate a full 180° tilt range. Our experience with patch-clamped membrane reconstructions convinced us that the micropipette would be a good configuration for our first attempt. Accordingly, we constructed a specimen stage for the Albany HVEM which was capable of tilting a micropipette through a full 180° angular range. We then computed the first 3D reconstruction from electron microscopy to contain a complete angular tilt range. The sample used in this reconstruction was two puff ball spores stuck to the end of a micropipette (Barnard *et al.*, forthcoming). The micrographs in Fig. 19 demonstrate that the full angular range was indeed sampled.

The general utility of the micropipette for computing tomographic 3D reconstructions with a full angular tilt range will be determined by how many samples can be prepared in such a configuration. We have just begun our investigation of new preparatory methods, and some of the approaches we are trying include

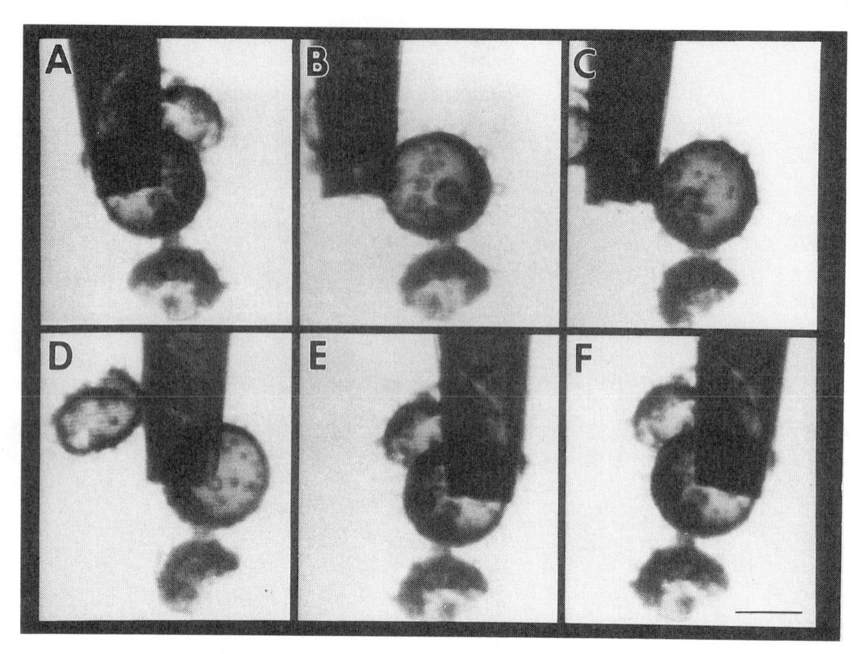

FIGURE 19. Achievement of a full angular tilt range in electron microscopy. The individual electron micrographs of the gallery were taken at (A) −90°, (B) −44°, (C) 0°, (D) +44°, and (E) +90°. (F) is the computed mirror image of (A). These views are part of a tilt series that was collected over an angular range of 180°, with a 2° angular increment, and with an accelerating voltage of 1.0 MV. The sample is puff ball spores attached to a glass micropipette, and the specimen stage was especially designed to make use of the cylindrical specimen geometry to achieve an unlimited angular tilt range. The images in the gallery have been digitized but not aligned. A comparison of (E) and (F) shows that the cumulative angular error of the tilt stage is small and that very little in plane rotation of the sample occurs through the tilt series. The latter indicates that the micropipette is nearly coaxial with the tilt axis. (From Barnard *et al.*, forthcoming.) Bar = 2.0 μm.

sucking samples into the micropipette along with a solution of negative stain; sucking embedding media with samples into the pipette; and using the micropipette to punch out cylindrical plugs from sections. It will probably be advantageous to place the sample at the end of, rather than into, the micropipette because the electron opacity of the glass wall of the micropipette is high and multiple scattering is probably occurring to some degree.

Regardless of the general utility of the present unlimited tilt stage, the configuration is already of great value for the computation of the patch-clamped reconstructions and for the development of restoration techniques (see Chapter 6). The reason for the latter is that the 3D reconstruction can be computed with a missing angular range by leaving out some of the data. Restoration can then be applied to this reconstruction, and the result compared to the reconstruction computed from the full data set. Presently one must rely on phantom objects for such tests, but even the best phantom objects are greatly oversimplified and notoriously poor indicators of how well the algorithm will perform with real data.

7. SUMMARY AND CONCLUSIONS

The chief advantage of the tomographic reconstruction method is its versatility. The examples discussed in this and the following chapter vary in size from 30 to over 5000 nm in diameter. Preparation methods range from the conventional techniques of negative stain and positively stained embedded sections to the more unusual techniques of selective staining and micropipette imaging. The versatility of tomography arises because it makes no assumptions about symmetry in the object and does not require identical units or subunits to be present. The lack of symmetry and identical unit structures is also the chief disadvantage of tomography because (1) there can be no signal-to-noise enhancement from averaging over identical units, and (2) all the different views required for the 3D reconstruction must be collected from a single object. These disadvantages limit the resolution that can be realized in tomographic reconstructions. For this reason the method is only used in cases where the more refined techniques are inapplicable.

The structures that have been successfully investigated by tomographic 3D reconstructions are generally one or two orders of magnitude larger than those investigated by the more refined methods (see Fig. 1). Higher accelerating voltages for the electron beam are generally required to form images of structures in this size range; the largest structures require an HVEM. In addition, the computation and analysis of 3D reconstructions requires more computing power and disk storage space as a result of the increased complexity of the structures. Suitable computers have only recently become readily affordable with the introduction of the latest generation of workstations. Finally, the increased complexity of the structures also makes interpretation of the 3D reconstructions more difficult. It is not unusual for more than 80% of an investigator's time to be spent in efforts to visualize and interpret a structure after the reconstruction is computed.

Because of the versatility of tomography and the complexity of the structures to which it is applied, each study presents the investigator with a unique set of problems. Sometimes, as in the case of the calcification of the turkey tendon, obtaining biological information is relatively straightforward. More often, as with the kinetochore, it is not. Often the investigator must create new approaches and a new set of techniques for analyzing and rendering a 3D reconstruction. However, the process becomes easier as more tomographic reconstruction projects are undertaken because we are accumulating a repertoire of approaches for 3D analysis. The interpretation will also be aided by the currently developing techniques for overcoming the effect of the missing angular range in the computation of the 3D reconstructions. These techniques include the construction of high-tilt stages, the development of an unlimited tilt stage, and the development of computational restoration methods such as POCS and maximum entropy.

Taken altogether, the versatility of tomography, the development of methods to overcome or minimize the effect of the missing angular range, the accumulation of strategies for 3D analysis, and the ready availability of modern computer workstations, indicate that tomography is rapidly becoming an important tool for 3D ultrastructural analysis.

ACKNOWLEDGMENTS

This work was supported by National Institutes of Health grants GM40165 and RR01219. I wish to thank Drs. J. Frank and W. J. Landis for helpful comments about the manuscript.

REFERENCES

Afzelius, B. (1959). Electron microscopy of the sperm tail: Results obtained with a new fixative. *J. Biophys. Biochem. Cytol.* **5**:269–278.

Allen, R. D. (1968). A reinvestigation of cross-sections of cilia. *J. Cell Biol.* **37**:825–831.

Amos, L. A. and Klug, A. (1974). Arrangement of subunits in flagellar microtubules. *J. Cell Sci.* **14**:523–549.

Avolio, J., Lebduska, S., and Satir, P. (1984). Dynein arm substructure and the orientation of arm-microtubule attachments. *J. Mol. Biol.* **173**:389–401

Barnard, D. P., McEwen, B. F., Frank, J., and Turner, J. N. (forthcoming). An unlimited-tilt stage for the high-voltage electron microscope.

Baumeister, W., Barth, M., Hegerl, R., Guckenberger, R., Hahn, M., and Saxton, W. O. (1986). Three-dimensional structure of the regular surface layer (HPI layer) of *Deinococcus radiodurans*. *J. Mol. Biol.* **187**:241–253.

Brenner, S., Pepper, D., Berns, M. W., Tan, E., and Brinkley, B. R. (1981). Kinetochore structure, dublication and distribution in mammalian cells: Analysis by human autoantibodies from scleroderma patients. *J. Cell Biol.* **91**:95–102.

Brinkley, B. R. and Stubblefield, E. (1966). The fine structure of the kinetochore of a mammalian cell in vitro. *Chromosoma* **19**:28–43.

Brinkley, B. R. and Stubblefield, E. (1970). Ultrastructure and interaction of the kinetochore and centriole in mitosis and meiosis. *Adv. Cell Biol.* **1**:119–185.

Brinkley, B. R., Valdivia, M. M., Tousson, A., and Balczon, R. D. (1989). The kinetochore: Structure and molecular organization, in *Mitosis: Molecules and Mechanisms* (J. S. Hyams and B. R. Brinkley, eds.), pp. 77–118. Academic Press, New York.

Brokaw, C. J. (1965). Non-sinusoidal bending waves of sperm flagella. *J. Exp. Biol.* **43**:155–169.

Brokaw, C. J. (1983). The constant curvature model for flagellar bending patterns. *J. Submicrosc. Cytol.* **15**:5–8.

Comings, D. E. and Okada, T. A. (1971). Fine structure of kinetochore in Indian Muntjac. *Exp. Cell Res.* **67**:97–110.

Crowther, R. A., DeRosier, D. J., and Klug, A. (1970). The reconstruction of a three-dimensional structure from projections and its application to electron microscopy. *Proc. R. Soc. London A* **317**:319–340.

Ellisman, M. H., Lindsey, J. D., Carragher, B. O., Kiyonaga, S. H., McEwen, L. R., and McEwen, B. F. (1990). Three-dimensional tomographic reconstructions of components of the Golgi apparatus imaged by selective staining and high voltage electron microscopy. *J. Cell Biol.* **111**:199a.

Fawcett, D. W. and Porter, K. R. (1954). A study of the fine structure of ciliated epithelia. *J. Morphol.* **94**:221–281.

Frank, J., McEwen, B., Radermacher, M., Turner, J. N., and Rieder, C. L. (1986). Three-dimensional tomographic reconstruction in high voltage electron microscopy, in *Proc. XIth Int. Congr. on Electron Microsc.*, pp. 1145–1150.

Frank, J., McEwen, B. F., Radermacher, M., Turner, J. N., and Rieder, C. L. (1987). Three-dimensional tomographic reconstruction in high voltage electron microscopy. *J. Electron Microsc. Tech.* **6**:193–205.

Gibbons, I. R. (1981). Cilia and flagella of eukaryotes. *J. Cell Biol.* **91**:107s–124s.

Gibbons, I. R. and Grimstone, A. V. (1960). On flagellar structure in certain flagellates. *J. Biophys. Biochem. Cytol.* **7**:697–716.

Glimcher, M. J. and Krane, S. M. (1968). The organization and structure of bone, and the mechanism of calcification, in *A Treatise on Collagen Biology* (G. N. Ramachandran and B. S. Gould, eds.), pp. 68–251. Academic Press, New York.

Goodenough, U. W. and Heuser, J. E. (1984). Structural composition of purified dynein proteins with in situ dynein arms. *J. Mol. Biol.* **180**:1083–1118.

Goodenough, U. W. and Heuser, J. E. (1985). Substructure of inner dynein arms, radial spokes, and the central pair/projection complex of cilia and flagella. *J. Cell Biol.* **100**:2008–2018.

Gorbsky, G. J., Sammak, P. J., and Borisy, G. G. (1987). Chromosomes move poleward in anaphase along stationary microtubules that coordinate disassembly from their kinetochore ends. *J. Cell Biol.* **104**:9–18.

Grimstone, A. V. and Klug, A. (1966). Observations on the substructure of flagellar fibres. *J. Cell sci.* **1**:351–362.

Hodge, A. J. and Petruska, J. A. (1963). Recent studies with the electron microscope on ordered aggregates of the tropocollagen macromolecule, in *Aspects of Protein Structure* (G. N. Ramachandran, ed.), pp. 289–300. Academic Press, New York.

Huang, B., Ramanis, Z., and Luck, D. J. L. (1982). Suppressor mutations in Chlamydomonas reveal a regulatory mechanism for flagellar function. *Cell* **28**:115–124.

Johnson, K. A. and Wall, J. S. (1983). Structure and molecular weight of the dynein ATPase. *J. Cell Biol.* **96**:669–678.

Kinnamon, J. C. (1990). High-voltage electron microscopy at the University of Colorado. *EMSA Bull.* **20**:115–122.

Klug, A. and Berger, J. E. (1964). An optical method for the analysis of periodicities in electron micrographs, with some observations on the mechanism of negative staining. *J. Mol. Biol.* **10**:565–569.

Klug, A., Crick, F. H. C., and Wyckoff, H. W. (1958). Diffraction by helical structures. *Acta Crystallogr.* **11**:199–213.

Koshland, D. E., Mitchison, T. J., and Kirschner, M. W. (1988). Poleward chromosome movement driven by microtubule depolymerization in vitro. *Nature* **331**:499–504.

Landis, W. J. (1985). Inorganic-organic interrelations in calcification: On the problem of correlating microscopic observation and mechanism, in *The Chemistry and Biology of Mineralized Tissues* (A. Veis, ed.), pp. 267–271. Elsevier, New York.

Landis, W. J. (1986). A study of calcification in leg tendon from the domestic turkey. *J. Ultrastruct. Mol. Struct. Res.* **94**:217–238.

Landis, W. J., Song, M. J., Leith, A., McEwen, L., and McEwen, B. F. (1990). Spatial relations between collagen and mineral in calcifying tendon determined by high voltage electron microscopy and tomographic 3D reconstruction. *J. Cell Biol.* **111**:24a.

Linck, R. W. and Langevin, G. L. (1982). Structure and chemical composition of insoluble filamentous components of sperm flagellar microtubules. *J. Cell Sci.* **58**:1–22.

Lindsey, J. D. and Ellisman, M. H. (1985a). The neuronal endomembrane system. I: Direct links between rough endoplasmic reticulum and the cis element of the Golgi apparatus. *J. Neurosci.* **5**:3111–3123.

Lindsey, J. D. and Ellisman, M. H. (1985b). The neuronal endomembrane system. II: The muliple forms of the Golgi apparatus. *J. Neurosci.* **5**:3124–3134.

Liu, Y., McEwen, B. F., Rieder, C. L., and Frank, J. (1990). Tomographic reconstructions of 0.25 μm thick serial sections of a mammalian kinetochore, in *Proc. XII Int. Congr. Electron Microscopy*, Vol. 1, pp. 476–477.

Luck, D. J. L. (1984). Genetic and biochemical dissection of the eukaryotic flagellum. *J. Cell Biol.* **98**:789–794.

Manton, I. and Clarke, B. (1952). An electron microscope study of the spermatozoid of Sphagnum. *J. Exp. Bot.* **3**:204–215.

Mastronarde, D. N., Wilson, C. J., and McEwen, B. F. (1989). Three-dimensional structure of intracellularly stained neurons and their processes revealed by HVEM and axial tomography. *Soc. Neurosci. Abstr.* **15**:256.

McEwen, B. F. and Frank, J. (1990). Application of tomographic 3D reconstruction methods to a diverse range of biological preparations, in *Proc. XII Int. Congr. Electron Microscopy*, Vol. 1, pp. 516–517.

McEwen, B. F., Radermacher, M., Rieder, C. L., and Frank, J. (1986). Tomographic three-dimensional reconstruction of cilia ultrastructure from thick sections. *Proc. Nat. Acad. Sci. USA* **83**:9040–9044.

McEwen, B. F., Rieder, C. L., and Frank, J. (1987a). Three-dimensional organization of the mammalian kinetochore. *J. Cell Biol.* **105**:207a.

McEwen, B. F., Rieder, C. L., Radermacher, M., Grassucci, R. A., Turner, J. N., and Frank, J. (1987b). The application of three-dimensional tomographic reconstruction methods to high-voltage electron microscopy, in *Proc. 45th Ann. Elect. Microsc. Soc. America*, Vol. 45, 570–573.

McEwen, B. F., Song, M. J., Ruknudin, A., Barnard, D. P., Frank, J., and Sachs, F. (1990). Tomographic three-dimensional reconstruction of patch-clamped membranes imaged with the high-voltage electron microscope, in *Proc. XII Int. Congr. Electron Microscopy*, Vol. 1, pp. 522–523.

Mitchison, T. J. (1988). Microtubule dynamics and kinetochore function in mitosis. *Ann. Rev. Cell Biol.* **4**:527–549.

Mitchison, T. J. (1989). Mitosis: Basic concepts. *Curr. Opin. Cell Biol.* **1**:67–74.

Nicklas, R. B. (1989). The motor for poleward chromosome movement in anaphase is in or near the kinetochore. *J. Cell Biol.* **109**:2245–2255.

Omoto, C. K. and Kung, C. (1980). Rotation and twist of the central-pair microtubules in cilia of Paramecium. *J. Cell Biol.* **87**:33–46.

Peachey, L. D. (1982). Three-dimensional structure of muscle membranes involved in the regulation of contraction in skeletal muscle fibers, *Cell Muscle Motility* **2**:221–230.

Pepper, D. A. and Brinkley, B. R. (1977). Localization of tubulin in the mitotic apparatus of mammalian cells by immunofluorescence and immunoelectron microscopy. *Chromosoma* **60**:223–235.

Pepper, D. A. and Brinkley, B. R. (1980). Tubulin nucleation and assembly in mitotic cells: Evidence for nucleic acids in kinetochores and centrosomes. *Cell Motility* **1**:1–15.

Radermacher, M. and Frank, J. (1984). Representation of three-dimensionally reconstructed objects in electron microscopy by surfaces of equal density. *J. Microsc.* **136**:77–85.

Rattner, J. B. (1986). The organization of the mammalian kinetochore. *Chromosoma* **93**:515–520.

Rattner, J. B. and Bazett-Jones, D. D. (1989). Kinetochore structure: Electron spectroscopic imaging of the kinetochore. *J. Cell Biol.* **108**:1209–1219.

Rieder, C. L. (1979). Ribonucleoprotein staining of centrioles and kinetochores in newt lung cell spindles. *J. Cell Biol.* **80**:1–9.

Rieder, C. L. (1982). The formation, structure, and composition of the mammalian kinetochore and kinetochore fiber. *Int. Rev. Cytol.* **78**:1–58.

Rieder, C. L. (1990). Formation of the astral mitotic spindle: Ultrastructural basis for the centrosome-kinetochore interaction. *Electron Microsc. Rev.* **3**:269–300.

Rieder, C. L. and Alexander, S. P. (1990). Kinetochores are transported poleward along a single astral microtubule during chromosome attachment to the spindle in newt lung cells. *J. Cell Biol.* **110**:81–95.

Ris, H. and Witt, P. L. (1981). Structure of the mammalian kinetochore. *Chromosoma* **82**:153–170.

Roos, U.-P. (1973). Light and electron microscopy of rat kangaroo cells in mitosis. II: Kinetochore structure and function. *Chromosoma* **41**:195–220.

Roos, U.-P. (1977). The fibrillar organization of the kinetochore and the kinetochore region of mammalian chromosomes. *Cytobiologie* **16**:82–90.

Ruknudin, A., Song, M. J., Auerbach, A., and Sachs, F. (1989). The structure of patch-clamped membranes in high voltage electron microscopy, in *Proc. 47th Ann. Electron Microsc. Soc. Am.*, Vol. 47, pp. 936–937.

Ruknudin, A., Song, M. J., and Sachs, F. (1991). The ultrastructure of patch-clamped membranes: A study using high voltage electron microscopy. *J. Cell Biol.* **112**:125–134.

Sachs, F. and Auerbach, A. (1984). The study of membranes using the patch clamp. *Ann. Rev. Biophys. Bioeng.* **13**:269–302.

Sachs, F. and Song, M. J. (1987). High voltage electron microscopy of patch-clamped membranes, in *Proc. 45th Ann. Electron Microsc. Soc. Am.*, Vol. 45, pp. 582–583.

Satir, P. (1974). The present status of the sliding microtubule model of ciliary motion, in *Cilia and Flagella* (M. A. Sleigh, ed.), pp. 131–142. Academic Press, New York.

Tilney, L. G., Bryan, J., Bush, D. J., Fujiwara, K., Mooseker, M. S., Murphy, D. B., and Snyder, D. H. (1973). Microtubules: Evidence for 13 protofilaments. *J. Cell Biol.* **59**:267–275.

Unwin, P. T. N. and Henderson, R. (1975). Molecular structure determination by electron microscopy of unstained crystalline specimens. *J. Mol. Biol.* **94**:425–440.

Warner, F. D. (1974). The fine structure of the ciliary and flagellar axoneme, in *Cilia and Flagella* (M. A. Sleigh, ed.), pp. 11–37. Academic Press, New York.

Williams, B. D., Mitchell, D. R., and Rosenbaum, J. L. (1986). Molecular cloning and expression of flagellar radial spoke and dynein genes of Chlamydomonas. *J. Cell Biol.* **103**:1–11.

Wilson, C. J. (1986). Postsynaptic potentials evoked in spiny neostriatal projection neurons by stimulation of ipsilateral and contralateral neocortex. *Brain Res.* **367**:201–213.

Wilson, C. J. (1987). Three-dimensional analysis of neuronal geometry using HVEM. *J. Electron Microsc. Tech.* **6**:175–183.

14

The Organization of Chromosomes and Chromatin

Christopher L. Woodcock

1. INTRODUCTION

The eukaryotic nucleus and its components are, in principle, ideal subjects for tomographic structural analysis. Within the nucleus, a number of crucial and closely regulated processes occur. These include DNA replication and chromatin assembly, RNA transcription, the processing and transport of mRNA and rRNA,

---

Christopher L. Woodcock • Department of Zoology and Program in Molecular and Cellular Biology, University of Massachusetts, Amherst, Massachusetts 01003

Electron Tomography: Three-Dimensional Imaging with the Transmission Electron Microscope, edited by Joachim Frank, Plenum Press, New York, 1992.

and chromosome condensation and decondensation. Many of these events have been studied in detail at the molecular level, yet very little is known about the three-dimensional (3D) spatial framework within which they occur. Electron tomography offers direct insight into these structural and functional aspects of nuclear organization.

However, applying electron tomography to nuclear structures is complicated at present by limitations in specimen preparation. As yet, there is no consensus as to ideal fixation conditions for nuclei, and small changes in the ionic milieu prior to or during fixation may have a dramatic effect on the ultrastructure. Also, methods that allow differentiation between nucleic acid and protein at the ultrastructural level are still in their infancy, imposing a limit to the interpretation of results. The asymmetric nature of many nuclear components precludes the use of image averaging, the benefits of which are amply illustrated in Chapter 15. Another consideration that is not a limitation in itself, but rather illustrates the amount of work that will be required for a thorough understanding of genetic structures, is that nuclear DNA ocurs as a DNA-protein complex with a number of hierarchical levels of structural organization ranging over several orders of magnitude from the 2.5-nm-diameter DNA helix to mitotic chromosomes measured in micrometers. As yet, the number of discrete hierarchical levels of DNA folding is unknown, but is an obvious primary goal of tomography investigations.

In this chapter, a brief review of current knowledge of nuclear components will be followed by a discussion of the preparative dilemmas faced by investigators. The three main foci of tomographic studies to data—metaphase chromosomes, interphase chromatin fibers, and transcription complexes—will then be considered in some detail. As will become clear, we are at the beginning of what promises to become a very active and fruitful area of investigation. As EM tomography becomes more accessible, a rapid accumulation of new information and the establishment of consensus structures can be anticipated. At present, the results of tomographic studies must be viewed as rather isolated data sets, each group of investigators having chosen a particular structure, set of preparation conditions, and level of resolution for his or her studies. These pioneering experiments set the stage for the future.

1.1. Composition and Ultrastructure of Chromatin and Chromosomes

Biochemical characterization of isolated nuclei has shown that, in most eukaryotes, the DNA is complexed with small, basic proteins, the *histones*. The fundamental unit of this complex is the nucleosome, a highly conserved structure in which DNA is wrapped twice around a core of eight histone molecules (van Holde, 1989). Nucleosomes are roughly disk-shaped, with a diameter of 11 nm and a height of 5.5 nm; their structure is known to a resolution of 0.7 nm from x-ray diffraction studies of nucleosome crystals (Richmond *et al.*, 1984). This has allowed a fairly detailed model of the DNA and histone components to be proposed, but the path of the protein backbones and the details of DNA-protein interaction are still unknown. A variable amount of "linker" DNA joins adjacent nucleosomes, and if nuclei are extensively dispersed, the "beads-on-a-string" conformation may be observed in the electron microscope (Olins and Olins, 1974). Dispersion, usually promoted by lowering the ionic strength of the medium, is critical for such observations, since nucleosomes are usually very closely packed in vivo. Under milder dispersion conditions, the DNA-histone complex is seen as a fiber, approximately 30 nm in diameter, containing nucleosomes (van Holde, 1989). The H1 class of histones is required for the formation and maintenance of 30-nm fibers; these histones are associated with linker DNA, and their location and interactions within chromatin are under active investigation. In the absence of any dispersive forces, the DNA-histone complex shows very little substructure, and individual 30-nm fibers often cannot be resolved.

Additional components found in the nucleus are other proteins, often termed nonhistone chromosomal proteins (NHCPs), and RNA complexed with specific ribonuclear proteins to form ribonucleoprotein (RNP) particles. The DNA-histone complex with associated bound NHCPs is collectively termed *chromatin.*

With conventional thin-section techniques, it is possible to differentiate a number of regions in the typical interphase nucleus, each of which has been assigned functional characteristics. For example, electron-dense areas, usually peripherally located (heterochromatin), contain transcriptionally inactive DNA, while the potentially active genes tend to occur at the boundary between heterochromatin and the less dense, more centrally located euchromatin (Hutchison and Weintraub, 1985). Also found in the "nucleoplasm" are darkly staining granules up to 50 nm in size that have been shown to be RNP particles. One or more nucleoli, often subdivided into fibrillar and granular zones are prominent nuclear structures whose functions in the synthesis of ribosomal RNA and the assembly of ribosomes are well established. In the typical interphase nucleus, these structures and regions are all that can be recognized in thin sections, and provide little guidance in understanding the complex events that must have been occurring at the time the nucleus was fixed.

During each cell cycle, nuclear DNA is replicated, and chromatin assembled on the daughter strands. Subsequently, the chromatin condenses into chromosomes that, under some conditions, are seen to consist largely of 30-nm chromatin fibers. The spatial arrangement of fibers within the chromosome is another aspect of nuclear structure that, despite considerable attention, is still the subject of debate. Following chromosome separation, nuclear membranes form around the chromosome masses, which then return to the interphase state. The biochemical and structural events accompanying chromosome condensation and decondensation are poorly understood.

One of the factors that has hampered ultrastructural studies of chromatin and chromosomes is the seemingly disordered, pleiomorphic nature of the structures involved. In a single section of a nucleus, or even in serial sections, it has been very difficult to grasp the 3D aspects of the material. Isolated 30-nm fibers have a well-defined morphology, and upon negative staining may yield images that hint tantalizingly at a basic underlying architecture. Yet the degree of order is low, and inadequately revealed in the 2D projection images obtained with conventional electron microscopy. For many nuclear components, the availability of direct 3D information is of critical importance.

1.2. Preparation of Chromatin and Chromosomes for Ultrastructural Study

For many subcellular structures, a consensus has been reached as to appropriate conditions under which the component may be prepared for ultrastructural investigation. Such a consensus has usually been reached using a number of approaches: thin sections, freeze-fracture, negative staining after isolation, etc. In the case of chromatin and chromosomes, there seems to be no single set of conditions that favor both preservation and structure determination.

A logical approach to in situ ultrastructural studies of nuclei and nuclear components is to select conditions under which, at the level of the light microscope, the in vivo morphology is retained. Buffers employing the polyamines spermine and spermidine, in addition to "physiological" levels of Na^+ and K^+, are effective for this purpose (e.g., Ruiz-Carrillo *et al.*, 1980). Under these conditions, the DNA-histone complex is very compact, and 30-nm fibers are not usually seen in conventional thin sections. The inability to resolve 30-nm chromatin fibers is probably due to the lack of contrast between fibers under these conditions and to the very contorted paths of the fibers producing complex superposition effects in thin sections. An analogous effect may be seen in the packing of DNA in phage heads where individual DNA fibers are generally not resolved. The possibility that chromatin fibers in vivo form a "fluid" of weakly interacting nucleosomes was suggested by McDowall *et al.* (1986), who found no evidence of 30-nm fibers in frozen hydrated sections of metaphase chromosomes. On the other hand, x-ray studies of whole cells and nuclei (Langmore and Schutt, 1980; Langmore and Paulson, 1983) show 30-nm–40-nm reflections that are interpreted as arising from the packing of chromatin fibers. Clearly, the status of the 30-nm fiber in native chromatin in an important topic to be addressed by electron tomography.

In isolated nuclei and chromosomes, 30-nm fibers become clearly resolved if the ionic strength is lowered appropriately, and if divalent and polyvalent counterions are low in concentration or absent. The conditions that allow visualization of 30-nm fibers also result in the swelling of nuclei, and the process may be considered as a relaxation of chromatin packing from its highly compact state to one in which fibers become sufficiently separated that they may be distinguished from each other. The difficulty of retaining the native nuclear morphology and observing a well-defined component such as the 30-nm fiber makes it essential to examine nuclei under a variety of conditions.

There are obvious advantages to isolating a subcellular component for study. In the case of chromatin, isolation is usually initiated by cleaving the DNA with a nuclease that cuts preferentially between nucleosomes and then allowing the fragments to be released into solution. Buffers designed to retain nuclear morphology do not allow adequate release, and it is necessary to select the same sort of conditions that favor 30-nm fiber visualization to effect this. If isolated chromatin is replaced in buffers containing the levels of polyamines or divalent cations needed to preserve nuclear morphology, precipitation occurs, and indeed the compact form of much of the chromatin found in vivo is so condensed that it may be appropriate to consider it a precipitate (e.g., Kellenberger, 1987). The 30-nm fibers in solution may be reversibly decondensed to an expanded chain of nucleosomes by lowering the ionic strength to about 10 mM monovalent ions (Thoma *et al.*, 1979), and for studying the arrangement of nucleosomes in the fiber it may be advantageous to examine different levels of condensation.

These considerations illustrate the difficulties in selecting conditions for examining the ultrastructure of chromatin and chromosomes. The problem is particularly acute for tomographic investigations, where (with current technology), a geat deal of time must be invested in a single reconstruction, and the exploration of a wide range of conditions is not feasible. In the absence of a consensus preparative milieu, each group of investigators has chosen conditions appropriate to their goals. This often makes comparison between the results of different groups difficult or impossible. On the other hand, the publication of data obtained using a wide range of conditions is essential for future studies.

1.3. Resolution Considerations

The purpose of this section to call attention to the criteria used to determine the limiting resolution of tomographic reconstructions of nuclear components, and to refer the reader to the appropriate detailed discussions elsewhere in the volume.

As discussed in Chapter 5, the limiting resolution (d) for the tomographic reconstruction of a spherical object diameter (D) from a series of N projections equally spaced between $+90°$ and $-90°$ is given by (Crowther *et al.*, 1970; Klug and Crowther, 1972)

$$d = \frac{\pi D}{N}$$

For nonspherical objects, D may be considered the maximum specimen distance traversed by the electron beam and is related to the specimen geometry, its orientation with respect to the tilt axis, and the angular tilt range (Radermacher, 1980; Frank *et al.*, 1987).

For the general case of a cylindrical fiber lying on a carbon support, D is dependent on the angle of the fiber with respect to the tilt axis (θ), the maximum tilt angle (α_{max}), and the width of the fiber measured perpendicular to the tilt axis (W). If $\theta < \alpha_{max}$, then

$$D = \frac{W}{\cos\theta}$$

and if $\theta \geqslant \alpha_{max}$, then

$$D = \frac{W}{\cos\alpha_{max}}$$

The most favorable orientation giving the best resolution occurs when the fiber lies parallel to the tilt axis, and D is equal to the fiber diameter. If the fiber is not circular in cross section, D is adjusted correspondingly using the same principle.

Similar considerations apply to the reconstruction of sections, thickness T

containing structures of width W, measured in the direction perpendicular to the tilt axis. If $W > T$, then

$$D = \frac{T}{\cos \alpha_{max}}$$

while if $W \leqslant T$, then

$$D = W$$

For complicated images, such as a section of a nucleus containing a number of chromatin fibers, the value of W may vary according to the distribution of material in the section. If an individual fiber remains separated from other material throughout the tilt range, then W is the fiber diameter (again, measured *perpendicular* to the tilt axis). If, however, other material becomes superimposed on the fiber at high tilts, then W is the combined diameter of the fiber plus overlapping material. These considerations apply to resolution in the plane of the specimen (normally referred to as the xy plane). As discussed in Chapter 5, the tilt range restriction leads to a poorer resolution in the z direction. In the case of a $\pm 60°$ tilt range, the z resolution is approximately twofold poorer than that calculated for xy.

For specimens in which the distribution of material requires that D be taken as the specimen thickness at the maximum tilt angle ($T/\cos \alpha_{max}$), there may be justification for relaxing the resolution limitation as defined above. For example, a tilt series that extends to $\pm 75°$ will result in a value of D (and hence resolution) that is almost twice that of a similar series that is restricted to $\pm 60°$. When low-pass filters based on these limits were applied to reconstructions derived from computer-generated projections of model images, it was clear that valid information was lost from the $\pm 75°$ case (C. L. Woodcock, unpublished results). As pointed out by Crowther *et al.* (1970), the resolution relationship allows for a worst case situation in which the images contain densely packed information. As the model images used in the simulations were rather simple combinations of lines and circles, filtering for the worst case situation was inappropriate. For real situations, there is at present no a priori measure of image complexity upon which decisions to relax resolution limits can be based, and it is therefore advisable to be conservative in filtering reconstructions and to avoid over-interpreting tomographic data.

2. *MITOTIC CHROMOSOMES*

At mitosis, chromatin is seen in its most compact and, in some ways, most ordered state. Specific genes map to specific locations (at the resolution of the light microscope) and show the linkage patterns expected of a linear organization. At the ultrastructural level, however, the fundamental architecture of the chromosome is still under debate. Over the past two decades, a variety of structures have been proposed, including specific folding patterns of 30-nm fibers (DuPraw, 1965), radially oriented loops of 30-nm fibers (Paulsen and Laemmli, 1977; Marsden and Laemmli, 1979; Adolph, 1980; Earnshaw and Laemmli, 1983), and large-scale

100-nm–200-nm-diameter "fibers" (Sedat and Manuelides, 1978; Zatsepina *et al.*, 1983). A helical model incorporating both the radial looping and 200-nm-fiber concepts has also been put forward (Rattner and Lin, 1985). The electron tomographic approach should, in principle, provide a rigorous test of the proposed models.

In the two published studies of chromosome structure using electron tomography, both the goals of the investigators and the starting material have been rather different. Belmont *et al.* (1986) concentrated on preserving the native compact state, and maximizing spatial resolution, while Harauz *et al.* (1987) used partially dispersed chromosomes, and focussed on following the paths of 30-nm fibers.

2.1. Compact Chromosomes

Belmont *et al.* (1986), have presented a very detailed comparative study of the structure of compact *Drosophila* mitotic chromosomes. The investigators took great care to preserve the native structure through the use of polyamine-containing buffers and, as far as possible, used similar buffer conditions both for preparing isolated chromosomes and for fixing chromosomes in situ. The material chosen was staged *D. melanogaster* embryos and colcemid-arrested Kc cells, the rationale for including the latter being that the chromosomes would be more compact than in the rapidly dividing embryos and better represent the general case of mitotic chromosomes.

To isolate chromosomes, embryos were homogenized in buffer containing the polyanions spermine and spermidine, and the detergents Brij 58 and digitonin, separated on sucrose gradients, and allowed to settle onto carbon-Formvar films. After adhesion, 2% glutaraldehyde was added and cross-linking allowed to proceed overnight before washing, dehydrating in ethanol, and critical point drying. The fiduciary markers for alignment were colloidal gold particles added to the grids before the chromosomes. For the in situ studies, similar buffer and fixation conditions were employed, followed by Epon embedding. After sectioning, uranyl and lead stains were applied, and gold particles adsorbed to both sides.

Electron microscopy of the critical-point-dried chromosomes was carried out at 1500 kV (Kratos/AED high-voltage EM), while the 120–200 nm sections were examined at 100 kV or 120 kV. The tilt series range was $\pm 60°$ and the increment 5° for both types of preparation. Negatives were digitized with an effective pixel size at the specimen level of 3.2 nm or 4.0 nm for isolated chromosomes, and 2.1 nm for sections. Alignment of the images was based on the positions of selected colloidal gold particles, as described by Lawrence in Chapter 8 and R-weighted back projection (Gordon and Herman, 1974; Ramachandran and Lakshminarayanan, 1971; Chapter 5) used to reconstruct the 3D volumes.

For the isolated critical-point-dried chromosomes, the reconstructions were filtered to a resolution of 20 nm, although the limiting resolution calculated according to the criteria of Crowther *et al.* (1970) was 33 nm. Similarly, the 10-nm limiting filter used for the reconstructions from sections was less than the calculated resolution by a factor of about 0.6. To enhance the very low contrast in the chromosome images (Fig. 1a, b), an algorithm (Peii and Lim, 1982) was used that

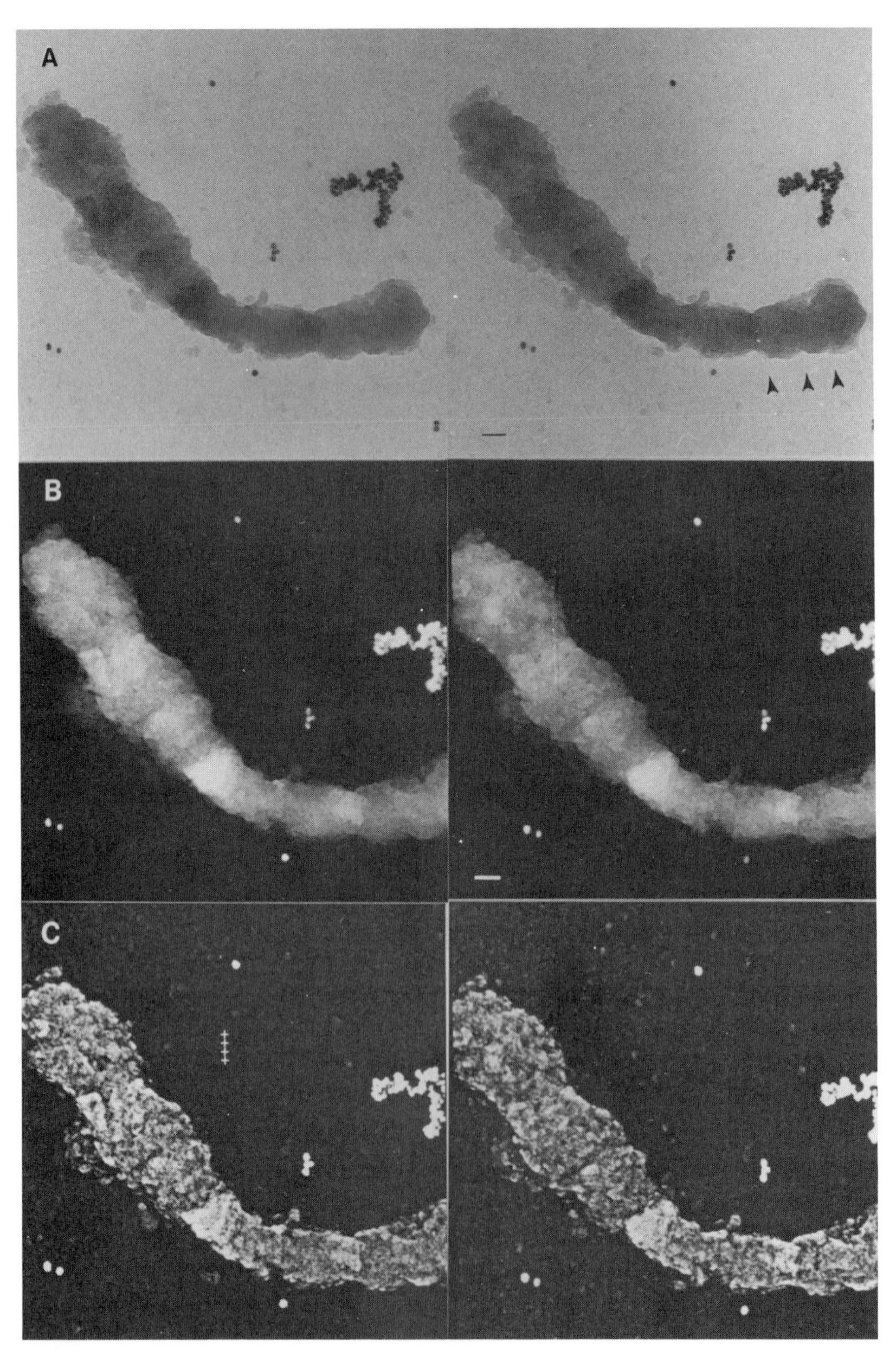

FIGURE 1. Stereopair of a critical-point-dried chromosome imaged at 1.5 MeV; tilt angles are +55° and +50°, with the tilt axis oriented vertically. (A) Prints made directly from the HVEM negatives. Grooves in the chromosome border delineating several continuous chromatin blocks are marked by arrowheads. (B) "Mass normalized" digitized images displayed on a 512 × 512 pixel, 256 gray-scale monitor in which the brightness of each pixel is proportional to the electron scattering from the projected path through the pixel. (C) Same image as in (B) after local contrast enhancement based on a box of size 21 × 21 pixels (67 × 67 nm). Bars in (A) and (B) represent 100 nm; crosses in (C) are separated by 32 nm; pixel size in (B) and (C) is 3.2 nm. (From Belmont *et al.*, 1986.)

defines local contrast as the difference between a pixel value and the mean of a box of surrounding pixels (local mean). New values for local contrast may then be selected and mapped into the image, giving a dramatic increase in contrast (Fig. 1C). Because user-selected factors are required for the enhancement algorithm, it is important to choose values that enhance the signal and reject the noise in an image. Belmont *et al.* (1986) stress that all features of the enhanced images could also be recognized in the original images. Nevertheless, a critical examination of the technique applied to micrographs with different levels of added "noise" would provide valuable guidelines for standardizing its future use.

Fig. 2 illustrates the structural complexity in the interior of a critical-point-dried chromosome as seen in slices of the reconstruction perpendicular to the chromosome axis. As an additional aid in interpreting the reconstructions, Belmont *et al.* (1986) rotated the 3D volume so that the chromosome axis was appropriately aligned, and then created stereo projections (Fig. 3) of selected regions. These views are of much greater assistance in understanding 3D structures than single slices.

From their reconstructions, Belmont *et al.* (1986) were able to distinguish four discrete levels of organization ranging from 12 to 80–100 nm within all three types of chromosome preparation, and an additional domain class of about 130 nm was found in the compact Kc chromosomes. Table 1 shows the results of their studies of compact *Drosophila* mitotic chromosomes. The smallest fiber observed is apparently the basic unit fiber and is seen nearly always as part of higher-order chromatin structures. One of the "domains" had a size of 28 nm, and presumably represents the 30-nm fiber that is the only structure seen under other conditions (see Section 2.2). Relationships between the different levels were difficult to establish, so that a hierarchical system in which the larger structures represent folding of the smaller could not be demonstrated clearly. Neither slices nor stereo-projections of the reconstructions provided any evidence for a central core, radial-loop structures, or any other ordered arrangement.

The difficulty in discerning any regular structural order in these compact chromosomes suggests a greater complexity than predicted from any current models. These results are especially compelling since very similar domain levels were found

TABLE 1. Widths of Fibers and Structural Domains Revealed in Tomographic Studies of Compact *Drosophila* Mitotic Chromosomes[a,b]

	Chromosome preparation and width of chromatids (nm)		
Fiber or domain width (nm)	Anaphase embryonic (~200)	CPD[c] embryonic (200–250)	Kc metaphase arrested (350–800)
12.0	√	√	√
24.0	++++	++	+
40–50	+	++++	++
80–100	+	++	++++
>130	–	–	√

[a] Modified from Belmont *et al.* (1986).
[b] The number of crosses corresponds to the relative prominence of particular size classes of structural domains within a given preparation.
[c] Critical point dried.

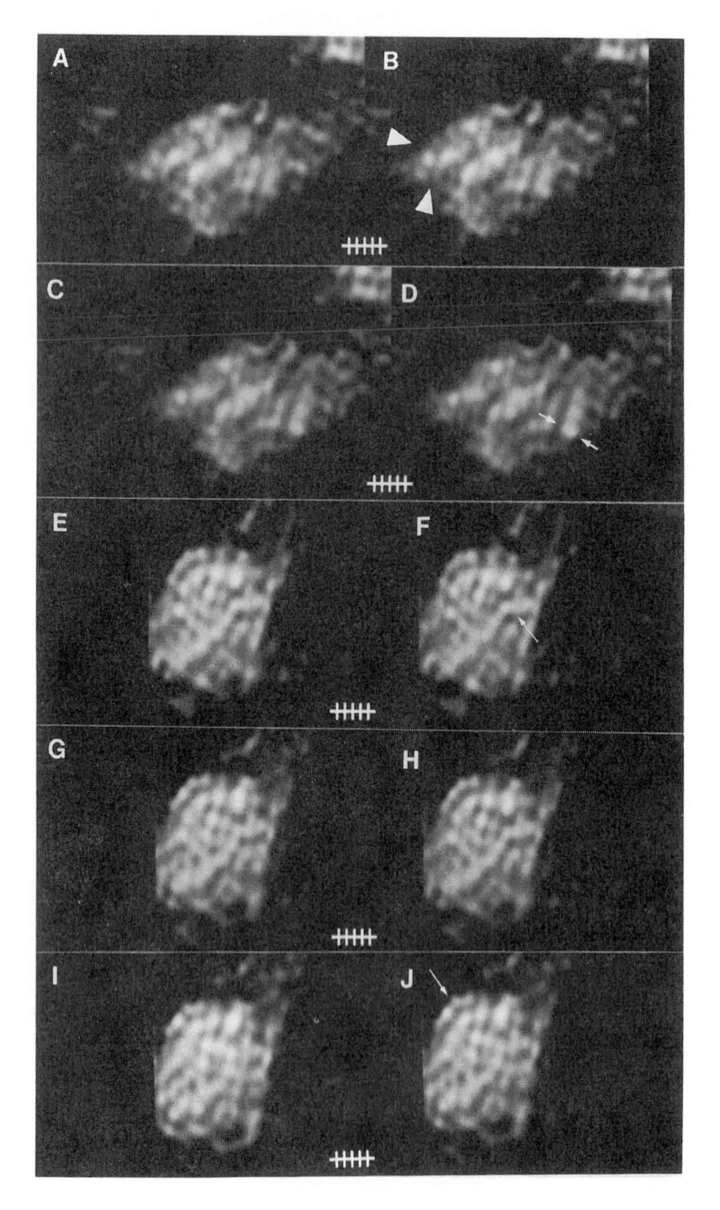

FIGURE 2. Selected slices of a tomographic reconstruction of a critical-point-dried chromosome. Two consecutive two-dimensional rotations, in different planes, were carried out on the reconstruction to place the chromosome axis roughly perpendicular to the direction of slicing. (A–D) Consecutive slices spaced 4.0 nm (1 pixel) apart. Arrowheads point to 80–100-nm structural domains; pairs of arrows delineate a structural domain 40 nm in diameter. (E–J) Consecutive slices from a region roughly 220 nm distant from that shown in (A–D). Long thin arrows point to fibers, roughly 24 nm in diameter. Crosses are separated by 20 nm (5 pixels). (From Belmont *et al.*, 1986.)

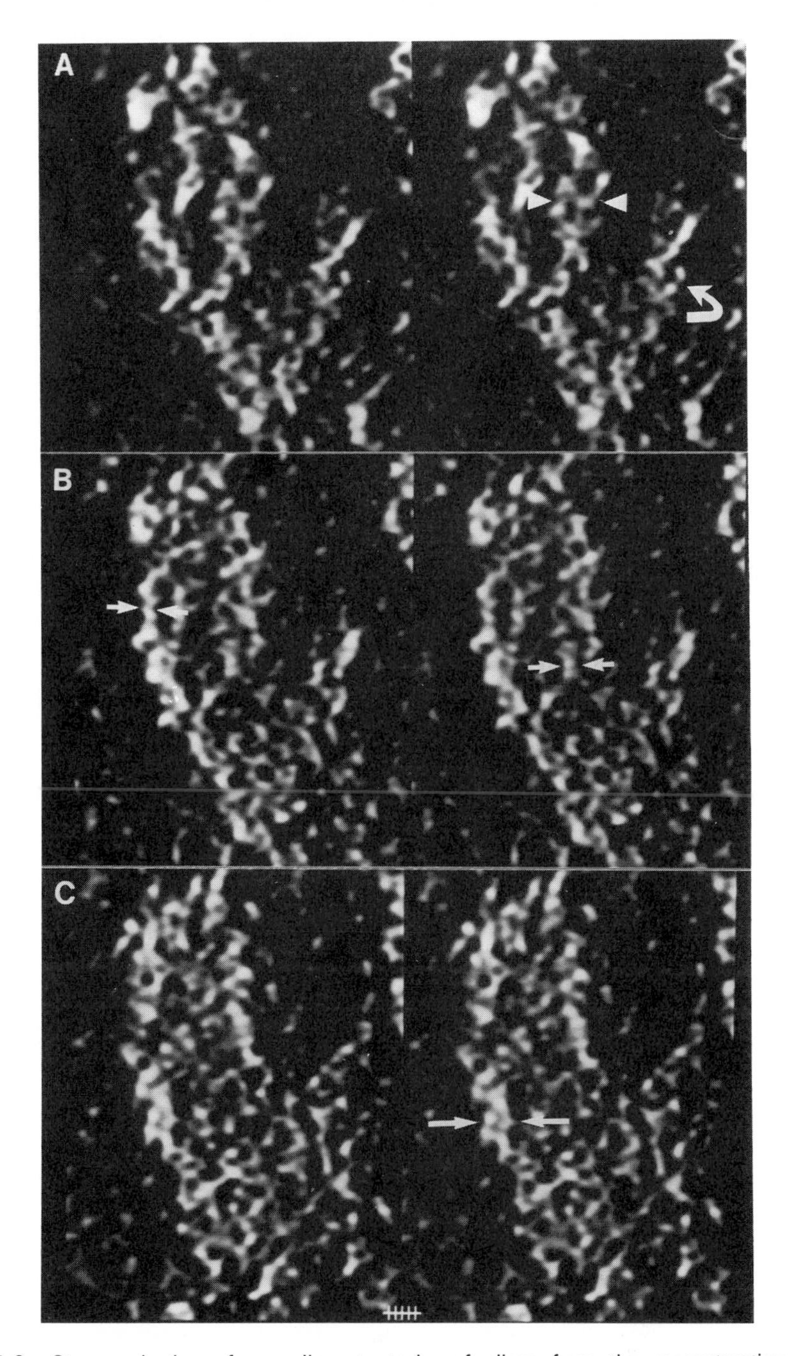

FIGURE 3. Stereoprojections from adjacent stacks of slices from the reconstruction of a metaphase chromosome embedded in situ in Epon. Each stack is 23.2 nm thick (11 slices). Short arrows in (B) point to fibers 24 nm in diameter; long arrows in (C) point to a 40–50-nm diameter structural domain. Arrowheads outline a region 80 nm wide present in (A) and (B); below it in (A) is another similar region, marked by a curved arrow. Crosses are separated by 10.5 nm (5 pixels). (From Belmont *et al.*, 1986.)

in chromosomes from different sources and with different preparation methods (Table 1); they provide a very solid basis upon which further work on partially dispersed chromosomes can be based.

2.2. Water-Spread Chromosomes

In the study of Harauz *et al.* (1987), lymphocyte cultures in metaphase arrest were lysed (no details given), spread on a water surface, and picked up on carbon-Formvar films to which polystyrene spheres (approximately 150 nm in diameter) had been added. The grids were then critical-point-dried from ethanol and examined unstained. Cross-linking with aldehydes was not used, and the chromosomes were unfixed in that sense; however, ethanol dehydration of unfixed DNA and DNA-protein complexes does introduce structural changes at the nucleosomal level (Woodcock *et al.*, 1976; Eickbusch and Moudrianakis, 1978).

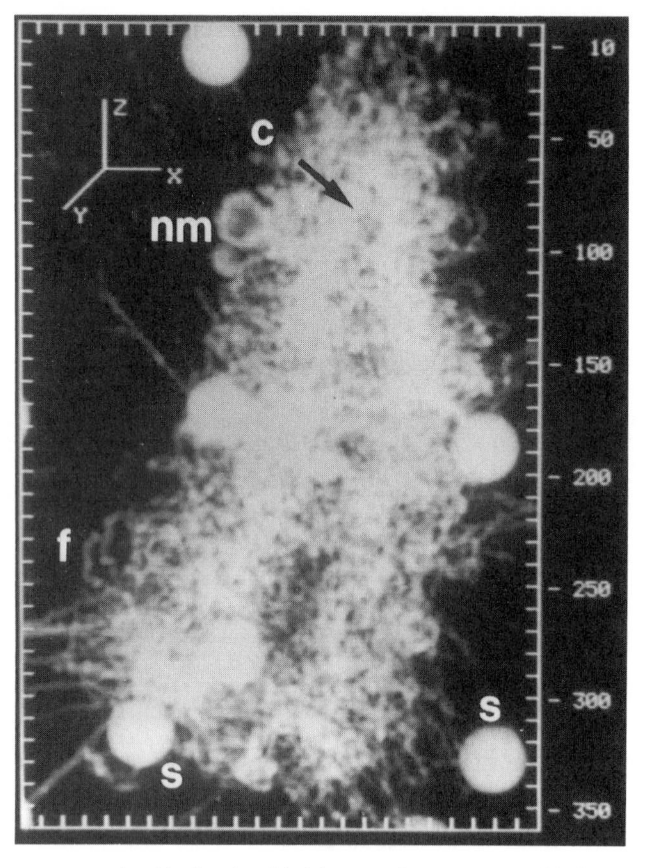

FIGURE 4. A water-spread, critical-point-dried human chromosome at 0° tilt angle (unfixed and unstained). The centromere region (c), a piece of contaminating nuclear membrane (nm), some 30 nm fibers (f), and polystyrene spheres (S) are indicated. The scales at the sides are spaced every 55 nm (10 pixels); the numbers along the sides refer to the line number of each 2D projection (also the slice number of the 3D reconstruction). (From Harauz *et al.*, 1987.)

The preparation of metaphase chromosomes by spreading on a water surface has been extensively used in the past (e.g., Bahr and Golomb, 1971) and, notwithstanding the exposure to very low ionic strength, the overall morphology is well maintained. Instead of the compact, smooth-edged structures seen after glutaraldehyde fixation in polyamine-containing buffers (Fig. 1; Belmont *et al.*, 1986), chromosomes appear as a mass of 30-nm fibers, some of which loop out from the periphery (Fig. 4).

Tilt data ($\pm 60°$, 3° increment) were collected at 100 kV and 12,000 × magnification, and digitized using a CCD camera with effective pixel size at the specimen level of 2.2 nm. Prior to alignment and reconstruction, the images were reduced 5-fold to give a final pixel size of 11 nm. Alignment was based on the polystyrene sphere positions, and reconstruction was by back-projection with "exact" filters. Previously, back-projection methods were extensively examined by the same group (van Heel and Harauz, 1986), and a filtering technique developed that constructs a filter function to match the actual angles of the tilt data set (Harauz and van Heel, 1986; Radermacher *et al.*, 1986; Chapter 5). With certain test images, this "*exact*" filter technique gives a substantial improvement in the fidelity of the reconstruction (Harauz and van Heel, 1986). In discussing the resolution of their chromosome

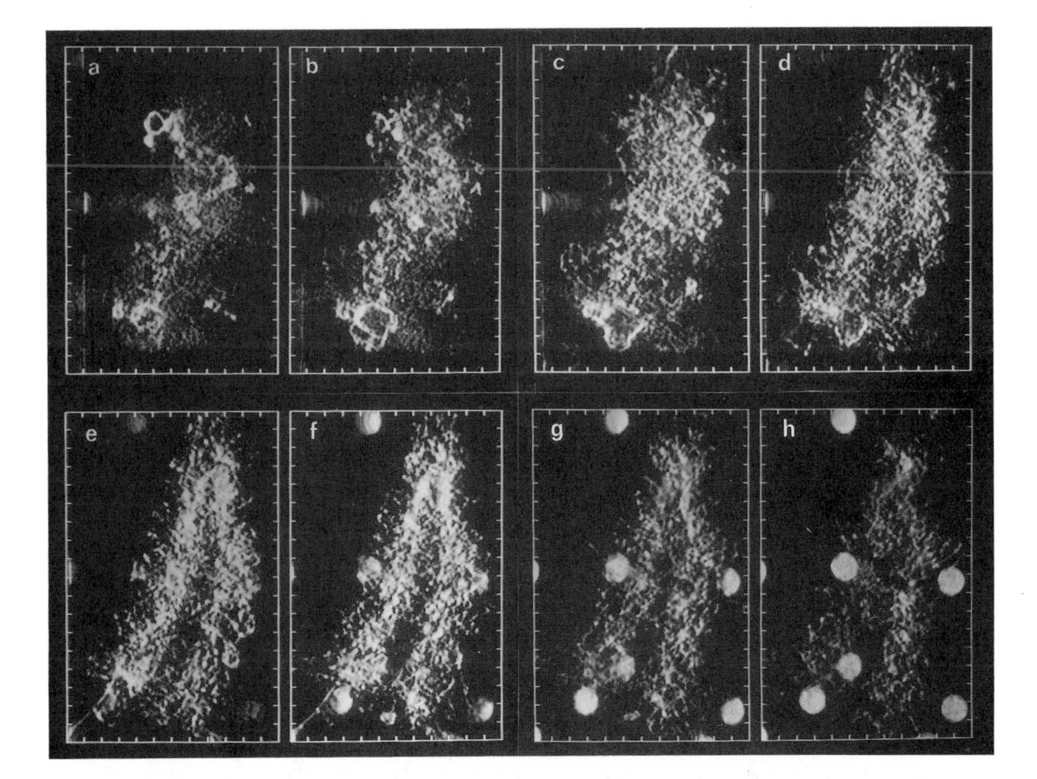

FIGURE 5. Selected slices (11 nm in thickness), parallel to the support film, through the reconstruction of the chromosome in Fig. 4. (a) is furthest away from the support, (h) closest. Density thresholding has been used to enhance the visualization of structural details. The scale bars are spaced 110 nm apart. (From Harauz *et al.*, 1987.)

reconstruction, Harauz *et al.* (1987) examined the point-spread function in the direction parallel to the electron beam (conventionally the z direction, but defined as the y direction in Harauz *et al.*, 1987). The missing wedge of information due to the tilt angle limitation resulted in a 2.2-fold elongation of a point image, yielding a resolution of 58 nm in the (conventional) z direction compared with 26 nm in the plane of the specimen. These values for the limiting resolution allowed 30-nm fibers to be resolved provided they did not approach each other too closely.

The reconstructed chromosome was about 0.8 μm thick compared with about 1.6 μm wide, and the surface adhering to the support film relatively flat (Harauz *et al.*, 1987). Some flattening of the structure is to be expected due to the surface tension of the water surface, and some may occur during electron irradiation (Luther *et al.*, 1988). Recently, Flannigan and Harauz (1989) applied the same preparative method to *Drosophila* polytene chromosomes, and found them to be severely flattened, a phenomenon they attributed to collapse during critical point drying.

The 30-nm fibers in the interior were clearly displayed in slices parallel to the support film (Fig. 5) and in shaded surface representations of the reconstruction (Fig. 6). In xz slices perpendicular to the long axis of the chromosome, the reconstruction artifacts make feature recognition and interpretation more difficult (Fig. 7).

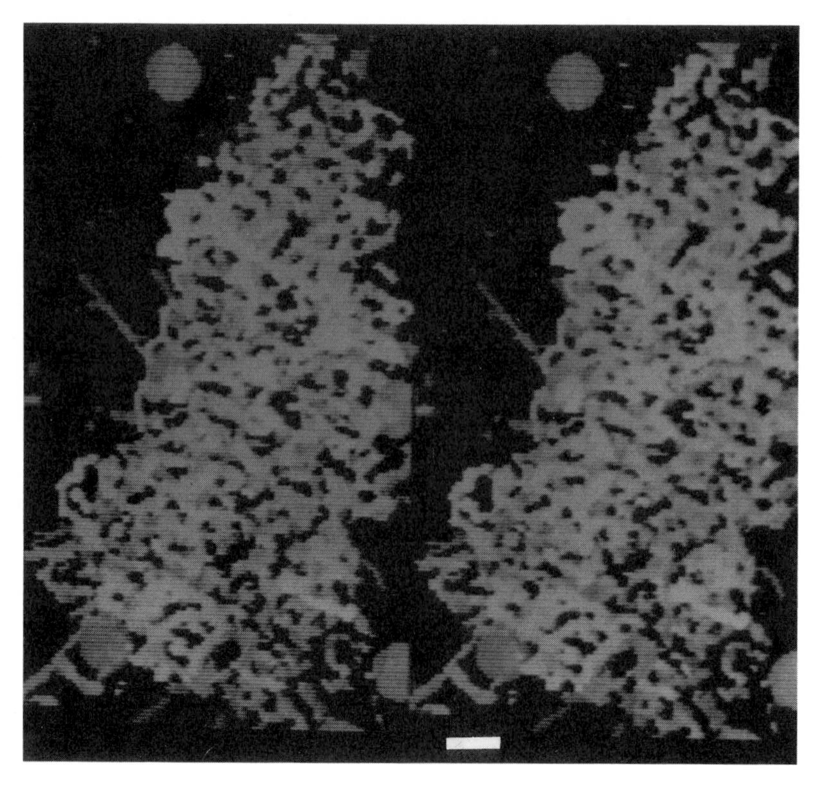

FIGURE 6. Surface representations of the reconstructed chromosome at viewing angles of 0° and 6° for stereo viewing. The bar represents 30 nm. (From Borland *et al.*, 1988.)

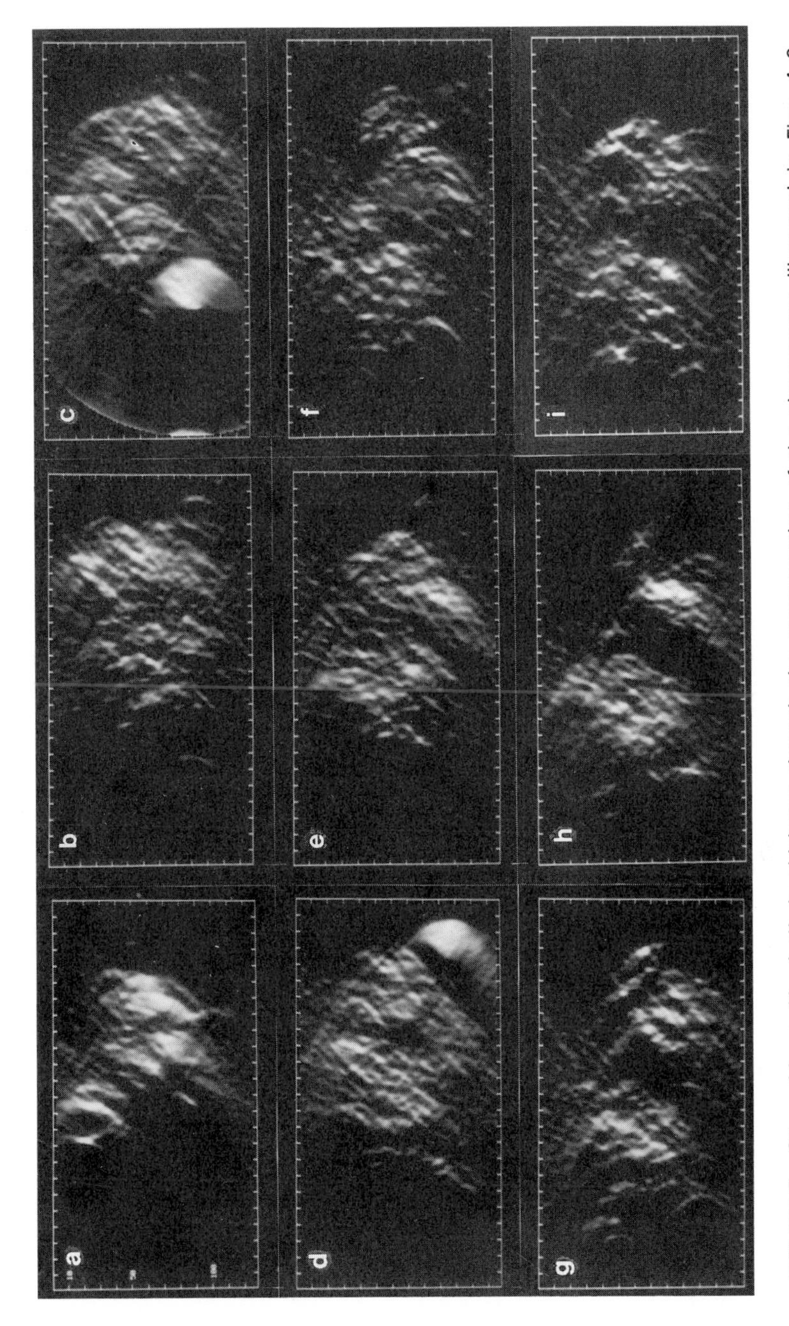

FIGURE 7. Slices 11 nm (1 pixel) in thickness through the reconstruction of the chromosome illustrated in Figs. 4–6 approximately perpendicular to the long axis of the chromosome. (From Harauz *et al.*, 1987.)

The presentation of *xz* views of a reconstruction provides a valuable assessment of the overall quality of the data that is not so evident from *xy* views.

Individual 30-nm fibers were traced through the chromosome using an automatic "fiber tracking" algorithm to map fiber segments in three dimensions (Borland *et al.*, 1988). From a given start point, the algorithm traced a path of maximum density through the reconstruction. The common pattern emerging from the tracing of over 200 segments of 30-nm fiber was of a looping path with a loop diameter between 100 and 350 nm. However, the random loop orientation and lack of either a dense or hollow chromatid core led the authors to reject radial-loop or helical coil models (Borland *et al.*, 1988). Rather, they suggested that clusters of variably oriented loops occur along the length of chromosomes, each cluster averaging 200 nm in length. Further support for this model was obtained by summing the total density along the length of a chromatid: the profile showed distinct maxima and minima spaced at 100 to 300-nm intervals (Borland *et al.*, 1988). The generality of this pattern can be tested using single 2D projections of chromosomes.

3. 30-NM CHROMATIN FIBERS

The integrity of the 30-nm fiber is maintained by interactions between the components, and, by analogy with other subcellular filamentous structures composed of subunits (e.g., microtubules), some regular arrangement of the subunits may be anticipated. Indeed, a number of investigators have revealed some degree of structural order in both isolated and in situ chromatin fibers (Woodcock *et al.*, 1984; Williams *et al.*, 1986; Woodcock and Horowitz, 1986; Horowitz *et al.*, 1990). However, such studies have shown only a limited degree of order, insufficient to define the structure, or even to distinguish unequivocally between the various models that have been proposed (Felsenfeld and McGhee, 1986). The tomographic approach, requiring no preconceptions about the structure, should be ideal in this case, at least for establishing nucleosome positions. Other information that is needed to fully define the structure, such as the path of linker DNA and the location of histone H1, requires the application of appropriate histochemical techniques in addition to tomography.

Chromatin fibers may be isolated from nuclei following treatment with a nuclease that cuts between nucleosomes, and then prepared for electron microscopy by negative staining. This procedure requires relaxing the compact in vivo chromatin conformation in order to release the fragments of fiber. Also, for in situ studies using thin sections, it is necessary to allow 30-nm fibers to separate in order to observe them. Unfortunately, it has not yet been possible to devise a single set of conditions that allows fibers to be observed in situ, and also cleaved and released. These factors must be borne in mind in interpreting structural studies on 30-nm chromatin fibers.

3.1. Isolated Fibers

In their pioneering studies using electron tomography, Subirana *et al.* (1983, 1985) examined both isolated negatively stained fibers from chicken erythrocyte nuclei, and in situ fibers in embedded nuclei from *Holotheria tubulosa* (sea cucumber) sperm. However, the bulk of the data and all of the micrographs presented by Subirana *et al.* (1985) are from the sections of whole nuclei; consequently, this work will be discussed in the section on in situ fibers.

More recently, Woodcock and McEwen (1988a, b; Woodcock *et al.*, 1990b) reconstructed segments of negatively stained chromatin fiber from *Necturus maculosus* (mud puppy) erythrocytes. This source of chromatin had previously been shown by Williams *et al.* (1986) to give relatively sharp x-ray reflections, and isolated fibers were seen to be partially ordered after negative staining. For the study of Woodcock and McEwen (1988a, b), chromatin fibers were released from nuclei after digesting with micrococcal nuclease in a buffer containing 80 mM monovalent ions. The fibers were fixed with 0.1% glutaraldehyde, applied to glow-discharged carbon films, and negatively stained with 1.5% sodium phosphotungstate pH 7.0 containing 0.015% glucose. Tilt data collection was at 100 kV and 30,000× magnification over a $\pm 70°$ tilt range with 5° increments, and digitization was carried out with an effective 1.1-nm pixel size at the specimen level, and 12-bit dynamic range. The fiduciary markers for alignment were provided by small stain-excluding specks, and the reconstruction performed by weighted back-projection.

Figure 8 shows a chromatin fiber (or group of fibers) in which two regions, denoted 1 and 2, have been reconstructed and examined in some detail (Woodcock and McEwen, 1988a, b; Woodcock *et al.*, 1990b). Both areas are of interest because they show the helixlike patterns noted in previous studies of isolated fibers (Woodcock *et al.*, 1984; Williams *et al.*, 1986). The orientations of the reconstructed segments with respect to the tilt axis are quite different, resulting in dramatic differences in limiting resolution (see Section 1.3). Taking into account the variations in fiber orientation, and the $\pm 70°$ tilt restriction, a low-pass Fourier filter with a radius of 1/3.3 nm was selected for filtering reconstruction from area 1, and a radius of 1/5.4 nm for area 2.

Initial examination of the reconstructed fiber showed that the cross section was elliptical rather than circular (Fig. 9). Three potential causes of this flattening were surface denaturation during adhesion to the carbon film, surface tension effects during negative staining, and electron beam damage. Of these, the latter evidently played an insignificant role since measurements from high-tilt images recorded with low electron doses indicated a similar amount of flattening (Woodcock, unpublished results). The influence of the support film on the structure was evident from the rather uniform stain exclusion at this z level (Fig. 10). In comparison, regions more distal, in z, from the support film showed prominent nucleosome-sized stain-exclusion regions (Fig. 10).

Two primary goals of the project were to locate the 3D positions of nucleosomes, and understand the basis of the helixlike patterns seen in 2D projections. Of these aims, the former proved to be the simpler. A 3D peak search with nearest-neighbor exclusion (similar to that used by Subirana *et al.*, 1985) was successfully used to identify potential nucleosome positions. Each peak position was displayed as an overlay on the appropriate slice of the reconstruction, and visually inspected for validity. Peak positions considered to be unlikely nucleosome sites because of their position, lack of definition, or proximity to other well-defined sites,

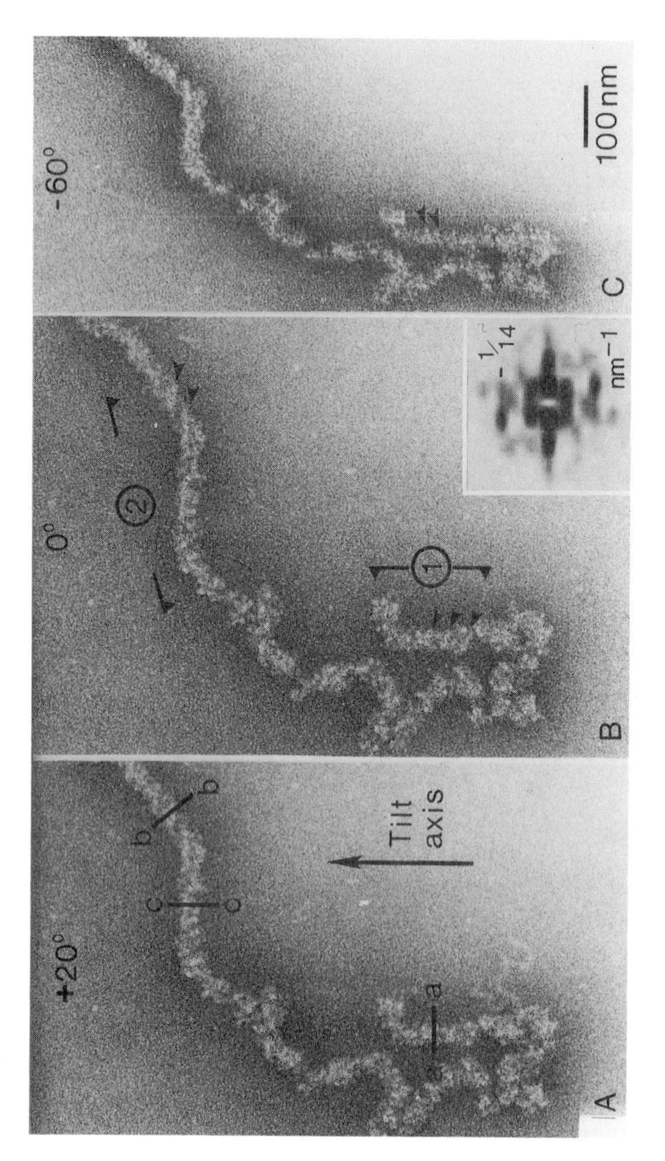

FIGURE 8. Isolated chromatin from *N. maculosus* erythrocyte nucleus stained with sodium phosphotungstate. (A)–(C) are selections from the ±70° tilt series. Regions marked 1 and 2 in (B) have been reconstructed and examined in detail. Insert in (B) is the power spectrum of area 1. (From Woodcock *et al.*, 1990b.)

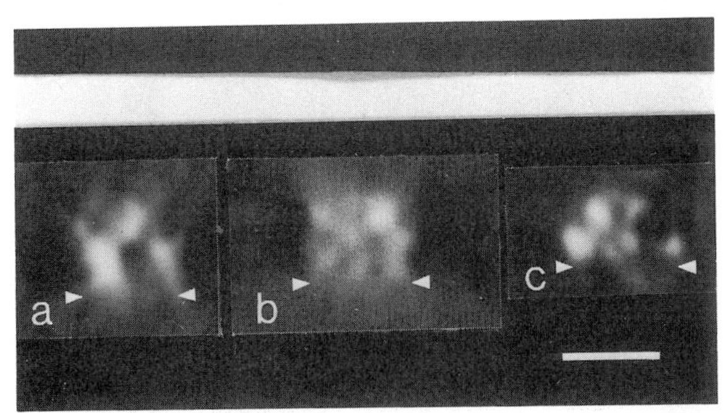

FIGURE 9. Vertical sections of the reconstruction perpendicular to the long axis of the fiber at the locations marked a–a, b–b, and c–c on Fig. 8A. Each represents a projection of an 11-nm length of fiber. Arrowheads show the approximate position of the support film. Bar = 30 nm. (From Woodcock *et al.*, 1990b.)

were eliminated. Using these rather subjective criteria, most of the stronger peaks were accepted as nucleosomes, the rejected ones proving to be external to the fiber. In area 1, 7.2 nucleosomes/11 nm were identified, consistent with direct STEM mass measurements (Woodcock *et al.*, 1984; Williams *et al.*, 1986; Graziano *et al.*, 1988). From the final peak lists, computer models were created in which 9-nm-diameter spheres were placed at each nucleosome location. Projections of this model matched projections of the complete 3D volume quite closely (Fig. 11), providing confidence that the general distribution of nucleosomes at the end of the negative staining procedure had been successfully extracted from the reconstruction. The most useful way to present this model was as a shaded surface representation viewed from different angles (Fig. 12), that could be presented as a "movie" loop. In comparison, shaded surface representations of the complete reconstruction were of little use: it was not possible to select a threshold that provided visual assistance in locating nucleosome positions or understanding the surface topology (Fig. 13).

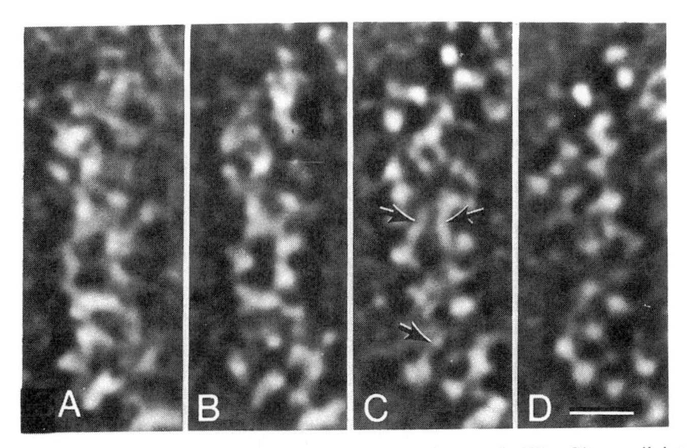

FIGURE 10. 5.5-nm-thick slices of the reconstruction of area 1 (Fig. 8) parallel to the support film. Each was produced by summing 5 adjacent *xy* slices of the reconstruction. Next to the support film (A), stain exclusion is diffuse, while furthest away (D), nucleosome-sized stain exclusions are prominent. Bar = 25 nm. (From Woodcock *et al.*, 1990b.)

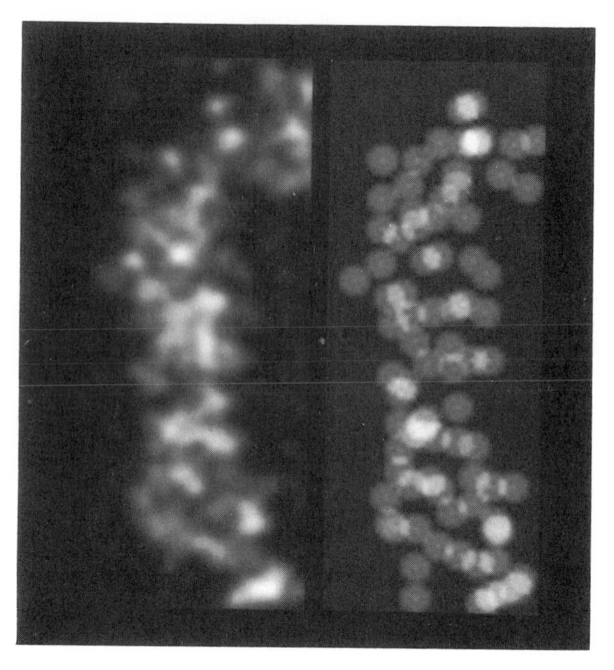

FIGURE 11. Comparison of the projection of the complete reconstruction of area 1 (left) with the corresponding projection of the sphere model (right). Each sphere is 9 nm in diameter. (From Woodcock *et al.*, 1990b.)

Considerable effort was spent exploring methods of displaying the reconstructions that would allow mental assimilation of the structure. One criterion for success was the ease with which the helixlike patterns seen in the low-tilt projections (Fig. 8) could be interpreted in 3D. A slice-by-slice approach (Fig. 10) proved to be ineffective, as did shaded surface representation (Fig. 13). More informative was a projection technique in which the maximum value encountered by a projection ray was placed in the final image (Tieman *et al.*, 1986). By selecting projection angles that resulted in stereo pairs, a surprisingly effective display of the 3D structure was obtained (Fig. 14). As computer hardware and software designed especially for rendering 3D volumes becomes more affordable and user friendly the problems of assimilating reconstruction data will diminish (see Chapters 10 and 11).

By using the various display options, Woodcock and McEwen (1988a, b) concluded that only in short fragments of fiber were true helical arrays of nucleosomes present. Oblique cross striations, as exhibited by region 1, could also be produced by the superimposition of stain exclusions at different levels along the *z* axis rather than by a true helix. This study showed clearly that nucleosomes were located peripherally in the 30-nm fiber, but did not disclose a uniform packing arrangement. It did, however, point out that more attention should be paid to adequate specimen preparation methods. An obvious first goal is to preserve the circular cross section of the fibers, and avoid the flattening (Kiselev *et al.*, 1990) that may dramatically alter nucleosome positions.

Lawrence *et al.* (1989) have also reconstructed a negatively stained 30 nm fiber,

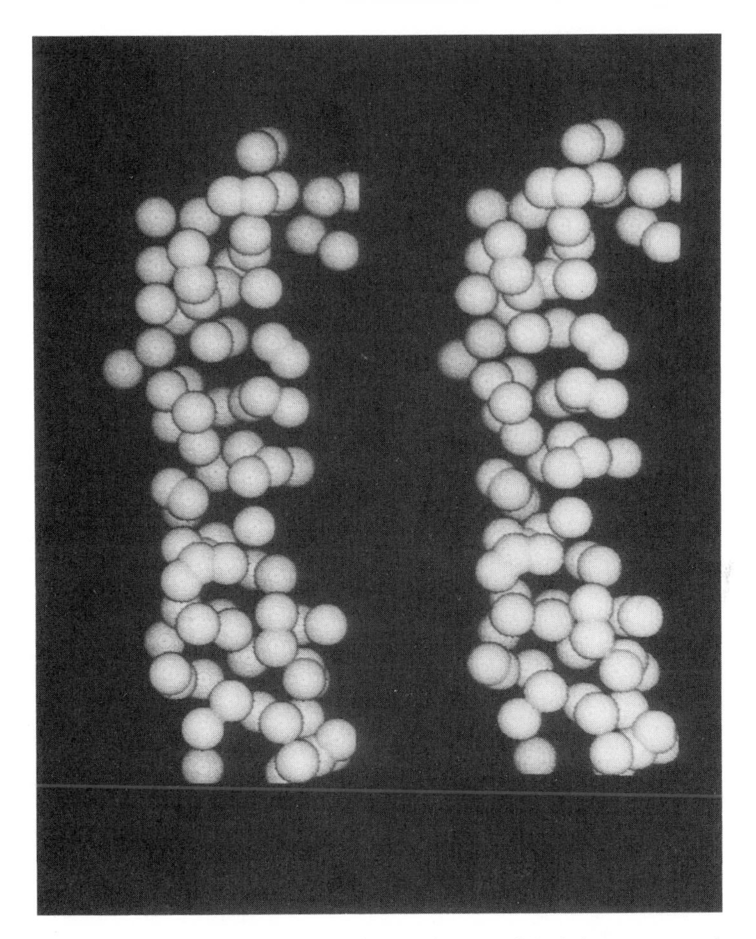

FIGURE 12. Shaded surface representation of the sphere model of the reconstruction of area 1 viewed at 0° (left) and 15° (right). The angular separation allows stereoviewing. (From Woodcock *et al.*, 1990b.)

although the goal of the work was to use chromatin as a test object for a maximum entropy reconstruction method, rather than to study chromatin structure. Fibers were isolated from *Parechinus angulosus* (sea urchin) sperm nuclei and fixed by dialysis against 1% formaldehyde in a buffer containing approximately 65 mM monovalent ions. Carbon grids were pretreated with alcian blue to provide a charged surface for adhesion, floated on a colloidal gold suspension (4–7-nm diameter), the chromatin preparation adsorbed, and methylamine tungstate used as the negative stain. Tilt series micrographs were taken at 120 kV and 30,000 × magnification over a $\pm 60°$ range and 6° tilt increment. Negatives were digitized at an effective pixel size of 0.83 nm. Alignment with gold bead positions was as described (Chapter 8), and the reconstruction carried out both by R-weighted back projection and by an implementation of the maximum entropy approach (see Chapter 5). The tilt data set, and a contour map of the reconstruction are shown in Figs. 15 and 16. A comparison of *xz* slices derived from the maximum entropy

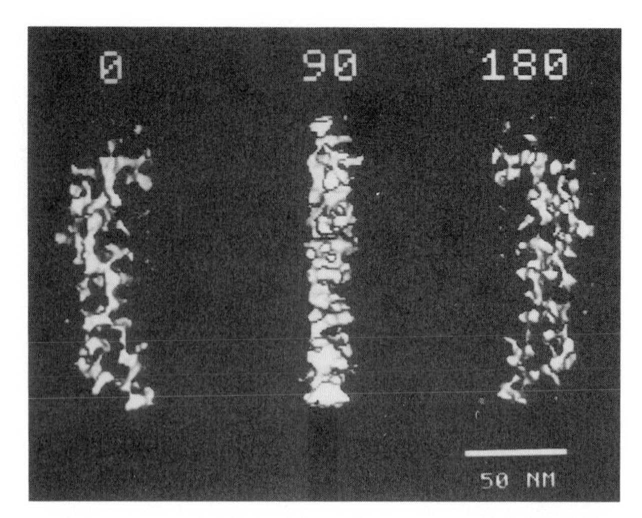

FIGURE 13. Shaded surface representation of the area 1 reconstruction viewed at angles of 0°, 90°, and 180°. The presentation is of little value in understanding the 3D structure of the fiber. (From Woodcock and McEwen, 1988a.)

reconstruction and weighted back projection is shown in Fig. 17. This maximum entropy implementation produces a clear improvement in the visual clarity of the slices, but as the authors point out, without prior knowledge of the structure, it is not possible to compare the veracity of the two methods. However, elsewhere in the paper (Lawrence *et al.*, 1989), reconstructions of a phantom test object showed that, with suitable input parameters, the maximum entropy methods gave a

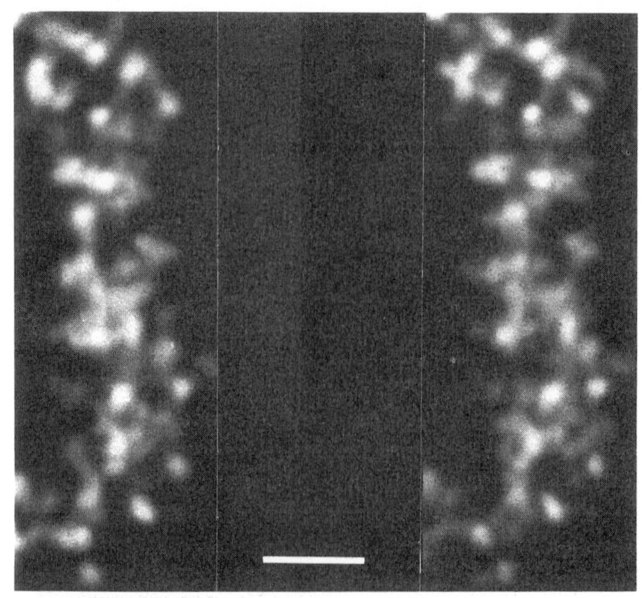

FIGURE 14. Stereo-pair of the reconstruction of area 1 produced by a ray-tracing technique. Bar = 25 nm. (From Woodcock *et al.*, 1990b.)

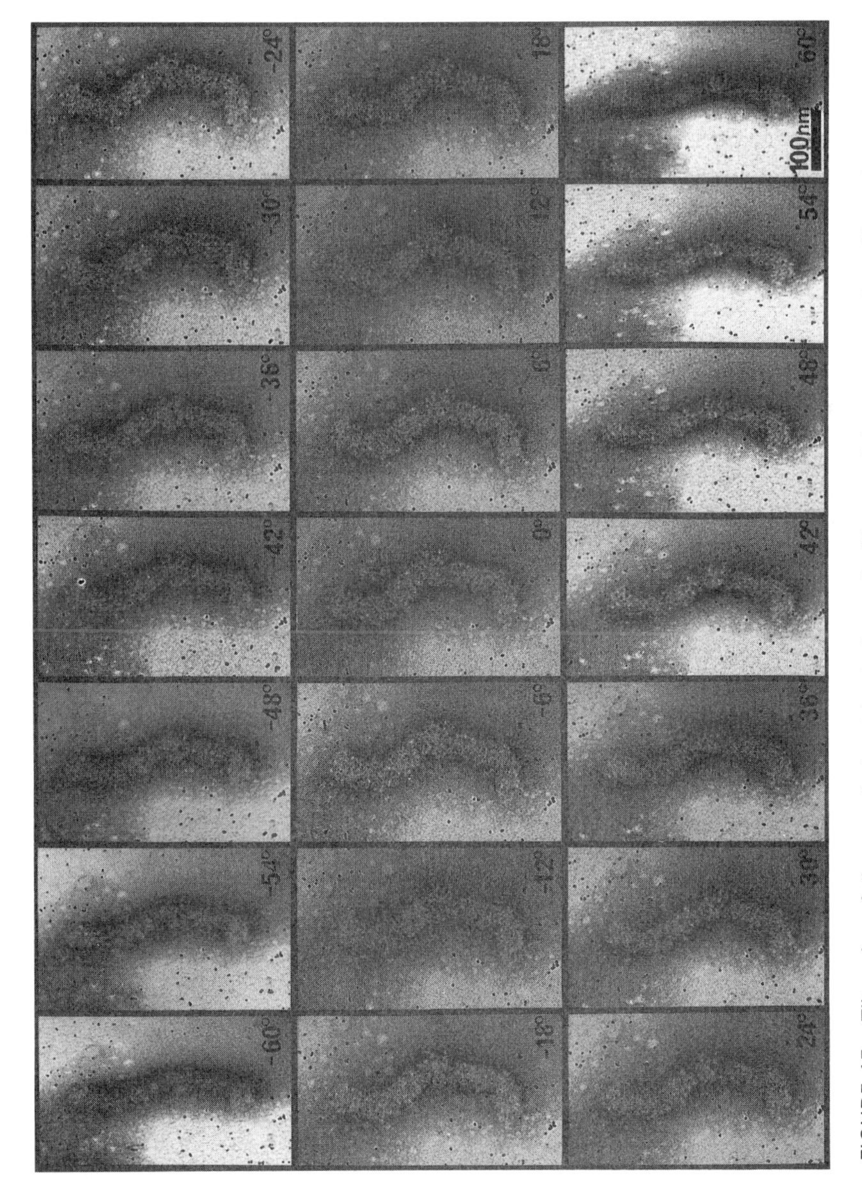

FIGURE 15. Tilt series of the negatively stained chromatin fiber used for reconstruction. (From Lawrence *et al.*, 1989.)

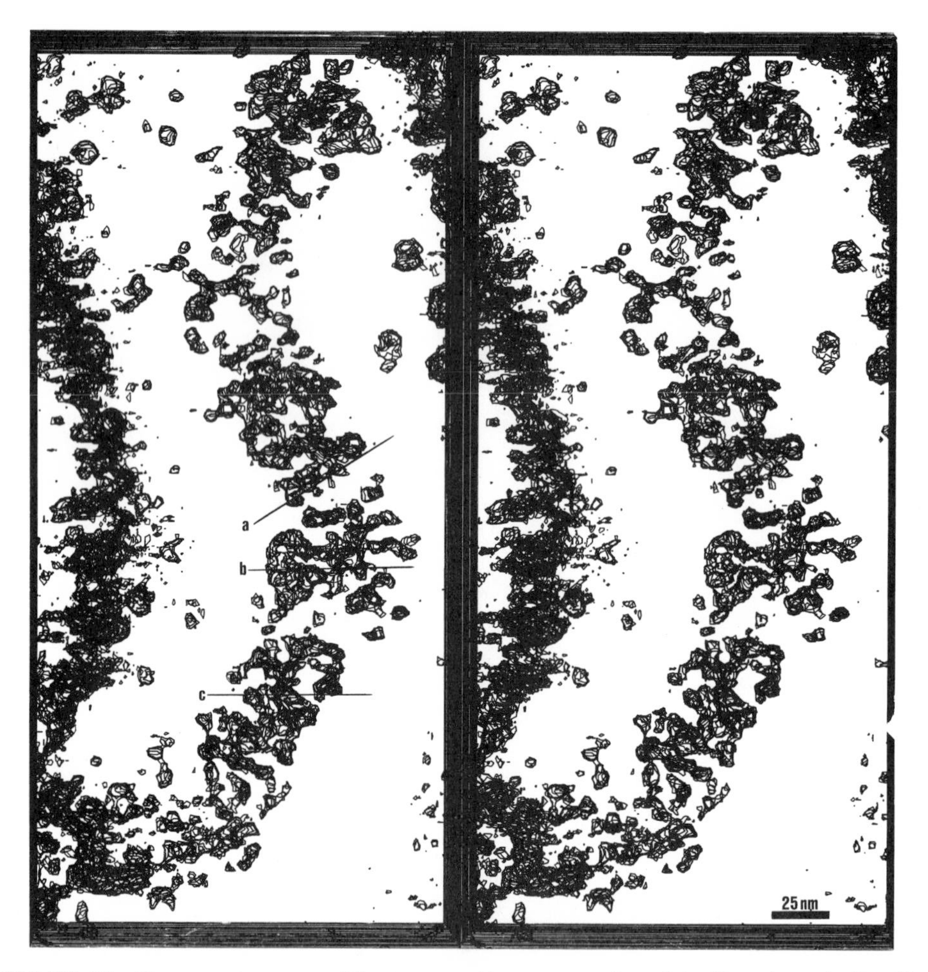

FIGURE 16. Stereo contour map of the tomographic reconstruction of the fiber shown in Fig. 15. The contour level was selected so that the shapes of objects corresponded with those seen in the gray-scale representations. (A)–(C) mark the locations of the vertical sections shown in Fig. 17. (From Lawrence *et al.*, 1989.)

reconstruction having a greater correlation coefficient with the original than did back-projection. The maximum entropy method was, moreover, more immune to random noise.

No ordered arrangement of nucleosomes was detected in the reconstruction, and, as in the studies on *Necturus* (Woodcock and McEwen, 1988a, b; Woodcock *et al.*, 1990b), flattening of the fiber was evident. One interesting feature of the slices perpendicular to the fiber axis (Fig. 17) is that the maximum entropy method accentuates the bright areas on either side of the fiber. Although these are locations at which heavy stain deposition is anticipated, there is also a reconstruction artifact caused by the restricted tilt range that contributes additional contrast at the same sites (Frank and Radermacher, 1986). In the reconstruction of Woodcock *et al.* (1990b), for example, it was found that the amount of stainlike contrast on either

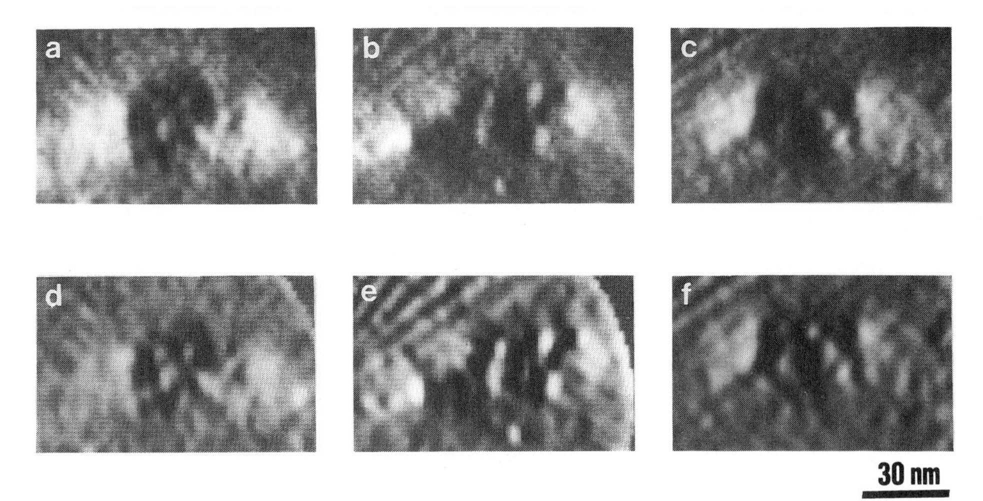

FIGURE 17. Vertical (*xz*) slices through the fiber at the locations marked in Fig. 16. For (a)–(c) the maximum entropy method was used, while (d)–(f) show the result obtained with weighted back-projection. Note that the contrast has been reversed with respect to Fig. 15. (From Lawrence *et al.*, 1989.)

side of the fiber was dependent on the fiber orientation. Where the orientation was perpendicular to the tilt axis, very little "stain" accumulation was observed (Fig. 9*c*). It will be important in future comparisons of reconstruction techniques to examine this phenomenon in more detail.

3.2. Studies on 30-nm Fibers in Sections of Nuclei

The work of Subirana *et al.* (1983, 1985) provided the first 3D view of a 30-nm chromatin fiber. Although reconstructions of four fibers, two from sections and two from negative stains were reported, only the section data from *H. tubulosa* sperm nuclei were described and illustrated in detail.

In order to separate the fibers, nuclei were allowed to swell in 0.25 M sucrose, 0.4 mM $CaCl_2$ before fixing in 2% glutaraldehyde and embedding. Tilt series were obtained at 38,000× magnification over a tilt range of $\pm 54°$ and increment of 6° for the reconstruction discussed here. Micrographs were digitized with a pixel size corresponding to 1.32 nm in the original specimen, and alignment used a cross-correlation procedure (Guckenberger, 1982) rather than the now more common fiduciary marker method. Reconstruction was by weighted back-projection.

Although nucleosomes were not readily visible in the original micrographs (Fig. 18), it was possible using a 3D peak search to detect nucleosome-sized stain densities in the reconstruction, and a model prepared by placing nucleosome-sized spheres at the peak positions (Fig. 19) showed a substantial match with the reconstruction (Fig. 20). The number of nucleosomes per unit length was low (4.8/10 nm) for a fully compact fiber with long linker (Williams *et al.*, 1986), perhaps due to the low ionic strength pretreatment of the nuclei. Again, no consistent pattern of 3D nucleosome positioning emerged. However, a 1D projection of the density along the fiber axis showed distinct maxima and minima with an interval

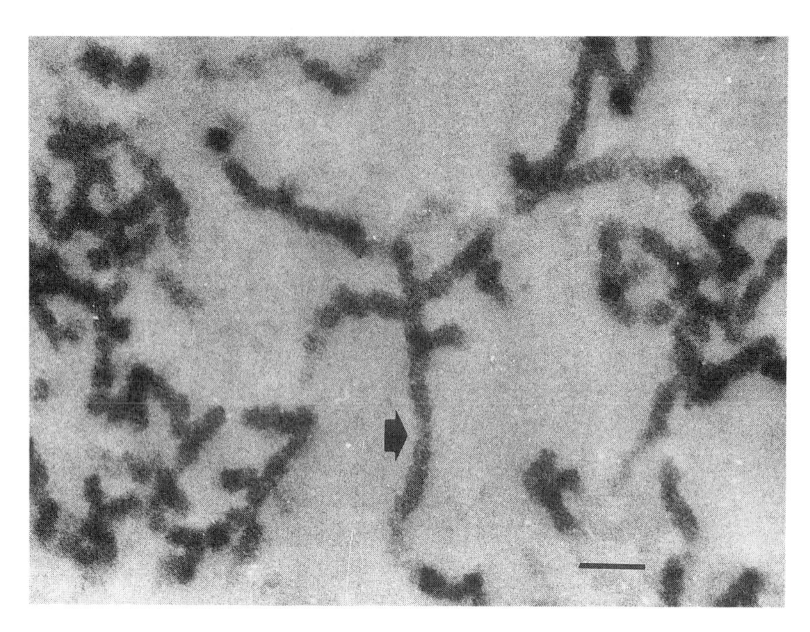

FIGURE 18. Section of sea cucumber sperm nucleus. The arrow denotes the fiber reconstructed in Figs. 19–21. The long axis of the fiber is approximately parallel to the tilt axis. Scale = 100 nm. From Subirana *et al.*, 1985.)

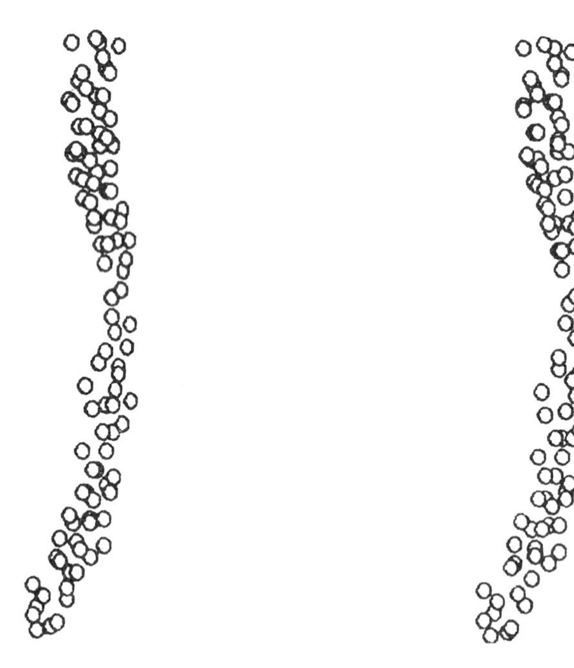

FIGURE 19. Stereo pair of sphere model of the reconstructed chromatin fiber, viewed perpendicular to the plane of the section. (From Subirana *et al.*, 1985.)

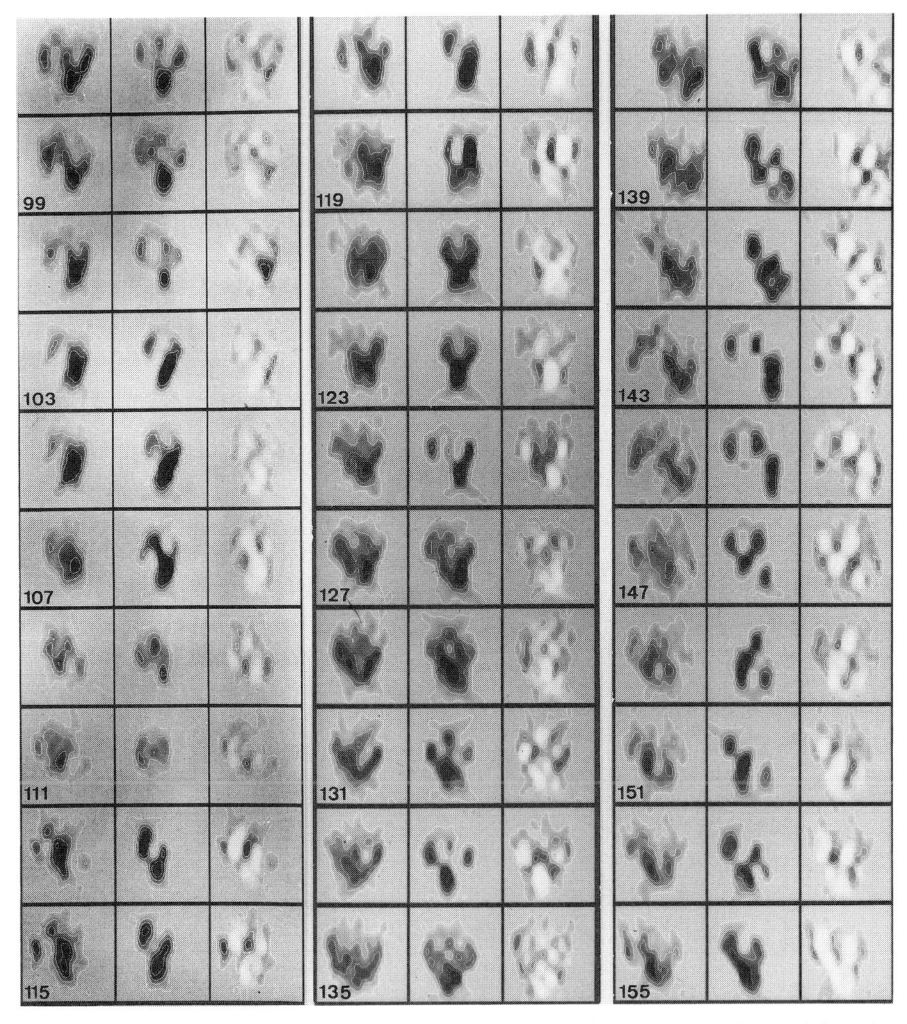

FIGURE 20. Vertical (xz) slices perpendicular to the tilt axis of the reconstruction and the sphere model. Each horizontal row of images shows at left, the reconstruction, in the center, the sphere model, and at right, the difference image. Numbers refer to the y coordinates of the slices. The side of each square corresponds to 43.4 nm. (From Subirana *et al.*, 1985.)

of about 10 nm, leading to the suggestion that the nucleosomes were arranged in disks, forming a layered structure. (But note that it is not necessary to carry out 3D reconstruction to test the generality of the phenomenon.) Subirana *et al.* (1985) also suggested how the individual nucleosomes in the fiber might be connected by linker DNA; however, the linker was not resolved in the reconstructions.

Tomographic reconstructions of chromatin fibers in situ were also described by Olins (1986), using the methods developed by the group at Oak Ridge, and used primarily for studies of active genes. Their unique methodology is discussed in Section 4.1. In the chromatin fiber study, chicken erythrocyte nuclei were prepared in two ways: slightly swollen nuclei provided an example of compact 30-nm fibers

(Fig. 21), and highly swollen nuclei provided unfolded fibers in which individual nucleosomes were clearly visible. Fig. 22 shows slices parallel to the plane of the section from the reconstruction of a region of Fig. 21 (contrast reversed) in which individual fibers are clearly resolved. In addition to the conventional uranyl acetate stain, the compact fibers were contrasted using the DNA-specific osmium ammine reaction (Derenzini *et al.*, 1982). These tomographic results were presented in the context of the general applicability of electron tomography to asymmetric structures, and details of the reconstructed fibers were not given. More recently, Olins *et al.* (1989b) have examined the osmium ammine reaction in detail, and made substantial improvements in its synthesis and usage. This contribution is likely to be extremely valuable in future tomographic studies of nuclear components (e.g., Woodcock *et al.*, 1990a).

Another tomographic study of sectioned chicken erythrocyte chromatin (Horowitz *et al.*, 1988) was initiated in order to maximize the benefits of improvements in fiber preservation achieved using low temperature methods (Horowitz *et al.*, 1990). Nuclei isolated at physiological ionic strength in the presence of spermine and spermidine, were fixed with 3% glutaraldehyde in a buffer containing 20 mM KCl, 20 mM Na cacodylate, pH 7.2, sprayed into liquid propane at −196 °C, freeze substituted with methanol at −90 °C, and embedded in the hydrophilic resin Lowicryl K11M at −50 °C. Sections with a nominal thickness of 100 nm were stained with uranyl and lead, carbon coated, treated with poly-L-lysine, and a suspension of 5-nm colloidal gold particles was applied. Tilt series were obtained at 150 kV (Philips EM 430) and 30,000 × magnification over a range of ±72° with a 1.5° increment. To reduce shrinkage during data collection, the sections were preirradiated for 15 min at the dose used for image capture. Also, the tilt series was built up of four interleaved 6°-increment rotations so that changes

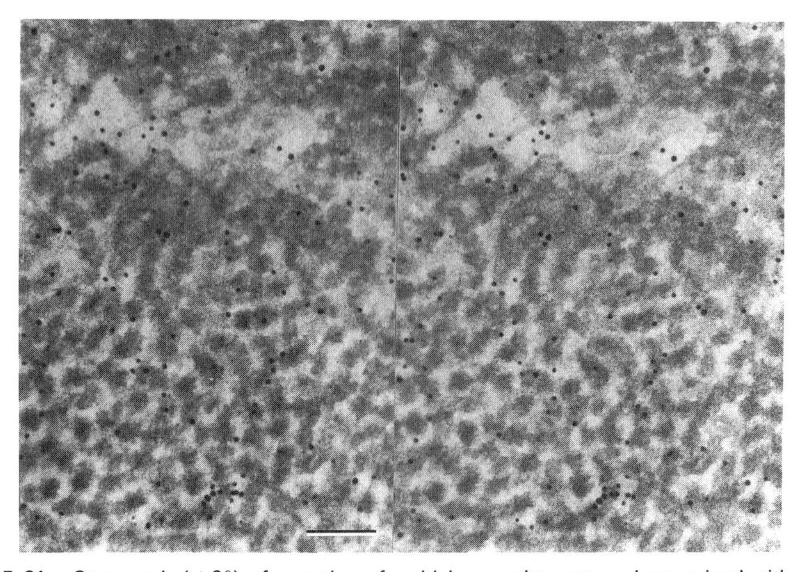

FIGURE 21. Stereo pair (±6°) of a section of a chicken erythrocyte nucleus stained with uranyl and lead salts. Closely packed 30-nm chromatin fibers in different orientations are seen. Bar = 100 nm. (From Olins, 1986.)

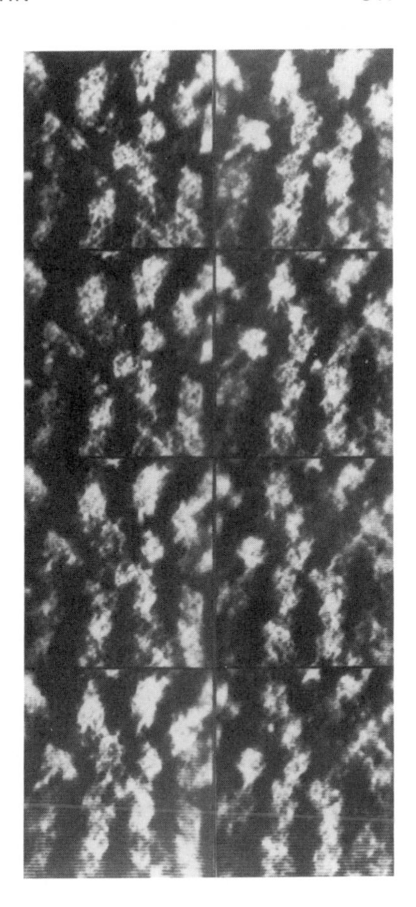

FIGURE 22. Slices parallel to the plane of the section (xy) from the reconstruction of a portion of the section seen in Fig. 21. Note that the contrast has been reversed. (From Olins, 1986.)

occurring over time would spread over the whole series. (A possible disadvantage of this approach is that any inaccuracies in tilt angle due to backlash in the goniometer are accentuated.) A novel aspect of data collection was the replacement of the photographic negative step with direct digitization, similar to the technique developed by the same group for recording images in light microscopy (Hiraoka *et al.*, 1987). For the electron microscopical implementation, a thinned YAG crystal coupled to a CCD camera (Photometrics Inc.) with a 368×576 photosensitive array was mounted below the standard film camera, and, after exposure, images read directly into a computer. Several microscope and camera operations, including goniometer tilt angle, camera readout, and correction of the images for pixel response and bad pixels, were under software control. For tilt data recording, the system developed by Agard and colleagues has the added advantage that after exposure, the electron beam may be blanked off, and the image inspected on a video display. If an alteration in focus is desirable, another image may be captured with minimal cost in electron dose. The effective pixel size at the specimen was 1.8 nm, and the dynamic range of the camera was 4096 levels of gray. Alignment and reconstruction followed the methods developed by Belmont *et al.* (1986), with slight modifications for CCD-derived images. Although complete 1.5° tilt series were collected, alignment of the complete series was often difficult, and in some cases it was advantageous to use only the first two passes through the tilt range, giving a tilt increment of 3°. This was presumably due to specimen warping or changes in microscope parameters during the extended time required to collect a 1.5° set. The behavior of Lowicryl sections during prolonged exposure to electrons is currently being investigated.

As discussed in Section 1.3, the limiting resolution for chromatin fibers in sections depends on the orientation with respect to the tilt axis, the degree of separation of individual fibers, and the maximum tilt angle. In the areas reconstructed (e.g., Fig. 23), some fibers were well separated, and the reconstructions filtered to eliminate spatial frequencies beyond 1/5 nm. In order to study regions of reconstruction with more closely packed fibers, which overlap when projected at high tilt angles, a more stringent filtering would be required.

Examination of the reconstructed volumes showed that although images recorded at the start and end of a tilt series showed no loss of specimen detail, and shrinkage in the plane of the sections was negligible, the reconstructions were about 50% thinner (in z) than anticipated. How much of this is due to shrinkage during electron exposure (Jesior, 1982; Berriman and Leonard, 1986; Luther *et al.*, 1988) and how much to the uncertainty of determining Lowicryl section thickness from interference colors is currently under investigation. It is also important to understand how z-axis shrinkage may affect the structure being studied. A simple linear reduction may be compensated, but a more complex phenomenon involving, for example, surface etching or preferential thinning of resin versus embedded material could have a significant impact on the reconstruction and its interpretation. The most direct way to address this question would be to carry out electron tomography on embedded specimens of known structure and dimensions (see Section 5.2).

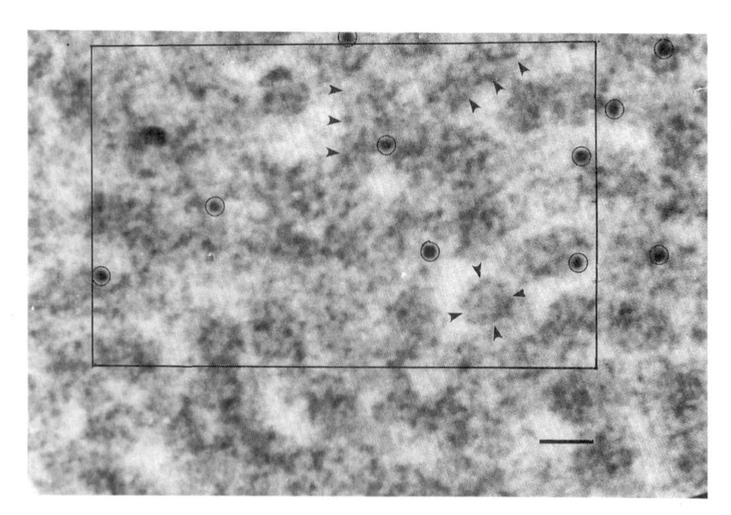

FIGURE 23. 0° tilt view of a section of nucleus recorded with a CCD camera as described in the text. Individual 30-nm fibers (some marked with arrowheads) are indistinct because of superposition in the ~100-nm-thick section. The black circles denote gold particles used as fiduciary markers, and the rectangle outlines a reconstructed region. Bar = 30 nm. (From C. L. Woodcock, R. A. Horowitz, and D. A. Agard, unpublished.)

The major goal of the investigation, to determine the 3D arrangement of nucleosomes within 30-nm fibers, was partially fulfilled. A slice-by-slice examination of the reconstrunctions was used to identify well-separated 30-nm fibers and to determine the trajectories of straight regions. The 3D volumes were then rotated so that transverse and longitudinal sections could be obtained by orthogonal slicing. Stereo pairs were also produced by projecting groups of slices at appropriate angles. In cross section, chromatin fibers appear as circular profiles with a staining pattern suggesting a radial peripheral arrangement of nucleosomes (Fig. 24*a*, *b*), an interpretation supported by central longitudinal views (Fig. 24*c*). Tangential longitudinal views (Fig. 24*c*) often show a regular pattern of staining indicating a close-packed arrangement of nucleosomes. A similar study using *Patiria miniata* (starfish) sperm nuclei has given almost identical results (Woodcock, C. L., Horowitz, R. A., and Agard, D. A., unpublished results).

The problem of adequately displaying these 3D volumes for mental assimilation is more acute with complex sections of nuclei than with the isolated, negatively stained fibers, and most of the common methods of reducing the complexity of the structure (e.g., shaded surface representations) are quite useless. For example, although a radial orientation of nucleosomes was clearly seen, an effective method for localizing and studying the 3D arrangement of putative nucleosome positions has not yet been devised. A promising approach to the problem is to use programs that enable 3D volumes to be rendered with variable levels of opacity and translucency, and viewed as rotating solids (see Chapter 10).

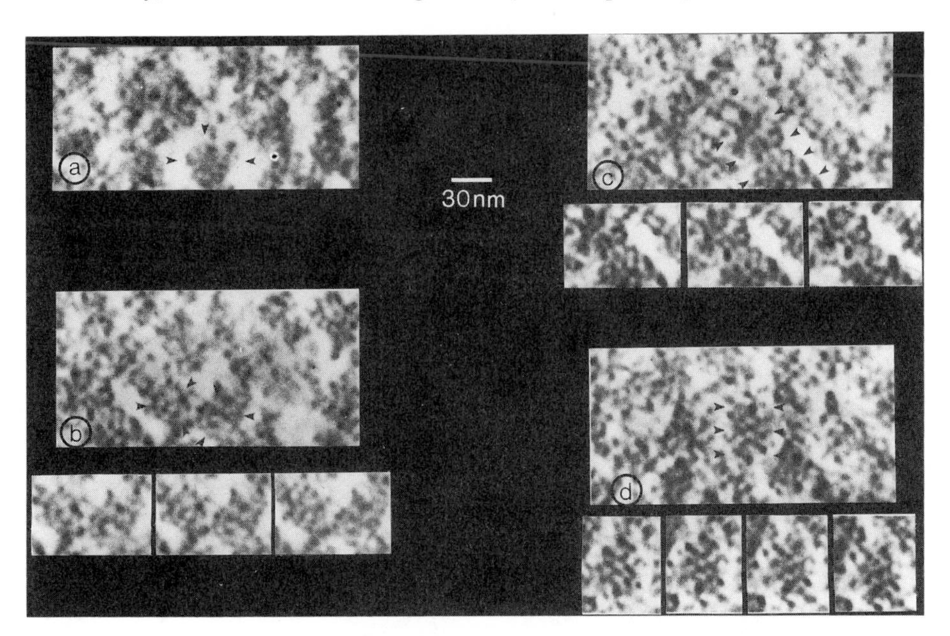

FIGURE 24. Single slices of the reconstruction of chicken erythrocyte nuclei oriented so as to give transverse sections (a, b), a central longitudinal section (c), and a tangential longitudinal section (d) of selected fibers (arrowheads). Inserts below (b–d) are subareas from adjacent 1.8-nm slices of reconstruction. The views support a peripheral location and radial orientation of nucleosomes in the 30-nm fiber. (From C. L. Woodcock, R. A. Horowitz, and D. A. Agard, unpublished.)

4. *TRANSCRIPTIONALLY ACTIVE CHROMATIN AND RNP PRODUCTS*

4.1. Transcription Complexes

The spatial organization of transcription is an intriguing and important topic. Very informative and aesthetically pleasing micrographs have been prepared by centrifuging highly dispersed nuclei with actively transcribing genes onto a suitable support film (Miller and Bakken, 1972). These two-dimensional views of transcription show the linear DNA strands with RNA polymerase molecules and nascent RNP chains attached. In cases where the nascent chains are well separated, nucleosomes are usually present. In vivo, the transcription complexes are in a much more compact form, and the mechanism by which the topical constraints associated with polymerase movement and RNP transport are overcome is of great interest.

Electron tomography provides an approach to the 3D organization of RNA synthesis, and has been effectively used to study transcription complexes and RNP particles. The system chosen by both groups that have worked in this area occurs on the polytene chromosomes of the salivary glands of the fourth instar larvae of *Chironomus tentans*. Intense transcription producing a 37 kb mRNA coding for a large secretory protein, results in "puffs" visible in the light microscope and known as Balbiani rings. Exposure of larvae to the drug pilocarpine increases the message production approximately 10-fold (Olins *et al.*, 1980).

Olins *et al.* (1983), in a landmark paper, reported the first tomographic results from this system, and have subsequently analyzed the transcription complexes in detail (Olins *et al.*, 1984, 1986). Salivary glands were fixed and embedded conventionally using glutaraldehyde fixation, osmium postfixation, and epoxy embedment (Olins *et al.*, 1980, 1982). Following uranyl and lead staining, sections 100 to 200 nm thick were coated with carbon, and colloidal gold particles were applied to both surfaces (Fig. 25). Tilt series micrographs were collected over a $\pm 60^\circ$ range and 5° to 10° increments at $37{,}000\times$ magnification. A vidicon camera was used for

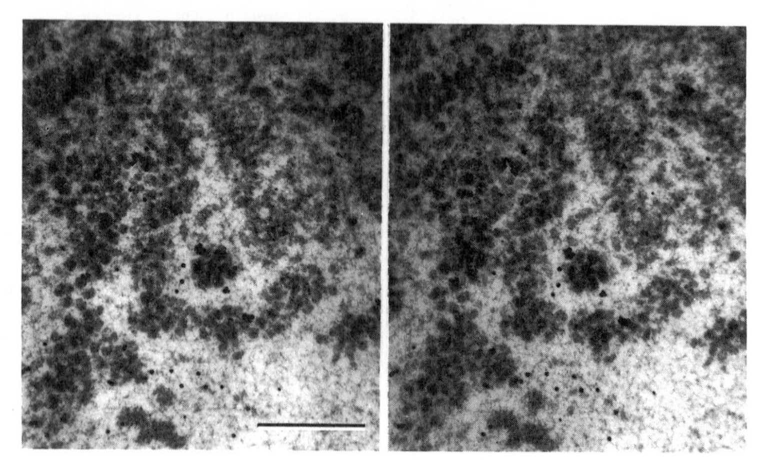

FIGURE 25. Stereo pair showing a typical example of the transcription complexes found in *Chironomus tentans* salivary glands. Bar = 500 nm. (From Olins *et al.*, 1984.)

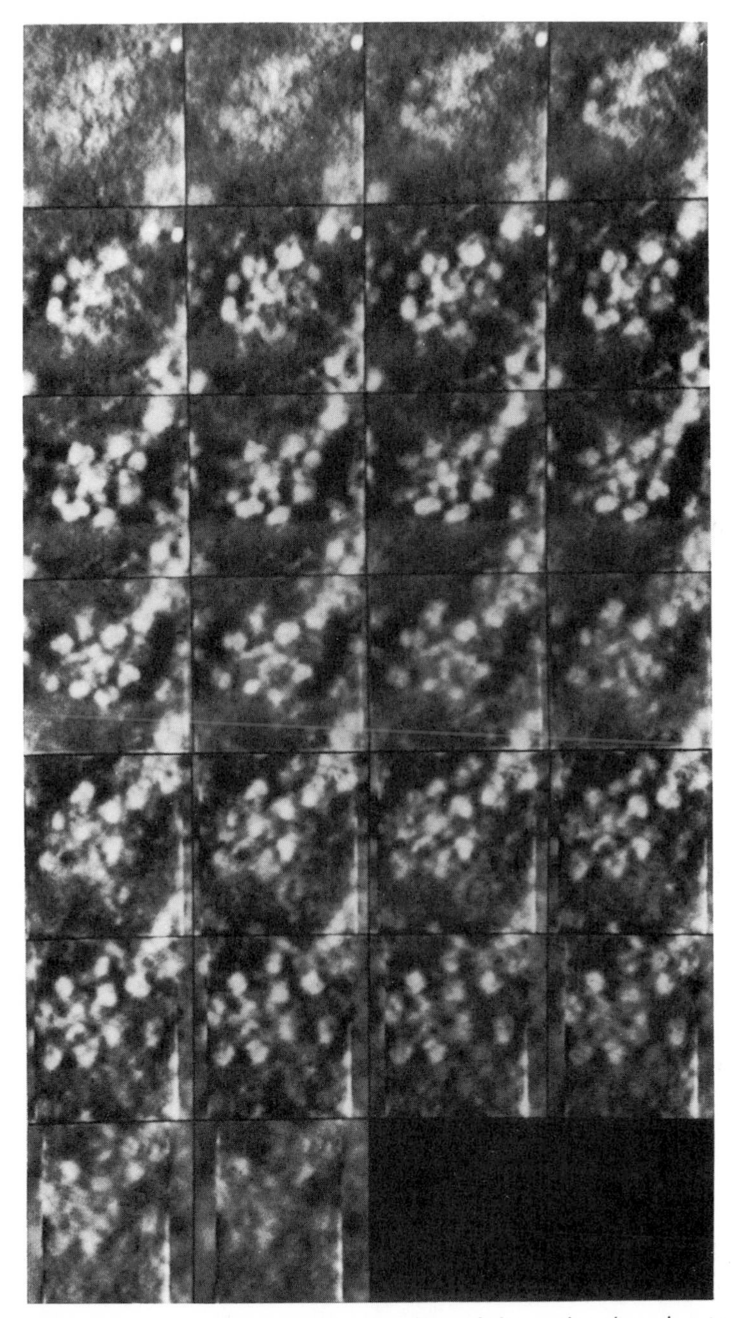

FIGURE 26. Consecutive segments parallel to the plane of the section through a transcription complex similar to that shown in Fig. 25. Individual RNP particles attached to the central axis by thin stalks are clearly seen. Each frame was created by summing two adjacent computed (xy) slices, and is 400 nm on a side and approximately 6 nm thick. (From Olins *et al.*, 1983.)

digitizing the micrographs, with pixels equivalent to about 3 nm at the specimen level. Least-squares analysis of gold bead positions was used to align the series and to provide accurate tilt angles and magnifications. In contrast to other tomographic studies of nuclear material, the reconstruction was carried out by assembling the 2D Fourier transforms of planes perpendicular to the tilt axis from the 1D transforms of the projections and then computing the inverse transform to reconstruct the real-space object (Dover *et al.*, 1981) (see Chapter 5).

Resolution values of 5 nm in the plane of the section, and 7.5 nm in the z direction were estimated (Olins *et al.*, 1983), although application of the Crowther *et al.* (1970) equation to a 5° tilt series of a transcription complex 200 nm in width yields a limiting resolution of about 17 nm. By examining slices of the reconstruction parallel to the plane of the section (Fig. 26), it was possible to trace the path of the fiberlike transcription complexes. Both the slices and balsa wood models created from them (Fig. 27) yielded clear images of the peripheral 40–50 nm RNP granules attached by stalks to the central chromatin axis.

A continuing aim of the work has been to measure the amount of DNA compaction that occurs during Balbiani ring transcription, a value that is related directly to the status of the DNA-histone during transcription (Olins *et al.*, 1984, 1986). It is clear from the size of the central axis that the chromatin is not in a 30-nm fiber conformation, but to distinguish between a fully extended DNA helix and a linear array of nucleosomes requires comparison of the length of the transcription complex with the known length of the gene. The most accurate measurements of the DNA axis length were based on reconstructions of thick (0.5–1.0 μm) sections recorded at 300 kV (Philips EM430), through which the path of the axis was tracked using a program specifically designed for this purpose. The resulting mean DNA compaction ratio of 2.69, S.D. 0.26, indicated a foreshortening amount that was intermediate between nucleosome-free DNA and a linear array of close-packed nucleosomes (Olins *et al.*, 1986). As the authors pointed out, some uncertainties still remain. An improved resolution of the reconstructions (in this

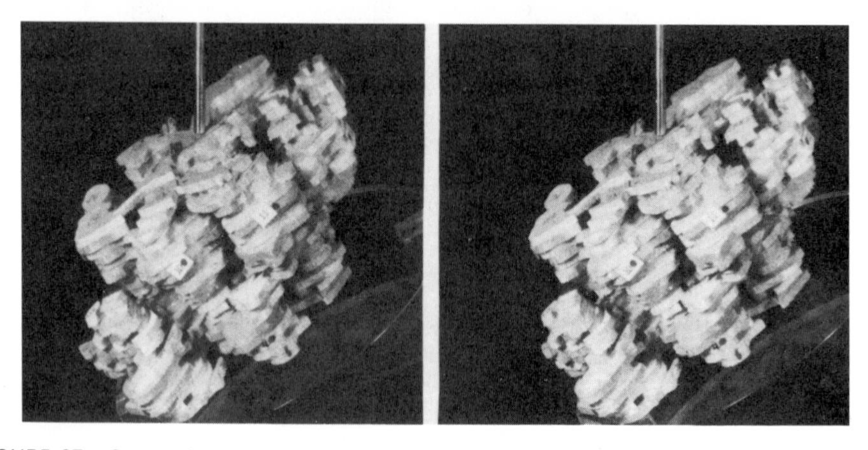

FIGURE 27. Stereo view of a balsa wood model of a portion of a Balbiani ring transcription complex. The model, derived from a tomographic reconstruction, has been oriented so that the loop axis is approximately vertical. (From Olins *et al.*, 1984.)

case estimated to be about 30 nm in the plane of the sections, and 45 nm in the *z* direction), and concomitant accuracy in tracking the axis and detecting the start and end points of transcription, would provide a more accurate value for the length of the axis.

The reconstructions of transcription complexes were also carefully checked for specific helical or other arrangements of the peripheral RNP granules (Olins *et al.*, 1984). Mapping granule positions with respect to distance along the axis and azimuthal angle showed that there were short regions over which a poorly defined helical arrangement could be observed.

In addition to providing the first glimpses of the spatial organization of transcription complexes, the group has introduced a number of technical innovations to electron tomography. One such is the use of a varifocal mirror to display 3D data sets (Levy *et al.*, 1987). In this instrument, a short-persistence phosphor is used to display a number of slices of a 3D image in synchronization with the 30-Hz raster. A reflection of this display is viewed in a mirror formed by a stretched mylar sheet. In synchrony with the display, the mirror is induced to undergo small changes in focal length by a loudspeaker placed behind it, causing the virtual image of the display to be elongated in the *z* direction. The viewer is thus given depth cues for seeing the structure in 3D (Fig. 28). The system may be made interactive for measurement and display operations. It is not yet clear whether this valuable, but rather expensive, tool will become more widespread as interest in tomography expands, or whether the newer generations of graphics workstations specifically

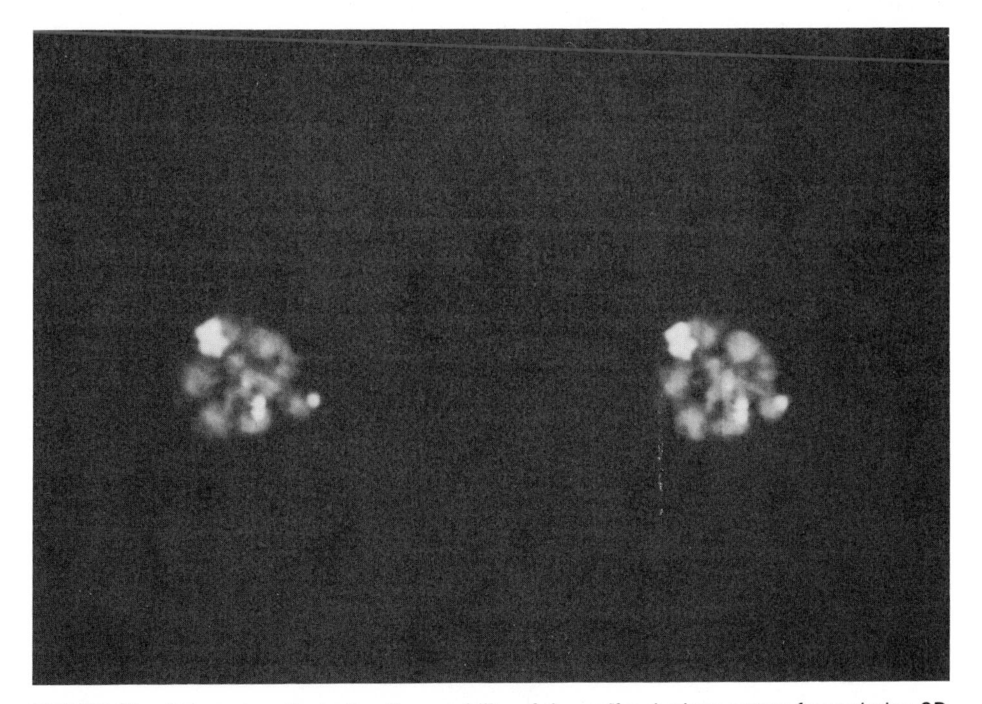

FIGURE 28. Stereo view illustrating the capability of the varifocal mirror system for analyzing 3D data. A portion of a Balbiani ring loop is seen end-on, with gold particles on both sides of the section. (From Levy *et al.*, 1987.)

designed for 3D image manipulation will provide a simpler and more cost-effective route for assimilating 3D structures.

More recently, Olins *et al.* (1989a) have explored the potential of energy-filtered imaging for electron tomography. One of the limitations in using conventional transmission electron microscopes for tomographic studies is that as the specimen thickness is increased, the proportion of inelastically scattered electrons increases. This effect is exacerbated by tilting, the effective specimen thickness increasing about threefold at a tilt angle of 70°. One solution is to use high-voltage or intermediate-voltage microscopes; another is to use a microscope in which specific energy loss windows can be selectively imaged (Ottensmeyer, 1986). Olins *et al.* (1989a) used sections approximately 0.5 μm thick of Balbiani ring material prepared as described above. After coating with colloidal gold particles, the sections were examined at 80 kV in the Zeiss EM 902 energy filtering instrument. The section thickness resulted in almost all transmitted electrons showing substantial energy loss: at 0° tilt, the energy loss spectrum showed a broad peak at about 125 eV, while at 40° the band had broadened, and the peak shifted to about 200 eV. There was a linear relationship between peak energy loss at a given tilt angle and t/t_0, where t_0 is the specimen thickness and t the effective thickness due to tilting. Based on theoretical considerations developed by Collieux *et al.* (1989), tilt series were collected at the peak energy loss value for each tilt angle, using a spectrometer slit with a width of 20 eV. Alignment and reconstruction were as discussed above (Olins *et al.*, 1983). Although the resolution of the reconstructions, based on a magnification of 6000 × and a pixel size of 10 nm^2, was quite low, the final result in terms of the structure of the transcriptional complexes was clearly comparable to that obtained with conventional microscopes, allowing the structure of the micrometer-long transcription complexes to be studied (Olins *et al.*, 1989a). The establishment of energy-filtered electron microscopy as a viable alternative to conventional or intermediate-voltage electron microscopy for collecting tomographic data is an important contribution; its future would be enhanced by the development of instruments equipped with eucentric goniometers and capable of higher accelerating voltages.

4.2. Ribonucleoprotein Particles

Once RNA synthesis is complete, the RNP complexes on the *C. tentans* Balbiani ring puffs are released into the nucleoplasm as conspicuous granules about 50 nm in diameter (Figs. 29, 30). The composition, morphology, and transport of these granules have been extensively studied by B. Daneholt and his colleagues. Electron tomography has been used to derive the 3D structure of the RNP granules in sections (Skoglund *et al.*, 1986) and after isolation (Wurtz *et al.*, 1990).

For the in situ study (Skoglund *et al.*, 1986), salivary gland material was fixed and embedded using conventional methodology, and sections were coated with 5–20-nm-diameter colloidal gold particles. The stained sections were observed at 60 kV and 20,000 × magnification, and tilt series were recorded over a ±60° range with 10° increment. Negatives were digitized with a pixel size corresponding to 1.25 nm at the specimen level, alignment used the gold particle positions, and reconstruction was by weighted back-projection. The resolution of 8–9 nm deter-

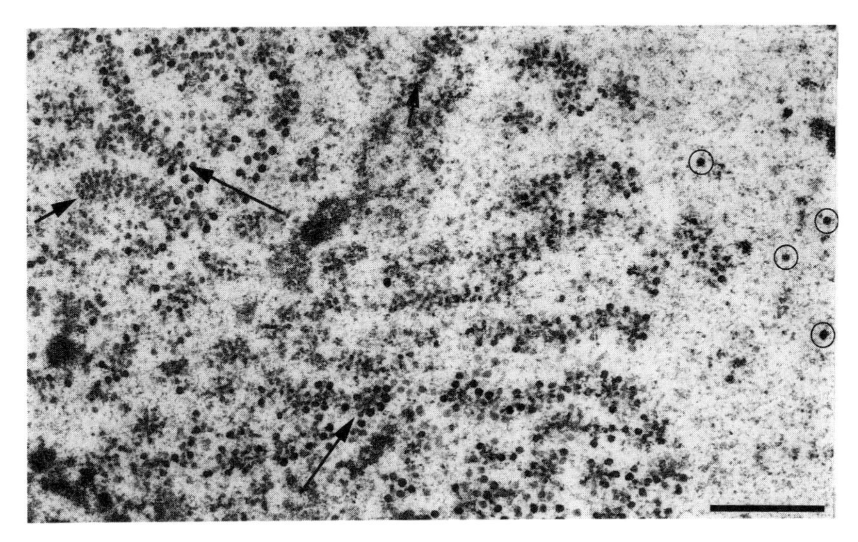

FIGURE 29. Section of a salivary gland nucleus of *Chironomus tentans* showing proximal (short arrows), and distal (long arrows) portions of transcription complexes. Released RNP granules are circled. Bar = 1 micrometer. (From Mehlin *et al.*, 1988.)

mined by the authors from characteristics of the reconstructed density maps is in accord with the formula of Crowther *et al.* (1970).

Since the goal of the experiments was to understand the overall shape of the granules rather than any internal structural detail, processing of the 3D volumes focused on delineating the edges of the structure. Balsa wood models of four particles were constructed from the density maps, the threshold between particle and background being calculated on the basis of significant deviations from the mean density (Fig. 31). To test for similarities among the reconstructed particles, a

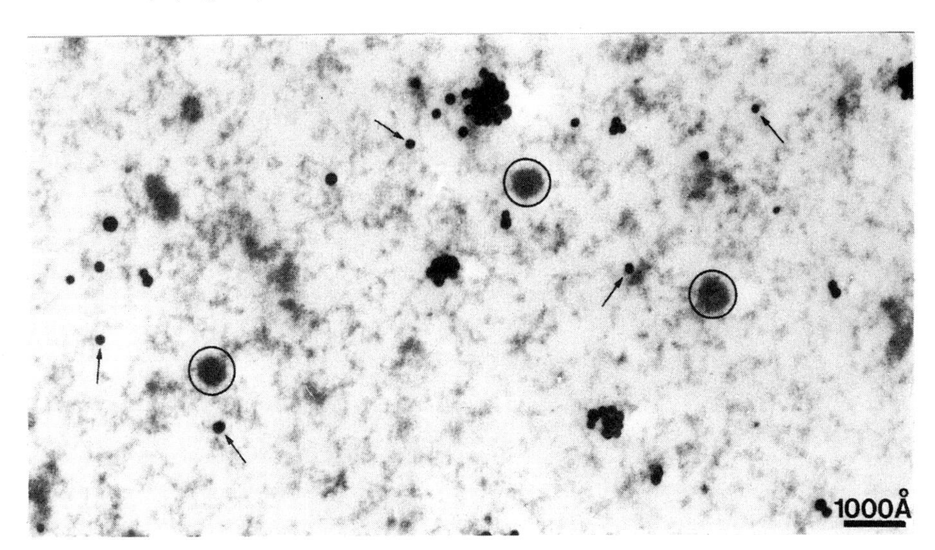

FIGURE 30. Detail of section showing RNP granules (circled) and adsorbed colloidal gold (arrows). (From Skoglund *et al.*, 1986.)

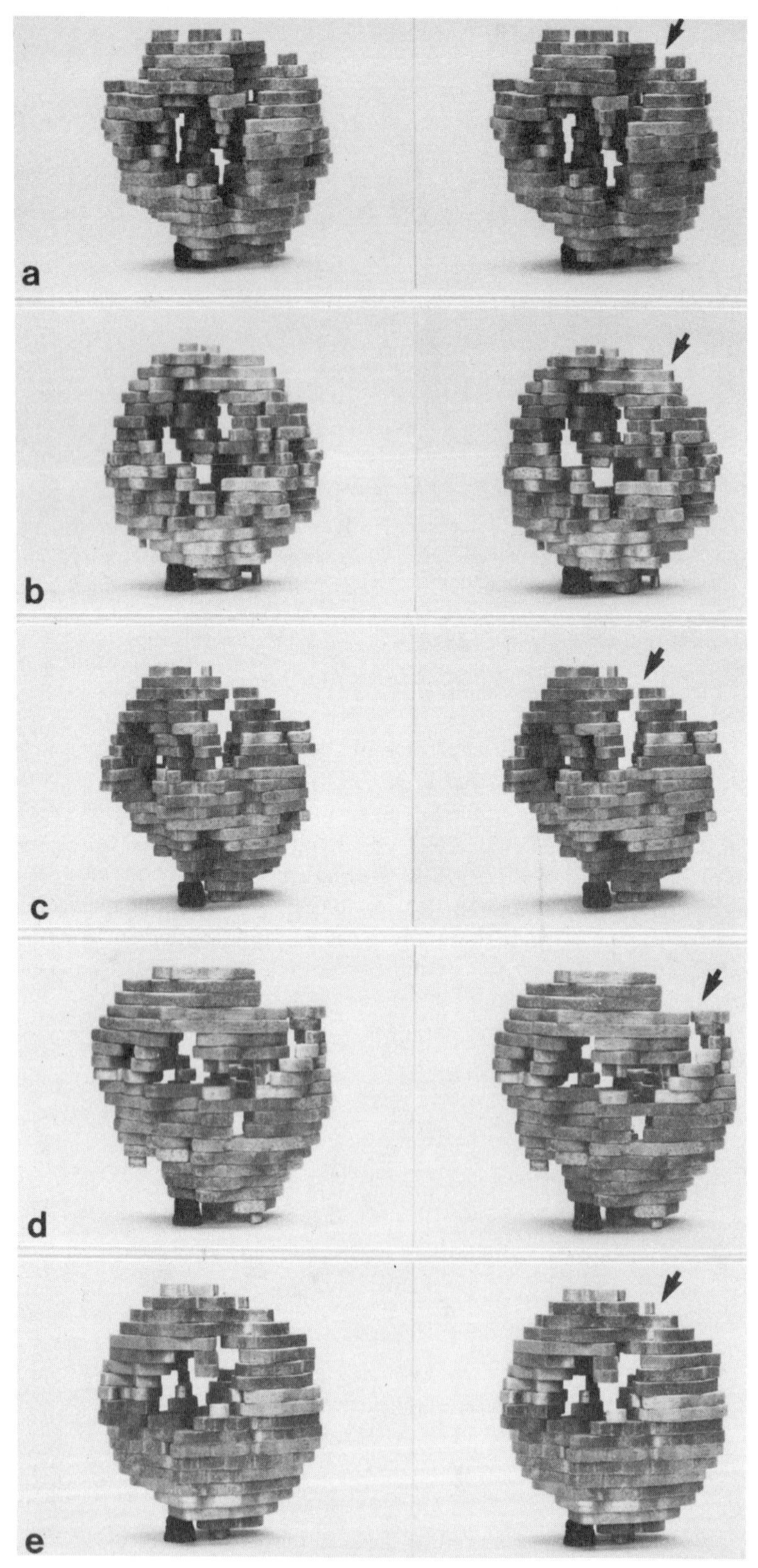

FIGURE 31. Stereo views of balsa wood models of four reconstructed RNP particles (a–d), aligned to a common orientation. (e) shows the computed average. Arrows denote the position of the slit in the toroid-like structure. (From Skoglund *et al.*, 1986.)

least-squares procedure was used to maximize the correlation coefficients between a reference particle and a second particle during 3D rotation and translation operations. By this method, well-defined maximal correlation coefficients of 0.75, 0.77, and 0.82 were obtained for three test particles with respect to the reference. The four particles were then oriented to place them in register and averaged (Fig. 31).

Earlier thin-section studies (Skoglund *et al.*, 1983) had indicated that the mature RNP granule was formed from an elongated RNP chain that is bent into a ring- or horseshoe-shaped conformation, assuming an almost spherical shape. Detailed examination of the averaged particle confirmed this view, and also suggested that four morphological domains were present (Skoglund *et al.*, 1986). It is difficult to confirm this interpretation from the model photographs presented (Fig. 31). Further tomographic investigations of the unfolding of the RNP granules as they are transported through the nuclear pores (Mehlin *et al.*, 1988) should be especially interesting.

In a recent characterization of isolated RNP granules, Wurtz *et al.* (1990) purified 300S particles from salivary cells and prepared tomographic reconstructions. A sucrose gradient fraction shown to contain 300S particles was diluted and layered onto 100 μL of buffered 0.5 M sucrose and 2% glutaraldehyde in a plastic microchamber. Then centrifugation was used to deposit the particles onto an EM grid at the bottom of the chamber. Subsequently, the grids were removed, washed with 0.4% Photoflo, and stained with ethanolic phosphotungstic acid. This method is based upon the technique described by Miller and Bakken (1972) for displaying actively transcribing genes. After staining, colloidal gold particles were added, and tomographic reconstruction was carried out as described above. It was found that the isolated particles shared the same basic morphology as the in situ ones. There were, however, some noticeable differences that can be ascribed to the method of preparation: after isolation, the granules were 20% smaller, had a less well-defined hollow center, and the portions facing the support film were somewhat distorted. The main goal of the tomography, to show that the 300S particles represented the Balbiani RNP granules, was clearly achieved.

5. CONCLUSIONS AND PROSPECTS

The past seven or so years have seen the very beginnings of what promises to be an increasingly informative and exciting tomographic exploration of the cell nucleus. Even with the limited number of studies to date, some common themes are emerging. Firstly, and in some senses rather disappointingly, electron tomography has not revealed any previously obscured structural order. On the contrary, for some components, such as mitotic chromosomes and 30-nm chromatin fibers, in which a long-range structural order was anticipated, these first tomographic results have suggested more complex internal architectures. A second, related theme is the problem of preserving these fragile components, a problem that becomes more acute for tomographic work than for 2D structural investigations. With no clear consensus for "best preparation method," the only way to understand the interrelationships between the different organizational levels of, for example, mitotic chromosomes will be to use a variety of preparation methods. In the case of the isolated 30-nm chromatin fiber, tomography has pointed out some obvious shortcomings of current preparation methods, and progress will require the exploration of alternative techniques to overcome or reduce these. In cases where the goals of the tomograhic study include detailed internal information, such as the substructure of compact chromosomes or 30-nm fibers in sections, interpretation becomes limited by the contrasting process. The presence of stain, whether from uranyl or lead salts or both, cannot be definitely correlated with a particular component. In this context, the development of stains that are specific for DNA (Derenzini *et al.*, 1982; Olins *et al.*, 1989b) becomes extremely important.

In order to make rapid progress, more groups of investigators than the present handful need to be attracted to the field. This will require a wider recognition of the importance of understanding the 3D spatial organization within the nucleus, and more importantly, a greater accessibility of tomographic methodology. Until recently, a daunting list of required expertises and instrumentation faced the prospective tomographer. Most critical was a thorough background in the theoretical underpinnings of tomography and computer image processing. It is now possible to obtain packages of programs that can be implemented by nonspecialists. As with all emerging technologies, there is the danger that use by the nonspecialist will lead to spurious results, but it is quite simple to set up a system of internal checks and standards to validate reconstructions. As discussed below, a set of standards would also serve to select the most appropriate reconstruction scheme for a specific project.

5.1. Instrumentation

In addition to the theoretical aspects of tomographic reconstruction, there are instrumental hurdles to be overcome by the beginning investigator. For tilt data recording, a eucentic goniometer stage is almost mandatory, and is available for most side-entry specimen holders. High-tilt holders are now commercially available that extend the angular range to $\pm 80^\circ$. For most specimens, the grid openings effectively restrict data collection to $\pm 70^\circ$, but the extra information in the z direction, as compared with a $\pm 60^\circ$ series, can be substantial (Baumeister *et al.*, 1986).

Even with modern tilt stages and holders, the collection and digitization of data remains a slow and tedious process, and skill is needed to keep a consistent focus through a tilt series. It is unfortunate that large-array (e.g., 1024×1024) CCD cameras with a wide dynamic range (Hiraoka *et al.*, 1987) are still not commercially available as options on electron microscopes. As described in Section 3.2, the system implemented by Agard and colleagues, in which a CCD camera is interfaced with a computer and software controls microscope parameters such as tilt angle, greatly simplifies the collection of a tilt series and reduces the electron dose absorbed by the specimen. More importantly, data collected directly from the microscope via a CCD camera is more accurate than that obtained with photographic film. The latter introduces inaccuracies through variations in development, drying, deviation from linearity in dose response, and background fog levels. Also, during digitization of a series of photographic negatives, the temporal stability of the light source limits the effective dynamic range. It is to be hoped that CCD systems suitable for the collection of tomographic data will soon become available as options on commercial electron microscopes.[†]

Once images are stored in the computer, alignment using gold bead positions may be partially automated, and most available programs provide a quantitative measure of alignment accuracy. This procedure, as well as subsequent reconstruction and processing, is greatly facilitated by image processing environments such as SPIDER (Frank *et al.*, 1981) or IMAGIC (van Heel and Keegstra, 1981) in which such housekeeping tasks such as file naming and creating lists of tilt angles for passing from one algorithm to the next can be carried out automatically. The use of standardized test images, discussed below in the context of assessing different reconstruction methods, would also be valuable for verifying the correct implementation of algorithms.

Although data collection, alignment, and reconstruction are exacting procedures, it is the last stages when the 3D structure must be assimilated mentally and the salient features extracted and presented that are, at present, the most time consuming. It is frustrating to have reconstructed a specimen but still be unable to comprehend it in 3D. Here, the rapid evolution of the power and storage capacity of moderately priced graphics workstations promises to simplify this stage of tomography. For example, with some currently available systems (such as Voxel-View by Vital Images Inc.), it is possible to roam interactively through a 3D structure in selected directions in real time, allowing the tracing of complex paths of fibers and other structures. Whole volumes may be rendered as semitransparent objects and viewed at selected angles or as a rotating solid.

In the ideal tomography laboratory, the microscope and computer should be integrated, the images coming from a CCD camera directly into the framework of a single image-processing system with sophisticated 3D display capabilities. Such systems are now feasible, and their commercial availability would make electron tomography accessible and attractive to a wider range of investigators.

5.2. Reconstruction Techniques

Even within the narrow limits of tomographic studies on nuclear components, a number of different approaches to reconstruction have been used. In addition to the most common method of R-weighted back-projection, exact filters (Harauz and van Heel, 1986; Radermacher, 1986), maximum entropy (Lawrence *et al.*, 1989), and Fourier reconstructions (Dover *et al.*, 1981) have been used. For the non-specialist, it would be very useful to establish a set of standard test objects and conditions for assessing the different methods. Very possibly, the best reconstruction method will differ according to the aims of the reconstruction project and the limitations of the data set. In electron tomography, as the stage is tilted, new regions of specimen become included in the image. The impact of this problem, which will vary with the type of specimen, should also be assessed with test images.

A set of internal standards would help to assess and minimize the potential problems caused by specimen shrinkage in the z direction. The extend of this effect, especially with plastic sections, has been recognized for some time (Jesior, 1982;

[†] *Note added in proof*: One such system has recently been introduced.

Berriman and Leonard 1986; Luther *et al.*, 1988). At best, when preirradiation is used to allow shrinkage to occur before data collection, both the specimen, and its resulting reconstruction are foreshortened along the *z* axis. On the other hand, if shrinkage occurs during the collection of a data set, or if it affects the specimen in a nonlinear way, the resulting reconstruction may be profoundly affected. An accurate determination of specimen thickness before and after preirradiation and data collection (from, e.g., gold bead positions) would help to document shrinkage, but an ultimate standardization would require the embedding and reconstruction of specimens of known shape and dimensions. The internal standards should match the size of the objects being reconstructed. Well-characterized viruses such as TMV might be suitable for, e.g., 30-nm fibers, while axoneme segments would be more appropriate for transcription complexes. Suspensions containing the standards could be added prior to fixation and serve to verify the quality and resolution of the reconstruction, as well as provide an internal length calibration.

In summary, many important topics relating to the spatial organization of the genetic material within the eukaryotic nucleus and pertinent to its functioning remain to be investigated. Electron tomography is a viable, and in some instances the only possible, method of approach. However, progress will be influenced by the rapidity with which current instrumentation can be integrated into a streamlined unit.

ACKNOWLEDGMENTS

I thank those colleagues who generously supplied prints from published and unpublished work, and A. S. Belmont, R. A. Horowitz, A. L. Olins, and D. E. Olins for critically reading the manuscript. Some of the unpublished work of the author was supported by NSF BBS-8714235 and NIH GM-43786.

REFERENCES

Adolph, K. W. (1980). Organization of chromosomes in mitiotic HeLa cells. *Exp. Cell Res.* **125**:95–103.

Bahr, G. F. and Golomb, H. M. (1971). Karyotyping of single human chromosomes from dry mass determined by electron microscopy. *Proc. Nat. Acad. Sci. USA* **68**:726–730.

Baumeister, W., Barth, M., Hegerl, R., Guckenberger, R., Hahn, M., and Saxton, W. O. (1986). Three-dimensional structure of the regular surface layer (HPI layer) of *Deinococcus radiodurans*. *J. Mol. Biol.* **187**:241–253.

Berriman, J. and Leonard, K. R. (1986). Methods for specimen thickness determination in electron microscopy. II:Changes in thickness with dose. *Ultramicroscopy* **19**:349–366.

Belmont, A. S., Sedat, J. W., and Agard, D. A. (1986). A three-dimensional approach to mitotic chromosome structure: evidence for a complex hierarchical organization. *J. Cell Biol.*, **105**:77–92.

Borland, L., Harauz, G., Bahr, G., and van Heel, M. (1988). Packing of the 30 nm chromatin fiber in the human metaphase chromosome. *Chromosoma (Berl)* **97**:159–163.

Collieux, C., Mory, C., Olins, A. L., Olins, D. E., and Tence, M. (1989). Energy filtered STEM imaging of thick biological sections. *J. Microsc.* **153**:1–21.

Crowther, R. A., DeRosier, D. J., and Klug, A. (1970). The reconstruction of a three-dimensional structure from projections and its application to electron microscopy. *Proc. R. Soc. London A* **317**:319–340.

Derenzini, M., Viron, A., and Puvion-Dutilleul, F. (1982). The Feulgen-like osmium-ammine reaction as a tool to investigate chromatin structure in thin sections. *J. Ultrastruct. Res.* **80**:133–147.

Dover, S. D., Elliott, A., and Kernagam, A. K. (1981). Three-dimensional reconstruction from images of tilted specimens: the paramyosin filament. *J. Microsc.* **122**:23–33.

DuPraw, E. J. (1965). Macromolecular organization of nuclei and chromosomes: a folded fibre model based on whole mount electron microscopy. *Nature* **206**:338–343.

Earnshaw, W. C. and Laemmli, U. K. (1983). Architecture of metaphase chromosomes and chromosome scaffolds. *J. Cell Biol.* **96**:84–93.

Eickbusch, T. H. and Moudrianakis, E. N. (1978). The compaction of DNA helices into either continuous supercoils or folded fiber rods and toroids. *Cell* **13**:295–306.

Felsenfeld, G. and McGhee, J. D. (1986). Structure of the 30 nm chromatin fiber. *Cell* **44**:375–377.

Flannigan, D. and Harauz, G. (1989). Three-dimensional reconstruction of a polytene chromosome segment from Drosophila melanogaster, in *Proc. Microscopical Soc. Canada*, Vol. XVI, pp. 90–91.

Frank, J., McEwen, B. F., Radermacher, M., Turner, J. N., and Rieder, C. L. (1987). Three-dimensional tomographic reconstruction in high voltage electron microscopy. *J. Electron Microsc. Tech.* **6**:193–205.

Frank, J. and Radermacher, M. (1986). Three-dimensional reconstruction of non-periodic macromolecular assemblies from electron micrographs, in *Advanced Techniques in Biological Electron Microscopy* (J. Koehler, ed.), pp. 1–72. Springer-Verlag, Berlin.

Frank, J., Shimkin, B., and Dowse, H. (1981). SPIDER—a modular software system of electron microscopy image processing. *Ultramicroscopy* **6**:343–358.

Gordon, R. and Herman, G. T. (1974). Three dimensional reconstruction from projections: a review of algorithms, *Int. Rev. Cytol.* **38**:111–151.

Graziano, V., Gerchman, S. E., and Ramakrishnan, V. (1988). Reconstitution of chromatin higher order structure from histone H5 and depleted chromatin, *J. Mol. Biol.* **203**:997–1007.

Guckenberger, R. (1982). Determination of the common origin in the micrographs of tilt series in three-dimensional electron microscopy, *Ultramicroscopy* **9**:167, 174.

Harauz, G., Borland, L., Bahr, G. F., Zeitler, E., and van Heel, M. (1987). Three-dimensional reconstruction of a human metaphase chromosome from electron micrographs, *Chromosoma* **95**:366–374.

Harauz, G. and van Heel, M. (1986). Exact filters for general geometry three dimensional reconstruction, *Optik* **73**:146–156.

Hiraoka, Y., Sedat, J. W., and Agard, D. A. (1987). The use of a charge-coupled device for quantitative optical microscopy of biological structures, *Science* **238**:36–41.

Horowitz, R. A., Giannasca, P. J., and Woodcock, C. L. (1990). Ultrastructural preservation of chromatin and nuclei: improvement with low temperature methods, *J. Microsc.* **157**:205–224.

Horowitz, R. A., Woodcock, C. L., Belmont, A. W., and Agard, D. A. (1988). 3D reconstruction of chromatin fibers in freeze-substituted nuclei. *J. Cell Biol.* **107**:313a.

Hutchison, N. and Weintraub, W. (1985). Localization of DNAase I-sensitive sequences to specific regions of interphase nuclei, *Cell* **13**:471–482.

Jesior, J-C. (1982). The grid sectioning technique: a study of catalase crystals, *EMBO J.* **1**:1423–1428.

Kellenberger, E. (1987). The compactness of cellular plasmas; in particular, chromatin compactness in relation to function, *TIBS* **12**:105–107.

Kiselev, N. A., Sherman, M. B., and Tsuprun, V. L. (1990). Negative staining of proteins. *Electron Microsc. Rev.* **3**:43–72.

Klug, A. and Crowther, R. A. (1972). Three-dimensional image reconstruction from the viewpoint of information theory. *Nature* **238**:435–440.

Langmore, J. P. and Paulson, J. R. (1983). Low-angle X-ray diffraction studies of chromatin structure in vivo and in isolated nuclei and metaphase chromosomes. *J. Cell Biol.* **96**:1120–1131.

Langmore, J. P. and Schutt, C. (1980). The higher order structure of chicken erythrocyte chromosomes in vivo. *Nature (London)*, **288**:620–622.

Lawrence, M. C., Jaffer, M. A., and Sewell, B. T. (1989). The application of the maximum entropy method to electron microscopic tomography. *Ultramicroscopy* **31**:285–302.

Levy, H. A., Margle, S. M., Tinnell, E. P., Durfee, R. C., Olins, D. E., and Olins, A. L. (1987). The varifocal mirror for 3-D display of electron microscope tomography. *J. Microsc.* **145**:179–190.

Luther, P. K., Lawrence, M. C., and Crowther, R. A. (1988). A method for monitoring the collapse of plastic sections as a function of electron dose. *Ultramicroscopy* **24**:7–18.

McDowall, A. W., Smith, J. M., and Dubochet, J. (1986). Cryo-electron microscopy of vitrified chromosomes in situ. *EMBO J.* **5**:1395–1402.

Marsden, M. P. F. and Laemmli, U. K. (1979). Metaphase chromosome structure: evidence for a radial loop model. *Cell* **17**:849–858.

Mehlin, H., Lonnroth, A., Skoglund, U., and Daneholt, B. (1988). Structure and transport of a specific premessenger RNP particle *Cell Biol. Int. Rep.* **12**:729–736.

Miller, O. L. Jr., and Bakken, A. H. (1972). Morphological studies on transcription *Acta Endocrinol. Suppl.* **168**:155–177.

Olins, A. L. (1986). Electron microscope tomography: 3-D reconstruction of asymmetric structures, in *Proc. 44th Ann. Mg. Electron Microscop. Soc. America* (G. W. Bailey, ed.), pp. 22–25. San Francisco Press, San Francisco.

Olins, A. L., Moyer, B., Sook-Hui, K., and Allison, D. P. (1989b). Synthesis of a more stable osmium ammine electron-dense DNA stain. *J. Histochem. Cytochem.* **37**:395–398.

Olins, A. L., and Olins, D. E. (1974). Spheroid chromatin units (nu bodies). *Science* **183**:330–332.

Olins, A. L., Olins, D. E., and Franke, W. W. (1980). Stereo-electron microscopy of nucleoli, Balbiani rings and endoplasmic reticulum in *Chironomus* salivary gland cells. *Eur. J. Cell Biol.* **22**:714–723.

Olins, A. L., Olins, D. E., and Lezzi, M. (1982). Ultrastructural studies of *Chironomus* salivary gland cells in different states of Balbiani ring activity. *Eur. J. Cell Biol.* **27**:161–169.

Olins, A. L., Olins, D. E., Levy, H. A., Durfee, R. C., Margle, S. M., and Tinnell, E. P. (1986). DNA compaction during intense transcription measured by electron microscope tomography. *Eur. J. Cell Biol.* **40**:105–110.

Olins, A. L., Olins, D. E., Levy, H. A., Durfee, R. C., Margle, S. M., Tinnell, E. P., Hingerty, B. E., Dover, S. D., and Fuchs, H. (1984). Modeling balbiani ring gene transcription with electron microscope tomography. *Eur. J. Cell Biol.* **35**:129–142.

Olins, A. L., Olins, D. E., Levy, H. A., Margle, S. M., Tinnell, E. P., and Durfee, R. C. (1989a). Tomographic reconstructions from energy-filtered images of thick biological sections. *J. Microsc.* **154**:257–265.

Olins, D. E., Olins, A. L., Levy, H. A., Durfee, R. C., Margle, S. M., Tinnell, E. P., and Dover, S. D. (1983). Electron microscope tomography: transcription in three dimensions. *Science* **220**:498–500.

Ottensmeyer, F. P. (1986). Elemental mapping by energy filtration: advantages, limitations, compromises. *Ann. NY Acad. Sci.* **483**:339–353.

Paulson, J. R. and Laemmli, U. K. (1977). The structure of histone depleted chromosomes. *Cell* **12**:817–818.

Peii, T. and Lim, J. S. (1982). Adaptive filtering for image enhancement. *Opt. Eng.* **21**:108–112.

Radermacher, M. (1980). *Dreidimensionale Rekonstruktion bei kegelformiger Kippung im Elektronenmikroskop.* Ph.D. thesis, Technical University Munich.

Radermacher, M., Wagenknecht, T., Verschoor, A., and Frank, J. (1986). A new 3-D reconstruction scheme applied to the 50S ribosomal subunit of *E. coli. J. Microsc.* **141**:RP1–RP2.

Ramachandran, G. N. and Lakshminarayanan, A. V. (1971). Three dimensional reconstruction from radiographs and electron micrographs: applications of convolutions instead of Fourier transforms. *Proc. Nat. Acad. Sci. USA* **68**:2236–2240.

Rattner, J. B. and Lin, C. C. (1985). Radial loops and helical coils coexist in metaphase chromosomes. *Cell* **42**:291–296.

Richmond, T. J., Finch, J. T., Rushton, B., Rhodes, D., and Klug, A. (1984). Structure of the nucleosome core particle at 7A resolution. *Nature (London)* **311**:532–537.

Ruiz-Carrillo, A., Puigdomench, P., Eder, G., and Lurz, R. (1980). Stability and reversibility of higher ordered structures of interphase chromatin: continuity of DNA is not required for maintenance of folded structure. *Biochemistry* **19**:2544–2554.

Sedat, J. and Manueledis, L. (1978). A direct approach to the architecture of eukaryotic chromosomes. *Symp. Quant. Biol. Cold Spring Harbor* **42**:331–350.

Skoglund, U., Anderson, K., Bjorkroth, B., Lamb, M. M., and Dancho, B. (1983). Visualization of the formation and transport of a specific hnRNP particle. *Cell* **34**:847–855.

Skoglund, U., Andersson, K., Strandberg, B., and Daneholt, B. (1986). Three-dimensional structure of a specific pre-messenger RNP particle established by electron microscope tomography. *Nature (London)* **319**:560–564.

Subirana, J. A., Munoz-Guerra, S., Aymami, J., Radermacher, M., and Frank, J. (1985). The layered organization of nucleosomes in 30 nm chromatin fibers. *Chromosoma* **91**:377–390.

Subirana, J. A., Munoz-Guerra, S., Radermacher, M., and Frank, J. (1983). Three dimensional reconstruction of chromatin fibers, *J. Biomol. Struct. Dyn.* **1**:705–714.

Thoma, F., Koller, T., and Klug, A. 1979. The salt-dependent superstructures of chromatin, and involvement of histone H1. *J. Cell Biol.* **83**:403–427.

Tieman, D. G., Murphey, R. K., Schmidt, J. T., and Tieman, S. B. (1986). A computer-assisted video technique for preparing high resolution pictures and stereograms from thick sections. *J. Neurol. Methods* **17**:231–245.

van Heel, M. and Harauz, G. (1986). Resolution criteria for three dimensional reconstruction, *Optik* **73**:119–122.

van Heel, M. and Keegstra, W. (1981). IMAGIC: a fast, flexible, and friendly image analysis software system, *Ultramicroscopy* **7**:113–130.

Van Holde, K. E. (1989). *Chromatin.* Springer-Verlag, N.Y.

Williams, S. P., Athey, B. D., Muglia, L. J., Schappe, R. S., Gough, A. H., and Langmore, J. P. (1986). Chromatin fibers are left-handed double helices with diameter and mass per unit length that depend on linker length. *Biophys. J.* **49**:223–235.

Woodcock, C. L. F., Frado L-L, Y., and Rattner, J. B. (1984). The higher order structure of chromatin: evidence for a helical ribbon arrangement. *J. Cell Biol.* **99**:42–52.

Woodcock, C. L. F. and Horowitz, R. A. (1986). Helical chromatin fibers in situ and after negative staining. *J. Cell Biol.* **103**:41a.

Woodcock, C. L., Horowitz, R. A., Bazett-Jones, D. A., and Olins, A. L. (1990a). Localization of DNA in chromatin using electron spectroscopic imaging and osmium ammine staining, in *Proc. XII Int. Congr. Electron Microscopy*, pp. 116–117. San Francisco Press.

Woodcock, C. L. and McEwen, B. F. (1988a). 3-D structure of negatively stained chromatin fibers, in *Proc. 46th Ann. Mg. Electron Microsc. Soc. America* (G. W. Bailey, ed.), pp. 168–168. San Francisco Press, San Francisco.

Woodcock, C. L. and McEwen, B. F. (1988b). Three dimensional reconstructions of negatively stained chromatin fibers. *J. Cell Biol.* **107**:312a.

Woodcock, C. L. F., Safer, J. P., and Stanchfield, J. E. (1976). Structural repeating units in chromatin. I Evidence for their general occurrence. *Exp. Cell Res.* **97**:101–110.

Woodcock, C. L., McEwen, B. F., and Frank, J. (1990b). Ultrastructure of chromatin. II 3D reconstruction of isolated fibers. *J. Cell Sci.* **99**:107–114.

Wurtz, T., Lonnroth, A., Ovchinnikov, L., Skoglund, U., and Daneholt, B. (1990). Isolation and partial characterization of a specific premessenger ribonucleoprotein particle. *Proc. Nat. Acad. Sci. USA* **87**:831–835.

Zatsepina, O. V., Polyakov, V. U., and Chentsov, Y. S. (1983). Chromonema and chromomere-structural units of mitotic and interphase chromosomes. *Chromosoma* **88**:91–97.

15

Three-Dimensional Reconstruction of Noncrystalline Macromolecular Assemblies

Terence Wagenknecht

1. INTRODUCTION

At present the only method available for determining the 3D molecular structure of most biological macromolecules is x-ray crystallography. However, some macromolecules associate to form large multisubunit complexes (e.g., viruses, ribosomes, multisubunit enzymes) for which crystals suitable for crystallographic studies are not easily obtained. Even in those cases where suitable crystals are available, there

Terence Wagenknecht • Wadsworth Center for Laboratories and Research, New York State Department of Health, Albany, New York 12201-0509; and Department of Biomedical Sciences, School of Public Health, State University of New York at Albany, Albany, New York 12222

Electron Tomography: Three-Dimensional Imaging with the Transmission Electron Microscope, edited by Joachim Frank, Plenum Press, 1992.

remain additional technical problems that will have to be solved before detailed structures can be determined (Yonath and Wittman, 1989). In this chapter I will describe recent structural determinations of noncrystalline biomacromolecular complexes by the technique of electron microscopy in conjunction with 3D reconstruction (electron tomography).

It should be emphasized at the outset that the smallest structural details (2–5 nm) that are visible in electron tomographic investigations of large biocomplexes are easily an order of magnitude larger than the near-atomic details resolved in x-ray crystallographic studies. Nevertheless, the information content of a determined structure depends on the size of the specimen as well as the resolution attained, and so low-resolution 3D reconstructions of large structures, such as ribosomes, do reveal many structural details, but not at the atomic level. Is low-resolution information useful in understanding the function of large macromolecular complexes? One of the goals of this chapter is to convince the reader that it is.

Electron tomography reveals the 3D distribution of density in the specimen, thereby removing the confusion arising from superposition of structural features that are present in the original images which show the specimen in projection along an axis normal to the plane of the micrograph. The reconstructed object can be viewed from arbitrary directions, and the spatial relationships of the structural details can be quantitatively assessed. Also, often internal structural features are revealed in addition to the surface topography. Study of the reconstruction in the light of known chemical and physical properties of the specimen will sometimes clarify existing hypotheses or lead to new hypotheses on the functioning of the specimen.

Usually much greater insight into function will be achieved when the technique of electron tomography is combined with other experimental approaches. For example, a functional site in a macromolecular complex could be labeled with a probe (e.g., a monospecific antibody) of sufficient size to be detected in a 3D reconstruction, thereby allowing its position on the surface of the complex to be determined precisely. Alternatively, specific components of the specimen might be selectively extracted or omitted in reassembly experiments conducted in vitro, and the effect of the missing component on overall structure determined (e.g., Carazo *et al.*, 1988). Some complexes exist in two or more conformational states, and these could be characterized separately by 3D reconstruction and then compared to reveal the nature of the structural rearrangements that occur when the two forms interconvert (Unwin and Ennis, 1984; Unwin *et al.*, 1988).

No attempt has been made here to review extensively all of the specimens to which the technique of electron tomography has been applied (for a discussion of reconstructions determined through 1984, the reader is referred to the review by Frank and Radermacher, 1986). The number of macromolecular complexes to which electron tomography has been applied is not large, and in some studies the specimen served mainly as a test object for the methodology and little attempt was made to glean information of biological significance from the reconstructions obtained. This situation is expected to change due to recent improvements in the methodology, the availability of the necessary software and lower computing costs. I will focus on recent electron tomographic studies in which a relatively new method of reconstruction was used (Radermacher *et al.*, 1987a, b; Radermacher, 1988). As will be discussed, this method is applicable to those macromolecular complexes that have well-defined structures such that averaging of data from images of many complexes is possible. Tomographic studies of ribosomal components will be emphasized, especially those of the large ribosomal subunit, which has been studied by several independent groups using different methodologies and specimen types.

This review does not discuss highly symmetric assemblies, such as icosahedral viruses and helical assemblages of subunits (e.g., tobacco mosaic virus), which, by virtue of their high symmetry, can be reconstructed three-dimensionally with little or no tilting of the specimen in the microscope (for reviews, see Stewart, 1988; Baker and Fuller, forthcoming). For reviews of recent tomographic studies of sub-cellular structures, such as chromatin and eukaryotic cilia and flagella, that cannot be reconstructed by techniques involving image averaging, see Chapters 13 and 14.

2. METHODOLOGY

Prior to 1987 the electron microscopic data used for tomographic reconstructions of macromolecular assemblies were of two types. In the first, and used most frequently, tilt series of the specimen (either negatively stained or embedded, sectioned, and stained) are collected by physically tilting the specimen about a single axis incrementally over as wide a range of angles as possible, usually within $\pm 60°$. Three-dimensional reconstructions are then computed individually for each particle analyzed (usually one or a few) using one of several 3D reconstruction algorithms. The major advantage of this single-axis tilting approach is that it is applicable to any type of specimen. The disadvantages of single-axis tilting are that macromolecular complexes suffer significant radiation damage when exposed to the electron doses required to collect the complete tilt series, and that the computational effort required to improve the signal-to-noise ratio by averaging many individual reconstructions can be prohibitive (Knauer *et al.*, 1983; Oettl *et al.*, 1983).

The second, not so well-developed scheme of data collection, involves the assignment of angles to naturally occurring orientations of the specimen (van Heel, 1983; Verschoor *et al.*, 1984; Crewe *et al.*, 1984). If necessary, the naturally occurring view could be supplemented with views obtained by limited tilting of the specimen grid. While this approach holds great promise, the main problem is that an objective and reliable method for determining precisely the angular relationships of the views has not yet been demonstrated for real biological specimens. Another problem, especially serious for negatively stained specimens adsorbed to a carbon film, is that if flattening distortions of the specimen have occurred, then it would no longer be correct to treat the views as representing differing orientations of identical objects.

In 1987 Radermacher *et al.* (1987a, b) described a new electron tomographic technique for reconstructing macromolecular specimens, and they applied the method to the 50S ribosomal subunit from *Escherichia coli.* Their technique is applicable to noncrystalline macromolecules that tend to adsorb to the specimen grid in one or more preferred orientations. While this requirement may seem restrictive, it is actually unusual for macromolecular complexes not to interact, at least to a degree, with the support film to produce preferential views. The 50S ribosomal subunit fulfills this criterion quite well as can be seen from Fig. 1 in which more than 90% of the subunits lie in a single orientation, which, in earlier work, had been named the "crown" view (Tischendorf *et al.*, 1974). In this study the ribosomal specimen was negatively stained by a method in which the subunits become "sandwiched" between two thin carbon films (Stöffler and Stöffler-Meilicke, 1983, 1988). Sandwiching of the specimen is highly desirable for application of the method of Radermacher *et al.* because the specimen becomes evenly stained, in contrast to the one-sided staining common in more conventional procedures for negative staining in which only one carbon film is present (Frank *et al.*, 1988).

Specimens exhibiting preferred orientations have the useful property that when the grid is tilted by an angle θ, the imaged macromolecules present many different viewing directions which can be determined precisely. The set of viewing directions form a conical tilt series of the specimen, with the object at the vertex, and the cone angle equal to the tilt angle. The azimuthal angles of the views are randomly distributed and are equivalent to the azimuthal orientations of the particles in the nontilted images (see Fig. 4 in Chapter 5), which can be determined by correlation methods (Frank *et al.*, 1978; Frank, 1980; Radermacher *et al.*, 1987b; Radermacher, 1988). The set of tilted images, following their alignment to a common origin, together with the azimuthal viewing angles obtained from the corresponding untilted images are all that is necessary to compute a 3D reconstruction using a modified weighted back-projection algorithm. Implicit in this method is the assumption that the macromolecular complexes are structurally identical. I refer to this technique of reconstruction as the method of Radermacher. For a detailed description of the method and its implementation the reader is referred to Chapter 5 and Radermacher *et al.* (1987b) and Radermacher (1988).

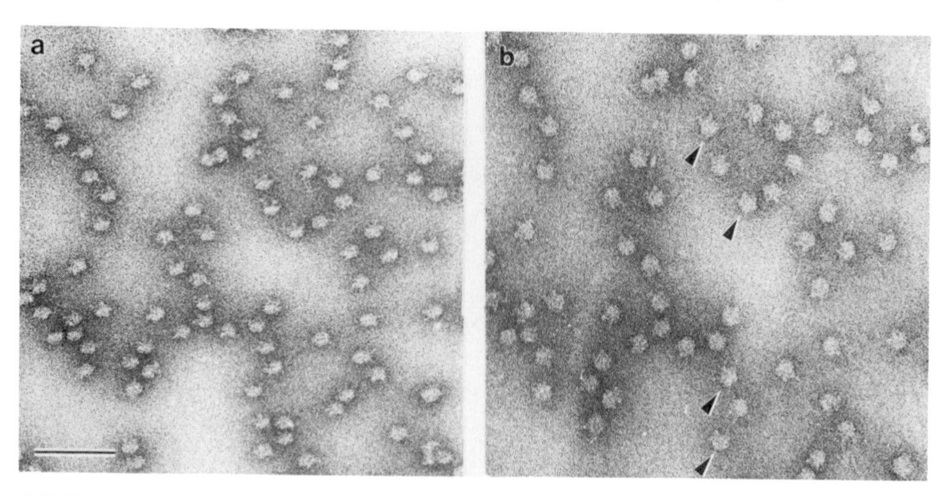

FIGURE 1. Electron micrograph of 50S ribosomal subunits from *E. coli* as an example of a supramolecular complex exhibiting a preferred orientation on the specimen grid. (a) Tilted specimen, recorded first, electron dose $<1000\ e/nm^2$). (b) Untilted specimen of same field as in (a). Note preponderance of "crown" orientations, a few of which are marked by arrows. (Reprinted with permission from Radermacher *et al.*, 1987, EMBO J. **6**:1107–1114.)

The method of Radermacher has several advantages over the other schemes outlined above. First, minimal electron exposure techniques can be employed because only micrographs of the specimen tilted maximally are used in the reconstruction scheme; the companion, untilted micrographs are used only to determine the azimuthal angles of the particles. Second, it is feasible to incorporate data from hundreds or even thousands of macromolecules into the reconstruction, which greatly increases the signal-to-noise ratio of the final reconstruction. Finally, because only particles exhibiting a single orientation are used, specimens that have become partially flattened in a homogeneous manner can be reconstructed without loss of resolution.

The example of the 50S ribosomal subunit discussed above in which nearly all of the subunits adhered to the grid in the same orientation is not typical of most macromolecules. More commonly, the images are much more heterogeneous due to differences among the macromolecules in orientation, preservation, staining, or structure. In such cases an objective method for sorting the images into homogeneous groups is needed. Multivariate statistical techniques such as correspondence analysis (van Heel and Frank, 1981; Frank and van Heel, 1982) and principal component analysis (Unser *et al.*, 1989) have proven to be powerful methods for sorting images in an objective manner, and as an aid in identifying the sources of the variations. In these analyses the images are treated as n-dimensional vectors (where n is the number of picture elements that comprise an image), and a coordinate transformation is computed such that the first axis of the new orthogonal coordinate system represents the direction of greatest interimage variance, the second represents the direction of next greatest variance, and so forth. For some specimens just the first two to four factors (axes) will account for a large percentage of the interimage variance and will be sufficient to describe the main groupings of images. For other specimens, in which the interimage variance is distributed over more factors, automatic classification techniques have been used to identify homogeneous groups of images (van Heel, 1984; van Heel and Stöffler-Meilicke, 1985; Frank *et al.*, 1988; Carazo *et al.*, 1990; Frank, 1990). Once a homogeneous group has been identified and characterized in untilted micrographs, the 3D electron tomographic method of Radermacher can be applied to the corresponding images in the tilted, companion micrographs, provided that the images comprising the group are present in sufficient numbers. For overviews of these image processing techniques the reader is referred to several recent reviews (Frank *et al.*, 1988; Frank, 1989).

3. PROKARYOTIC RIBOSOMES

No other specimen has been investigated more thoroughly by electron tomographic methods than the bacterial ribosome. Ribosomes are the cellular organelles whose function is to synthesize proteins from the constituent amino acids as specified by the genetic code imprinted in messenger RNA. The chemical composition of ribosomes is well known, but the 3D structure of the ribosome remains a formidable problem. In eubacteria, ribosomes consist of two subunits, a large (50S) subunit and a small (30S) subunit. The 50S subunit (M_r 1.45×10^6) is composed of 35 different proteins and two molecules of ribosomal RNA (rRNA) known as 5S rRNA (120 nucleotides in *E. coli*) and 23S rRNA (2904 nucleotides). The smaller 30S subunit (M_r 850,000) comprises 21 proteins and one molecule of 16S rRNA (1542 nucleotides). All 54 proteins have had their amino acids sequences determined for *E. coli* (Wittman-Liebold, 1984), and the nucleotide sequences of the three rRNAs are also known (for reviews see Erdmann *et al.*, 1985; Noller, 1984). The rRNA and ribosomal proteins account for 65% and 35%, respectively, of the ribosome's mass.

Impressive progress has been made recently in x-ray crystallographic analysis of ribosomal specimens (for review see Yonath and Wittman, 1989), but the task of determining the 3D structures at high resolution is still probably years away. In lieu of x-ray crystallography, electron microscopy has made important contributions to understanding the basic 3D architecture of the bacterial ribosome. Several groups have attempted to determine the gross shapes of the ribosome and its subunits by direct visual analysis of electron micrographs of negatively stained ribosomal subunits and intact ribosomes (reviewed by Wittman, 1983). Ribosomal particles tend to lie on the specimen grid in a few easily recognized orientations having distinctive shapes. By tilting the specimens in the microscope, the relative orientations of the characteristic views could be roughly estimated. Additionally, antibodies specific for individual ribosomal proteins have proven to be useful markers in the analyses (Stöffler and Stöffler-Meilicke, 1986; Oakes *et al.*, 1986).

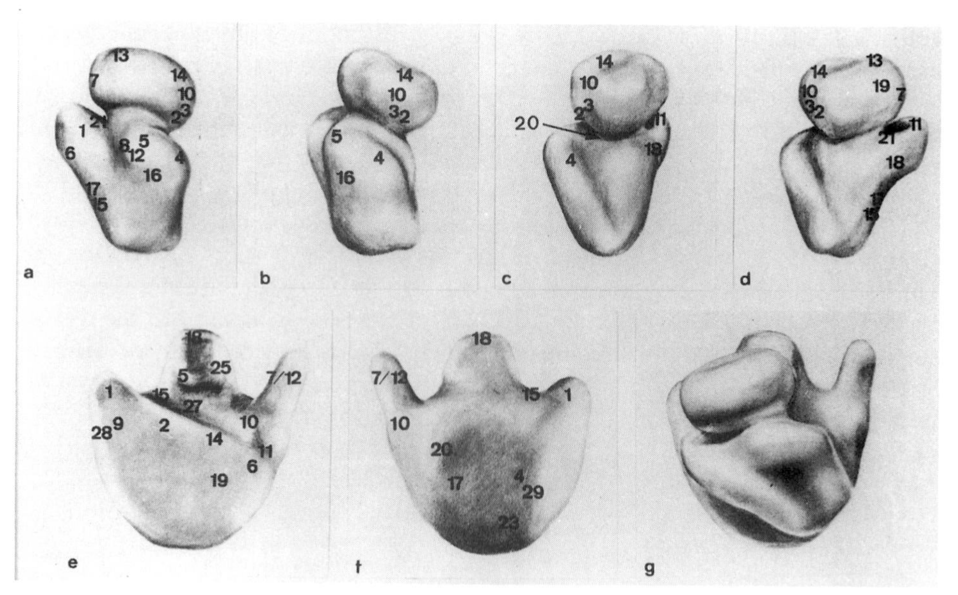

FIGURE 2. Visually derived model of *E. coli* ribosome and subunits. The model shown is adapted from Stöffler-Meilicke and Stöffler (1990 with permission, copyright American Society for Microbiology), but the somewhat different models proposed by Lake (Oakes *et al.*, 1986) or Vasiliev and colleagues (Vasiliev *et al.*, 1983) could just as well have been shown. (a)–(d), views of the 30S subunit, (e), (f) front and back views, respectively, of the 50S subunit, (g) 70S ribosome viewed from 30S side. The numbers refer to the locations of ribosomal proteins as determined by immunoelectron microscopy, except for S20 (S refers to a small subunit protein, L to a large subunit protein) which was localized by neutron scattering (Capel *et al.*, 1987).

These methods, which are really a qualitative form of electron tomography (using the human brain, instead of a computer to reconstitute the shape from the projection data), led to models for the prokaryotic ribosome and its subunits that were, not surprisingly, somewhat different. Over the years the various models proposed have partially converged such that currently most references to the visual models cite one of several similar models, one of which is shown in Fig. 2. Although the visually derived models have been of great value in serving as structural frameworks for interpreting the results of numerous chemical and physical studies on ribosomes, their utility is severely limited by the low level of structural detail present and the subjective nature of the process by which they were derived.

3.1. The Large (50S) Ribosomal Subunit

The 50S subunit has been studied more intensively by electron tomographic methods than any other specimen. Three independent groups, each using different tomographic methods have reported 3D reconstructions of 50S subunits from electron micrographs of negatively stained, noncrystalline specimens (Oettl *et al.*, 1983; Hoppe *et al.*, 1986; Radermacher *et al.*, 1987a, 1987b; Vogel and Provencher, 1988). Also, the large subunit has been reconstructed by the more conventional electron crystallographic method from negatively stained 2D crystalline sheets (Lake, 1985; Yonath *et al.*, 1987). Unless the visually derived models are grossly incorrect, it seems reasonable to expect that the 3D reconstructions will appear as refined versions of the visual models (Fig. 2*e*, *f*) in which additional morphological details are resolved. As will become apparent, the tomographically derived reconstructions fulfill this expectation. I will describe first results from our own laboratory, which has produced the most detailed models, and then compare these with reconstructions determined elsewhere.

Figure 3 shows a model of the 50S subunit based upon a 3D reconstruction determined by the method of Radermacher *et al.* discussed in Section 2. The reproducible resolution of the reconstruction was estimated by several criteria to be 2–3 nm (Radermacher *et al.*, 1987a, b).* Clearly the reconstructed subunit is in basic agreement with the currently favored versions of the visually derived models described above. Specifically, the subunit consists of a main, roughly hemispherical body which bears three projecting structures known as the *stalk*, *central protuberance*, and *L1 shoulder*. One qualitative difference between the reconstructed subunit and the visual models is that the region of the main body labeled PL in Fig. 3*a*, which is a region that interacts with the small 30S subunit in the complete ribosome, is

* There is no universally accepted method for quantifying the resolution of tomographically determined 3D reconstructions. The two commonly used methods are the differential phase residual (Frank *et al.*, 1981; Radermacher *et al.*, 1987b; Radermacher, 1988) and Fourier ring correlation (Saxton and Baumeister, 1982; van Heel *et al.*, 1982) methods which, unfortunately, give different values for the reproducible resolution. The latter always indicates better resolution than the former. When a range of resolutions is cited in the text (e.g., 2–3 nm), the first value was determined by Fourier ring correlation and the second by the differential phase residual method. If only one value is cited, it is the differential phase residual value. Resolutions quoted in the literature for 3D reconstructions determined from 3D crystals tend to be liberal estimates, and correspond more closely to Fourier ring correlation estimates than to differential phase residual estimates (see also Unser *et al.*, 1987).

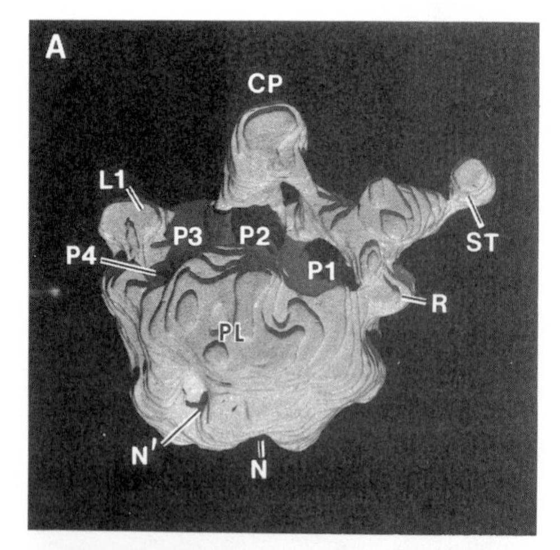

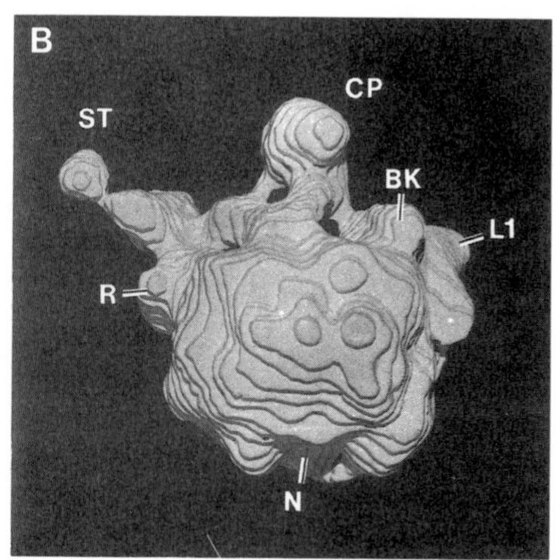

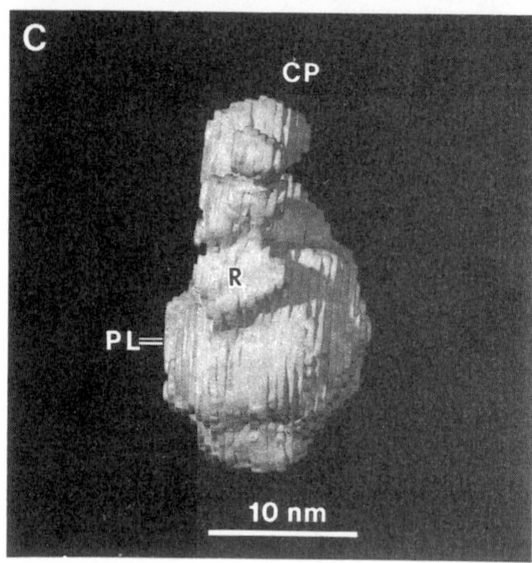

FIGURE 3. Three-dimensional solid models of the *E. coli* 50S ribosomal subunit determined by electron tomography according to Radermacher *et al.* (1987a, b). (a), (b), (c) views from the front (interface), back, and side respectively. Abbreviations: CP, central protuberance; BK, back knob; N, N', notch and minor notch; L1, L1 ridge or shoulder; P1-P4, pockets 1-4 which define the interface canyon; PL, plateau; R, ridge; ST, L7/L12 stalk. (Reprinted with permission from Radermacher *et al.*, 1987, EMBO J. **6**:1107–1114.)

somewhat convex in overall shape rather than flat or concave. More importantly, many structural details are resolved in the 3D reconstruction that were not present in any of the visually derived models.

The largest, and perhaps most significant, of the new features is a large stain-filled furrow located just below the central protuberance and extending across the face of the subunit from the base of the stalk to the L1 shoulder region (labeled P1-P4 in Fig. 3). This groove, which is about 16 nm in length, 4 nm in width and up to 4 nm deep in some regions, has been named the *interface canyon* because it is located on the face of the 50S subunit that binds the smaller 30S subunit to form the complete ribosome. The limitations inherent in the visually derived models are underscored by the observation that none of those models has a feature resembling the interface canyon, despite its large overall dimensions. Interestingly, the interface canyon lies in a region of the subunit that contains the major functional sites of the subunit, including the peptidyl transferase active site, which has been mapped by the immunoelectron microscopy technique (see Oakes *et al.*, 1986, for a review) to a region just below and possibly to the left of the central protuberance (subunit viewed as in Fig. 3*a*) and the sites where elongation factors G and Tu bind, near the base of the stalk.

The peptidyl transferase region lies in a region of the 3D reconstruction (Fig. 3*a*) near the pocket of the interface canyon labeled P2. Intriguingly, subpocket P2 forms a stain-filled hole at its base that extends through the subunit to the back side (the hole on the back side is visible in Fig. 3*b* just to the left of the "back knob" (BK) feature). It was hypothesized that this hole represents the site where the nascent, elongating polypeptide escapes from the peptidyl transferase active site (Radermacher *et al.*, 1987a). Independent evidence supporting this interpretation has come from recent immunoelectron microscopy experiments designed to map the location of the nascent polypeptide chain in the ribosome (Ryabova *et al.*, 1988). Antibodies specific for nascent peptide were found to bind at two main sites. One site was at the position of the hole on the back of the three-dimensionally reconstructed subunit (Fig. 3*b*), and the second site was further down on the back of the subunit near a location that had been mapped in earlier studies (Bernabeu and Lake, 1982). Surprisingly, little antibody bound between the two sites on the subunit's back, a pattern which was interpreted by Ryabova *et al.* as indicating that the polypeptide might lie in a channel on or just beneath the surface of the subunit. The tomographically reconstructed subunit appears to be consistent with this interpretation because it contains an extended region of high density (i.e., penetrated by stain) on the back of the subunit leading from the vicinity of the P2 hole toward an indentation on the back, the location of which corresponds to the upper bound of the lower site identified in the immunoelectron microscopy study. This region is not visible in the model as depicted in Fig. 3*b* because the density level selected for this representation was too low. The path of this channel or groove becomes apparent, however, at a slightly higher density threshold, as llustrated in Fig. 4 (see

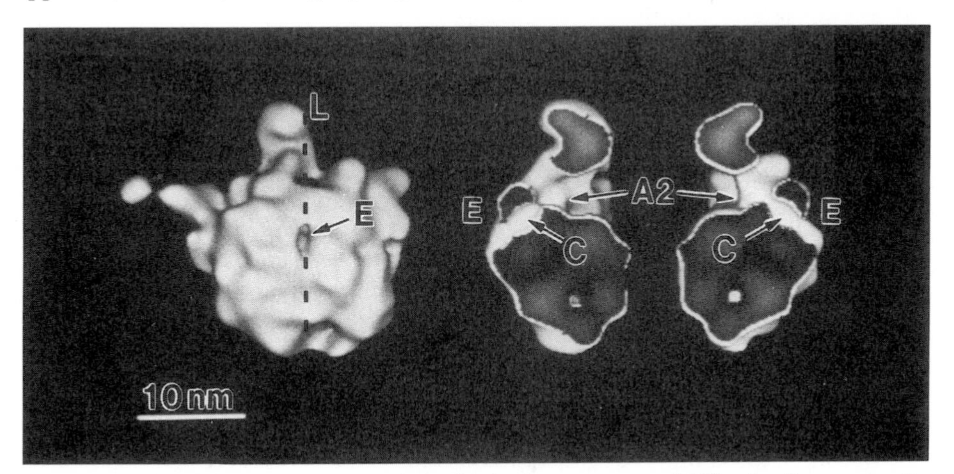

FIGURE 4. Possible path of nascent polypeptide in the reconstructed 50S subunit reconstruction of Radermacher *et al.* (1987). Abbreviations: L, line lying in plane along which the subunit was sliced to produce the two cutaway views shown to the right; C, stained channel leading from interface side to back of subunit; E, putative exit site for nascent peptide; C2, subpocket 2 of the interface canyon (labeled P2 in Figures 3 and 5). (Illustration provided by M. Radermacher, adapted from Frank *et al.*, 1990).

also Fig. 5 in Radermacher *et al.*, 1987a). The tentative identification of this channel as the path taken by the exiting nascent polypeptide illustrates the power of electron tomography in conjunction with other experimental approaches to reveal structural details that can be related to the functioning of complex macromolecular assemblies.

Probably the most intricate and detailed structural features present in the 3D reconstruction of the 50S subunit occur at the junction of the central protuberance and the main body (Fig. 3*a*, *b*). A network of three stain-excluding bridges connects the central protuberance to the body of the 50S subunit. Two of the bridges define the upper half of the hole P2.

A somewhat surprising characteristic of the reconstruction is the variability in density present throughout the volume, which suggests that the negative stain penetrates deep into the interior regions of the subunit. What this corresponds to as far as the internal structure is concerned is not known at present, but positive staining of rRNA has been suggested as an explanation.

Probably the most thoroughly studied ribosomal proteins are L7 and L12, which together form the stalk feature of the large subunit (Fig. 2*e*, *f*). These two proteins, which differ from one another only in the acetylation of the amino terminus of L7, are highly conserved throughout the plant and animal kingdoms, a fact indicative of their importance in the functioning of the ribosome. The L7 and L12 polypeptides exist as a heterodimer in solution and are apparently dimeric in the 50S ribosomal subunit as well. A model for the L7/L12 stalk based largely upon x-ray crystallographic studies has been proposed in which the L7 and L12 are arranged in a parallel side-by-side manner with the carboxy terminal regions forming a domain that joins the stalk to the subunit (Liljas, 1982). A thin flexible region of peptide joins the amino-terminal and carboxy-terminal globular domains. The bilobed appearance of the L7/L12 stalk in the 3D reconstruction shown in Fig. 3 is consistent with this interpretation.

3.2. L7/L12-Depleted Large Ribosomal Subunits

Unlike all of the other ribosomal proteins, which are present in just one copy per ribosome, there are two copies of the L7/L12 pair per 50S subunit. Several experimental observations suggest that only one pair forms the stalk, and that the other dimer might be located between the stalk and the central protuberance (Traut *et al.*, 1986; Moller and Maassen, 1986). In order to clarify this issue, electron tomography, again using the method of Radermacher, was applied to 50S subunits that had been depleted of all L7/L12 by extraction with ethanol/NH_4Cl (Carazo *et al.*, 1988). The results of the reconstruction are presented in Fig. 5.

Removal of the L7/L12 proteins, which constitute less than 2% of the total mass of the large subunit, produced a number of unexpected effects on the 3D architecture of the subunit. Two major conformational states were identified for the L7/L12-depleted subunits by application of multivariate statistical and classification techniques to the aligned images obtained from the untilted specimen. Surprisingly, the main difference between the two conformations as determined from the 3D reconstructions (Fig. 5*c*, *d*) was not in the region neighboring the stalk but on the opposite side of the subunit in the vicinity where the ribosomal protein L1 is

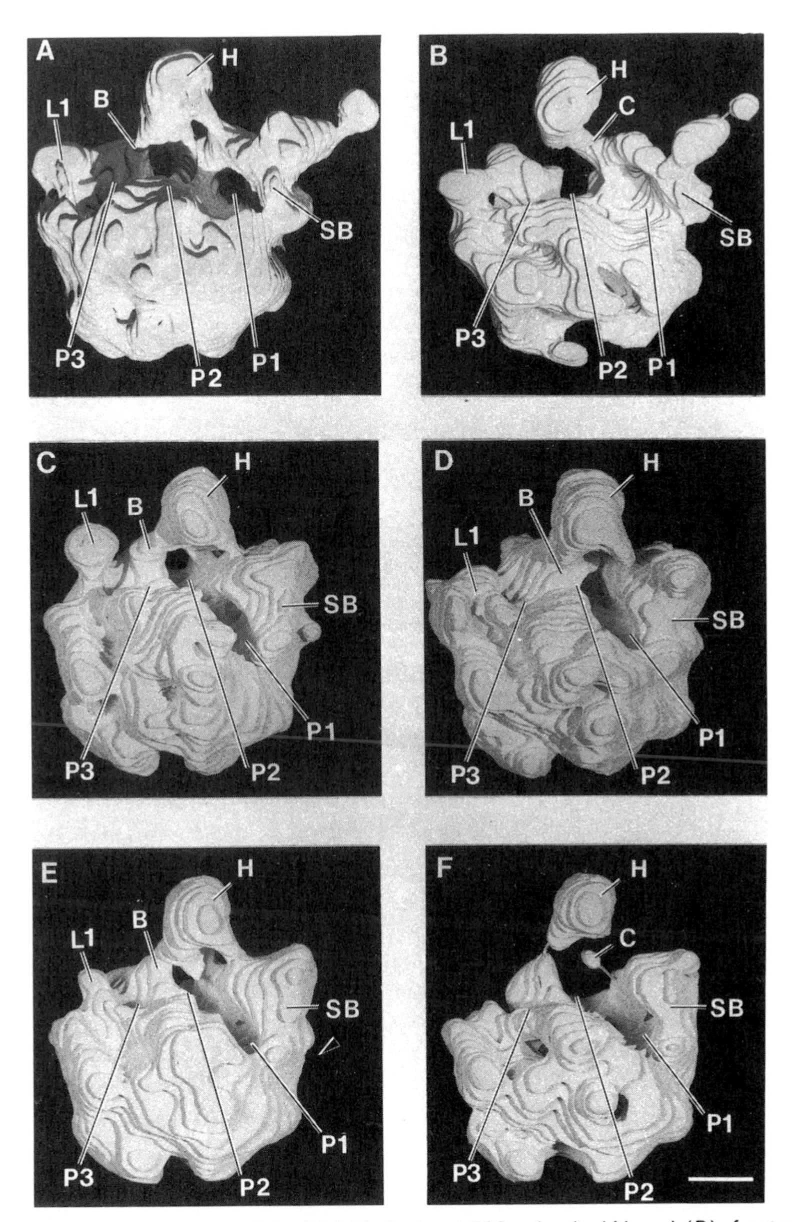

FIGURE 5. 3D reconstruction of the L7/L12-depleted 50S subunit. (A) and (B): front views of the reconstruction of the intact subunit (same reconstruction as in Fig. 3) at two different contour levels. Shown here for purposes of comparison. (C) and (D): front views of two reconstructions determined for different subsets of images identified in micrographs of the L7/L12 depleted subunits. Note absence of stalk on right hand side. (E) and (F): reconstructions at two density threshold levels of L7/L12-depleted subunits obtained by combining the images contributing to the two reconstructions in (C) and (D). Abbreviations: H, head (central protuberance); B, bridge; L1, L1 ridge or shoulder; P1, P2, P3, three of the pockets forming the interface canyon; SB, stalk base region. (From Carazo *et al.*, 1988 with permission of Journal of Molecular Biology; copyright 1988 by Academic Press, Inc. [London] Limited.)

located. This region of the subunit is more than 15 nm away from the site where the L7/L12 stalk attaches to the subunit. In one of the conformations there is an upward projecting mass in the L1 region (Fig. 5*c*), whereas in the other this mass is less pronounced and appears to project to the left (Fig. 5*d*). Additional structural rearrangements in this region of the subunit are also evident.

Both conformations of the stalkless subunit differ similarly from that of the intact subunit in the region where the stalk would normally bind. In the subunits depleted of L7/L12 the stalk base region appears to have collapsed into the interface canyon and to have moved closer to the central protuberance. Also, the rightmost bridge of density connecting the central protuberance to the body of the subunit is reduced in size and density relative to that in the intact subunit (compare Figs. 5*a* with 5*e* and 5*b* with 5*f*). These observations could be viewed as being consistent with the hypothesis that one of the L7/L12 dimers lies somewhere between the stalk base and the central protuberance, although the complexity of the structural rearrangements associated with removal of L7/L12 precludes an unambiguous conclusion to be made on this point.

Both the large-scale structural changes observed when the L7/L12 proteins were removed from the 50S subunit and the occurrence of the two conformational states were completely unexpected. At the very least, the structural rearrangements indicate that the 50S subunit is not a rigid structure. It remains to be determined whether the two conformations have a physiological significance. It will be interesting to determine whether the selective removal of other ribosomal components also causes such widespread conformational changes in the subunit.

3.3. Other Reconstructions of the 50S Ribosomal Subunit

The first electron tomographic reconstructions of the *E. coli* 50S subunit were determined from single-axis tilt series of negatively stained subunits (Oettl *et al.*, 1983; Hoppe *et al.*, 1986). Subunits in both the "crown" and another orientation, the "kidney" view, were reconstructed. The surface topography of these reconstructions is not as well defined as it is in the reconstructions described above, probably due to radiation damage and much less extensive averaging of data from different particles. Nevertheless the reconstructions that were determined for subunits in the crown orientation by single-axis tilting are in overall agreement with the reconstructions determined by Radermacher's method discussed above. For example, the interface canyon is clearly present in both reconstructions. The surprisingly strong contrast variations present in the internal regions of the subunit also appeared similarly in the two reconstructions. Reconstructions determined from particles that adsorbed to the grid in the kidney orientation also displayed the interface canyon and several other surface features that were present in the crown-view reconstructions, but there were also significant differences in overall shape. It is not clear to what extent these differences were caused by differential distortions of the subunits arising from their very different orientations on the specimen grid; incomplete embedding of the subunits in the negative stain, especially of the subunit in the kidney orientation which would be thicker in the direction normal to the grid than the crown-oriented subunit, could also account for the differences between the reconstructions obtained for these two views.

A third tomographic reconstruction of the 50S ribosomal subunit, in this case from *Bacillus stearothermophilus*, has been determined by yet a third method (Vogel and Provencher, 1988). This reconstruction combined data from a tilt series of seven subunits, not all of which were crown views; consequently there was no missing cone in the 3D Fourier transform from which the reconstruction was computed. Although the resolution was rather low, the overall shape of the subunit is similar to that found in the reconstructions of the *E. coli* subunit. The interface canyon does not appear to have been resolved in this reconstruction, and a region of high density was found in the subunit's interior which was not seen in the reconstructions of the *E. coli* subunit.

The overall agreement among the independently obtained tomographic reconstructions of the large ribosomal subunit is reassuring, and their consistency with the visually derived models, is a useful result. The results from immunoelectron microscopy studies (Stöffler and Stöffler-Meilicke, 1986) can be directly transferred from the visually derived models to the 3D reconstructions.

Probably the most worrisome potential problem in the interpretation of 3D reconstructions determined by electron tomography of negatively stained specimens concerns the degree to which the specimen was flattened or otherwise distorted when it was dried down on the carbon foil of the microscope grid. It is well known that the amount of flattening varies widely for different specimens and with different staining procedures. Fortunately, ribosomes are relatively resistant to flattening (Kellenberger *et al.*, 1982), an observation that is supported by the reconstructions of the 50S subunits themselves which have dimensions in the direction of flattening (i.e., normal to the carbon support film) that differ little from values expected from independent determinations.

Three-dimensional reconstructions of the 50S subunit determined from 2D crystalline specimens might provide another means to evaluate the structural integrity of isolated subunits on the specimen grid. Macromolecular assemblies in a crystalline state are thought to be protected, at least partially, from the deleterious effects of drying and staining. A preliminary reconstruction of the *E. coli* subunit clearly showed the three main protruding features (stalk, central protuberance, L1 shoulder) and an overall shape consistent with that of the visually derived models and the tomographically reconstructed subunits, but little additional fine structure was resolved (Lake, 1985; Oakes *et al.*, 1986). More recently, another 3D reconstruction of the 50S subunit, this one from *B. stearothermophilus*, was described in which the overall structure is somewhat different from the visual models and all other 3D reconstructions (Yonath *et al.*, 1987; Yonath and Wittman, 1989). For example, although three or four projecting structures were present in the reconstruction of Yonath *et al.*, the overall shape was sufficiently different from previous models that it was not even possible to identify the central protuberance, stalk, or L1 shoulder. Furthermore, a tunnel of about 10 nm length and 2.5 nm diameter, not detected in other reconstructions, traverses the subunit. Based upon its dimensions, the tunnel might correspond to the right-hand half of the interface canyon of the tomographic reconstruction shown in Fig. 5*a*, *c*, which leads to the back side of the subunit via the pocket P2. The apparent structural differences between the *B. stearothermophilus* reconstruction reported by Yonath *et al.* and the reconstructions of the 50S subunit from *E. coli* appear not to be explicable as species related in that electron microscopy studies and electron tomography of the isolated *B. stearothermophilus* subunit (Vogel and Provencher, 1988) show the same overall structure as the *E. coli* subunit. The reasons for the dissimilarity of the reconstruction of Yonath *et al.* (1987) from all other models of the 50S subunit remain puzzling.

3.4. 70S Prokaryotic Ribosome

Verschoor *et al.* (1986) showed that the complete ribosome from *E. coli* assumes several orientations in negatively stained preparations which are amenable to correlation alignment and image averaging. A class of images known as "overlap" views (Lake, 1976) was found to occur with sufficient frequency to be reconstructed in three dimensions by electron tomography (Wagenknecht *et al.*, 1989a; Carazo *et al.*, 1989). The overlap views as a class are heterogeneous, due in large part to orientational variability, and it was necessary to apply multivariate statistical analyses to identify more homogeneous subsets of images. However, the resulting reduction in the number of ribosomal images used to determine the reconstructions was probably responsible for the lower resolutions obtained, typically about 4 nm, relative to those obtained for the 50S subunit. Nevertheless a number of new structural details were resolved in the reconstructions (Fig. 6).

Views of the reconstructed ribosome from the side bearing the small (30S) subunit show that a large portion of the interface canyon is exposed on the right-hand side of the 50S subunit (Fig. 6*b*). The exposed portion is large enough to accommodate at least two transfer RNA (tRNA) molecules in a configuration that allows their aminoacyl ends to approach the putative location of the peptidyl transferase center located below the central protuberance, while still allowing the anticodon loops to contact the 30S subunit at the junction of the head and body which is where the mRNA decoding site is located (Gornicki *et al.*, 1984; Wagenknecht *et al.*, 1988b). This arrangement of the tRNA's closely matches a model proposed previously by Spirin (1983) on the basis of structural and biochemical considerations. The reconstruction does not eliminate other models for the tRNA locations, but is merely more consistent with an arrangement of the type depicted. It should now be possible to map directly the locations of the tRNA molecules by reconstructing ribosome-tRNA complexes by electron tomography.

A few minor, but possibly significant, differences between the 50S portion of the ribosome reconstruction and the independently reconstructed 50S subunit are apparent. Both the central protuberance and the stalk are tilted slightly forward (i.e., toward the interface side of the subunit) in the ribosome relative to their orientations in the subunit alone. It is not known, however, whether these movements are characteristic of the native structures or are induced by the conditions of electron microscopy (e.g., interactions with the carbon supporting film), but in either case they indicate that these portions of the subunit are deformable. Interestingly, another 3D reconstruction of the 70S subunit, determined by electron crystallography (Arad *et al.*, 1987), also showed evidence of similar structural rearrangements in the 50S subunit (Yonath and Wittman, 1989).

The shape of the 30S subunit in the reconstructed 70S ribosome is remarkably similar to its shape in visually derived models (compare Figs. 6*b* and 2*g*). Unfor-

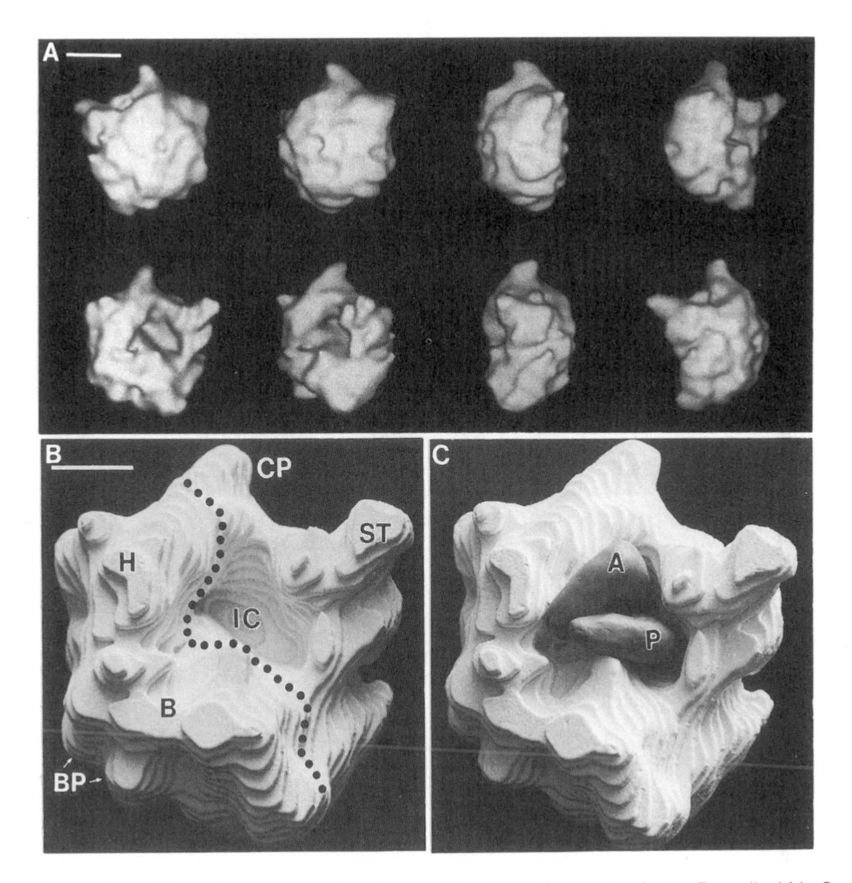

FIGURE 6. Three-dimensional reconstruction of the 70S ribosome from *E. coli*. (A) Surface representations of a gallery of views related by 45 rotations about vertical axis. Scale bar, 10 nm. (B) Styrofoam model of front view (overlap orientation). Dashed line indicates approximate boundary between the small (to the left) and large subunits. Scale bar, 5 nm. (C) As in (B) but with scale models of two tRNA molecules (labeled A and P) inserted into interface canyon. Abbreviations: H, head of 30S subunit; B, main body of 30S subunit; BP, bilobed platform of 30S; IC, interface canyon of 50S subunit; ST, stalk of 50S subunit, CP, central protuberance of 50S. (From Wagenknecht *et al.*, 1989a).

tunately the surface of the 30S subunit that faces the 50S subunit was not resolved in the reconstruction, and so a full comparison of the subunit's shape with that of the visually derived models (Fig. 2*a–d*) cannot be made. The failure to resolve the interface is probably due to insufficient resolution, which is lowest in the required direction because the grid cannot be tilted to 90° (missing cone problem, see Chapter 5), and to partial flattening or collapse of the subunits onto each other during drying (see below).

Unfortunately, the isolated 30S subunit has proved difficult to reconstruct by the method of Radermacher because it exhibits many different image types, presumably reflecting different orientations. Several reconstructions of the isolated

30S subunit by electron tomography have been reported, either from data collected by single-axis tilting in the microscope (Knauer *et al.*, 1983) or from data obtained without experimental tilting by assigning viewing angles to the various naturally occurring orientations using multivariate statistical techniques and other information as a guide (van Heel, 1983; Verschoor *et al.*, 1984). Both of these approaches have their disadvantages, and a consensus on the overall shape of the 30S subunit has not yet been reached.

Although the orientational heterogeneity of the overlap class of images generally complicates the analysis, this behavior can also be useful. For example, a number of reconstructions were determined for the 70S ribosome in the overlap range of views that differed mainly by a rotation about a single axis. Comparisons of these reconstructions provided clues on how the shape of the ribosome depended on its orientation on the grid (Carazo *et al.*, 1989). Over a range of about 25° the reconstructions showed only minor structural variations, but at one extreme of this range the separation of the two subunits increased slightly. This suggests that some collapse of the two subunits onto one another did indeed occur during preparation for electron microscopy. Additional support for this possibility comes from the appearance of "lateral view" images of the 70S ribosome in which the two subunits are positioned more or less side by side on the grid as opposed to being one atop the other as in the overlap views (Verschoor *et al.*, 1986). A 3D reconstruction determined from crystalline ribosomes from *B. stearothermophilus* also shows substantially more space between the subunits (Arad *et al.*, 1987; Yonath and Wittman, 1989). Finally, a recently described preliminary reconstruction of the 70S ribosome determined from a noncrystalline, frozen hydrated specimen also shows considerably more separation of the two subunits (Penczek *et al.*, 1990).

4. EUKARYOTIC RIBOSOMES

More recently, electron tomographic reconstructions have been determined for the eukaryotic small (40S) ribosomal subunit (Verschoor *et al.*, 1989) and the complete (80S) ribosome from rabbit reticulocytes (Verschoor, 1989; Frank *et al.*, 1990; Verschoor and Frank, 1990). The eukaryotic ribosome (M_r ~4,200,000) is considerably larger and much less is known about its 3D structure and biochemistry than for its eubacterial counterpart (M_r ~2,500,000). For example, only a few ribosomal proteins have been mapped on the surface of the eukaryotic small subunit by immunoelectron microscopy, whereas all 21 small subunit proteins have been mapped in the *E. coli* subunit. Visually derived models have been described for the eukaryotic ribosomal subunits, but the agreement among them is not as good as for the prokaryotic ribosome. Nevertheless, as will become apparent, some morphological features are conserved between eukaryotes and prokaryotes, presumably reflecting universality of function. Thus, a comparison of 3D reconstructions of ribosomes from the two kingdoms might allow the identification of conserved functional sites as well as regions in the structures which have changed through evolution.

4.1. Small 40S Ribosomal Subunit

The eukaryotic small (40S) ribosomal subunit, unlike the eubacterial 30S subunit, lies on the specimen grid in just two principal orientations, making it an excellent specimen for single-particle image-processing techniques. As for the 30S subunit, the 40S subunit is an elongate object (25 nm) consisting of two main parts, a main body connected to a smaller head, and it lies on the specimen grid with its long axis approximately parallel to the support film. The two main orientations, named L (left) and R (right) lateral views (Boublik and Hellmann, 1977; Frank *et al.*, 1982), are related to each other by a rotation of about 160° about the particle's long axis. Thus far, only the L view has been reconstructed (resolution 3.8 nm) from images selected with the aid of multivariate statistical analysis, again used to reduce the heterogeneity among the images that arise primarily from "rocking" motions about the particle's long axis (Verschoor *et al.*, 1989).

Figure 7 shows a gallery of surface representations of the reconstructed subunit related by successive 20° rotations about the long axis. Landmark features, identified previously from two-dimensional analyses of L- and R-type images, such as the "beak," "back lobes," and "feet," are prominent in the 3D reconstruction. At several locations the negative stain has penetrated the surface of the subunit to reveal rather large clefts and channels, quite different from the mostly smooth surfaces of the visual models. Of most interest, insofar as functional implications are concerned, is the region at the junction of the head and body, where decoding of mRNA is thought to occur by analogy with the prokaryotic small subunit. Here there is a large trough, especially prominent on the face of the subunit to which the large subunit binds (the interface side) to form the 80S ribosome (320° view on Fig. 7). This trough is clearly large enough to contain simultaneously the anticodon regions of two rRNA molecules such as are thought to be present in the standard two-site model of the peptide chain elongation. At the base of the trough there is a channel (2–3 nm diameter) leading to the exterior side (i.e., the side facing the cytoplasm in the complete ribosome) of the subunit at a position just above the upper lobe feature. One cannot help speculating that such an intriguing structural feature is important in the functioning of the ribosome; perhaps, the channel conducts the mRNA from the subunit:subunit interface to the exterior surface. Further tomographic studies of 40S subunits complexed with mRNA or analogues thereof could resolve issues such as this.

Regulation of translation occurs mainly at the stage of initiation, a process involving the 40S subunit, mRNA, initiator tRNA, GTP, and at least nine additional proteins called initiation factors. Two of the initiation factors, eIF-3 and eIF-2, have been mapped on the surface by electron microscopy (Emanuilov *et al.*, 1978; Lutsch *et al.*, 1985; Bommer *et al.*, 1988). Factor eIF-3, a large platelike complex, has been mapped to a position which, in our reconstruction, would place it across the two back lobes, adjacent to, or overlapping with, the channel feature in the 3D reconstruction. Factor eIF-3 is necessary for binding of the ternary complex (eIF-2-GTP-Met-$tRNA_f$) to the 40S subunit, and might also be involved in the subsequent binding of mRNA. Factor eIF-2 binds to one of the lateral faces of the subunit near the head-body junction, but due to uncertainties as to the absolute handedness of the subunit in that study (Bommer *et al.*, 1988) it was

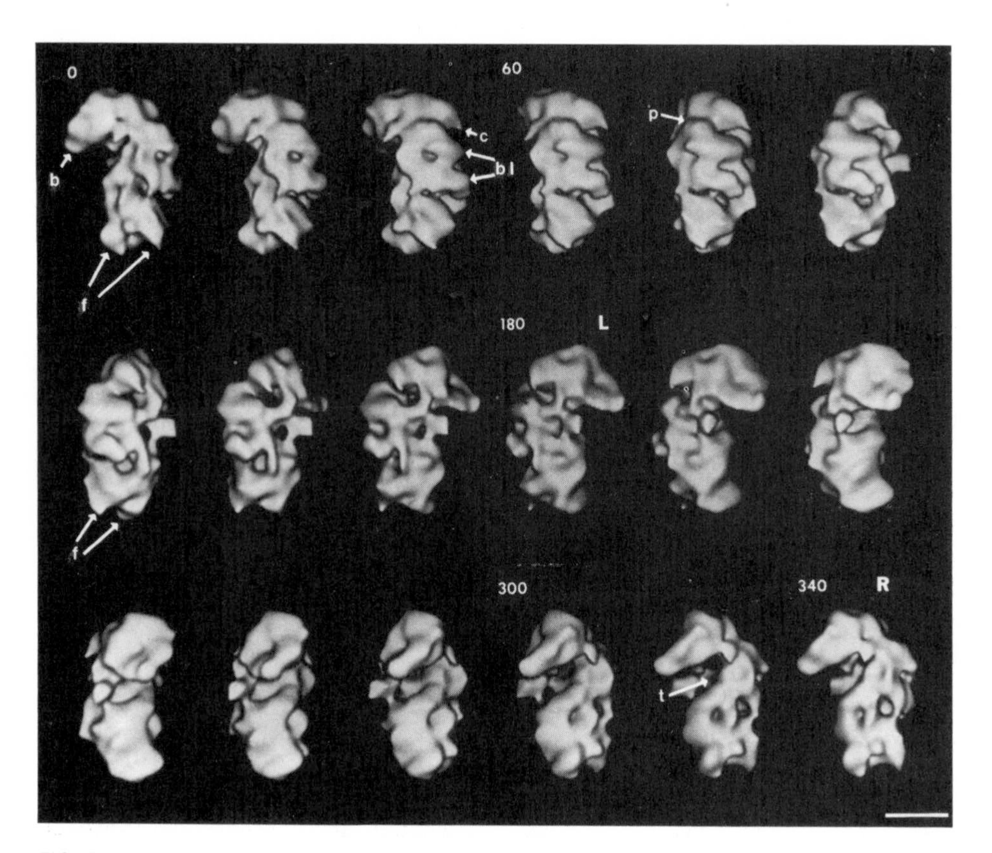

FIGURE 7. Surface representations of reconstructed eukaryotic small ribosomal subunit from rabbit reticulocytes. The views are related by successive 20° rotations about a vertical axis. Views or the L and R orientations are indicated. The surface that interacts with the large ribosomal subunit is represented by the 320° view and is characterized by a troughlike feature (t) which narrows to a channel that appears to extend all the way through to the exterior surface of the subunit (see 140°–180° views). Other abbreviations: b, beak; c, crest; bl, bilobed platform; f, basal lobes or feet. Scale bar, 10 nm. (From Verschoor *et al.*, 1989, with permission of Journal of Molecular Biology; copyright 1989 by Academic Press, Inc. [London] Limited.)

not clear whether the interface or exterior face was involved. Clearly the 3D reconstruction of the 40S subunit will be of great value in the interpretation of future results from immunoelectron microscopy and biochemical studies.

4.2. The 80S Ribosome

The eukaryotic ribosome lies on the specimen grid in a number of orientations one of which, termed the left-featured frontal view, has been reconstructed by 3D electron tomography method to a resolution of 3.7 nm (Verschoor, 1989; Verschoor and Frank, 1990; Frank *et al.*, 1990). As the methodology for reconstructing noncrystalline macromolecules by the method of Radermacher has evolved, it has become possible to analyze micrographs of macromolecules in which the images show considerable heterogeneity. Specifically, for the 80S ribosome, only 10–20%

of the images were left-featured frontal views, and within this class there was some heterogeneity which was detected and reduced to acceptable levels by multivariate statistical analysis.

A gallery of surface representations of the reconstruction is shown in Fig. 8. Using the previously reconstructed 40S subunit as a guide, the relationship of the large and small subunits to one another is clearly evident. Major features of the 40S subunit, such as the beak and back lobes, are most visible in the 0°, 20°, 140°, 200°, and 340° views. A striking difference of this reconstruction from that of the 70S prokaryotic ribosome discussed in Section 3.4 is that there is much more space between the small and large subunits in the eukaryotic ribosome. This may be due partly to an orientation of the 80S ribosome on the specimen grid that is more favorable for revealing intersubunit detail. In left-featured frontal views, the two subunits are in a nearly side-by-side arrangement as opposed to the piggy-back arrangement of the overlap view used to reconstruct the 70S ribosome (Section 3.4); in the latter the intersubunit regions would be susceptible to compression during air drying. The large amount of space between the upper portions of the subunits comprising the 80S ribosome is located where the major events in translation occur, and appears to be adequate to allow interactions with the non-ribosomal macromolecules involved in protein synthesis. Extensive intersubunit interactions appear to be restricted to the lower half of the subunits where in the reconstruction the subunits appear to be fused over an extended area.

Until now little has been known about the overall shape of the 60S subunit. As revealed by the reconstruction of the 80S ribosome, it appears somewhat different from the prokaryotic 50S subunit, although several features appear to be conserved, such as the central protuberance (CP in 160° view in Fig. 8) and the interface canyon (IC in 160° view) which may be more shallow in the 60S as compared to the 50S subunit. The 60S subunit appears to have a more ellipsoidal overall shape than the 50S subunit, although the differing directions of flattening in the 70S and 80S specimens could have exaggerated this difference. A structure analogous to the stalk of the 50S subunit is not resolved in the 80S reconstruction, but the reason for this is not clear at present. A stalk feature is occasionally visible in individual ribosomes of untilted specimen.

It appears that, in general, conservation of substructure in the eukaryotic versus prokaryotic ribosomal reconstructions is stronger in the subunit-subunit interface regions, whereas the exterior regions are more variable.

Three-dimensional reconstructions of eukaryotic ribosomes have been determined by electron crystallographic methods for both negatively stained (Unwin, 1977) and frozen hydrated specimens (Milligan and Unwin, 1986). Due to their lower resolution (5–9 nm), it is difficult to compare these with Verschoor's tomographically determined reconstruction of the 80S ribosome. However, one point deserves mention. A low-density tunnel some 15–20 nm in length was detected in the reconstruction of the frozen hydrated ribosomes, and it was hypothesized to represent a conduit through which the growing polypeptide passed during its exit from the ribosome. A tunnel is not resolved in the tomographic reconstruction shown in Fig. 8 despite its higher resolution. It could be argued that this is somehow due to the different specimen preparation techniques employed—negative staining versus nonstained, frozen hydrated. On the other hand, the intersubunit

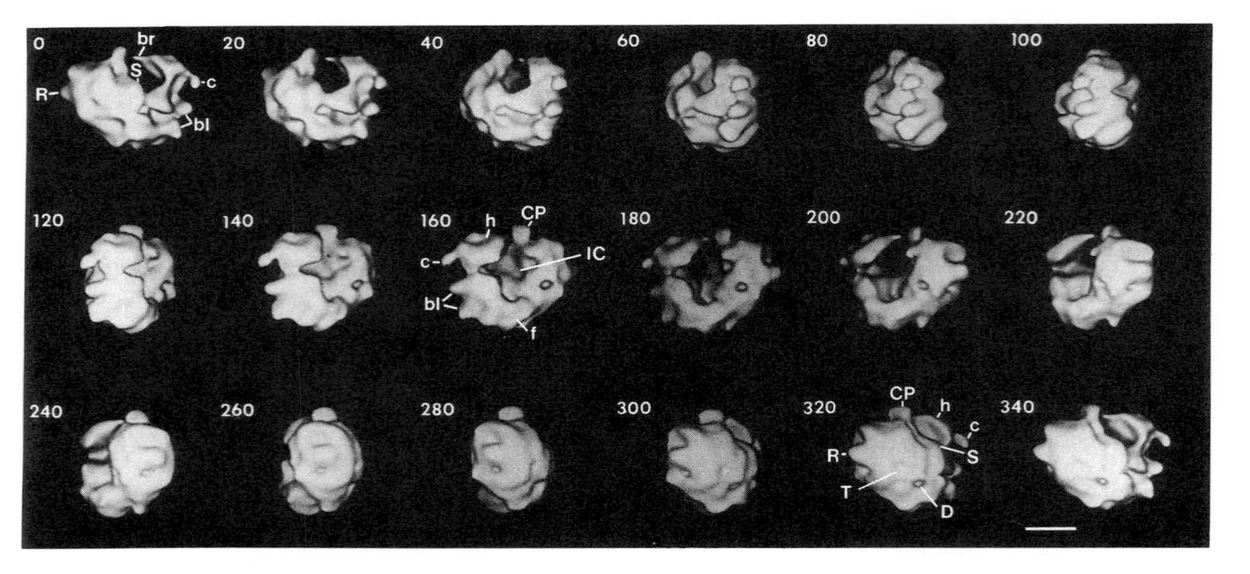

FIGURE 8. Surface representations of reconstructed eukaryotic ribosome. The views are related by successive 20° rotations about a vertical axis. Abbreviations for 40S feature: L, R refer to the 40S subunit in L or R orientation; h, head; c, crest; br, bridge; bl, back lobes, f, feet or basal lobes. Abbreviations for 60S subunit: CP, central protuberance; IC, interface canyon; R, stalk-base ridge; S, shoulder; D, dimple possibly associated with the exit domain of nascent chain; T, trough possibly associated with exit domain. Scale bar, 10 nm. (From Verschoor and Frank, 1989, with permission of Journal of Molecular Biology; copyright 1990 by Academic Press, Inc. [London] Limited.)

gap, the most prominent feature of the tomographic reconstruction, was apparently not present in the reconstructions of the frozen hydrated specimen. An alternative interpretation of the feature identified as a tunnel, which appears likely based upon comparisons with the reconstruction of Verschoor (1989), is that it corresponds to the intersubunit gap identified in the tomographic reconstruction.

5. HEMOCYANIN FROM ANDROCTONUS AUSTRALIS (SCORPION)

Hemocyanins are copper-containing proteins which function as oxygen carriers in arthropods and molluscs. They are not related to the more familiar hemoglobins that perform the same function in mammals. Hemocyanins vary among species in their quaternary structures and protein compositions. In the arthropod *Androctonus australis*, for example, the hemocyanin complex comprises 24 polypeptides, 8 of which are different but closely related (M_r ~75,000). The complete assembly is made up of four hexamers of which there are two types differing in the composition of their peptides. The two types of hexamer associate to form dodecamers, and, finally, two dodecamers form the complete hemocyanin molecule as shown in Fig. 9 (for a review

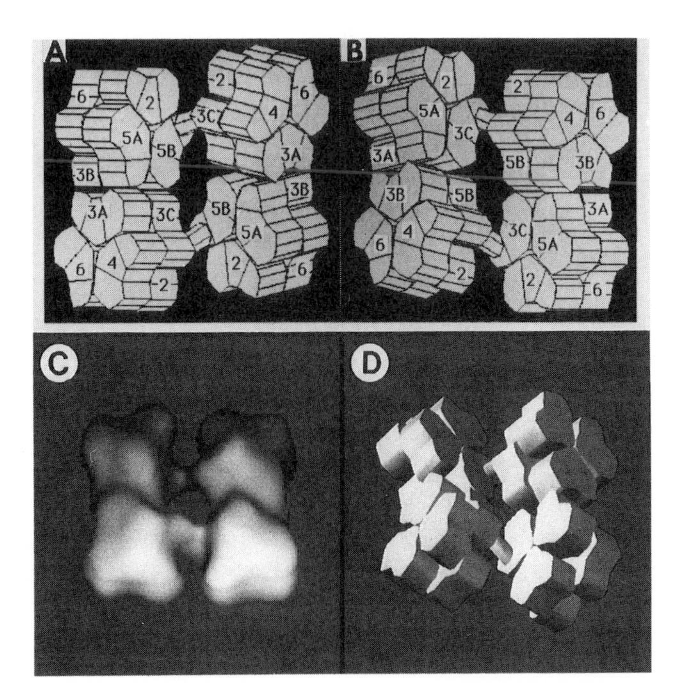

FIGURE 9. Model of quaternary structure of *A. australis* hemocyanin as determined by Lamy *et al.* (1981, 1985). The complex is shown in an orientation known as the "top" view which occurs frequently in the electron microscope. (a), (b) show views of the top and bottom surfaces referred to as "flop" and "flip" respectively (reprinted with permission from Boisset *et al.*, 1988). (c) Surface representation of a tomographic 3D reconstruction determined by the method of Radermacher. (d) A model in the same orientation. (c, d adapted from Boisset *et al.*, 1990; with permission of Journal of Molecular Biology; copyright 1990 by Academic Press, Inc. [London] Limited.)

see Lamy *et al.*, 1985b). The model in Fig. 9*a* shows the individual hemocyanin molecules as kidney-shaped objects in accordance with the high-resolution structure of the homomeric hexamer of another arthropod, *Panulirus interruptus*, determined by x-ray crystallography (Gaykema *et al.*, 1984).

As illustrated by Fig. 9, the eight different polypeptides (named 2, 3A, 3B, 3C, 4, 5A, 5B, 6) occupy defined positions in the 24-mer which have been mapped by the technique of immunoelectron microscopy in an elegant series of studies by Lamy and co-workers (Lamy *et al.*, 1981, 1985a, 1985b). In order to improve the precision of epitope localizations, these same workers have begun to use monoclonal antibodies and single-particle image-processing technology in their most recent studies (Boisset *et al.*, 1988). Electron tomography will be essential in these studies, and its feasibility has already been demonstrated by the determination of the 3D architecture of the *A. australis* hemocyanin in the absence of bound antibody. This reconstruction will serve as a reference and control for future reconstructions of hemocyanin to which subunit-specific antibodies are bound. Figure 9*c* shows the results of reconstructing images in which the hemocyanin was in the top orientation (as in Fig. 9*a*, *b*). Although the boundaries of the individual monomeric subunits are not resolved at the resolution attained (3.5–4.9 nm), a detailed analysis of the reconstruction reveals 24 stain-excluding centers corresponding to the twenty-four 75-kD subunits in the positions predicted by the quaternary structure that was deduced by immunoelectron microscopy and model building. Extensive use of multivariate statistical and classification techniques was necessary in these studies in order to extract reasonably homogeneous sets of particles with the result that the sets contained insufficient images to realize the potential resolution permitted by the quality of the images. The use of larger data sets should result in improvements in the resolution. The orientation of the two bridges of density connecting subunits 3C and 5B (Fig. 9*c*) which join the two dodecamers to each other, is very clearly resolved and agrees with predictions made previously. It is anticipated that future studies of immunocomplexes should allow the mapping of epitopes on the surface the hemocyanin complex with precisions of 1 nm or better (Lamy *et al.*, 1985b).

6. SARCOPLASMIC CALCIUM RELEASE CHANNEL/FOOT STRUCTURE

A major outstanding problem in the physiology of contraction in striated muscle is to how nerve impulses arriving at the neuromuscular junction induce muscle contraction, a process referred to as excitation-contraction (E-C) coupling. It is known that the plasma membrane of the muscle becomes depolarized in E-C coupling, and that the immediate cause of contraction is an increase of Ca^{2+} in the cytoplasm, but the intervening steps have remained elusive (for recent review see Fleischer and Inui, 1989). One of the key components in the process, the channel responsible for releasing the Ca^{2+} stored in the sarcoplasmic reticulum into the cytoplasm has recently been isolated and characterized (Inui *et al.*, 1987; Lai *et al.*, 1988; Imagawa *et al.*, 1987). Surprisingly, this calcium release channel, also referred to as the ryanodine receptor, was found to be equivalent morphologically to a structure known as the *junctional foot*, which had been observed previously in electron micrographs of sectioned muscle tissue (Franzini-Armstrong, 1970). The junctional feet span the gap between two major membrane systems of muscle: the sarcoplasmic reticulum, analogous to the endoplasmic reticulum in other cell types, and the transverse tubules, which are contiguous with the plasma membrane and formed from it by invagination (and therefore become depolarized during E-C coupling). Based upon its location in muscle, the calcium release channel/foot

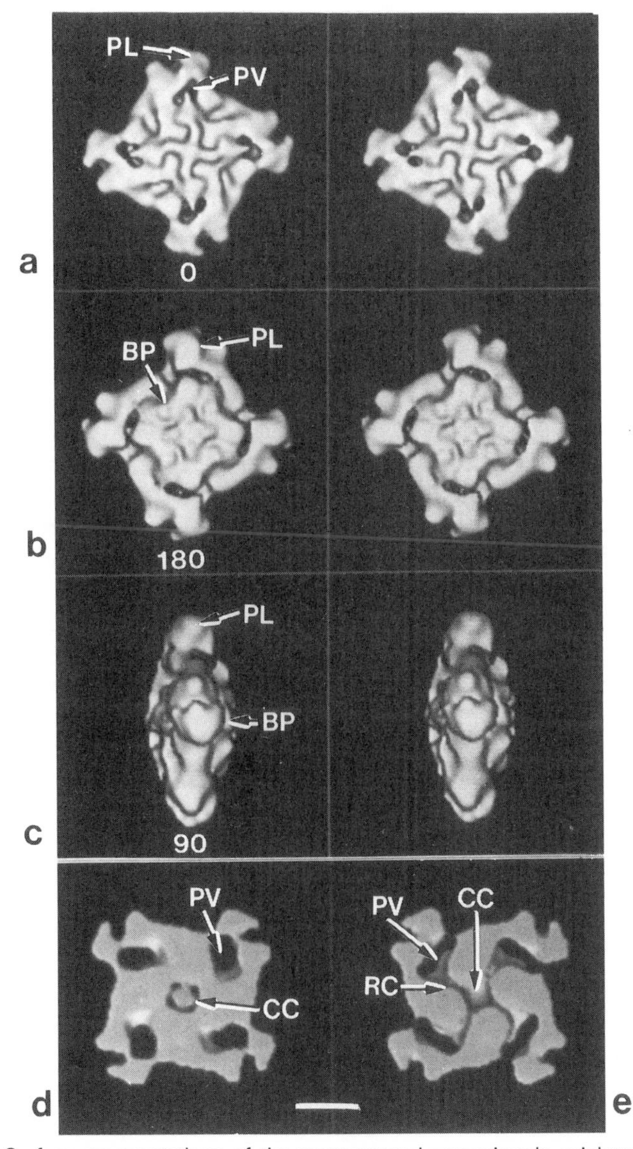

FIGURE 10. Surface representations of the reconstructed sarcoplasmic calcium release channel/foot structure (JCC). (a)–(c), stereo pairs of the JCC. (a), face that interacts with transverse tubule and cytoplasm; (b), face in contact with sarcoplasmic membrane; (c), side view. (d), (e) reconstruction sliced open to reveal internal staining. (From Wagenknecht *et al.*, 1989b with permission of Nature [London]; copyright 1989 Macmillan Magazines, Ltd.)

structure, hereafter referred to as the junctional channel complex (JCC), might be involved in sensing the depolarization of the plasma membrane in addition to its function in releasing calcium from sarcoplasmic stores.

The isolated JCC exists as a tetramer of identical polypeptides and has a total molecular mass of 2.3×10^6, making it the largest ion channel known. Upon adsorption to the carbon foil of an electron microscope grid and after negative staining, the complexes invariably assume an orientation exhibiting fourfold symmetry and an overall square shape. A detailed analysis showed that all of the images were of the same handedness, indicating that the JCC adsorb to the grid exclusively by one of the two faces. These properties make the JCC an ideal specimen for the application of the electron tomographic method of Radermacher (Wagenknecht *et al.*, 1989b).

Surface representations of the reconstructed JCC are shown in Fig. 10. One of the two faces possesses a projecting platform (BP in Fig. 10*b*) and it is likely that it, or some part of it, inserts into the sarcoplasmic membrane (Wagenknecht *et al.*, 1989b). Presumably one or more Ca^{2+}-conducting channels are present in this part of the complex, but they are not resolved in the reconstruction, probably due to limited resolution (3.0–4.0 nm). The outer regions of the complex, which would be located in the cytoplasm of the muscle cell in situ, contain stain-filled cavities (PV, peripheral vestibules) that appear to extend to the surface of the molecule on both faces (Fig. 10*a*, *b*). In Fig. 10*d*, *e* the reconstruction is displayed as if it had been sliced into complementary halves along a direction perpendicular to the fourfold symmetry axis. These cutaway views reveal two major stain-accessible features, a central cavity and four radial channels which connect the central cavity to the peripheral vestibules. This pattern of staining suggests a possible pathway, shown schematically in Fig. 11, for Ca^{2+} flow when the JCC is in the open state. Calcium ions are shown entering the central cavity via the putative membrane traversing channel(s) associated with the basal platform, and traversing the radially running channels to the peripheral vestibules from which the cytoplasm would be accessible.

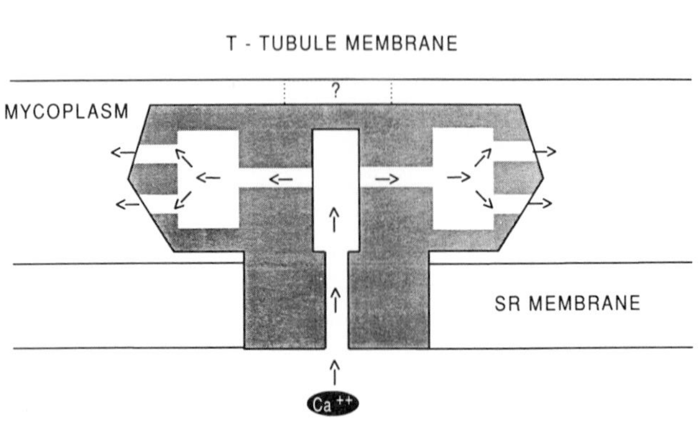

FIGURE 11. Schematic interpretation of the reconstructed JCC showing possibly pathway taken by released calcium ions.

An intriguing possibility suggested by the model is that channel gating could take place in the narrow radial channels, well away from the transmembrane portion of the complex; if the gates in the four radial channels operate independently of one another, then this could account for the results of electrophysiological studies on reconstituted JCC in which four substates of conductance were often observed (Smith *et al.*, 1988; Liu *et al.*, 1989). Further electron tomographic studies of the JCC at higher resolution and in defined open and closed states should aid in testing such speculations.

7. ELECTRON TOMOGRAPHY: PERSPECTIVE AND OUTLOOK

The 3D reconstructions discussed demonstrate that electron tomography is an effective alternative to electron crystallography when applied to negatively stained, macromolecular complexes. Electron tomographic techniques are just beginning to be applied to unstained specimens, and consequently, it is not known whether resolutions significantly better than 2 nm will be achievable, as has been attained for bacteriorhodopsin by 3D dimensional electron crystallography (Henderson and Unwin, 1975). On the other hand, the tomographic methods work well on very large biological complexes, which generally are the most difficult specimens to study by x-ray or electron crystallographic techniques. Because time need not be spent finding conditions for growing crystals, progress in determining reconstructions tomographically is relatively rapid. For example, it should be feasible to systematically characterize chemically modified specimens (e.g., labeled with monospecific antibody or other macromolecular ligand) for the purpose of localizing sites of structural or functional importance on the surface of macromolecular assemblies.

Currently, the resolution-limiting factors in electron tomography are specimen related. Specimen preservation is of great concern, even in low-resolution studies where the negative staining technique is used. Isolated negatively stained macromolecules are subject to distortions upon adsorbing to the specimen grid and during drying in air. Some macromolecular assemblies are more resistant to flattening distortions than others (Kellenberger *et al.*, 1982). Crystalline specimens are believed to be more resistant to these effects by virtue of the intermolecular contacts present, but they are not more resistant to specimen thinning caused by electron exposure (Berriman and Leonard, 1986). The latter effect can be severe, causing reductions in specimen thickness of 25% or more, if minimal electron beam exposure techniques are not employed, or if extensive tilt series are collected from the same area of the specimen.

It is apparent that better specimen preparation techniques are required in order for electron tomographic methods to be exploited fully. In this regard, the use of unstained, frozen hydrated cryoelectron microscopy shows great promise. Taylor and Glaeser (1976) introduced this methodology which was later greatly simplified (Lepault *et al.*, 1983; Dubochet *et al.*, 1988). In this technique, macromolecules, either in suspension or adsorbed to a carbon film, are rapidly frozen while still embedded in a thin film of water or buffer. The specimen grids are transferred to a special cryoholder for electron microscopy, all the while maintaining the temperature below about −140°C. The rapid freezing results in a vitreous, structurally featureless form of ice in which macromolecular structure can be preserved to near-atomic resolution, although retrieving the high-resolution data by electron microscopy is not trivial. Because there is no stain to preserve the molecules, it is essential to record micrographs under conditions of minimal electron exposure, and to generate contrast it is necessary to defocus the objective lens of the microscope. Obviously, there are practical limits to be considered in adjusting these parameters. Although frozen hydrated specimens generally show no tendency to flatten when suspended in ice, it is not yet known to what degree macromolecules are distorted by adsorption to a carbon film.

Presently, there appear to be no impediments to determining 3D reconstructions of frozen hydrated macromolecules at moderate resolution (>2 nm) by the electron tomographic method of Radermacher; it has already been shown that the correlation procedures required for rotational and translational alignment can be applied to frozen hydrated 50S ribosomal subunits (Wagenknecht *et al.*, 1988a) and to the sarcoplasmic calcium release channel/foot structure (Radermacher *et al.*, 1990) discussed in Sections 3.1 and 6, respectively. Recently a 3D reconstruction of the *E. coli* ribosome has been reported (Penczek *et al.*, 1990; Frank *et al.*, 1991). Tomographic techniques, currently being developed, in which naturally occurring orientations are identified and used (van Heel, 1987; Goncharov *et al.*, 1987) might also be well suited to frozen hydrated specimens. Because of the susceptibility of frozen hydrated macromolecules to radiation damage, tomographic reconstructions from data obtained from experimental tilt series are probably not feasible, except in low-resolution studies or in cases where the specimen has sufficient symmetry such that only tilts over a restricted range are required, as for example, in the case of clathrin cages (Vigers *et al.*, 1986a, b).

Ever more examples of subcellular supramolecular complexes are being discovered and characterized. Accumulating evidence suggests that many fundamental cellular processes including DNA replication, mRNA synthesis and processing, protein biosynthesis, and major metabolic pathways are catalyzed by ordered assemblies of the requisite enzymes (Srere, 1987). Many of these assemblies are labile and difficult to isolate, much less crystallize for structural studies. Electron tomography, especially applied to images of frozen hydrated complexes, may be the only practical approach for determining the native 3D architecture of such complexes.

ACKNOWLEDGMENTS

I thank Joachim Frank and Adriana Verschoor for discussions and comments. This work was supported in part by NIH grant R29 GM38161.

REFERENCES

Arad, T., Piefke, H. S., Weinstein, A., Yonath, A., and Wittman, H. G. (1987). Three-dimensional image reconstruction from ordered arrays of 70S ribosome. *Biochimie* **69**:1001–1006.

Baker, T. S. and Fuller, S. Forthcoming.

Bernabeu, C. and Lake, J. A. (1982). Nascent polypeptide chains emerge from the exit domain of the large ribosomal subunit: immune mapping of the nascent chain. *Proc. Nat. Acad. Sci. USA* **79**:3111–3115.

Berriman, J. and Leonard, K. R. (1986). Methods of specimen thickness determination in electron microscopy. II: Changes in thickness with dose. *Ultramicroscopy* **19**:349–366.

Boisset, N., Frank, J., Taveau, J. C., Billiald, P., Motta, G. J., Sizaret, P. Y., and Lamy, J. (1988). Intramolecular localization of epitopes within an oligomeric protein by immunoelectron microscopy and image processing. *Proteins: Struct. Funct. Genet.* **3**:161–183.

Boisset, N., Wagenknecht, T., Radermacher, M., Frank, J., and Lamy, J. N. (1990). Three-dimensional reconstruction of native *Androctonus australis* hemocyanin. *J. Mol. Biol.* **216**:743–760.

Bommer, U.-A., Lutsch, G., Behlke, J., Stahl, J., Nesytova, N., Henske, A., and Bielka, H. (1988). Shape and location of eukaryotic initiation factor eIF-2 on the 40S ribosomal subunit of rat liver. *Eur. J. Biochem.* **172**:653–662.

Boublik, M. and Hellman, W. (1978). Comparison of *Artemia salina* and *Escherichia coli* ribosome structure by electron microscopy. *Proc. Nat. Acad. Sci. USA* **75**:2829–2833.

Capel, M. S., Engelmann, D. M., Freeborn, B. R., Kjeldgaard, M., Langer, J. A., Ramakrishnan, V., Schindler, D. G., Schneider, D. K., Schoenborn, B. P., Sillers, I.-Y., Yabuki, S., and Moore, P. B. (1987). A complete mapping of the proteins in the small ribosomal subunit of *Escherichia coli. Science* **238**:1403–1406.

Carazo, J. M., Rivera, J. M., Zapata, E. L., Radermacher, M., and Frank, J. (1990). Fuzzy sets-based classification of electron microscopy images of biological macromolecules with an application to ribosomal particles. *J. Microsc.* **157**:187–203.

Carazo, J. M., Wagenknecht, T., and Frank, J. (1989). Variations of the three-dimensional structure of the *Escherichia coli* ribosome in the range of overlap views: An application of the methods of multicone and local single-cone three-dimensional reconstruction. *Biophys. J.* **55**:465–477.

Carazo, J. M., Wagenknecht, T., Radermacher, M., Mandiyan, V., Boublik, M., and Frank, J. (1988). Three-dimensional structure of the 50S *Escherichia coli* ribosomal subunits depleted of proteins L7/L12. *J. Mol. Biol.* **201**:393–404.

Crewe, A. V., Crewe, D. A., and Kapp, O. H. (1984). Inexact three-dimensional reconstruction of a biological macromolecule from a restricted number of projections. *Ultramicroscopy* **13**:365–372.

Dubochet, J., Adrian, M., Chang, J.-J., Homo, J.-C., Lepault, J., McDowall, A. W., and Schultz, P. (1988). Cryo-electron microscopy of vitrified specimens. *Q. Rev. Biophys.* **21**:129–228.

Emanuilov, I., Sabatini, D. D., Lake, J. A., and Freienstein, C. (1978). Localization of eukaryotic initiation factor 3 on native small ribosomal subunits. *Proc. Nat. Acad. Sci. USA* **75**:1389–1393.

Erdmann, V. A., Wolters, J., Huysmans, E., and De Wachter, R. (1985). Collection of published 5S, 5.8S and 4.5S ribosomal RNA sequences. *Nucl. Acids Res.* **13**:r105–r153.

Fleischer, S. and Inui, M. (1989). Biochemistry and biophysics of excitation-contraction coupling. *Ann. Rev. Biophys. Biophys. Chem.* **18**:333–364.

Frank, J. (1980). The role of correlation techniques in computer image processing, in *Topics in Current Physics* (P. W. Hawkes, ed.), Vol. 13, pp. 187–222. Springer-Verlag, Berlin.

Frank, J. (1989). Image analysis of single macromolecules. *Electron Microsc. Rev.* **2**:53–74.

Frank, J. (1990). Classification of macromolecular assemblies studied as "single particles." *Q. Rev. Biophys.* **23**:281–329.

Frank, J., Bretaudiere, J. P., Carazo, J. M., Verschoor, A., and Wagenknecht, T. (1988). Classification of images of biomolecular assemblies: A study of ribosomes and ribosomal subunits of *Escherichia coli. J. Microsc.* **150**:99–115.

Frank, J., Goldfarb, W., Eisenberg, D., and Baker, T. S. (1978). Reconstruction of glutamine synthetase using computer averaging. *Ultramicroscopy* **3**:283–290.

Frank, J. and Radermacher, M. (1986). Three-dimensional reconstruction of nonperiodic macromolecular assemblies from electron micrographs, in *Advanced Techniques in Electron Microscopy III* (J. K. Koehler, ed.), pp. 1–72. Springer-Verlag, New York.

Frank, J., Radermacher, M., Wagenknecht, T., and Verschoor, A. (1988). Studying ribosome structure by electron microscopy and computer image processing. *Methods Enzymol.* **164**:3–35.

Frank, J. and van Heel, M. (1982). Correspondence analysis of aligned images of biological particles. *J. Mol. Biol.* **161**:134–137.

Frank, J., Verschoor, A., and Boublik, M. (1981). Computer averaging of electron micrographs of 40S ribosomal subunits. *Science* **214**:1353–1355.

Frank, J., Verschoor, A., and Boublik, M. (1982). Multivariate statistical analysis of ribosome electron micrographs: L and R lateral views of the 40S subunit from HeLa cells. *J. Mol. Biol.* **161**:107–137.

Frank, J., Verschoor, A., Radermacher, M., and Wagenknecht, T. (1990). Morphologies of eubacterial and eucaryotic ribosomes as determined by Three-dimensional electron microscopy, in *The Ribosome: Structure, Function and Evolution* (W. Hill, A. Dahlberg, R. Garrett, P. B. Moore, D. Schlessinger, and J. R. Warner, eds.), pp. 107–113. Am. Soc. Microbiol., Washington DC.

Frank, J., Penczek, P., Grassucci, R., and Srivastava, S. (1991). Three-dimensional reconstruction of the *Escherichia coli* ribosome in ice: The distribution of ribosomal RNA. *J. Cell. Biol.* **115**:595–605.

Franzini-Armstrong, C., 1970. Studies of the triad. I: Structure of the junction in frog twitch fibers. *J. Cell Biol.* **47**:488–499.

Gaykema, W. P. J., Hol, W. G. J., Vereijken, J. M., Soeter, N. M., Bak, H. J., and Beintema, J. J. (1984). A structure of copper containing oxygen carrier protein *Panulirus interruptus* hemocyanin. *Nature* **309**:23–29.

Goncharov, A. B., Vainshtein, B. K., Ryskin, A. I., and Vagin, A. A. (1987). Three-dimensional reconstruction of arbitrarily oriented particles from their electron photomicrographs. *Sov. Phys. Crystallogr.* **32**:504–509.

Gornicki, P., Nurse, K., Hellman, W., Boublik, M., and Ofengand, J. (1984). High resolution localization of the tRNA anticodon interaction site on the *Escherichia coli* 30S ribosomal subunit. *J. Biol. Chem.* **259**:10493–10498.

Henderson, R. and Unwin, P. N. T. (1975). Three-dimensional model of purple membrane obtained by electron microscopy. *Nature* **257**:28–32.

Hoppe, W. (1981). Three-dimensional electron microscopy. *Ann. Rev. Biophys. Bioeng.* **10**:563–592.

Hoppe, W., Oettl, H., and Tietz, H. R. (1986). Negatively stained 50S ribosomal subunits of *Escherichia coli. J. Mol. Biol.* **192**:291–322.

Imagawa, T., Smith, J. S., Coronado, R., and Campbell, K. P. (1987). Purified ryanodine receptor from skeletal muscle sarcoplasmic reticulum is the Ca^{++}-permeable pore of the calcium release channel. *J. Biol. Chem.* **262**:16636–16643.

Inui, M., Saito, A., and Fleischer, S. (1987). Purification of the ryanodine receptor and identity with feet structures of junctional terminal cysternae of sarcoplasmic reticulum from fast skeletal muscle. *J. Biol. Chem.* **262**:1740–1747.

Kellenberger, E., Haner, M., and Wurtz, M. (1982). The wrapping phenomenon in air-dried and negatively stained specimens. *Ultramicroscopy* **9**:139–150.

Knauer, V., Hegerl, R., and Hoppe, W. (1983). Three-dimensional reconstruction and averaging of 30S ribosomal subunits of *Escherichia coli* from electron micrographs. *J. Mol. Biol.* **163**:409–430.

Lai, F. A., Erickson, H. P., Rousseau, E., Liu, Q.-Y., and Meissner, G. (1988). Purification and reconstitution of the calcium release channel from skeletal muscle. *Nature* **331**:315–319.

Lake, J. A. (1976). Ribosome structure determined by electron microscopy of *Escherichia coli* small subunits, large subunits and monomeric ribosomes. *J. Mol. Biol.* **105**:131–159.

Lake, J. A. (1985). Evolving ribosome structure: Domains in archaebacteria, eubacteria, eocytes and eukaryotes. *Ann. Rev. Biochem.* **54**:507–530.

Lamy, J., Bijlholt, M. M. C., Sizaret, P. Y., Lamy, J. N., van Bruggen, E. F. J. (1981). Quaternary structure of scorpion (*Androctonus australis*) hemocyanin: localization of subunits with immunological and electron microscopy. *Biochemistry* **20**:1849–1856.

Lamy, J. N., Lamy, J., Billiald, P., Sizaret, P. Y., Cave, G., Frank, J., and Motta, G. (1985a). An approach to direct intramolecular localization of antigenic determinants in *Androctonus australis* hemocyanin with monoclonal antibodies by molecular immunoelectron microscopy. *Biochemistry* **24**:5532–5542.

Lamy, J. N., Lamy, J., Sizaret, P. Y., Billiald, P., and Motta, G. (1985b). Quaternary structure of arthropod hemocyanin, in *Respiratory Pigments in Animals, Structure-Function Relations* (J. N. Lamy, J. P. Truchot, and R. Gille, eds.), pp. 73–86. Springer-Verlag, Berlin.

Lepault, J., Booy, F. P., and Dubochet, J. (1983). Electron microscopy of frozen biological suspensions. *J. Microsc.* **129**:89–102.

Liljas, A. (1982). Structural studies of ribosomes. *Progr. Biophys. Molec. Biol.* **40**:161–228.

Liu, Q.-Y., Lai, F. A., Rousseau, E., Jones, R. V., and Meissner, G. (1989). Multiple conductance states of the purified calcium release channel complex from skeletal sarcoplasmic reticulum. *Biophys. J.* **44**:415–424.

Lutsch, G., Benndorf, R., Westermann, P., Behlke, J., Bommer, U.-A., Bielka, H. (1985). On the structure of native small ribosomal subunits and initiation factor eIF-3 isolated from rat liver. *Biomed. Biochim. Acta* **44**:K1–K7.

Milligan, R. A. and Unwin, P. N. T. (1986). Location of exit channel for nascent protein in 80S ribosome. *Nature* **319**:693–695.

Moller, W. and Maassen, J. A. (1986). On the structure, function, and dynamics of L7/L12 from *Escherichia coli* ribosomes, in *Structure, Function and Genetics of Ribosomes* (B. Hardesty and G. Kramer, eds.), pp. 309–325. Springer-Verlag, New York.

Noller, H. F. (1984). Structure of ribosomal RNA, *Ann. Rev. Biochem.* **53**:119–162.

Oakes, M., Henderson, E., Scheinman, A., Clark, M., and Lake, J. A. (1986). Ribosome structure, function, and evolution: Mapping ribosomal RNA, proteins, and functional sites in three dimensions, in *Structure, Function and Genetics of Ribosomes* (B. Hardesty and G. Kramer, eds.), pp. 47–67. Springer-Verlag, New York.

Oettl, H., Hegerl, R., and Hoppe, W. (1983). Three-dimensional reconstruction and averaging of 50S ribosomal subunits of *Escherichia coli* from electron micrographs. *J. Mol. Biol.* **163**:431–450.

Penczek, P., Srivastava, S., and Frank, J. (1990). The structure of the 70E E. coli ribosome in ice, in *Proc. XIIth Int. Congr. for Electron Microscopy* (L. D. Peachey and D. B. Williams, eds.), pp. 506–507. San Francisco Press, San Francisco.

Provencher, S. W. and Vogel, R. H. (1988). Three-dimensional reconstruction from electron micrographs of disordered specimens. I: Method, *Ultramicroscopy* **25**:209–222.

Radermacher, M. (1988). Three-dimensional reconstruction of single particles from random and non-random tilt series. *J. Electron Microsc. Tech.* **9**:359–394.

Radermacher, M., Wagenknecht, T., Grassucci, R., Frank, J., Saito, A., Inui, M., Chadwick, C., Reif, S., and Fleischer, S. (1990). Native architecture of the calcium channel/foot structure from skeletal muscle. *Biophys. J.* **57**:501a.

Radermacher, M., Wagenknecht, T., Verschoor, A., and Frank, J. (1987a). Three-dimensional structure of the large ribosomal subunit from *Escherichia coli. EMBO J.* **6**:1107–1114.

Radermacher, M., Wagenknecht, T., Verschoor, A., and Frank, J. (1987b). Three-dimensional reconstruction from single-exposure random conical tilt series applied to the 50S ribosomal subunit of *Escherichia coli. J. Microsc.* **146**:113–136.

Ryabova, L. A., Selivanova, O. M., Barabov, V. I., Vasiliev, V. D., and Spirin, A. S. (1988). Does the channel for nascent peptide exist inside the ribosome? *FEBS Lett.* **226**:255–260.

Saxton, W. O. and Baumeister, W. (1982). The correlation averaging of a regularly arranged bacterial cell envelope protein. *J. Microsc.* **127**:127–138.

Srere, P. A. (1987). Complexes of sequential metabolic enzymes. *Ann. Rev. Biochemistry* **56**:89–124.

Smith, J. S., Imagawa, T., Ma, J., Fill, M., Campbell, K. P., and Coronado, R. (1988). Purified ryanodine receptor from rabbit skeletal muscle is calcium-release channel of sarcoplasmic reticulum. *J. Gen. Physiol.* **92**:1–26.

Spirin, A. S. (1983). Location of tRNA on the ribosome. *FEBS (Fed. Eur. Biochem. Soc.) Lett.* **156**:217–221.

Stewart, M. (1988). Computer image processing of electron micrographs of biological structures with helical symmetry. *J. Electron Microsc. Tech.* **9**:325–358.

Stöffler, G. and Stöffler-Meilicke, M. (1983). The ultrastructure of macromolecular complexes studied with antibodies, in *Modern Methods in Protein Chemistry* (H. Tschesche, ed.), pp. 409–455. de Gruyter, New York.

Stöffler, G. and Stöffler-Meilicke, M. (1986). Immunoelectron microscopy of *Escherichia coli* ribosomes, in *Structure, Function and Genetics of Ribosomes* (B. Hardesty and G. Kramer, eds.), pp. 28–46. Springer-Verlag, New York.

Stöffler, G. and Stöffler-Meilicke, M. (1988). Localization of ribosomal proteins on the surface of ribosomal subunits from *Escherichia coli* using immunoelectron microscopy. *Methods Enzymol.* **164**:503–520.

Taylor, K. A. and Glaeser, R. M. (1976). Electron microscopy of frozen hydrated biological specimens. *J. Ultrastruct. Res.* **55**:448–456.

Tischendorf, G. W., Zeichhardt, H., and Stöffler, G. (1974). Determination of the location of proteins L14 L17, L18, L19, L22, L23 on the surface of the 50S ribosomal subunit of *Escherichia coli* by immune electron microscopy. *Mol. Gen. Genet.* **134**:187–208.

Traut, R. R., Tewari, D. S., Sommer, A., Gavino, G. R., Olson, H. M., and Glitz, D. G. (1986). Protein topography of ribosomal functional domains: effects of monoclonal antibodies to different epitopes in *Escherichia coli* protein L7/L12 on ribosome function and structure, in *Structure, Function and Genetics of Ribosomes* (B. Hardesty and G. Kramer, eds.), pp. 286–308. Springer-Verlag, New York.

Unser, M., Trus, B. L., and Steven, A. C. (1987). A new resolution criterion based on spectral signal-to-noise ratio. *Ultramicroscopy* **23**:39–52.

Unser, M., Trus, B. L., and Steven, A. C. (1989). Normalization procedures and factorial representations for classifications of correlation-aligned images: a comparative study. *Ultramicroscopy* **30**:299–310.

Unwin, P. N. T. (1977). Three-dimensional model of membrane-bound ribosomes obtained by electron microscopy. *Nature* **269**:118–122.

Unwin, P. N. T. and Ennis, P. D. (1984). Two configurations of a channel-forming membrane protein. *Nature* **307**:609–613.

Unwin, P. N. T., Toyoshima, C., and Kubalek, E. (1988). Arrangement of the acetylcholine receptor subunits in the resting and desensitized states determined by cryoelectron microscopy of crystallized torpedo postsynaptic membranes. *J. Cell Biol.* **107**:1123–1138.

van Heel, M. (1983). Three-dimensional reconstruction of the 30S E. coli ribosomal subunit, in *Proc. 41st Ann. Meet. Electron Microsc. Soc. Am.* (G. W. Bailey, ed.), pp. 460–461. San Francisco Press.

van Heel, M. (1984). Multivariate statistical classification of noisy images (randomly oriented biological macromolecules). *Ultramicroscopy* **13**:165–184.

van Heel, M. (1987). Angular reconstitution: a posteriori assignment of projection directions for 3D reconstructions. *Ultramicroscopy* **21**:111–124.

van Heel, M. and Frank, J. (1981). Use of multivariate statistics in analyzing the images of biological macromolecules. *Ultramicroscopy* **6**:187–194.

van Heel, M., Keegstra, W., Schutter, W., and van Bruggen, E. F. (1982). In *The Structure and Function of Invertebrate Respiratory Proteins, Life Chemistry Reports*, Suppl. 1 (E. J. Wood, ed.), EMBO Workshop Leeds, pp. 69–73.

van Heel, M. and Stöffler-Meilicke, M. (1985). Characteristic views of *E. coli* and *B. Stearothermophilus* 30S ribosomal subunits in the electron microscopy. *EMBO J.* **4**:2389–2395.

Vasiliev, V. D., Selivanova, O. M., Baranov, V. I., and Spirin, A. J. (1983). Structural study of translating 40S ribosomes from *Escherichia coli. J. Electron Microsc.* **155**:167–172.

Verschoor, A. (1989). *Morpho-structural Studies of the Eukaryotic Ribosome: Three-Dimensional Reconstructions from Single-Particle Electron Microscopic Specimens*, Ph.D. thesis, State University of New York at Albany, Albany, NY.

Verschoor, A. and Frank, J. (1990). Three-dimensional structure of the mammalian cytoplasmic ribosome. *J. Mol. Biol.* **214**:737–749.

Verschoor, A., Frank, J., Radermacher, M.. Wagenknecht, T., and Boublik, M. (1984). Three-dimensional reconstruction of the 30S ribosomal subunit from randomly oriented particles. *J. Mol. Biol.* **178**:677–698.

Verschoor, A., Frank, J., Wagenknecht, T., and Boublik, M. (1986). Computer-averaged views of the 70S monosome from *Escherichia coli. J. Mol. Biol.* **187**:581–590.

Verschoor, A., Zhang, N. Y., Wagenknecht, T., Obrig, T., Radermacher, M., and Frank, J. (1989). Three-dimensional reconstruction of mammalian 40S ribosomal subunit. *J. Mol. Biol.* **209**:115–126.

Vigers, G. P. A., Crowther, R. A., and Pearse, B. M. F. (1986a). Three-dimensional structure of clathrin cages in ice. *EMBO J.* **5**:529–534.

Vigers, G. P. A., Crowther, R. A., and Pearse, B. M. F. (1986b). Location of the 100 kD-50 kD accessory proteins in clathrin coats. *EMBO J.* **5**:2079–2085.

Vogel, R. H. and Provencher, S. W. (1988). Three-dimensional reconstruction from electron micrographs of disordered specimens. II: Implementation of results. *Ultramicroscopy* **25**:223–240.

Wagenknecht, T., Carazo, J. M., Radermacher, M., and Frank, J. (1989a). Three-dimensional reconstruction of the ribosome from *Escherichia coli. Biophys. J.* **55**:455–464.

Wagenknecht, T., Frank, J., Boublik, M., Nurse, K., and Ofengand, J. (1988b). Direct localization of the tRNA-anticodon interaction site on the *Escherichia coli* 30S ribosomal subunit by electron microscopy and computerized image averaging. *J. Mol. Biol.* **203**:753–760.

Wagenknecht, T., Grassucci, R., and Frank, J. (1988a). Electron microscopy and computer image averaging of ice-embedded large ribosomal subunits from *Escherichia coli*. *J. Mol. Biol.* **199**:137–147.

Wagenknecht, T., Grassucci, R., Frank, J., Saito, A., Inui, M., and Fleischer, S. (1989b). Three-dimensional architecture of the calcium channel/foot structure of sarcoplasmic reticulum. *Nature* **338**:167–170.

Wittman, H. G. (1983). Architecture of prokaryotic ribosomes. *Ann. Rev. Biochem.* **52**:35–65.

Wittman-Liebold, B. (1984). Primary structure of *Escherichia coli* ribosomal proteins. *Adv. Prot. Chem.* **36**:56–78.

Yonath, A., Leonard, K. R., and Wittman, G. H. (1987). A tunnel in the large ribosomal subunit revealed by three-dimensional image reconstruction. *Science* **236**:813–816.

Yonath, A. and Wittman, H. G. (1989). Challenging the three-dimensional structure of ribosomes. *Trends Biochem. Sci.* **14**:329–335.

Index